Power Line Worker

Level One

Trainee Guide

Prentice Hall

Boston Columbus Indianapolis New York San Francisco Upper Saddle River
Amsterdam Cape Town Dubai London Madrid Milan Munich Paris Montreal Toronto
Delhi Mexico City São Paulo Sydney Hong Kong Seoul Singapore Taipei Tokyo

National Center for Construction Education and Research
President: Don Whyte
Director of Product Development: Daniele Stacey
Power Line Worker Project Manager: Daniele Stacey
Production Manager: Tim Davis
Quality Assurance Coordinator: Debie Ness
Desktop Publishing Coordinator: James McKay
Production Specialist: Laura Wright
Editors: Chris Wilson, Debie Ness

Writing and development services provided by Topaz Publications, Liverpool, NY
Lead Writer/Project Manager: Thomas Burke
Desktop Publisher: Joanne Hart
Art Director: Megan Paye
Permissions Editors: Andrea LaBarge, Tonia Burke
Writers: Thomas Burke, Troy Staton, Pat Vidler, Gerald Shannon, John Tianen, Darrell Wilkerson

Pearson Education, Inc.
Editorial Director: Vernon R. Anthony
Senior Product Manager: Lori Cowen
Senior Managing Editor: JoEllen Gohr
Senior Project Manager: Steve Robb
AV Project Manager: Janet Portisch
Operations Supervisor: Deidra M. Skahill
Art Director: Jayne Conte
Cover Photo: Tim Davis
Director of Marketing: David Gesell
Executive Marketing Manager: Derril Trakalo
Senior Marketing Coordinator: Alicia Wozniak
Full-Service Project Management: Michael B. Kopf, S4Carlisle Publishing Services
Printer/Binder: LSC Communications
Cover Printer: LSC Communications
Text Fonts: Palatino and Univers

Credits and acknowledgments for content borrowed from other sources and reproduced, with permission, in this textbook appear at the end of each module.

ISBN-13: 978-0-13-257109-8
ISBN-10: 0-13-257109-9

16 2022

Preface

To the Trainee

Welcome to your first year of training in power line work. If you are training under an NCCER Accredited Training Program Sponsor, you have successfully completed *Power Industry Fundamentals* and are well on your way to more skill-specific training.

Power Line Worker Level One addresses the fundamental aspects of power line work, including safety, electrical theory, climbing techniques, aerial framing and rigging, and operating utility service equipment.

When you successfully complete this skill-specific training, you will join the ranks of thousands of men and women whose primary responsibility is to provide and restore electricity to millions of businesses and residences across the nation.

Trained power line workers are expected to be in high demand, as more than 20% of the current workforce will retire by the next decade. And the demand will only intensify as the nation struggles to modernize an aging energy infrastructure.

In this training, you will learn that electricity travels great distances, through transmission lines, substations, and distribution lines. Electricity travels both overhead and underground. Depending upon the area of power line work, you could be working at heights greater than 250 feet or at depths in excess of 50 feet. In power line work, variety and opportunity await those with the skills, interest, and willingness to learn.

We wish you success as you progress through this training program. Should you have any comments on how NCCER might improve upon this textbook, please complete the User Update form located at the back of each module and send it to us. We will always consider and respond to input from our customers.

We invite you to visit NCCER's website at **www.nccer.org** for information on the latest product releases and training, as well as online versions of the *Cornerstone* newsletter and Pearson's Contren® product catalog.

Your feedback is welcome. You may email your comments to **curriculum@nccer.org** or send general comments and inquiries to **info@nccer.org**.

NCCER Standardized Curricula

The NCCER is a not-for-profit 501(c)(3) education foundation established in 1996 by the world's largest and most progressive construction companies and national construction associations. It was founded to address the severe workforce shortage facing the industry and to develop a standardized training process and curricula. Today, NCCER is supported by hundreds of leading construction and maintenance companies, manufacturers, and national associations. The NCCER Standardized Curricula was developed by NCCER in partnership with Pearson Education, Inc., the world's largest educational publisher.

Some features of NCCER's Curricula are as follows:

- An industry-proven record of success
- Curricula developed by the industry for the industry
- National standardization providing portability of learned job skills and educational credits
- Compliance with the Office of Apprenticeship requirements for related classroom training *(CFR 29:29)*
- Well-illustrated, up-to-date, and practical information

NCCER also maintains a National Registry that provides transcripts, certificates, and wallet cards to individuals who have successfully completed a level of training within a craft in NCCER's Standardized Curricula. *Training programs must be delivered by an NCCER Accredited Training Sponsor in order to receive these credentials.*

Special Features

In an effort to provide a comprehensive, user-friendly training resource, we have incorporated many different features for your use. Whether you are a visual or hands-on learner, this book will provide you with the proper tools to get started in the power line worker industry.

Introduction

This page is found at the beginning of each module and lists the Objectives, Performance Tasks, Trade Terms, and Required Trainee Materials for that module. The Objectives list the skills and knowledge you will need in order to complete the module successfully. The Performance Tasks give you the opportunity to apply your knowledge to the real world duties that power line workers perform. The list of Trade Terms identifies important terms you will need to know by the end of the module. Required Trainee Materials list the materials and supplies needed for the module.

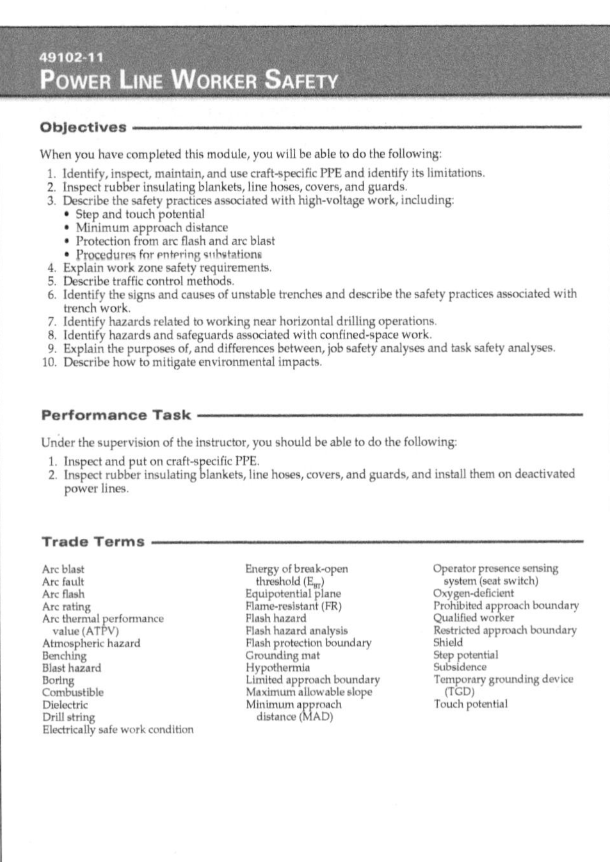

49102-11

POWER LINE WORKER SAFETY

Objectives

When you have completed this module, you will be able to do the following:

1. Identify, inspect, maintain, and use craft-specific PPE and identify its limitations.
2. Inspect rubber insulating blankets, line hoses, covers, and guards.
3. Describe the safety practices associated with high-voltage work, including:
   - Step and touch potential
   - Minimum approach distance
   - Protection from arc flash and arc blast
   - Procedures for entering substations
4. Explain work zone safety requirements.
5. Describe traffic control methods.
6. Identify the signs and causes of unstable trenches and describe the safety practices associated with trench work.
7. Identify hazards related to working near horizontal drilling operations.
8. Identify hazards and safeguards associated with confined-space work.
9. Explain the purposes of, and differences between, job safety analyses and task safety analyses.
10. Describe how to mitigate environmental impacts.

Performance Task

Under the supervision of the instructor, you should be able to do the following:

1. Inspect and put on craft-specific PPE.
2. Inspect rubber insulating blankets, line hoses, covers, and guards, and install them on deactivated power lines.

Trade Terms

Arc blast
Arc fault
Arc flash
Arc rating
Arc thermal performance value (ATPV)
Atmospheric hazard
Benching
Blast hazard
Boring
Combustible
Dielectric
Drill string
Electrically safe work condition
Energy of break-open threshold (E_{BT})
Equipotential plane
Flame-resistant (FR)
Flash hazard
Flash hazard analysis
Flash protection boundary
Grounding mat
Hypothermia
Limited approach boundary
Maximum allowable slope
Minimum approach distance (MAD)
Operator presence sensing system (seat switch)
Oxygen-deficient
Prohibited approach boundary
Qualified worker
Restricted approach boundary
Shield
Step potential
Subsidence
Temporary grounding device (TGD)
Touch potential

Color Illustrations and Photographs

Full-color illustrations and photographs are used throughout each module to provide vivid detail. These figures highlight important concepts from the text and provide clarity for complex instructions. Each figure reference is denoted in the text in *italic type* for easy reference.

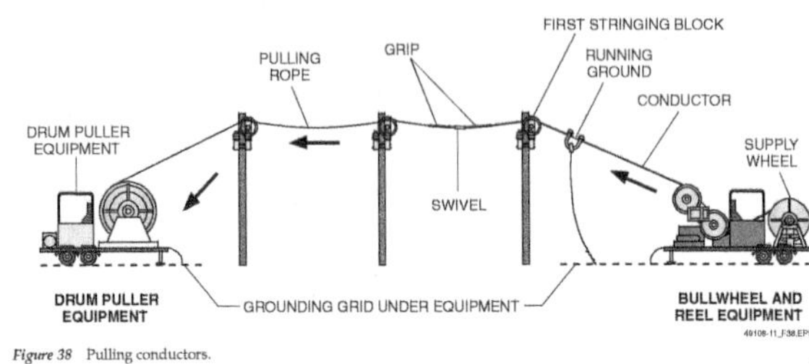

Figure 38 Pulling conductors.

Notes, Cautions, and Warnings

Safety features are set off from the main text in highlighted boxes and are organized into three categories based on the potential danger of the issue being addressed. Notes simply provide additional information on the topic area. Cautions alert you of a danger that does not present potential injury but may cause damage to equipment. Warnings stress a potentially dangerous situation that may cause injury to you or a co-worker.

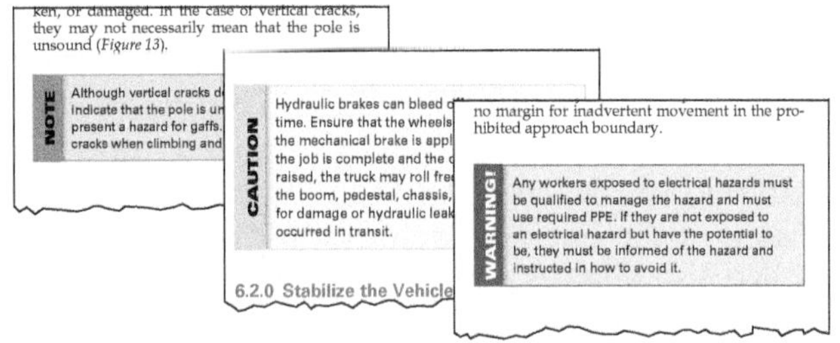

Did You Know?

The Did You Know? features offer hints, tips, and other helpful bits of information from the trade.

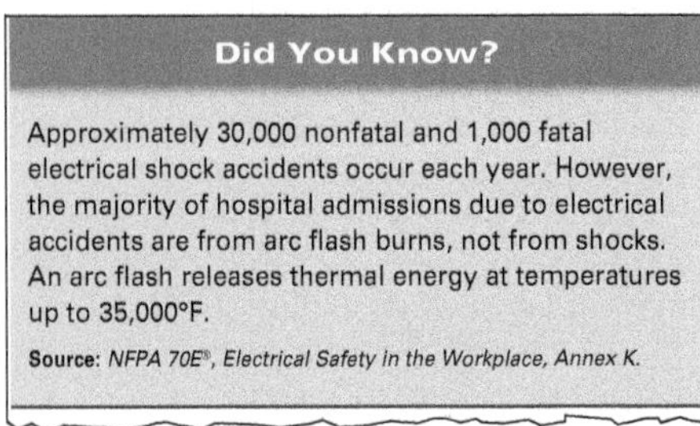

Did You Know?

Approximately 30,000 nonfatal and 1,000 fatal electrical shock accidents occur each year. However, the majority of hospital admissions due to electrical accidents are from arc flash burns, not from shocks. An arc flash releases thermal energy at temperatures up to 35,000°F.

Source: *NFPA 70E®, Electrical Safety in the Workplace, Annex K.*

On Site

On Site features provide a head start for those entering the electrical transmission and distribution fields by presenting technical tips and professional practices from power line workers on a variety of topics. On Site features often include real-life scenarios similar to those you might encounter on the job site.

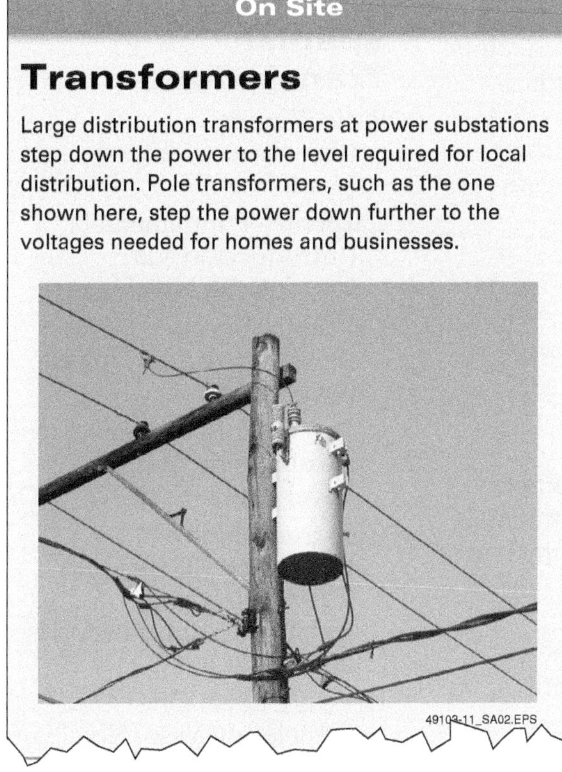

On Site

Transformers

Large distribution transformers at power substations step down the power to the level required for local distribution. Pole transformers, such as the one shown here, step the power down further to the voltages needed for homes and businesses.

49103-11_SA02.EPS

Think About It

Think About It features use "What if?" questions to help you apply theory to real-world experiences and put your ideas into action.

Think About It

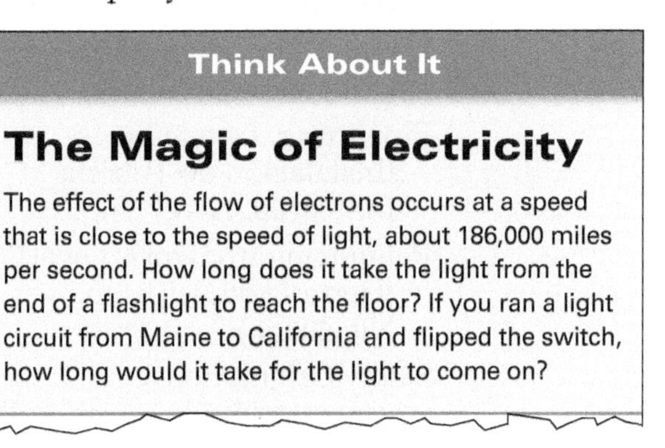

The Magic of Electricity

The effect of the flow of electrons occurs at a speed that is close to the speed of light, about 186,000 miles per second. How long does it take the light from the end of a flashlight to reach the floor? If you ran a light circuit from Maine to California and flipped the switch, how long would it take for the light to come on?

Step-by-Step Instructions

Step-by-step instructions are used throughout to guide you through technical procedures and tasks from start to finish. These steps show you not only how to perform a task but also how to do it safely and efficiently.

trically safe work condition. Most companies have detailed written procedures for performing this task.

Step 1 Determine whether any other crews are working on the circuit. All distribution lines are treated as if they are energized unless your team has performed a de-energizing procedure. If two or more crews are working on the same lines or equipment, each crew must independently perform a de-energizing procedure and apply their own lockout/tagout devices to energy controls.

Step 2 Designate one qualified member of the crew as the employee in charge of the electrical clearance.

Trade Terms

Each module presents a list of Trade Terms that are discussed within the text and defined in the Glossary at the end of the module. These terms are denoted in the text with blue bold type upon their first occurrence. To make searches for key information easier, a comprehensive Glossary of Trade Terms from all modules is located at the back of this book.

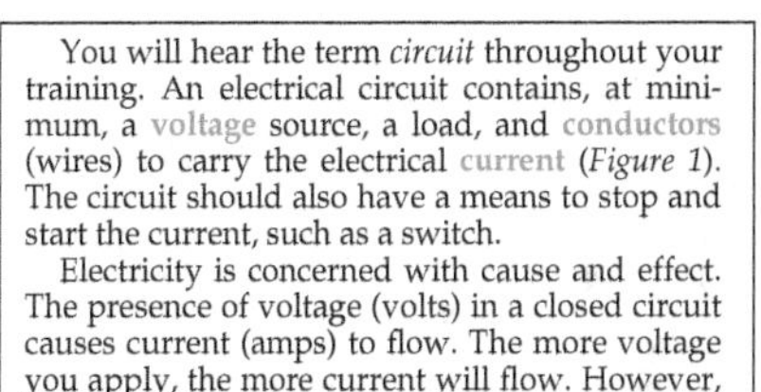

You will hear the term *circuit* throughout your training. An electrical circuit contains, at minimum, a **voltage** source, a load, and **conductors** (wires) to carry the electrical **current** (*Figure 1*). The circuit should also have a means to stop and start the current, such as a switch.

Electricity is concerned with cause and effect. The presence of voltage (volts) in a closed circuit causes current (amps) to flow. The more voltage you apply, the more current will flow. However, the amount of current flow is also determined by how much **resistance**, in **ohms** (Ω), the load offers

Review Questions

Review Questions are provided to reinforce the knowledge you have gained. This makes them a useful tool for measuring what you have learned.

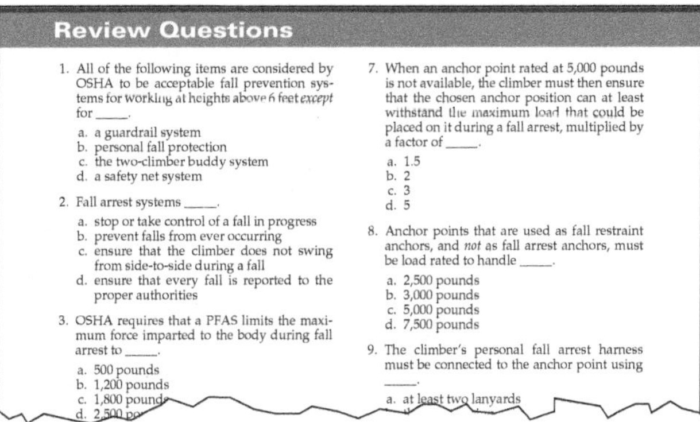

Review Questions

1. All of the following items are considered by OSHA to be acceptable fall prevention systems for working at heights above 6 feet *except* for _____.
   a. a guardrail system
   b. personal fall protection
   c. the two-climber buddy system
   d. a safety net system
2. Fall arrest systems _____.
   a. stop or take control of a fall in progress
   b. prevent falls from ever occurring
   c. ensure that the climber does not swing from side-to-side during a fall
   d. ensure that every fall is reported to the proper authorities
3. OSHA requires that a PFAS limits the maximum force imparted to the body during fall arrest to _____.
   a. 500 pounds
   b. 1,200 pounds
   c. 1,800 pound
   d. 2,500 po

7. When an anchor point rated at 5,000 pounds is not available, the climber must then ensure that the chosen anchor position can at least withstand the maximum load that could be placed on it during a fall arrest, multiplied by a factor of _____.
   a. 1.5
   b. 2
   c. 3
   d. 5
8. Anchor points that are used as fall restraint anchors, and *not* as fall arrest anchors, must be load rated to handle _____.
   a. 2,500 pounds
   b. 3,000 pounds
   c. 5,000 pounds
   d. 7,500 pounds
9. The climber's personal fall arrest harness must be connected to the anchor point using _____.
   a. at least two lanyards

Contren® Curricula

NCCER's training programs comprise more than 80 construction, maintenance, pipeline, and utility areas and include skills assessments, safety training, and management education.

Boilermaking
Cabinetmaking
Carpentry
Concrete Finishing
Construction Craft Laborer
Construction Technology
Core Curriculum: Introductory Craft Skills
Drywall
Electrical
Electronic Systems Technician
Heating, Ventilating, and Air Conditioning
Heavy Equipment Operations
Highway/Heavy Construction
Hydroblasting
Industrial Coating and Lining Application Specialist
Industrial Maintenance Electrical and Instrumentation Technician
Industrial Maintenance Mechanic
Instrumentation
Insulating
Ironworking
Masonry
Millwright
Mobile Crane Operations
Painting
Painting, Industrial
Pipefitting
Pipelayer
Plumbing
Reinforcing Ironwork
Rigging
Scaffolding
Sheet Metal
Signal Person
Site Layout
Sprinkler Fitting
Tower Crane Operator
Welding

Green/Sustainable Construction

Building Auditor
Fundamentals of Weatherization
Introduction to Weatherization
Sustainable Construction Supervisor
Weatherization Crew Chief
Weatherization Technician
Your Role in the Green Environment

Energy

Introduction to the Power Industry
Introduction to Solar Photovoltaics
Introduction to Wind Energy
Power Industry Fundamentals
Power Generation Maintenance Electrician
Power Generation I&C Maintenance Technician
Power Generation Maintenance Mechanic
Power Line Worker
Solar Photovoltaic Systems Installer

Pipeline

Control Center Operations, Liquid
Corrosion Control
Electrical and Instrumentation
Field Operations, Liquid
Field Operations, Gas
Maintenance
Mechanical

Safety

Field Safety
Safety Orientation
Safety Technology

Management

Fundamentals of Crew Leadership
Project Management
Project Supervision

Supplemental Titles

Applied Construction Math
Careers in Construction
Tools for Success

Spanish Translations

Basic Rigging (Principios Básicos de Maniobras)
Carpentry Fundamentals (Introducción a la Carpintería, Nivel Uno)
Carpentry Forms (Formas para Carpintería, Nivel Trés)
Concrete Finishing, Level One (Acabado de Concreto, Nivel Uno)
Core Curriculum: Introductory Craft Skills (Currículo Básico: Habilidades Introductorias del Oficio)
Drywall, Level One (Paneles de Yeso, Nivel Uno)
Electrical, Level One (Electricidad, Nivel Uno)
Field Safety (Seguridad de Campo)
Insulating, Level One (Aislamiento, Nivel Uno)
Ironworking, Level One (Herrería, Nivel Uno)
Masonry, Level One (Albañilería, Nivel Uno)
Pipefitting, Level One (Instalación de Tubería Industrial, Nivel Uno)
Reinforcing Ironwork, Level One (Herreria de Refuerzo, Nivel Uno)
Safety Orientation (Orientación de Seguridad)
Scaffolding (Andamios)
Sprinkler Fitting, Level One (Instalación de Rociadores, Nivel Uno)

Acknowledgments

This curriculum was revised as a result of the farsightedness and leadership of the following sponsors:

Baltimore Gas & Electric
Cianbro Corporation
Gaylor, Inc.
MasTec, Inc.
Oneonta Job Corps Academy
Pumba Electric LLC
Quanta Services Inc.
Southeast Lineman Training Center
Vision Quest Academy

This curriculum would not exist were it not for the dedication and unselfish energy of those volunteers who served on the Authoring Team. A sincere thanks is extended to the following:

James Anthony
Robert Groner
O'Neil Boivin
David Brzozowski
Kurt Gastel
Larry Harvey
Joe Holley
Craig Hopkins
Gordon Johnson
Mark Lagasse
L. J. LeBlanc
James McGowan
Scott Mitchell
David Powell
Michael A. Roedel
Jonathan Sacks
James "Shane" Smith

A final note: This book is the result of a collaborative effort involving the production, editorial, and development staff at Pearson Education, Inc., and the National Center for Construction Education and Research. Thanks to all of the dedicated people involved in the many stages of this project.

NCCER Partners

American Fire Sprinkler Association
Associated Builders and Contractors, Inc.
Associated General Contractors of America
Association for Career and Technical Education
Association for Skilled and Technical Sciences
Carolinas AGC, Inc.
Carolinas Electrical Contractors Association
Center for the Improvement of Construction Management and Processes
Construction Industry Institute
Construction Users Roundtable
Construction Workforce Development Center
Design Build Institute of America
Merit Contractors Association of Canada
Metal Building Manufacturers Association
NACE International
National Association of Minority Contractors
National Association of Women in Construction
National Insulation Association
National Ready Mixed Concrete Association
National Technical Honor Society
National Utility Contractors Association
NAWIC Education Foundation
North American Technician Excellence
Painting & Decorating Contractors of America
Portland Cement Association
SkillsUSA
Steel Erectors Association of America
The Manufacturers Institute
U.S. Army Corps of Engineers
University of Florida, M.E. Rinker School of Building Construction
Women Construction Owners & Executives, USA

Contents

Note: *NFPA 70*®, *National Electrical Code*®, and *NEC*® are registered trademarks of the National Fire Protection Association, Inc., Quincy, MA 02269. All *National Electrical Code*® and *NEC*® references in this module refer to the 2011 edition of the *National Electrical Code*®.

49110-11

Rigging

Explains how to select and use rigging equipment. Covers common rigging equipment and rigging methods that are likely to be used by power line workers. Also covers hand signals and other methods of communication between the rigger and the crane operator. (12.5 Hours)

49111-11

Setting and Pulling Poles

Provides instructions for the storage, loading, and transport of wooden utility poles. Includes use of the digger derrick to dig the hole and install the pole. Also covers pole removal using a hydraulic jacking device. (20 Hours)

49112-11

Trenching, Excavating, and Boring Equipment

Covers the use and maintenance of trenching equipment, backhoe/loaders, and horizontal directional drilling equipment for the installation of direct-buried power lines. Includes a review of safety guidelines related to buried utilities. (7.5 Hours)

49113-11

Introduction to Electrical Test Equipment

Introduces the basic test equipment used by electrical workers to test and troubleshoot electrical circuits. Also covers specialized line worker test equipment, including the high-voltage detector, phase rotation tester, megohmmeter, phasing stick, and hi-pot tester. (7.5 Hours)

Glossary

Index

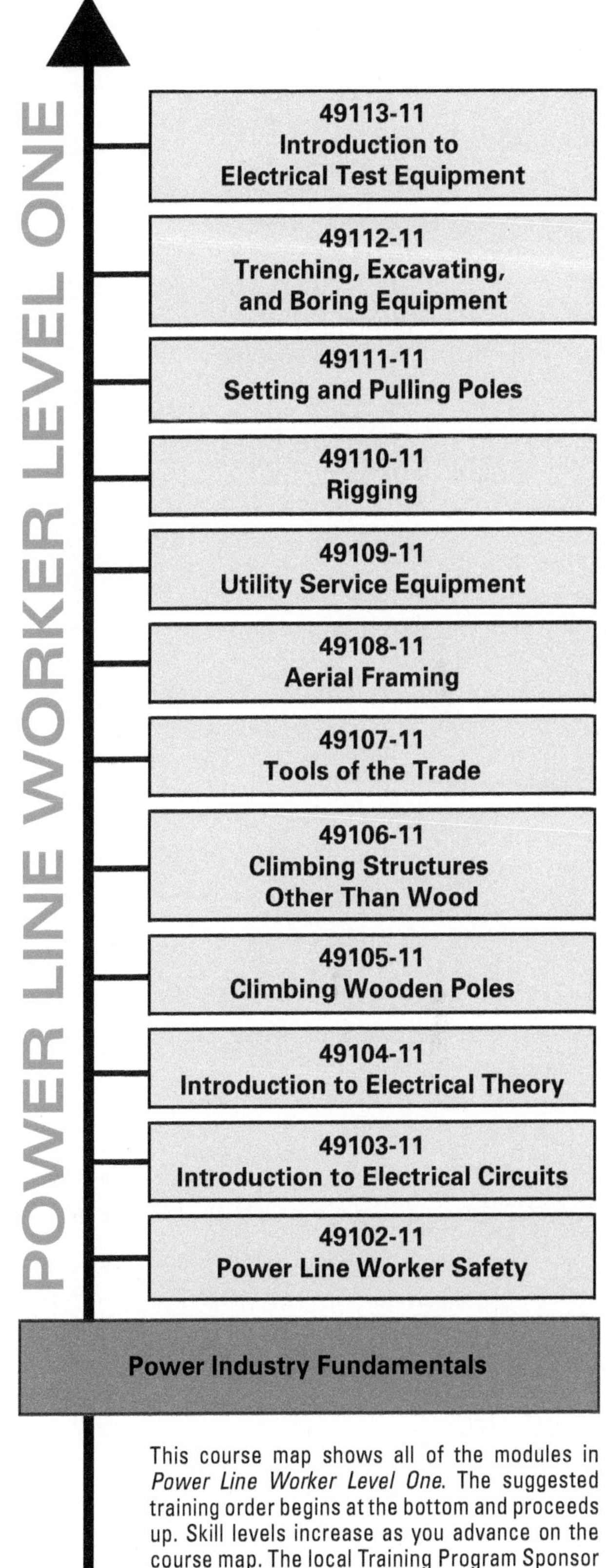

This course map shows all of the modules in *Power Line Worker Level One*. The suggested training order begins at the bottom and proceeds up. Skill levels increase as you advance on the course map. The local Training Program Sponsor may adjust the training order.

Power Line Worker Safety

49102-11

Trainees with successful module completions may be eligible for credentialing through NCCER's National Registry. To learn more, go to **www.nccer.org** or contact us at **1.888.622.3720.** Our website has information on the latest product releases and training, as well as online versions of our *Cornerstone* newsletter and Pearson's Contren® product catalog.

Your feedback is welcome. You may email your comments to **curriculum@nccer.org**, send general comments and inquiries to **info@nccer.org**, or use the User Update form at the back of this module.

V.1 7/11

49102-11

Power Line Worker Safety

Objectives

When you have completed this module, you will be able to do the following:

1. Identify, inspect, maintain, and use craft-specific PPE and identify its limitations.
2. Inspect rubber insulating blankets, line hoses, covers, and guards.
3. Describe the safety practices associated with high-voltage work, including:
   - Step and touch potential
   - Minimum approach distance
   - Protection from arc flash and arc blast
   - Procedures for entering substations
4. Explain work zone safety requirements.
5. Describe traffic control methods.
6. Identify the signs and causes of unstable trenches and describe the safety practices associated with trench work.
7. Identify hazards related to working near horizontal drilling operations.
8. Identify hazards and safeguards associated with confined-space work.
9. Explain the purposes of, and differences between, job safety analyses and task safety analyses.
10. Describe how to mitigate environmental impacts.

Performance Task

Under the supervision of the instructor, you should be able to do the following:

1. Inspect and put on craft-specific PPE.
2. Inspect rubber insulating blankets, line hoses, covers, and guards, and install them on deactivated power lines.

Trade Terms

Arc blast
Arc fault
Arc flash
Arc rating
Arc thermal performance value (ATPV)
Atmospheric hazard
Benching
Blast hazard
Boring
Combustible
Dielectric
Drill string
Electrically safe work condition
Energy of break-open threshold (E_{BT})
Equipotential plane
Flame-resistant (FR)
Flash hazard
Flash hazard analysis
Flash protection boundary
Grounding mat
Hypothermia
Limited approach boundary
Maximum allowable slope
Minimum approach distance (MAD)
Operator presence sensing system (seat switch)
Oxygen-deficient
Prohibited approach boundary
Qualified worker
Restricted approach boundary
Shield
Step potential
Subsidence
Temporary grounding device (TGD)
Touch potential

Contents

Topics to be presented in this module include:

Figures and Tables

Figures and Tables (*continued*)

1.0.0 Introduction

Electrical line workers perform a vital role in providing electrical power to homes, businesses, schools, and hospitals. A dependable supply of electricity is seldom noticed, but an interrupted supply is always noticed.

A power line worker's primary duty is to safely install and maintain the electrical transmission and distribution system. Safety is the first priority.

On the job, you must follow safety policies and procedures. It is important to understand that these documents establish the minimum standards required. At any time, your supervisor, a co-worker, or you may decide to use a higher standard. A higher standard is acceptable, but a lower standard is never acceptable.

As a line worker, it is your responsibility to perform your duties safely and to ensure that your co-workers perform their jobs safely. To stay safe on the job, you need to be able to identify hazardous aspects of your job. Then you need to know how to take measures to protect yourself and others against these hazards while safely performing your job.

To an inexperienced worker, this can be an overwhelming task. You will learn the basics of job safety during this course, but knowledge alone will not protect you and your co-workers. You will need to apply this knowledge effectively, and that takes time and experience.

Initially, you will work with seasoned workers who will help you gain experience. During this time, it is important to be a good co-worker. To learn, you need to listen carefully to instructions, observe other workers performing a job, and stay alert to any hazards.

1.1.0 Industry Standards

Electrical line work is covered by a number of industry and government standards. Among them are the following:

- *ASTM F1506, Standard Performance Specification for Flame-Resistant Textile Materials for Wearing Apparel for Use by Electrical Workers Exposed to Momentary Electric Arc and Related Thermal Hazards*
- *ASTM F1505, Standards for Insulated and Insulating Hand Tools*
- *ASTM D1048, Standard Specification for Rubber Insulating Blankets*
- *ASTM F479, Standard Specification for In-Service Care of Insulating Blankets*
- *NFPA 70®, National Electrical Code® (NEC®)*
- *NFPA 70E®, Standard for Electrical Safety in the Workplace*
- *OSHA Standard 29, Part 1910, Subpart R, Section 269*
- *National Electrical Safety Code® (NESC®)*, supplied by the IEEE provides requirements for electrical installations
- *ASTM Z87.1 Standards for Safety Glasses*
- *OSHA Standard 29, Section 1910.147* identifies conditions for simple lockout procedures for non-electrical work
- *ANSI/ASSE Z87.1 – 2003, Occupational and Educational Personal Eye and Face Protection Devices*

Virtually everything done on the job and every piece of equipment used is governed by a standard. Many of these standards are referred to by name and title in this module. At this point in your career, it is not as important for you to memorize the name and number of a standard as it is for you to be aware of its existence.

One exception to this is the OSHA standards. OSHA sets minimum safety requirements for your job in *OSHA Standard 29, Part 1910, Subpart R, Section 269*. This standard specifically covers electric power generation, transmission, and distribution systems. Most companies have a safety program that will include safety rules, policies, and procedures that will meet or exceed all minimum safety regulations established by OSHA. OSHA standards can be found on the OSHA web site (**www.osha.gov**). While studying this module, go onto the OSHA web site and look at some of the requirements for your job.

Your instructor will be able to show you examples of other industry specifications that apply to your job. As you progress at your workplace, you will be expected to learn more about industry specifications and how to use them. Your employer will have copies in their reference library.

1.2.0 Line Worker Safety

It is part of your job to know how and when to use all of your employer's safety rules, policies, and procedures. It will take time to become familiar with all safety guidelines, but until then you can use the following basic rules of safety to stay safe:

- Use tools, equipment, and personal protective equipment the way they were designed.
- Wear a hardhat and safety equipment at all times where required.
- Immediately correct or report to a supervisor all unsafe conditions.
- Perform only tasks for which you have been trained.
- Work toward understanding company safety rules and policies.

- Take responsibility for yourself and your co-workers.
- Inspect tools, PPE, and equipment daily before use and never use damaged or unsafe equipment.
- Follow your employer's policy for tagging out, repairing, or discarding unserviceable tools and equipment.
- Get involved with your company safety program.
- When in doubt, stop and ask before proceeding!

2.0.0 Introduction to Electrical Power and Hazards

Even when electrical equipment is working properly, it can create hazardous situations for untrained workers. When a fault occurs, the results can be catastrophic. It is important to have a basic understanding of the principles of electricity and its potential hazards so that you can protect yourself.

2.1.0 Electrical Power

In order to work safely around electrical power, you need to keep this principle in mind: when voltage has a complete path, current flows. *Figure 1* shows a simple lamp circuit with a dimmer switch. When the switch is set to Off as shown in *Figure 1A*, no current flows so the lamp does not glow. When the switch is set to On, current flows and the lamp glows as shown in *Figure 1B*. A simple dimmer switch is used to represent variable resistance. When more resistance is placed in the circuit, there is less current flow, and the lamp dims. As resistance is removed from the current path, the current flow increases and the lamp brightens.

A complete electrical circuit requires at least two conductors. One conductor allows current to flow from the source, while the second conductor allows it to flow back to the source. One of the characteristics of the earth is that it conducts electricity, so another way to complete a circuit is to use the earth as one of the conductors (*Figure 2*).

> **On Site**
>
> **Fuses**
>
> Did you ever wonder about how a fuse works? A fuse is used to protect a circuit against overcurrent conditions and is made from a thin strip of metal. Some fuses are so small they are little more than a fine strand of wire. As current flows through the fuse, the metal gets hot. If too much current flows, the fuse melts, which stops current flow. That's a blown fuse.

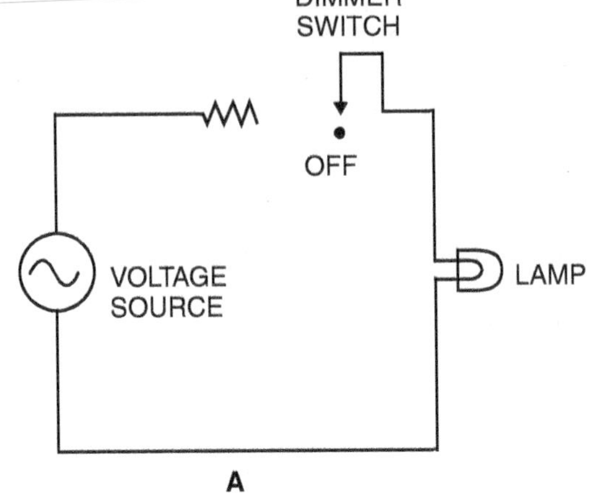

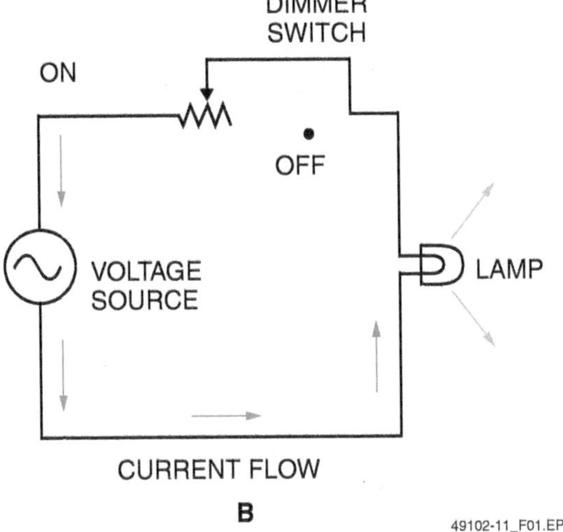

Figure 1 Simple electrical circuit.

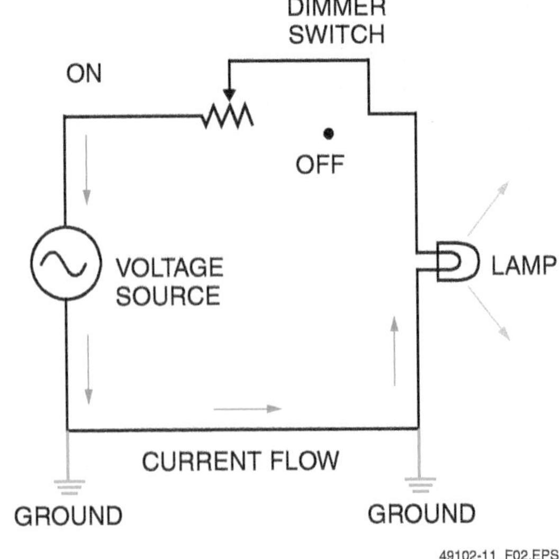

Figure 2 Earth ground.

Virtually all electrical circuits use the earth as a reference. To complete the current path, a copper rod is driven deep into the earth and used as the reference. This point is commonly called ground.

2.2.0 Electrical Hazards

Electrical hazards are caused by current flowing through unintended paths. These paths can be any conductive object, such as metal rods, fences, or your body. Anyone working on or near electrical equipment may encounter one of the following electrical hazards:

- Shock hazard
- Step potential and touch potential
- Arc flash and arc blast

2.2.1 Shock Hazard

In an electrical distribution system, the path of current flow is made up of the electrical power lines, transformers, bus bars, and other electrical devices. Current travels along its intended path, so the circuit is safe unless contact is made with an energized part. When you contact ground (or grounded objects) and an energized part, you complete a circuit and voltage and current will flow through you. When a fault occurs, unintended paths of current flow may develop, causing normally de-energized parts to become energized. This condition is known as a ground fault.

The damage from an electrical contact depends on the amount of current flowing through the person, which is a function of voltage and body resistance, as well as the parts of the body receiving current flow. Current flow through the chest cavity (heart) and the head (brain) is the most lethal. *Table 1* shows the effects of current on the human body.

A 9V battery has a typical current flow of 5 to 15mA. As shown in *Table 1,* a minor shock of 5mA can come from something as small as a 9V battery. The human body can easily resist this magnitude of electrical shock, but even a shock at this level can cause an involuntary movement away from the source. This can result in injuries from jumping back from the electrical shock source.

When the current is between 6mA and 30mA, the shock causes loss of muscular control. This may result in the worker falling from an elevated position or may cause the worker to fall into a more dangerous electrical source. So even though the shock itself may have no lasting effects, the results of the shock may cause serious injuries.

As the current increases beyond 20mA, muscle contractions can prevent the victim from pulling away from the current source. At 50mA, respiratory paralysis may result in suffocation. Current levels above 150mA may cause the heart to beat in an abnormal rhythm. This condition is fatal unless the heart rhythm is corrected using a medical device called a defibrillator. Current levels of 4A or more may stop the heart, resulting in death unless immediate first aid is provided. Remember, voltage doesn't kill; current kills.

Other effects of electrical shock include entry and exit wounds from high-voltage contact, and thermal burns from current flow of a few amps and greater. Thermal burns are often not apparent at first; however, the tissue in the current path may be destroyed, causing the skin to die from the inside over time.

Current will follow all paths back to its source. Given the premise that current follows all paths to ground, you can prevent yourself from being shocked by being electrically isolated from ground. This means that no part of your body touches a ground point. Another way to prevent electrical shock is to create an equipotential plane where you are working.

Table 1 Effects of Current on the Human Body

Current Value	Typical Effects
1mA	Perception level. Slight tingling sensation.
5mA	Slight shock. Involuntary reactions can result in other injuries.
6 to 30mA	Painful shock, loss of muscular control.
50 to 150mA	Extreme pain, respiratory arrest, severe muscular contractions. Death possible.
1,000mA to 4,300mA	Ventricular fibrillation, severe muscular contractions, nerve damage. Typically results in death.

49102-11_T01.EPS

Did You Know?

Approximately 30,000 nonfatal and 1,000 fatal electrical shock accidents occur each year. However, the majority of hospital admissions due to electrical accidents are from arc flash burns, not from shocks. An arc flash releases thermal energy at temperatures up to 35,000°F.

Source: *NFPA 70E®, Electrical Safety in the Workplace, Annex K.*

2.2.2 Step and Touch Potentials

The earth has resistance from one point to another. The amount of resistance depends on the type of soil in the area, the amount of moisture in the soil, and the presence of underground pipes and lines. In addition, when working with very high voltages, and because of the earth's resistance, it is possible for a voltage to develop between two points of the earth.

A good example of this occurs in switchyards and substations. Switchyards and substations are used to isolate lines and equipment, increase and decrease voltage for further transmission, and to reduce voltage for local distribution. The ground grid around these facilities is buried deep in the earth and extends several feet outside of the fence perimeter. It is bonded to the fence at frequent intervals. The intent of the ground grid is to eliminate any voltage potential that may develop. In addition, the grid provides a path to the earth for system neutral and dissipates lightning and switching surges. When potentials do develop in the earth, it is usually during a fault situation. These potentials are called step potential and touch potential (*Figure 3*).

Step potential is the voltage between the feet (usually about 3.28 feet or 1 meter in length) of a person standing near an energized grounded object. It is equal to the difference in voltage between two points at different distances from the electrode. The substation yard is covered with several inches of crushed gravel to provide a level of insulation or high resistance between a person and earth so a step potential can't develop.

Touch potential is the voltage between the energized object being touched and ground. The ground path may be through any other body part of the person touching the object. For example, when you touch an object with one hand and a grounded object with the other, the current path is through the chest cavity so current flows through the heart; when you touch an object with one hand and are standing on a grounded surface, the current path is from your hand through your body to your feet. The touch potential can be very high when the object is grounded far from the place where the person is in contact with it. As shown in the figure, there is also a transferred touch voltage where the difference in potential is great enough to create an arc between the grounded object and the person.

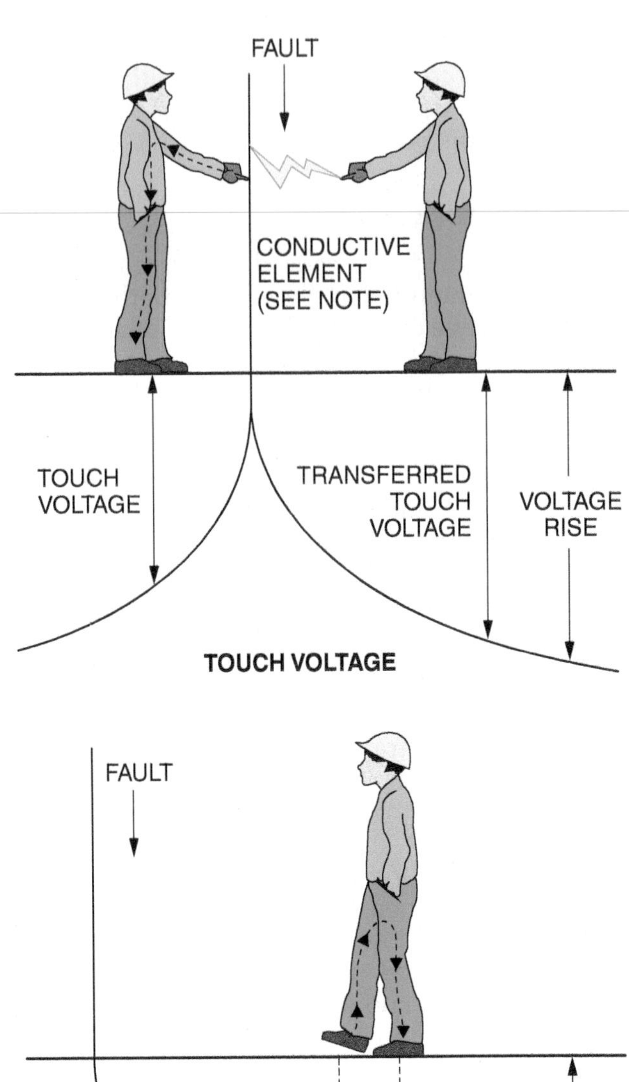

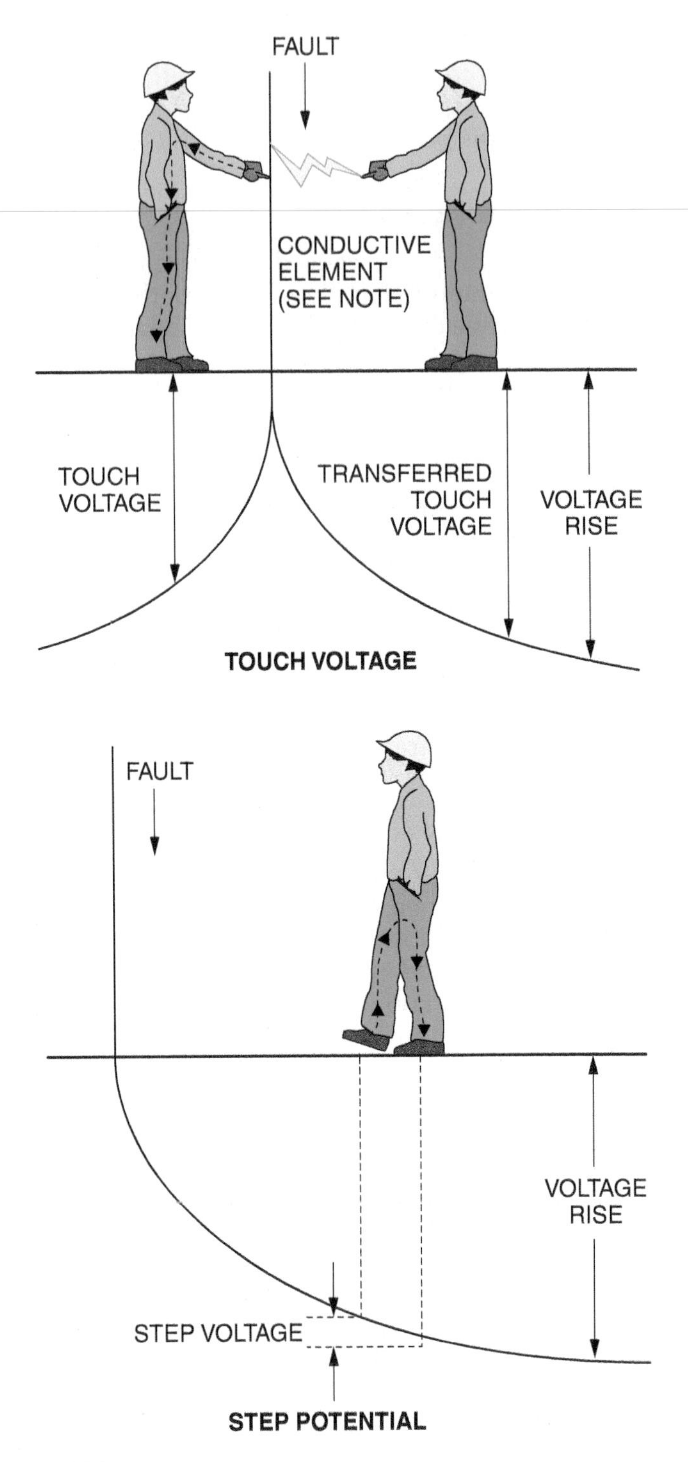

Figure 3 Step and touch potential.

2.2.3 Arc Flash and Arc Blast

Air is not typically a good conductor of electricity. However, there are certain conditions in which current can jump through the air to reach another point and complete the path of current flow. This is called an arc. Arcs can happen anywhere there is voltage.

In the winter, when the air is dry, you may cause an arc when you touch a metal object and discharge static electricity. When this happens,

the electrical charge can be seen passing through the air from one point to the other. That visible arc is called an arc flash. The static produces a small amount of current flow that can shock you when you discharge it. While the shock from static electricity is uncomfortable, it is not dangerous because the voltage source is static. Once it is discharged, the source is gone.

When working around high voltage, an arc can be very dangerous because there is a constant and powerful source behind it that allows a tremendous amount of current flow. The current is called **arc fault** current and it causes an arc flash (*Figure 4*). Depending on the source, an arc flash can release an enormous amount of thermal energy to the point that an explosion occurs. This is called an arc blast and it is comparable to a dynamite blast. It can destroy nearby equipment, melt metals, and cause serious injuries from flying shrapnel and molten metal.

3.0.0 Electrical Safety

The equipment used to transfer electrical power from the power generating plant to the customer is designed to eliminate arcs and to make equipment safe. However, nothing is foolproof. The grounding and bonding systems in substations carry electric current into the earth under both normal and fault conditions. For that reason, a broken or disconnected grounding or bonding connection can be a lethal hazard to a worker completing the path to ground under fault conditions or when induced voltage is possible. It can also result in loss of service and equipment damage.

The first step in protecting yourself is to recognize when a hazard exists and then to avoid it. There is always the potential for an electrical shock when working around any level of voltage. It is safest to work around de-energized equipment. Since this is not always possible, it is necessary to establish hazard boundaries and understand how to use protective devices and equipment to lessen your risk when working around energized equipment.

Whenever possible, de-energize equipment before you begin working on it. Once equipment is de-energized, lockout/tagout devices must be attached to the source to prevent someone from unknowingly energizing the circuit. In addition, a **temporary grounding device (TGD)** must be used to protect workers from electrical shock hazards should the circuit somehow become energized.

Finally, a number of tools and personal protective equipment (PPE) offer protection. Some types of PPE are specially designed and rated for electrical hazard protection, while others are used for more general hazards. For example, safety shoes can protect your toes from falling objects, but electrical safety shoes can also help protect you from electrical shock. Insulated tools are essential in this type of work.

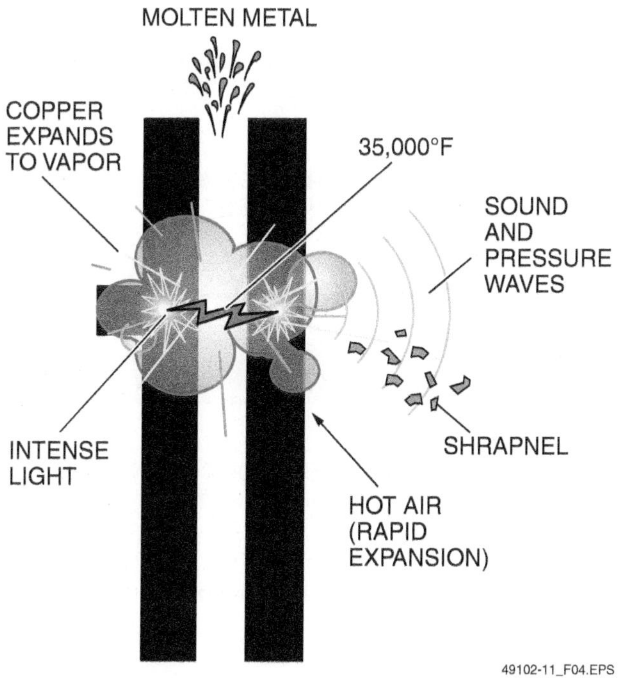

MOLTEN METAL
COPPER EXPANDS TO VAPOR
35,000°F
SOUND AND PRESSURE WAVES
SHRAPNEL
INTENSE LIGHT
HOT AIR (RAPID EXPANSION)
49102-11_F04.EPS

Figure 4 Arc flash.

3.1.0 Hazard Boundaries

OSHA regulations address safety issues in the workplace, including electrical safety issues. While OSHA is a federal agency and its regulations are law, OSHA also relies on national consensus standards for certain requirements. For electrical safety, OSHA recognizes several standards from the National Fire Protection Association (NFPA). *NFPA 70®, National Electrical Code®* (*NEC®*), provides requirements for electrical installations. *NFPA 70E®, Standard for Electrical Safety in the Workplace,* provides practical safe working requirements related to hazards arising from the use of electricity. The additional PPE that *NFPA 70E®* requires also applies to generation station and substation work. Always follow and maintain company safety policies.

Special safety procedures are required when working on or near circuits with voltage levels of more than 50V line-to-line. Maintaining a safe distance is the best protection against electrical hazards.

Only a qualified worker may work unsupervised in areas with unguarded, uninsulated energized lines or equipment operating at 50 volts or more. An unqualified worker must remain well out of the danger zone at specified distances as follows:

- There is no safe distance specified for voltages less than 50 volts
- 10 feet (3.05 meters) for 50 to 750 volts
- 20 feet (6.1 meters) for greater than 750 volts

Industry standards have established specific limits of approach to exposed energized parts. These limits are for personal protection and are called approach boundaries. Hazard boundaries include a flash protection boundary, shock protection boundary, and minimum approach distance (MAD). Only qualified persons are allowed within flash and shock protection boundaries and MADs. Unqualified personnel must be trained to stay away from potentially dangerous electrical equipment and processes. *Figure 5* shows a diagram comparing flash and shock protection boundaries. When working inside these boundaries, special PPE, tools, and other equipment must be used.

When working with electrical equipment, assume that the equipment is energized until you have personally verified otherwise. You should assume that all electrical equipment presents potential shock hazards, flash hazards, and blast hazards regardless of voltage sources until you have verified otherwise.

Ideally, work on or near electrical equipment would always be performed with no electrical power applied (also known as an electrically safe work condition), but that is not always possible. Therefore, equipment installations are analyzed by specially trained workers to identify electrical hazards and to establish hazard boundaries.

Every possible electrical hazard within a work area must be analyzed and documented. This is called a flash hazard analysis. These areas must be clearly marked with appropriate signs (see *Figure 6*) indicating the hazard as well as information about what PPE must be worn in the area. Specific PPE for a given situation is based on the information gathered from the analysis of a given hazard. That documented data includes all the electrical hazards (arc flash, blast, and shock). After all hazards have been documented, all personnel (qualified and unqualified) working in the area must be trained to recognize and avoid the identified hazards. Only qualified persons using all required PPE are allowed to enter and work inside the electrical hazard boundaries.

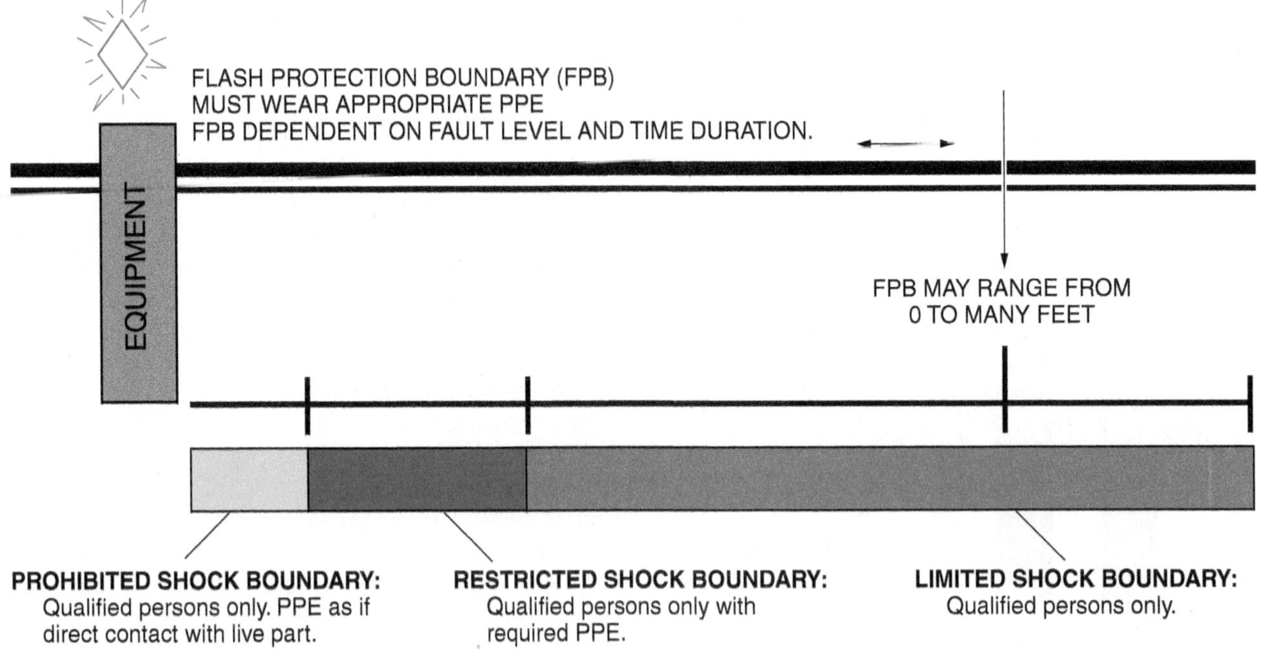

Figure 5 Approach limits.

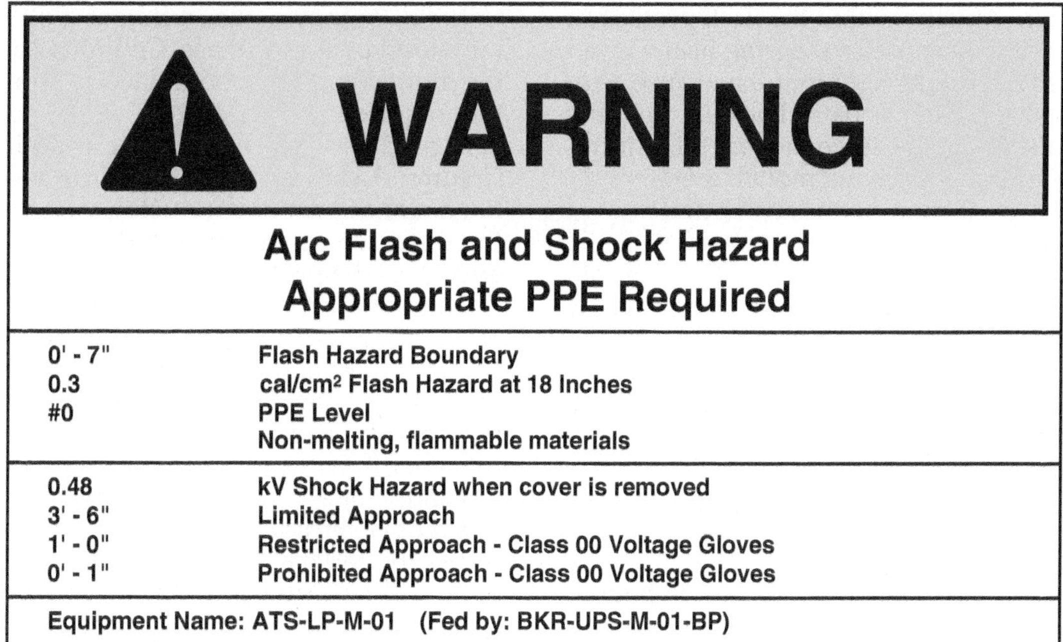

WARNING
Arc Flash and Shock Hazard
Appropriate PPE Required
0' - 7"
Flash Hazard Boundary
0.3
cal/cm2 Flash Hazard at 18 Inches
#0
PPE Level
Non-melting, flammable materials
0.48
kV Shock Hazard when cover is removed
3' - 6"
Limited Approach
1' - 0"
Restricted Approach - Class 00 Voltage Gloves
0' - 1"
Prohibited Approach - Class 00 Voltage Gloves
Equipment Name: ATS-LP-M-01 (Fed by: BKR-UPS-M-01-BP)
49102-11_F06.EPS

Figure 6 Electrical hazard warning sign.

3.1.1 Flash Protection Boundary

When an arc flash hazard is present, an arc flash protection boundary must be established. This boundary is determined by how far away a person would need to be located to avoid receiving serious burns in the event of an arc flash. Anyone within the flash protection boundary is exposed to the possibility of both second-degree (blistering) and third-degree (tissue-destroying) burns should arcing occur.

Workers inside the arc flash protection boundary must wear flame-retardant clothing, including underwear. The heat released during arcing can be at temperatures high enough to melt fibers in clothing made from acetate, nylon, polyester, and rayon, which would increase injuries a worker may receive. OSHA prohibits any clothing made from these fibers (fully or blends) unless the fibers have been specially treated to resist burning and melting.

Depending on the situation, an arc flash protection boundary might be within a shock protection boundary or outside of (exceed) a shock protection boundary. Many electrical safety programs establish both the flash protection boundary and the outer shock protection boundary at whatever distance is greater, as determined by the hazard analysis.

When an electrical fault causes an arc flash (*Figure 7*), the explosion produces both a fireball and a shock wave extending away from the arc

Figure 7 Arc flash.

flash location. Anyone within the flash protection boundary will be exposed to searing heat as well as an extremely bright light that may cause temporary loss of vision and pain. The heat from arc flashes is often hot enough to melt metal fixtures, causing a hazard from flying molten metal.

The flash protection boundary is determined for thermal energy, but blast accompanies the arc flash. The blast creates a shock wave that can blow equipment apart and blow people away from the blast. Shrapnel, toxic gases, and copper vapor explode in all directions. The blast also creates sound waves that can damage hearing. The amount of current flowing through the fault affects the size of the arc flash.

3.1.2 Shock Protection Boundaries

There are three electrical shock protection boundaries or limits of approach. *NFPA 70E®* identifies the following electrical shock boundaries:

- Limited approach boundary
- Restricted approach boundary
- Prohibited approach boundary

Table 2 shows the shock protection approach boundaries to exposed energized parts listed. Column 1 shows the different voltage levels, measured phase-to-phase. Columns 2 and 3 cover the limited approach boundary. Column 2 shows the required distance from an exposed movable conductor (such as an overhead line), while Column 3 shows the distance from an exposed fixed circuit part or conductor. Column 4 covers the restricted approach boundary, and Column 5 addresses the prohibited approach boundary.

The exposed energized component can be a wire or a mechanical component inside the electrical equipment. All boundary distances are measured from that point. When establishing boundaries, exposed movable conductors are treated differently than exposed fixed circuit parts or conductors.

Limited approach boundary – This is a shock protection boundary at a specified distance from an exposed energized part that can be crossed only by qualified persons. Unqualified persons may cross a limited approach boundary for purposes of on-the-job learning and when escorted by a qualified person. All unqualified personnel in the area must be made aware of the hazards and warned not to cross the boundary.

Restricted approach boundary – This is a shock protection boundary that, due to its proximity

Table 2 Approach Boundaries for Shock Protection [Data from *70E® Table 130.2(C)*]

Nominal System Voltage Range, Phase to Phase[2]	Limited Approach Boundary[1]		Restricted Approach Boundary[1] Includes Inadvertent Movement Adder	Prohibited Approach Boundary[2]
	Exposed Movable Conductor[3]	Exposed Fixed Circuit Part		
Less than 50V	Not specified	Not specified	Not specified	Not specified
50V to 300V	3.05m (10 ft 0 in)	1.07m (3 ft 6 in)	Avoid contact	Avoid contact
301V to 750V	3.05m (10 ft 0 in)	1.07m (3 ft 6 in)	304.8mm (1 ft 0 in)	25.4mm (0 ft 1in)
751V to 15kV	3.05m (10 ft 0 in)	1.53m (5 ft 0 in)	660.4mm (2 ft 2in)	177.8mm (0 ft 7 in)
15.1kV to 36kV	3.05m (10 ft 0 in)	1.83m (6 ft 0 in)	787.4mm (2 ft 7in)	254mm (0 ft 10 in)
36.1kV to 46kV	3.05m (10 ft 0 in)	2.44m (8 ft 0 in)	838.2mm (2 ft 9in)	431.8mm (1 ft 5 in)
46.1kV to 72.5kV	3.05m (10 ft 0 in)	2.44m (8 ft 0 in)	1.0m (3 ft 3 in)	660mm (2 ft 2 in)
72.6kV to 121kV	3.25m (10 ft 8 in)	2.44m (8 ft 0 in)	1.29m (3 ft 4 in)	838mm (2 ft 9 in)
138kV to 145kV	3.36m (11 ft 0 in)	3.05m (10 ft 0 in)	1.15m (3 ft 10 in)	1.02m (3 ft 4 in)
161kV to 169kV	3.56m (11 ft 8 in)	3.56m (11 ft 8 in)	1.29m (4 ft 3 in)	1.14m (3 ft 9 in)
230kV to 242kV	3.97m (13 ft 0 in)	3.97m (13 ft 0 in)	1.71m (5 ft 8 in)	1.57m (5 ft 2 in)
345kV to 362kV	4.68m (15 ft 4 in)	4.68m (15 ft 4 in)	2.77m (9 ft 2 in)	2.79m (8 ft 8 in)
500kV to 550kV	5.8m (19 ft 0 in)	5.8m (19 ft 0 in)	3.61m (11 ft 10 in)	3.54m (11 ft 4 in)
765kV to 800kV	7.24m (23 ft 9 in)	7.24m (23 ft 9 in)	4.84m (15 ft 11 in)	4.7m (15 ft 5 in)

Note: For arc flash protection boundary, see *70E® Section 130.3(A)*.

1. See definition in *70E® Article 100* and text in *70E® Section 130.2(D)(2)* and *Annex C* for elaboration.
2. For single-phase systems, select the range that is equal to the system's maximum phase-to-ground voltage multiplied by 1.732.
3. A condition in which the distance between the conductor and a person is not under the control of the person.The term is normally applied to overhead line conductors supported by poles.

49102-11_T02.EPS

to exposed energized parts, requires the use of shock protection techniques and equipment when crossed. The restricted approach boundary may be crossed only by qualified persons using the required PPE (*Figure 8*) and authorized by an energized electrical work permit. Work within the restricted approach boundary requires that rubber insulating equipment be used. It also requires the use of insulated tools for voltages of 1,000V and below, or live-line tools above 1,000V. The restricted approach boundaries include an added safety margin to compensate for inadvertent movement of the worker. The interaction of exposed energized parts and test equipment or hand tools is the cause of many shock and arc flash incidents when the tool or test lead becomes part of the circuit path. Some estimates are that 75 percent or more of arc incidents begin in this way.

Prohibited approach boundary – This is a shock protection boundary at a specified distance from an exposed energized part. Work within this boundary is considered the same as making contact with the energized part. Any part of the body crossing the prohibited approach boundary must be suitably insulated and protected. There is no margin for inadvertent movement in the prohibited approach boundary.

Figure 8 Worker using appropriate PPE inside a restricted approach boundary.

WARNING!

Any workers exposed to electrical hazards must be qualified to manage the hazard and must use required PPE. If they are not exposed to an electrical hazard but have the potential to be, they must be informed of the hazard and instructed in how to avoid it.

3.1.3 Minimum Approach Distance

Energized transmission lines represent another electrical hazard to line workers. OSHA has established MADs for working in the area of these lines. A MAD specifies the closest distance from any energized component that an unprotected worker may approach. It specifically refers to any unprotected body part of a worker or any part of a conductive object such as a tool carried by a worker. To violate the MAD, a worker must wear some type of protective insulation device or use special live-line tools. Protective insulation devices can include insulated gloves and sleeves, hoods, and suits. They can also include devices that are attached to equipment, such as rubber blankets, hose covers, and guards. The minimum distance is based on voltage levels found at the equipment. The MAD for qualified workers in areas with unguarded, uninsulated energized lines is shown in *Table 3*.

NOTE

When you wear an insulating device, keep in mind that only the protected body part may violate the MAD. For example, when wearing insulating gloves, only your hands and the parts of your arms that are covered by the gloves can be inside the MAD. Uncovered parts of arms must remain outside the MAD.

3.2.0 De-Energized Equipment

Whenever possible, de-energize equipment before working on it. Most companies have written guidelines that you must follow to de-energize any equipment. It is important to follow all guidelines because many circuits receive power from multiple sources. To place equipment in an electrically safe work condition, all sources must be removed.

Table 3 Energized Line Work Minimum Approach Distance

Kilovolts (AC)	Phase-to-Ground	Phase-to-Phase
0.05 to 1.0	Avoid contact	Avoid contact
1.1 to 15.0	2'-1" (0.64m)	2'-2" (0.66m)
15.1 to 36.0	2'-4" (0.72m)	2'-7" (0.77m)
36.1 to 46.0	2'-7" (0.77m)	2'-10" (0.85m)
46.1 to 72.5	3'-0" (0.90m)	3'-6" (1.05m)
72.6 to 121	3'-2" (0.95m)	4'-3" (1.29m)
138 to 145	3'-7" (1.09m)	4'-11" (1.50m)
161 to 169	4'-0" (1.22m)	5'-8" (1.71m)
230 to 242	5'-3" (1.59m)	7'-6" (2.27m)
345 to 362	8'-6" (2.59m)	12'-6" (3.80m)
500 to 550	11'-3" (3.42m)	18'-1" (5.50m)
765 to 800	14'-11" (4.53m)	26'-0" (7.91m)

From *OSHA 29 CFR 1910.269 Electric Power Generation, Transmission, and Distribution*

Once equipment is de-energized, lockout/tagout devices must be attached to all sources to prevent someone from unknowingly energizing the circuit. Then you must verify that no power is applied to the circuit using a voltmeter. Finally, temporary grounding devices must be placed at the area where the work is performed to protect workers from hazardous differences of electrical potentials should the circuit somehow become energized.

The following is a general procedure based on the requirements of *OSHA Standard 29, Section 1910.269* for placing equipment into an electrically safe work condition. Most companies have detailed written procedures for performing this task.

Step 1 Determine whether any other crews are working on the circuit. All distribution lines are treated as if they are energized unless your team has performed a de-energizing procedure. If two or more crews are working on the same lines or equipment, each crew must independently perform a de-energizing procedure and apply their own lockout/tagout devices to energy controls.

Step 2 Designate one qualified member of the crew as the employee in charge of the electrical clearance.

Step 3 Determine whether the source control is accessible to those not on the work crew. If so, the control must be rendered inoperable (removed or locked or otherwise disabled) during the performance of work.

Step 4 De-energize the section of line or equipment. Only the designated employee in charge has the authority to request that the system operator de-energize a section of line or equipment.

Step 5 Open, disable, or otherwise render inoperable all switches, connections, jumpers, and other devices through which known sources of energy may be supplied to the lines and/or equipment being serviced. Lockout and tagout controls to indicate that employees are at work.

Step 6 Open, disable, or otherwise render inoperable all automatic and remotely controlled switches that could apply power to the lines and/or equipment being serviced. Apply tags that prohibit operation of the disconnected device as well as indicate that employees are at work.

Step 7 After the electrical source(s) has been removed, test the lines and/or equipment with an approved voltage meter to ensure that they are de-energized.

Step 8 Install protective grounds as required.

Step 9 After all of these steps have been performed and the lines are found to be de-energized, the lines and equipment may be treated as de-energized.

While work is being performed and the employee in charge needs to transfer duties to another employee due to an emergency, the employee's supervisor must inform the system operator. The workers on the work crew must also be informed of the transfer. The new employee in charge is then responsible for the clearance.

When work has been completed, the employee in charge will perform the following to release a clearance:

- Notify employees on the crew that the clearance is to be released.
- Determine that all employees in the crew are clear of the lines, equipment, and tools.
- Determine that all protective grounds installed by the crew have been removed.
- Report this information to the system operator.
- Release the clearance.

The person releasing a clearance must be the same person that requested the clearance, unless responsibility has been transferred as described above. Clearances may be released and the circuits re-energized only when the following conditions have been met:

- All protective grounds have been removed.
- All crews working on the lines or equipment have released their clearances.

- All employees are clear of the lines and equipment.
- All protective tags have been removed from a given point of disconnection.

3.3.0 Procedures for Entering a Substation

As described in previous sections, switchyards and substations are constructed with a massive grounding grid buried deep in the earth under them. High fencing with locked entries surround them. The fencing is there to keep unauthorized personnel out. The fences also have warning signs posted all around. The fencing and warning signs are specifically designed to keep personnel away from the high voltage equipment inside the fence.

To reduce the risk of shock, the grounding grid extends well beyond the fence in order to protect anyone who might come too close. Inside the fencing, walking areas are covered with gravel, which adds a layer of protective insulation.

Only trained, qualified, and authorized personnel are allowed inside the fence surrounding a switchyard or substation. They must be wearing all the approved PPE for such work. That includes all applicable arc flash equipment if work must be performed within specified distance of a potential arc source. All unqualified workers must be kept outside the fence and outside the hazard areas. Anyone entering a substation should review all the flash and shock boundaries associated with the switchyard or substation that is to be entered. Workers entering a switchyard or substation may also need signed work permits before entering the work areas.

4.0.0 Protective Equipment

Protective equipment includes PPE along with task-specific tools and devices that are used to help keep workers safe while working on energized and de-energized equipment. It is important to understand the uses and limitations of equipment. Remember, safety equipment is designed to keep you safe as you perform your job. It is not a substitute for safe working practices, and it is not a reason to bypass safety devices or procedures. Further, all protective equipment has electrical ratings and limits. It is up to you to know your equipment's ratings and limitations and to use the correct PPE.

Your job requires the use of insulated or insulating hand tools; live-line tools such as insulated hot sticks with various tools attached; temporary grounding devices; and other protective equipment. These devices are manufactured according to industry standards. A standard will specify the ratings of equipment within the standard, the markings identifying equipment conforming to the standard, and the manufacturer testing requirements. Other standards address requirements for in-service care, inspection, and testing. For example, *ASTM F479, Standard Specification for In-Service Care of Insulating Blankets* identifies care and maintenance for insulating blankets.

4.1.0 Personal Protective Equipment

You will be using PPE while performing your job. PPE is designed to help you to stay safe when you are performing your job. Almost all of the equipment that you will use will have an electrical rating. Electrically rated protective equipment has been designed and tested to meet electrical safety requirements set forth in industry standards. Some electrical and non-electrical safety equipment may appear similar. Before you use any equipment, be sure that you know and understand its rating and limitations.

In addition to wearing the correct PPE, you should remove all jewelry and avoid wearing decorative metal belt buckles.

National consensus standards, such as ASTM and ANSI, have specific requirements for the inspection, testing, and maintenance of various types of PPE. For PPE not covered by a specific standard, consult the manufacturer's literature supplied with the equipment.

The main items of PPE are the hard hat, eye protection, and work gloves. When working on a job site, employees may also be required to wear a reflective vest.

On Site

PPE

Many people hear the phrases *electrical safety* or *hazard* and automatically think of PPE. That is understandable because you have been trained to use PPE to protect yourself. However, PPE is not your first line of defense against electrical hazards. PPE comes into play only when all other possible measures have been taken to eliminate the need to be exposed to an electrical hazard, and is your last line of defense against injury or death.

4.1.1 Hard Hats

There are three electrical classes for hard hats. They have the following electrical insulation ratings:

- Class E (electrical) tested to withstand 20,000 volts
- Class G (general) tested to withstand 2,200 volts
- Class C (conductive) no electrical protection

Line workers wear Class E hard hats. Always inspect your hard hat before using it. Never use a hard hat that has damaged webbing or a cracked outer shell. Always adjust your hard hat to fit properly before you wear it. Replace hard hats according to company policy and manufacturer's schedule.

4.1.2 Footwear

There are several types of footwear that may be worn for line work. Regular safety shoes are worn to protect feet and toes. Electrical safety shoes have high-resistance soles that help isolate you from ground and reduce the severity of an electrical shock should you make contact with an energy source. They are rated at 14,000 volts at 60 hertz. These soles retain their resistance only when they are clean and dry. Wet, dirty, or oil-soaked soles conduct electricity and should not be worn.

Dielectric overshoes (*Figure 9*) provide more insulation than regular electrical safety shoes. The soles are tested to 20,000 volts under varying conditions. This type of footwear is used when there is a possibility of step potential. There are many styles of these shoes. Some soles are designed with a special tread and heel to improve safety while climbing poles. Some dielectric overshoes have a safety toe box and arch supports. They are available in both a shoe and boot style.

Climbing boots (*Figure 10*) are heavy-duty boots that provide line workers with good calf, arch, and ankle support and protection. They have deep heels for more secure pole climbing. They also have non-slip walking surfaces and oil-resistant soles that provide some protection from electrical shock. These boots are designed for protection when climbing poles. This footwear has minimal protection from electrical shock and should not be worn in the place of electrical safety shoes or dielectric overshoes.

4.1.3 Chaps

Chaps are leg coverings that are worn over regular clothing to increase leg protection (*Figure 11*). Line workers often wear chaps when working around thick vegetation and with chainsaws.

49102-11_F09.EPS

Figure 9 Dielectric overshoes.

49102-11_F10.EPS

Figure 10 Lineman climbing boots.

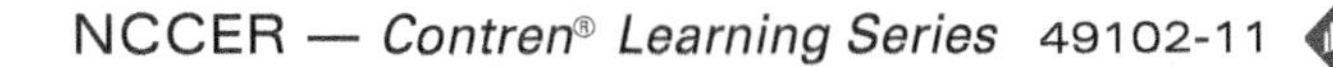

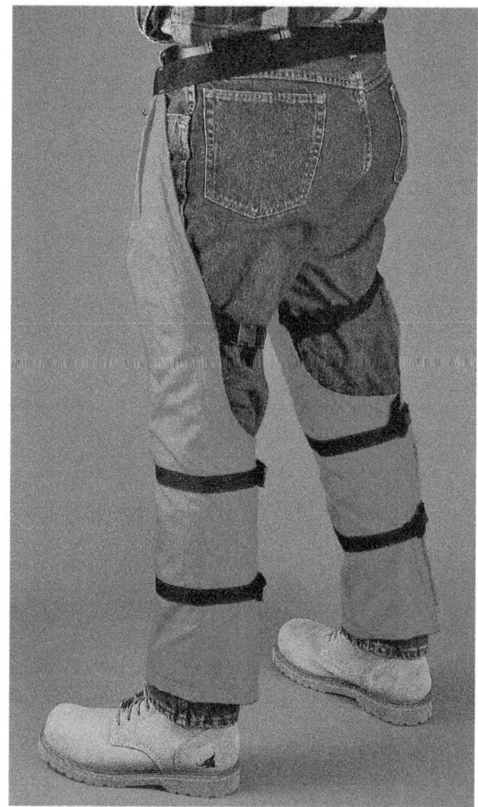

Figure 11 Protective chaps.

Typical chaps can be made from heavy fabric or leather, while more rugged chaps are made of Kevlar® or ballistic material.

4.1.4 Eye Protection

Standard safety glasses with or without side shields are designed to protect your eyes from flying debris. Arc flash goggles (*Figure 12*) are designed to protect your eyes from the intense light of an arc flash and from flying debris. Arc flash goggles do not protect the face from the heat of an arc flash. Full face shields are required in some situations. Glasses must meet *ANSI Z87.1* requirements.

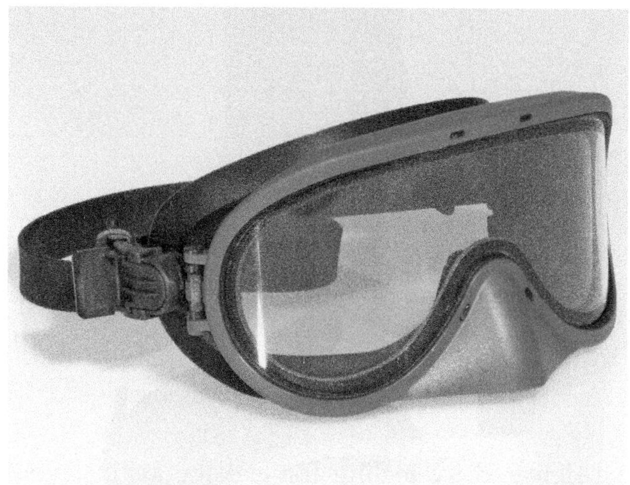

Figure 12 Arc flash goggles.

4.1.5 Hearing Protection

Hearing protection may be the insert type (earplug) or the cover-up type (ear muff) or a combination of both. Although not offered as arc-rated components, conventional earplugs have not been shown to melt or increase the chance of injury. OSHA requires a hearing protection program if noise levels exceed 85 Dba over an eight (8) hour period.

4.1.6 Hand and Arm Protection

Leather work gloves are worn to protect your hands from scrapes and cuts. Rubber insulating gloves are used to provide the hands with insulation from voltages. Rubber insulating sleeves are used to provide the arms with insulation from voltages. Rubber insulating devices must be rated for the voltages present. *Figure 13* shows the voltage ratings, labeling, and color codes for gloves. The higher the class number, the higher the voltage rating is. The voltage rating is marked on a color-coded label near the wrist of the glove. The glove color itself does not identify the voltage class. Rubber protective equipment has two voltage ratings: rated voltage and maximum use voltage.

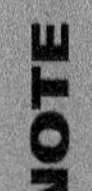

NOTE

The use of Classes 00 through 4 used for gloves has no relation to the hazard risk categories used to classify levels of arc flash hazard described in the next section.

Rubber insulating gloves are intended to be used only under their protective leathers (*Figure 14*). In those rare cases when a task requires dexterity not possible with leathers on, the leathers may be removed only to complete that portion of the task. Gloves that have been used without leathers can only be used at half of their voltage rating until passing an electrical test at the test voltage for the original rating. Cotton glove liners can be used to increase wearer comfort.

Class Color	Proof Test Voltage AC/DC	Max. Use Voltage AC/DC	Insulating Rubber Glove Label
00 Beige	2,500/10,000	500/750	10 SALISBURY ANSI / ASTM MADE IN U.S.A. D120 CLASS 00 TYPE I MAX USE VOLT 500V AC
0 Red	5,000/20,000	1,000/1,500	10 SALISBURY ANSI / ASTM MADE IN U.S.A. D120 CLASS 0 TYPE I MAX USE VOLT 1000V AC
1 White	10,000/40,000	7,500/11,250	10 SALISBURY ANSI / ASTM MADE IN U.S.A. D120 CLASS 1 TYPE I MAX USE VOLT 7500V AC
2 Yellow	20,000/50,000	17,000/25,000	10 SALISBURY ANSI / ASTM MADE IN U.S.A. D120 CLASS 2 TYPE I MAX USE VOLT 17000V AC
3 Green	30,000/60,000	26,500/39,750	10 SALISBURY ANSI / ASTM MADE IN U.S.A. D120 CLASS 3 TYPE I MAX USE VOLT 26500V AC
4 Orange	40,000/70,000	36,000/54,000	10 SALISBURY ANSI / ASTM MADE IN U.S.A. D120 CLASS 4 TYPE I MAX USE VOLT 36000V AC

49102-11_F13.EPS

Figure 13 Voltage rating codes of gloves.

WARNING!

Do not wear watches, rings, or bracelets under rubber gloves. Sharp edges can puncture gloves.

Rubber insulating gloves must be maintained in a safe and reliable condition, stored properly, and visually inspected before each use. Visually inspect rubber insulating equipment before use each day and whenever damage is suspected. Remove rubber gloves from the protective leathers for inspection and inspect both the rubber gloves and the leathers. Look for any grit or wire bits that can damage the gloves. *Figure 15* identifies common problems to look for during inspection. Turn rubber gloves and leathers inside out during inspection. Lightly stretch rubber blankets and gloves to look for small cracks or checking.

Rubber insulating gloves must also undergo an air test and/or a water test before each use and whenever damage is suspected. The gloves and sleeves may be inflated with a portable inflator (*Figure 16*). Gloves may be inflated by flipping the glove and capturing air by rolling the cuff upward. Inflate gloves to no more than half-again (150 percent) of their normal size. Air leaks may be detected by feeling for leakage against your cheek or looking for dust puffs when glove dust is used. On a job site, it may be necessary to find a sheltered area in which to perform the air test.

49102-11_F14.EPS

Figure 14 Rubber insulating gloves and leather.

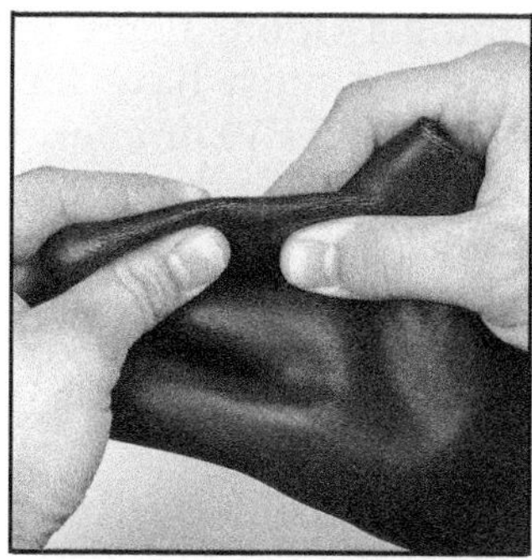

CRACKING & CUTTING
This type of damage is caused by prolonged folding or compressing.

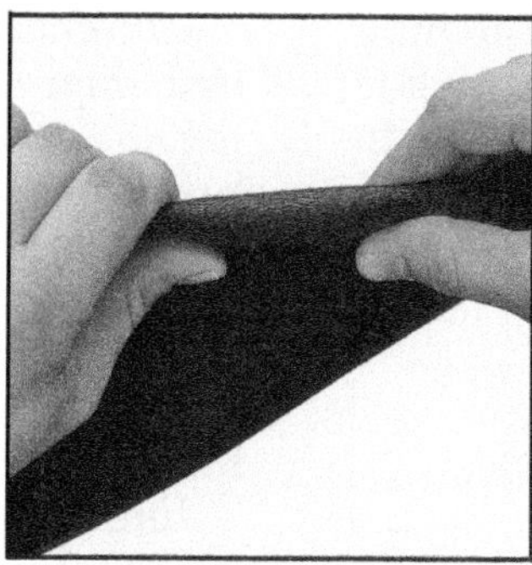

UV CHECKING
Storing in areas exposed to prolonged sunlight causes UV checking.

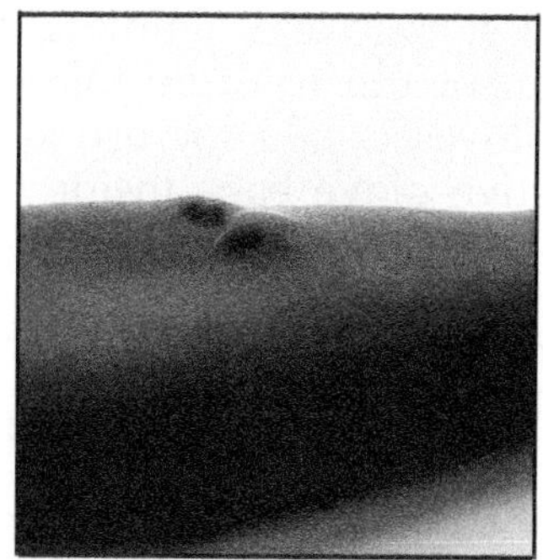

CHEMICAL ATTACK
This photo shows swelling caused by oils and petroleum compounds.

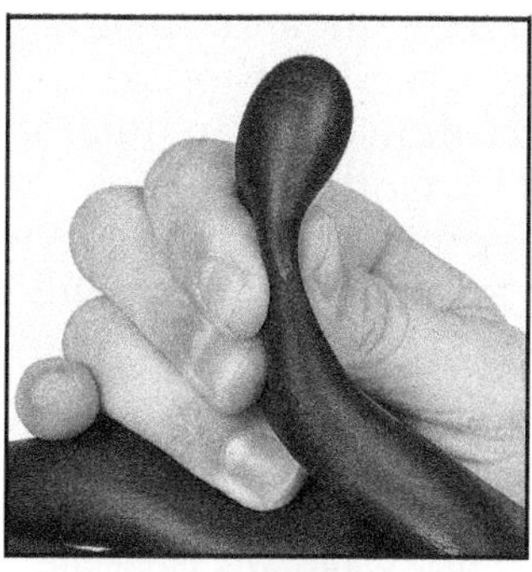

SNAGS
Damage shown is due to wood or metal splinters or other sharp objects.

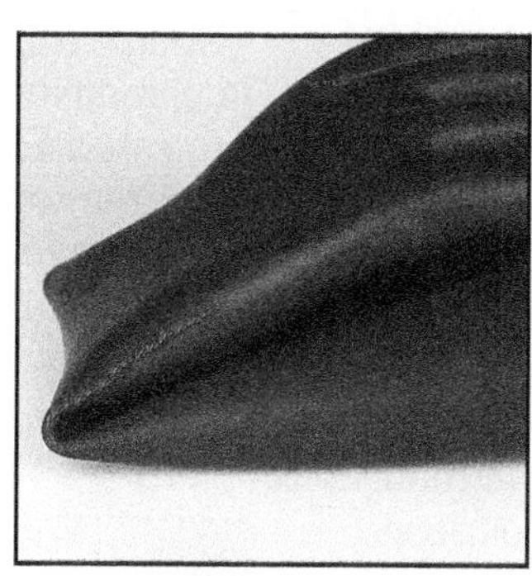

AVOID FOLDING ELECTRICAL GLOVES
The strain on rubber at a folded point is equal to stretching the glove to twice its length.

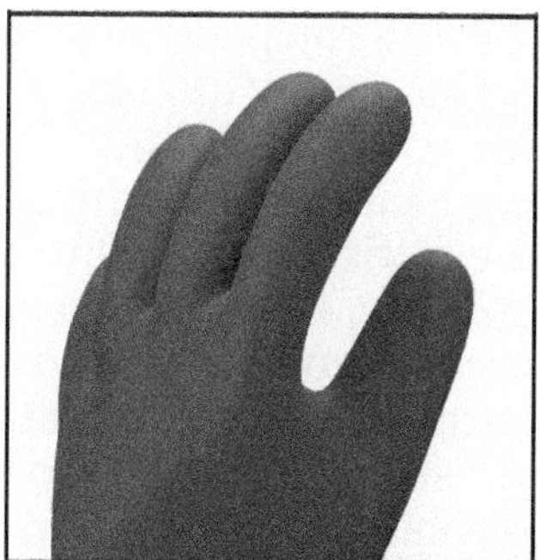

AVOID STORING INSIDE OUT
Storing reversed gloves strains the rubber severely and promotes ozone cutting.

49102-11_F15.EPS

Figure 15 Glove inspection.

49102-11_F16.EPS

Figure 16 Rubber glove on a glove inflator.

The water test is performed by filling the glove halfway with water and rolling the cuff to check for leaks.

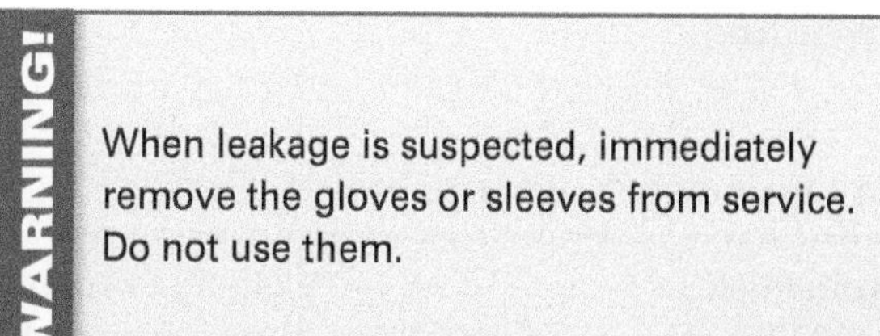

> **WARNING!**
>
> When leakage is suspected, immediately remove the gloves or sleeves from service. Do not use them.

Gloves and other rubber insulating materials can be damaged by petroleum products and hand lotions. Some types are also affected by sunlight and ozone. When rubber insulating equipment is contaminated by oils or grease, wipe off the contamination immediately. As soon as practical, wash the item with an approved cleaning product such as mild dishwashing liquid. Much damage to rubber insulating products can be avoided by careful storage and handling. Rubber insulating equipment must not be folded or stored inside out. Gloves issued for use should be stored in protective glove bags inside their leathers with cuffs down.

WARNING!

Ensure that rubber insulating gloves and leathers fit the hand properly and that the leathers are the same class as the glove. Failure to do so will not provide the expected level of protection.

4.1.7 Flame-Resistant Clothing

Electrical arcs can generate a tremendous amount of heat. This energy can ignite some materials and cause a fire. It is important for your safety to wear flame-resistant (FR) clothing when working on or near electrical equipment. Although FR material can ignite, it will not continue to burn after the source of ignition is removed. Clothing made from cotton and other natural fibers is naturally flame resistant. Clothing made of synthetic material is prohibited where employees are exposed to electrical hazards. When exposed to extreme heat, synthetic clothing melts to the skin, causing further injury.

To protect yourself from heat-related injuries, you should always report to work wearing clothing made of natural fiber (long sleeved shirts and long pants) and footwear that is non-conductive and high enough to protect the ankle area not covered by the pants. Many companies provide fire-resistant coveralls that can be worn over regular natural fiber clothing. In some locations, FR clothing is standard daily wear. In other locations, FR clothing and other PPE must be readily available when required.

Commercial FR clothing is more flame resistant than street clothing and must have an arc thermal performance value (ATPV) or energy of break-open threshold (EBT) rating that provides adequate protection against the expected hazards. These values are determined by how long the material can withstand thermal energy before it breaks down. *Figure 17* shows some typical arc flash PPE that is also fire resistant.

Information on the electrical hazard warning label will tell you what type of PPE you will need to wear when you are in the flash hazard boundary. *Table 4* shows the hazard/risk categories (HRC), characteristics, and arc rating value of clothing and PPE items. HRC categories go from 0 through 4, with 4 being the most flame-resistant. The arc rating is given in J/cm^2 and cal/cm^2; this is a measure of energy over 1 cm^2 of skin. Only arc-rated FR clothing is allowed as an outer layer for Categories 1 to 4. FR clothing must be worn properly for maximum protection. Garments with sleeves rolled up or fasteners undone will not provide overall protection.

NOTE

Before attempting a task wearing a flash suit and hood, first put on all required PPE and practice working in similar lighting and space conditions in a safe area. Try using the tools you will need for the job while wearing the gloves and their protectors. Think about the location of the planned job. Is it a tight work area? Is there adequate lighting?

Each rated garment has an ATPV or E_{BT} rating on its tag. The tag also provides instructions for washing. Many manufacturers recommend

Table 4 Protective Clothing Characteristics

Hazard/Risk Category	Clothing Description (Typical number of clothing layers is given in parentheses)	Required Minimum Arc Rating of PPE (cal/cm^2)
1	FR shirt and FR pants or FR coveralls (1 layer)	4
2	Cotton underwear plus FR shirt and FR pants (1 or 2 layers)	8
3	Cotton underwear plus FR shirt and FR pants plus FR coveralls, or cotton underwear plus two FR coveralls (2 or 3 layers)	25
4	Cotton underwear plus FR shirt and FR pants plus mulitlayer flash suit (3 or more layers)	40

49102-11_T04.EPS

Figure 17 Typical arc flash/flame-resistant PPE.

home laundering using a mild detergent. Home laundering is recommended, as commercial laundering may degrade the protective properties of the material. The use of fabric softeners is generally not recommended. Check with the manufacturer to determine the expected lifespan of the FR rating. Some protective clothing is designed for a single use, while others retain their rating for up to 18 months or longer with weekly home laundering.

WARNING!

All flame-resistant PPE apparel must be inspected prior to each use. Look for tears, worn spots, and contamination such as oil or grease. If it is contaminated or damaged, do not use it. Turn it in for cleaning, repair, or replacement. Garments with holes or rips are not reliable. FR clothing may be repaired using the same material as the clothing.

Labels, embroidered emblems, or logos are not recommended. When used, they must be attached to FR clothing in accordance with *ASTM F1506, Standard Performance Specification for Flame-Resistant Textile Materials for Wearing Apparel for Use by Electrical Workers Exposed to Momentary Electric Arc and Related Thermal Hazards*. Never sew on any label, emblem, or logo to your FR apparel yourself.

In the past, electrical arc flash protective equipment had a reputation for being hot, heavy, and uncomfortable, restricting both movement and vision. In addition, some protective apparel causes low visibility or lack of manual dexterity. The obvious danger of uncomfortable or over-protective PPE is that workers will avoid using it. The comfort of FR apparel has improved greatly due to the use of lighter materials, increased breathability, and higher insulating values. For example, when comparing the face shield of a 10-year-old arc flash hood with one of the same class purchased last year, light transmission and visibility are greatly improved (*Figure 18*). FR garments may be layered for additional protection.

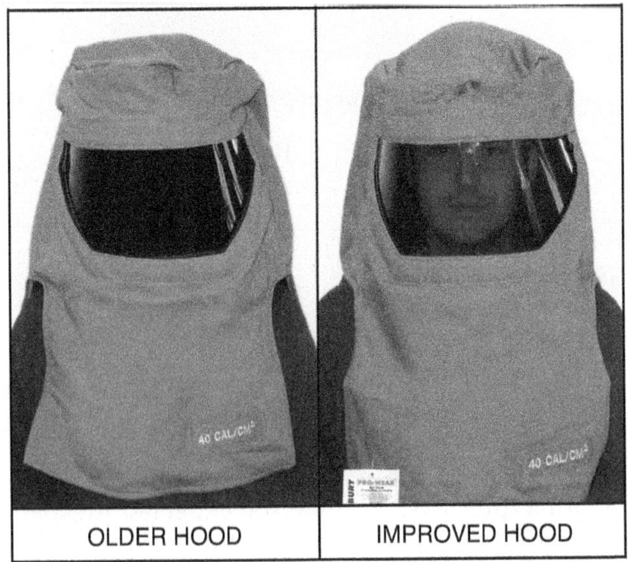

49102-11_F18.EPS

Figure 18 Newer arc hoods offer improved comfort and visibility.

On Site

Arc Suppression Blankets

Arc suppression blankets are made out of ballistic material that provides a barricade to deflect potential arc hazards. It is secured using clips through the side loops.

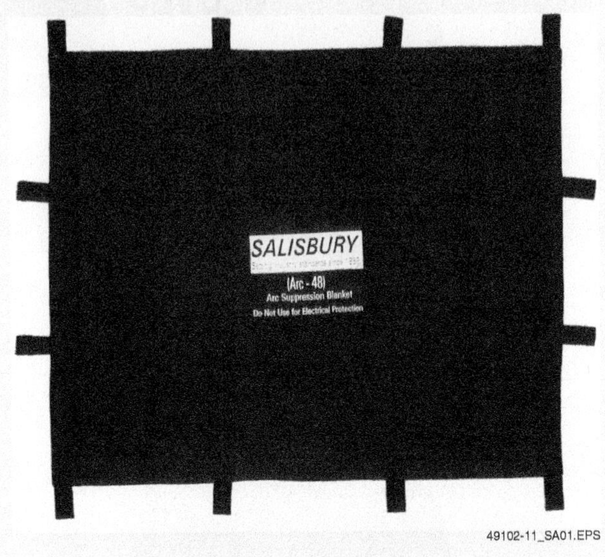

49102-11_SA01.EPS

Arc-resistant rain suits are available for use in inclement weather (*Figure 19*). These types of rain suits have a higher degree of flame resistance than cotton or other natural fiber clothing, but they have varying ATPV ratings. Always check the manufacturer's specification for more detailed information.

4.1.8 Face Protection

An arc-rated face shield with chin protector attached to a hard hat may provide adequate protection for the neck, head, and face areas for HRC 1 and 2 exposures. Combining the face shield with an arc-rated balaclava (*Figure 20*) or an overall hood can provide protection for HRC 2 or greater exposure.

4.2.0 Lockout/Tagout

Lockout/tagout devices (*Figure 21*) are used to prevent power from being turned on while work is being performed on equipment. *OSHA Standard 29, Section 1910.269* specifies the requirements for the electrical energy control procedures involving power distribution lines. *NFPA 70E®* specifies three basic electrical energy control

49102-11_F19.EPS

Figure 19 Arc-resistant rain suit.

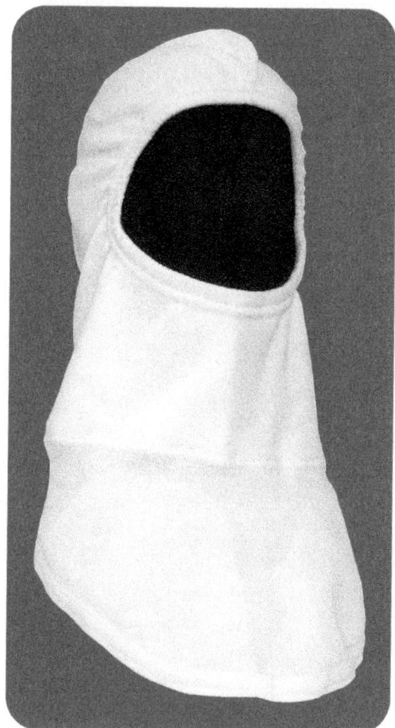

49102-11_F20.EPS

Figure 20 Balaclava (sock hood).

ELECTRICAL LOCKOUT

PNEUMATIC LOCKOUT

49102-11_F21.EPS

Figure 21 Lockout/tagout devices.

procedures involving lockout/tagout devices. They are as follows:

- *Individual qualified employee control procedure* – Certain types of minor work by a qualified person may be allowed without placement of a lockout device under specific conditions. The energy control procedures of your employer may or may not recognize individual or exclusive employee control procedures.

- *Simple lockout/tagout procedure* – Any lockout involving only qualified persons de-energizing a single set of conductors or electrical source for the sole purpose of performing work within the limited approach boundary is considered a simple lockout/tagout. It is important to recognize that unqualified personnel may not work under this procedure. *OSHA Standard 29, Section 1910.147* identifies conditions for simple lockout procedures for non-electrical work. Those procedures may be used by any authorized employee. The energy control procedures of your employer may or may not distinguish between simple lockout for electrical and simple lockout for other purposes.
- *Complex lockout/tagout procedure* – When any of the following conditions are present, a complex lockout procedure is required:
  - Multiple energy sources
  - Multiple crews or multiple crafts
  - Multiple locations
  - Multi-shift work

The intent of the detailed requirements of a complex lockout is to require a written execution plan and to assign specific responsibility and leadership. Logistics involving multiple sources to be locked out and/or multiple employees and crews who must place personal locks under this policy must be defined. Procedures for group lockout devices or use of lockboxes are required.

4.3.0 Temporary Protective Grounds

Temporary protective grounds (TPGs) are placed after equipment has been de-energized. Their purpose is to protect workers from electrical shock caused by hazardous differences of potential that might occur if power is induced in the circuit by lightning strikes or accidental restoration of power. These devices place ground potential at the place where personnel are working. The effect is to short any unintended voltage to ground and provide an area of equipotential.

TPGs (*Figure 22*) are required to conduct the maximum fault current expected at a location for the time necessary for protective devices to operate to clear the fault. The clamps must be able to conduct the required fault current and create a mechanical connection strong enough to withstand the magnetic forces generated during a fault. Failure of the temporary grounding equipment or resistance in connections or conductors could expose workers to lethal shock hazard in the event of lightning strike or accidental re-energizing of the circuit.

As with all protective equipment, temporary grounding equipment must be inspected before each use. When equipment is in place for more than one day, inspect the grounding equipment before each shift. Do not remove and replace it, as doing so creates an additional hazard. Visual inspection includes examining for dirt or corrosion on contact surfaces, as well as damage to conductor jackets. Also check to see whether any conductor strands have pulled out at a ferrule, and look for loose connections between the ferrule and clamp.

WARNING!

Use only labeled tools complying with *ASTM F1505* or other accepted standards. Taping, shrink tubing, or plastic dipping of tools will not provide the same level of protection as properly rated tools.

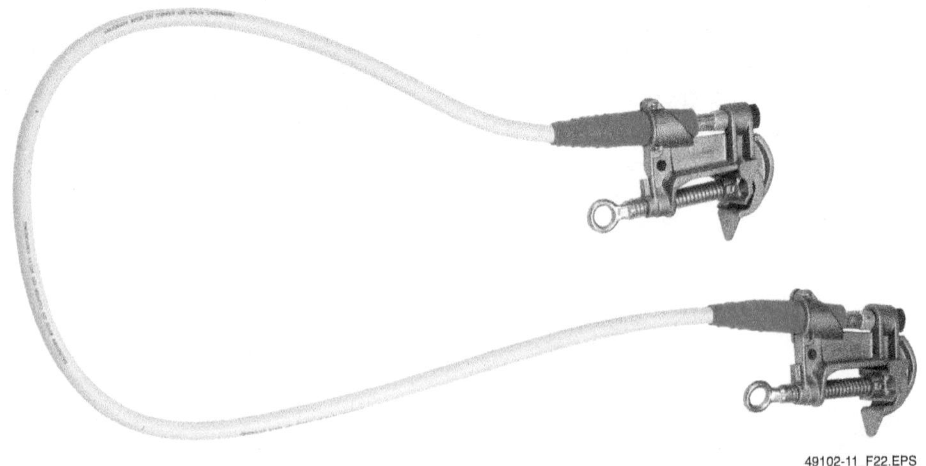

Figure 22 TPGs.

The application and removal of temporary grounds is considered work on energized parts and is hazardous. There are many instances of electrical shock events and arc flash events during the installation or removal of temporary grounding. Procedures and required PPE must be used per the written policies of the employer. Temporary grounds must be sized to limit the voltage potential across a worker's body to 100V for a 15-cycle event or 75V for a 30-cycle event. This is typical for work on overhead lines. It is critical that there be no added impedance in the ground path due to loose connections or surface contamination.

WARNING!

In overhead distribution, conductors from an energized line may blow into or fall onto conductors of a de-energized line that may be on the same structure or crossing under it. Switching errors may also re-energize a circuit that has been tagged out.

Temporary grounding is connected by first clamping one end of the jumper to the ground source and then clamping the other end of the jumper to the line or phase using a hotstick. Subsequent jumpers are connected from the grounded phase to the other lines or phases. When temporary grounding is applied, each ground should be identified and logged by location. Always uncoil grounds completely.

Removal is in reverse order, with the last connection removed being the first connection to ground. Be sure to account for all ground jumpers.

Store temporary grounds neatly rolled and hung up or placed in storage bags. Identify each temporary ground set or jumper to allow traceability of use and testing. Inspect grounding equipment periodically while in storage. Perform electrical testing at a minimum of once per year and after every instance of carrying fault current. The testing may be done with a high current source or a micro-ohmmeter. The acceptance criterion is impedance through the clamps and conductor that will not result in excessive voltage potential across the worker during a fault.

Figure 23 shows live-line work using a hot stick to place temporary grounds on an overhead line and create an electrically safe work condition for a lineman standing above the clamp-on grounding bar cluster. The lineman is not exposed to shock hazard because his body is within an equipotential plane created through grounding.

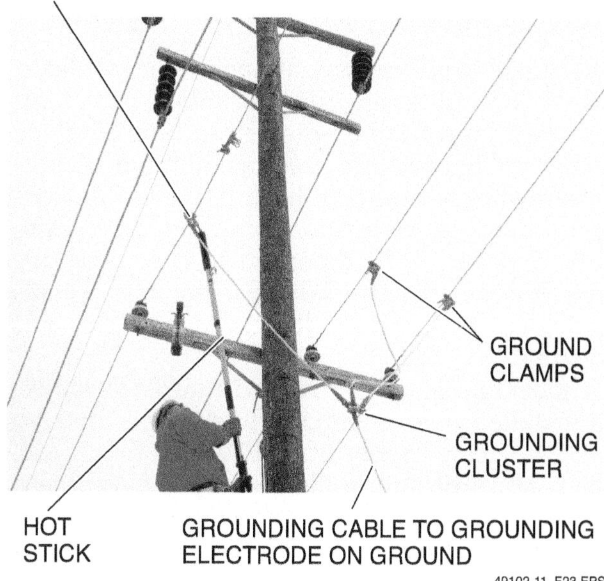

Figure 23 Hot stick in use.

4.4.0 Live-Line Tools

Requirements for live-line tools are found in *OSHA Standard 29, Section 1910.269(j)*. Tools made of fiberglass-reinforced plastic (FRP) must be designed and constructed to withstand test voltage of 100,000V per foot of length for five minutes per *ASTM F711, Standard Specification for Fiberglass-Reinforced Plastic (FRP) Rod and Tube Used in Live-Line Tools*. Live-line tools must be maintained as follows:

- Wipe the tools clean and visually inspect them before use for any defects that could affect the insulating quality or mechanical integrity of the tool.
- Remove the tools from service for electrical testing and additional inspection if any defect or contamination is visible after wiping.
- Remove the tools from service at least once every two years for electrical testing and additional visual inspection.
- Handle the tools with care.
- Store the tools in a protective sleeve or tube to prevent damage and protect against condensation and dust.

CAUTION

Tools stored in a cool, dry environment must be allowed to acclimate to the work environment before use. Condensation on a tool may reduce or eliminate its insulating qualities.

4.5.0 Other Tools and Protective Equipment

Other protective equipment, including insulated and insulating tools, ropes, rubber insulating equipment, protective shields, and physical or mechanical barriers are used in situations involving exposure to electrical hazards. These devices are used in all elements, including dust, rain, snow, and sun. It is important to clean and dry these devices after use. Before using any insulating device, inspect it for dirt, holes or cracks, and moisture.

Rubber line hoses, covers, and guards (*Figure 24*) are installed over energized line connections so that work can be performed within the MAD. Like rubber insulating gloves and sleeves, these devices have specific voltage ratings and must be selected based on the voltage level in the work zone. These devices must be periodically tested, cleaned, and inspected for serviceability. Voltage and air tests are the most common methods of testing.

Rubber blankets (*Figure 25*) are used to cover oddly shaped connections. They are held in place with non-conductive clamps. Blankets should be rolled and stored in a blanket rollup or canister intended for that purpose. Store clamps in a separate container. Rubber blankets are stretched by laying them flat, folding them diagonally, and rolling in each direction on both sides of the blanket. Rubber blankets can be tested using specialized equipment, such as the blanket tester shown in *Figure 26*. Do not use tape to secure blankets for storage.

49102-11_F24.EPS

Figure 24 Rubber line hoses, covers, and guards in place.

5.0.0 Traffic Control

Line workers often work on or near public roads where moving traffic presents additional hazards. When the work site is located near traffic, your supervisor will carefully study the area and make various plans to ensure that the job can

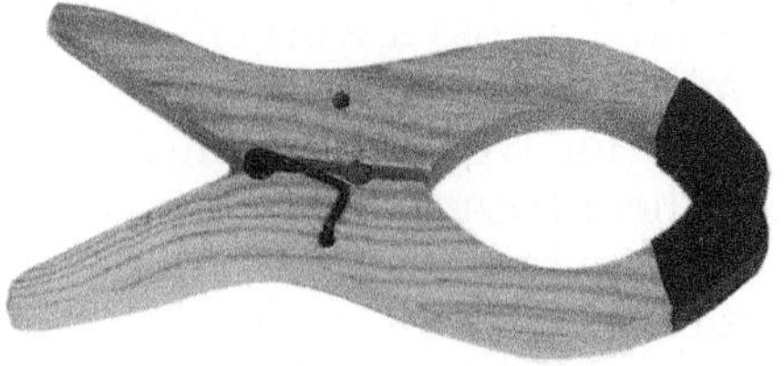

BLANKET CLAMP PIN
USED TO HOLD BLANKET IN PLACE

RUBBER BLANKET WITH
STORAGE ROLLUP

49102-11_F25.EPS

Figure 25 Rubber blanket.

49102-11_F26.EPS

Figure 26 Rubber blanket tester.

be completed safely and efficiently. You need to stay alert for hazards when working on or near a roadway.

When the normal use of a road or sidewalk is disrupted, temporary traffic control (TTC) measures must be used. TTC measures help to ensure that motorists, pedestrians, and bicyclists can safely use the road and sidewalks while you safely perform your job. The best-known TTC measures are those familiar orange or orange/white striped barricades, cones, and signs.

When a job is scheduled or long-term, specially trained workers create maintenance of traffic (MOT) plans to keep traffic moving smoothly and to allow work to be completed. They also decide about the need for and position of TTC devices. The Federal Highway Administration publishes a manual that states the basic principles for changing the flow of traffic at a work site. It also explains how to design and use traffic control devices. This manual is called the *Manual on Uniform Traffic Control Devices (MUTCD) for Streets and Highways*. It is used for all streets and highways open to the public regardless of type or class or the *public* agency having control of the road. The *MUTCD* clearly states requirements for all aspects of MOT plans, as well as specifying the design of signs, barricades, and other channelizing devices. No other signs or devices may be used on a public roadway.

Figure 27 shows a typical TTC zone. The TTC zone consists of the following:

- *Advanced warning area* – This area alerts drivers that they are approaching a work zone.
- *Transition area* – This area allows traffic ample space to move to the next lane.
- *Buffer space* – This area provides additional protection to workers in the work space should any vehicle fail to move to the next lane.
- *Work space* – This area gives workers a safe area to perform their duties. The work space should be large enough to contain workers and equipment and to permit work to be safely completed. Workers should not stray from this space.
- *Second buffer space* – This area provides additional protection to workers in the work space should any vehicle move to the work lane early.

Figure 28 shows a typical TTC zone on a curve. Notice that these figures show that traffic gradually merges from the right lane into the left. The transition is long enough to alert drivers of the upcoming work zone and provide enough time for the drivers to move into the next lane. When the work area is right after a curve, the transition area starts before the beginning of the curve to alert drivers of the approaching work area.

The TTC zones shown in *Figures 27* and *28* provide for a safe work area and an orderly flow of traffic. However, the work you perform near a roadway will often be urgent, so there will be little time to make a MOT plan. At times like these, it is best to adhere to some of the principles behind a formal MOT plan, such as providing drivers with ample warning of the approaching work zone and using standard channelizing devices (*Figure 29*). Keep in mind that for a traffic control device to be useful, it must meet the following five basic requirements:

- Fulfill a need
- Attract attention
- Have a clear and simple meaning
- Command respect from road users
- Give enough time for proper response

TTC measures and associated devices are used to provide for safe and orderly movement of traffic through or around work zones and to protect workers, responders to traffic accidents, and equipment. At the same time, the TTC zone must permit workers to complete the project in a reasonable time.

Flaggers may be used in the TTC zone to direct traffic around the work zone. Flaggers are responsible for the public safety, so they must be trained in safe traffic control practices. Flaggers must keep themselves and workers safe, as well as direct traffic using signs, paddles, or flags. Should you need to be a flagger, you will need to do the following:

- First, select a position where you are clearly visible to approaching drivers and at a distance that allows drivers ample time to safely stop their vehicles. Notice in *Figure 28* that the flagger is positioned before the curve.
- Before beginning work, identify a safe escape route should a driver fail to follow your directions.
- Clearly, firmly, and politely give specific instructions.
- Stay alert and move quickly to avoid danger.
- Use signaling devices such as paddles and flags to provide clear direction to drivers.
- Apply safe traffic control practices, sometimes in stressful or emergency conditions.
- Identify unsafe traffic situations and warn workers in time to avoid injury.
- Wear highly visible safety apparel. This is usually a vest that is either fluorescent orange-red or fluorescent yellow-green. The vest needs to be visible at a minimum of 1,000 feet.
- Use a flashlight or battery-powered lantern during any period of low visibility.

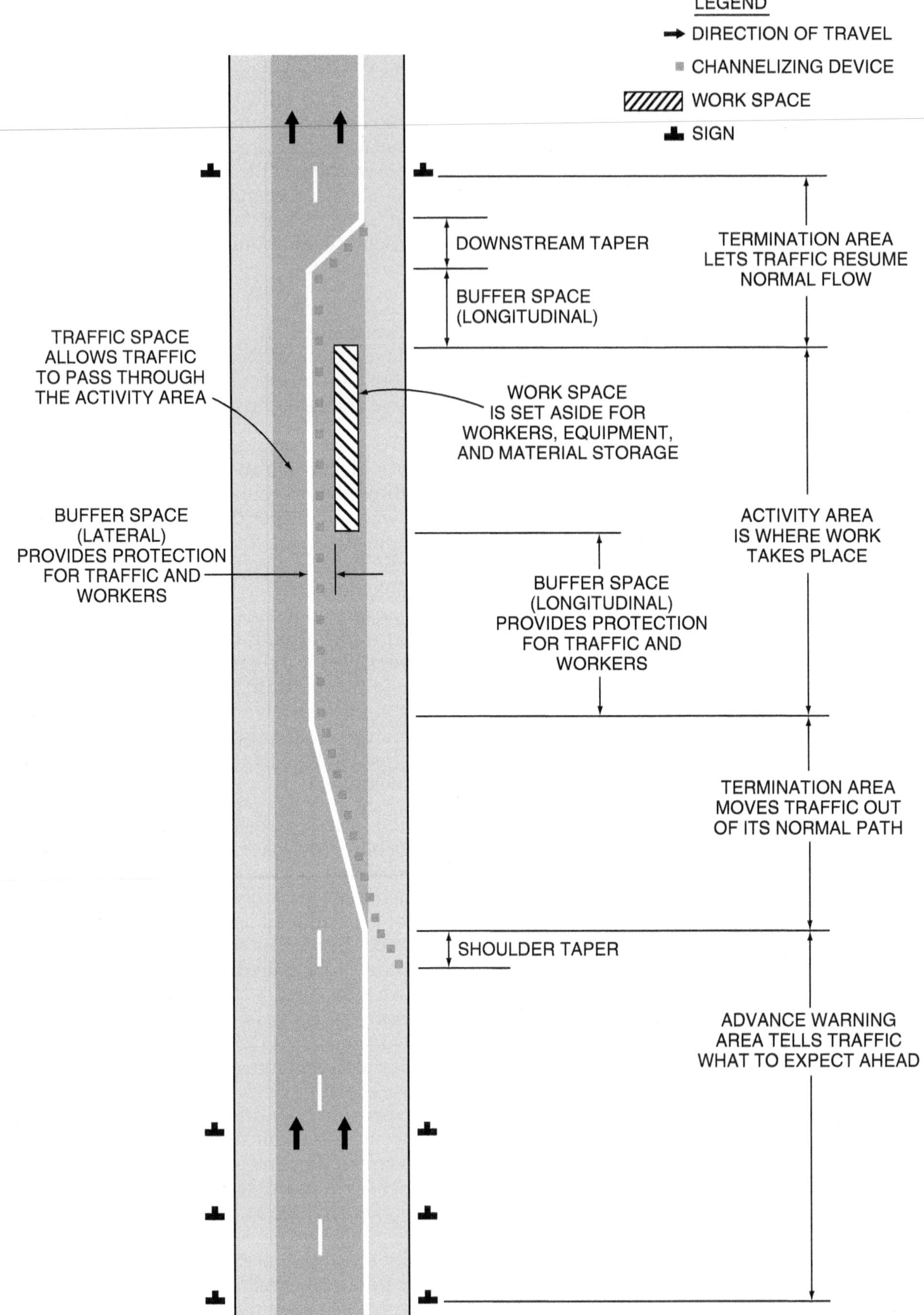

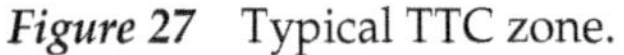

Figure 27 Typical TTC zone.

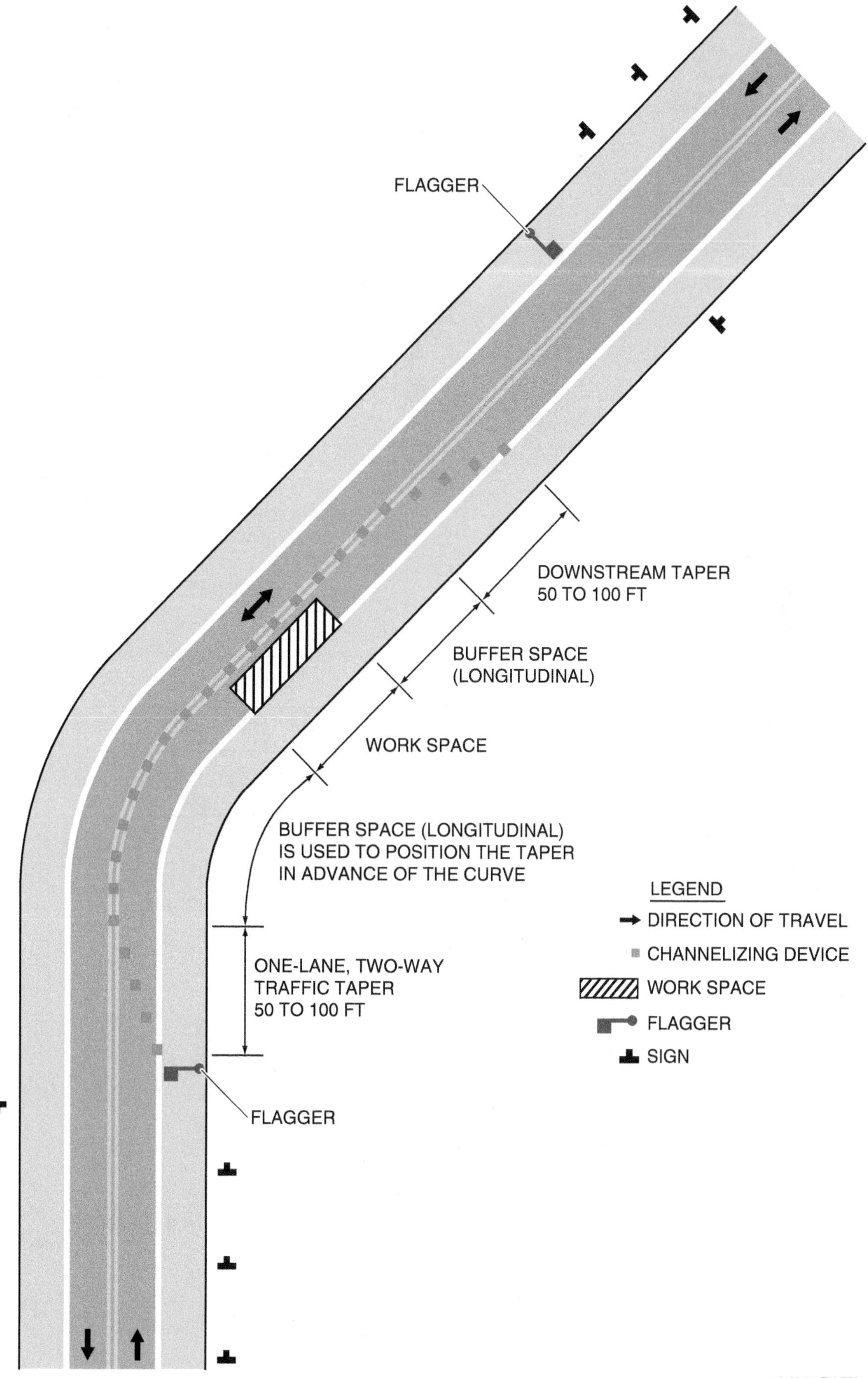

Figure 28 Typical TTC zone on a curve.

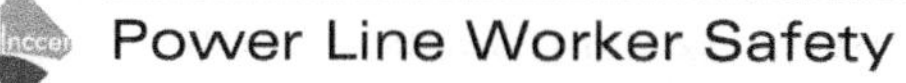

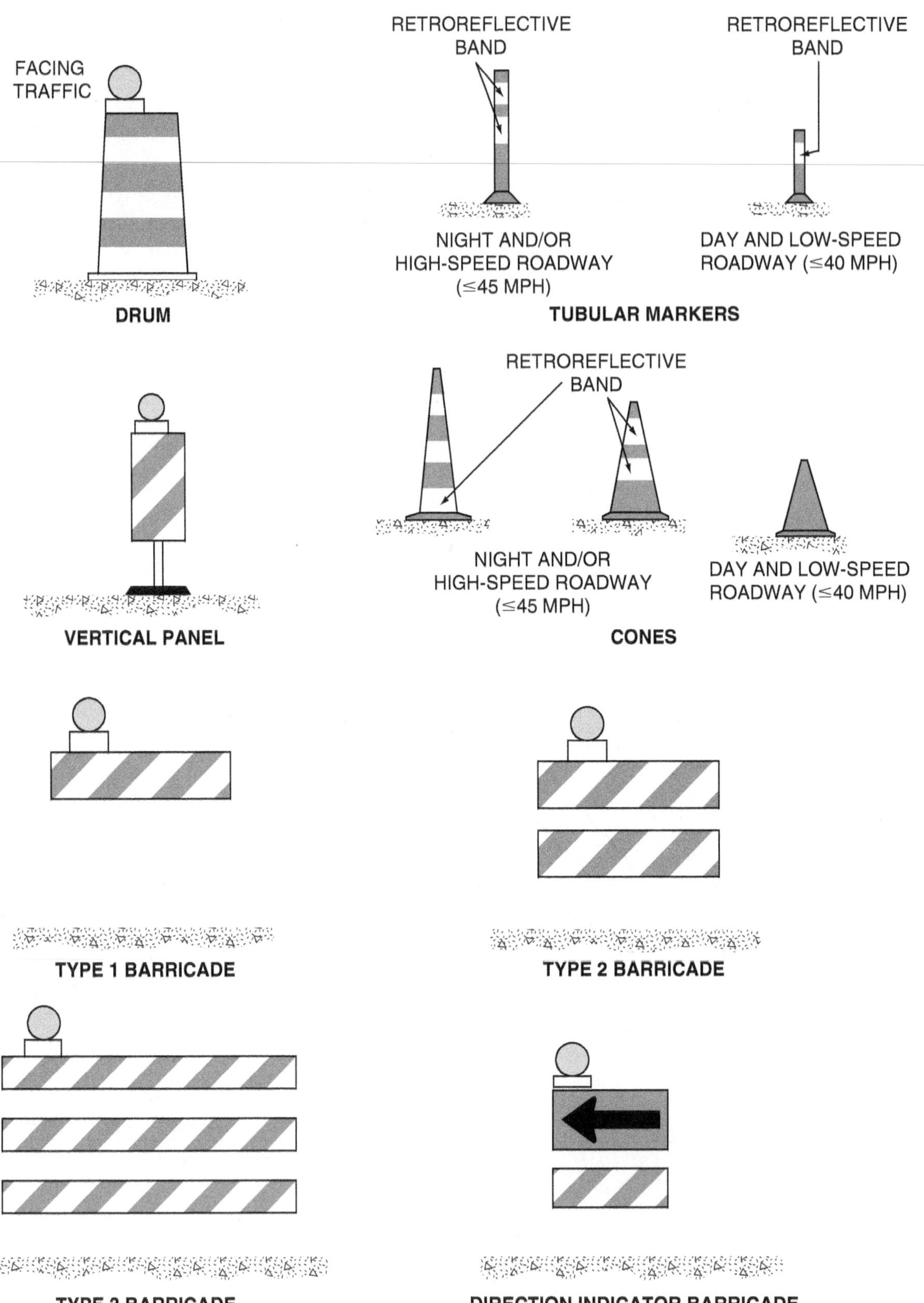

Figure 29 Common channelizing devices.

- Use a stop/slow paddle or an approved red flag to direct drivers.
- When using a flag, it should always be unfurled. To stop traffic, hold the flag staff at shoulder level. To allow traffic to pass, point the flag staff toward the ground. To slow traffic, wave the flag up and down from shoulder level and down toward your foot (*Figure 30*).

Figure 30 Flagger hand signals.

6.0.0 Trenching Safety

As you learned in the *Basic Safety* module of the Core Curriculum, a trench is a narrow excavation made below the surface of the ground in which the depth is greater than the width and the width does not exceed 15 feet. Power lines are often placed in trenches (*Figure 31*). Safety is a crucial aspect of any excavation job, so it is important that you understand how to work safely around trenches.

Trenching operations are planned and performed under the guidance of a company-assigned qualified person. This person is a specially trained worker who has knowledge of trenching principles, knows how to counteract hazardous conditions, and can keep workers safe. In addition, the qualified person will have knowledge of, and know how to apply, OSHA guidelines.

6.1.0 Preparation

Before any trenching begins, existing underground utility hazards must be identified. Of course, this includes electrical utilities as well as communications, gas, water, and sewage utilities.

Figure 31 Electrical lines in a trench.

The most common way of identifying these hazards is to use established services such as One-Call. This service provides information about buried utilities in the area. This service must be contacted at least 72 hours before doing any excavations. They will send someone to locate all buried utilities at the job and mark the locations with color-coded flags (*Table 5*). If this type of service does not exist in your area, all local utility companies must be contacted to identify buried utility lines in the area.

After the utilities have been located, the final step to preparation is to perform a physical inspection of the site to determine any other potential hazards or problems. When inspecting the job site, look for the following:

- Notices for buried utilities
- Storm drains and sewers
- Overhead power lines
- Gas or water meters
- Junction boxes
- Mailboxes
- Light poles
- Manhole covers and catch basins
- Sunken ground

6.2.0 Trenching Hazards

During trenching operations, soil is removed from the ground. This can cause extreme pressures to be placed on the trench walls, and may cause them to collapse. The weight from workers and equipment around the trench places an additional load on its walls that could cause them to collapse. Vibrations from machinery can cause the nearby soil to settle, which can also cause the walls to collapse. Cave-ins are the most common hazard in trenching work. The collapse of trench walls can instantly bury workers. To counter these problems, trench walls are reinforced with shoring equipment. As an alternative to shoring, sloping of the trench walls can be used in some instances.

Table 5 APWA Underground Color Codes

Color	Meaning
Red	Electric power lines, cables, conduit and lighting cables
Yellow	Gas, oil, steam, petroleum, or gaseous material
Orange	Communication, alarm or signal lines, cables or conduits
Blue	Potable water
Green	Sewers and drain lines
Purple	Reclaimed water, irrigation and slurry lines
White	Proposed excavation
Pink	Temporary survey markings

Workers could be buried when one or both edges of the trench cave in or when one or both walls slide in or shear away and collapse. One cubic foot of soil weighs between 75 and 125 pounds, depending on composition and moisture content. One cubic yard is the size of a 3' × 3' × 3' cardboard box. It is a small amount of space, yet one cubic yard of soil can weigh more than 3,000 pounds. That's the weight of a small car and more than enough weight to seriously injure or kill a worker.

Failure of unsupported trench walls is not the only cause of burial. Tons of dirt can be dumped on the workers if the excavated earth slides into the trench. Such slides occur when the spoil pile is placed too close to the edge of the trench or when the ground beneath the pile gives way. There must be a minimum of 2 feet between the trench wall and pile of excavated soils as well as any tools, equipment, and materials.

The following conditions will likely lead to a trench cave-in. If you notice any of these conditions, immediately inform your supervisor. These conditions are listed in order of seriousness:

- Disturbed soil from previously excavated ground
- Trench intersections where large corners of earth can break away
- A narrow right-of-way causing equipment and traffic to travel too close to the edge of the trench
- Vibrations from nearby equipment and traffic
- Water from underground sources that causes soil to become saturated, and therefore unstable
- Drying of exposed trench walls that causes the natural moisture that binds together soil particles to be lost
- Inclined layers of soil dipping into the trench, causing layers of different types of soil to slide one upon the other and cause the trench walls to collapse

6.3.0 Trenching Safety Guidelines

When working around any excavation or trench, you are responsible for personal safety. You are also responsible for the safety of others in the work trench. The following guidelines must be enforced to ensure everyone's safety:

- Be alert. Watch and listen for possible dangers.
- Never enter an excavation without the approval of the qualified person on site.

- Never operate your machinery above workers in an excavation.
- Ensure that the OSHA-approved qualified person inspects the excavation daily for changes in the environment, such as rain, frost, or severe vibration from nearby heavy equipment.
- Stay alert for other machinery and stay clear of any vehicle that is being loaded.
- Keep machinery away from trenches and the excavated soil at least 2 feet from the edge of the trench.
- Stop work immediately if there is any potential for a cave-in. Make sure any problems are corrected before starting work again.
- Use shoring, trench boxes, benching, or sloping for excavations and trenches over 5 feet deep.

Some trenches that you will work around will be narrow and shallow and this will decrease the dangers but not eliminate them. Safety precautions must be exercised at all times to prevent injury to yourself and others. Some of the hazards you may encounter during a trenching operation include the following:

- Cave-ins due to trench failure
- Falls and cave-ins from workers too close to the trench edge
- Flooding from broken water or sewer mains
- Electrical shock from striking buried electrical cable
- Toxic liquid or gas leaks from nearby facilities or pipes
- Motor traffic if the excavation site is near a highway
- Falling dirt or rocks from an excavator bucket

6.4.0 Indications of an Unstable Trench

A number of stresses and weaknesses can occur in an open trench. For example, increases or decreases in moisture content can affect the stability of a trench. There are a number of signs that can indicate that a trench is about to fail. These conditions are shown in *Figure 32*.

Tension cracks usually form one-quarter to one-half of the way down from the top of a trench. Sliding or slipping may occur as a result of tension cracks. In addition to sliding, tension cracks can cause toppling. Toppling occurs when the trench's vertical face shears along the tension crack line and topples into the trench. Subsidence is when pressure on the surface of the ground around the trench stresses the trench walls and causes the wall to bulge. If uncorrected, this condition can cause wall failure and trap workers in the trench or topple equipment near the trench. Bottom heaving is caused by downward pressure created by the weight of adjoining soil. This pressure causes a bulge in the bottom of the cut. These conditions can occur even when shoring and shielding are properly installed.

Another indication of an unstable trench is boiling. Boiling is when water flows upward into the bottom of the cut. A high water table is one of the causes of boiling. Boiling can happen quickly and can occur even when shoring or trench boxes are used. If boiling starts, stop what you are doing and leave the trench immediately.

When you are working near a trench, stay alert to changes. Watch for developing cracks, moisture, or small movements in and around the trench walls. Alert your supervisor and co-workers to any changes you notice in the walls or ground surrounding the excavation, because they can be an early indication of a more severe condition. Never move your equipment near a trench when you think there may be a problem. The weight of the equipment could cause the trench to collapse.

6.5.0 Making the Trench Safe

There are several ways to make a trenching site safer. Trenches are often reinforced with shoring or shielding systems. Other times trench walls are cut at an angle away from the trench floor to relieve pressure and avoid cave-ins. This is called sloping. Shoring, shielding, and sloping are different methods used to protect workers and equipment. It is important that you recognize the differences between them.

CAUTION

Protective systems are designed for even loads of earth. Heaving and squeezing can place uneven loads on the shielding system and may stress particular parts of the protective system.

Shoring in a trench is placed against the trench walls to support them and prevent their movement and collapse. Shoring does more than provide a safe environment for workers in a trench. Because it restrains the movement of trench walls, shoring also stops the shifting of surrounding soil, which may contain buried utilities or on which sidewalks, streets, building foundations, or other structures are built.

Trench shields, also called trench boxes (*Figure 33*), are placed in unshored trenches to protect personnel from trench wall collapse. They provide no support to trench walls or surrounding soil, but for specific depths and soil conditions, will

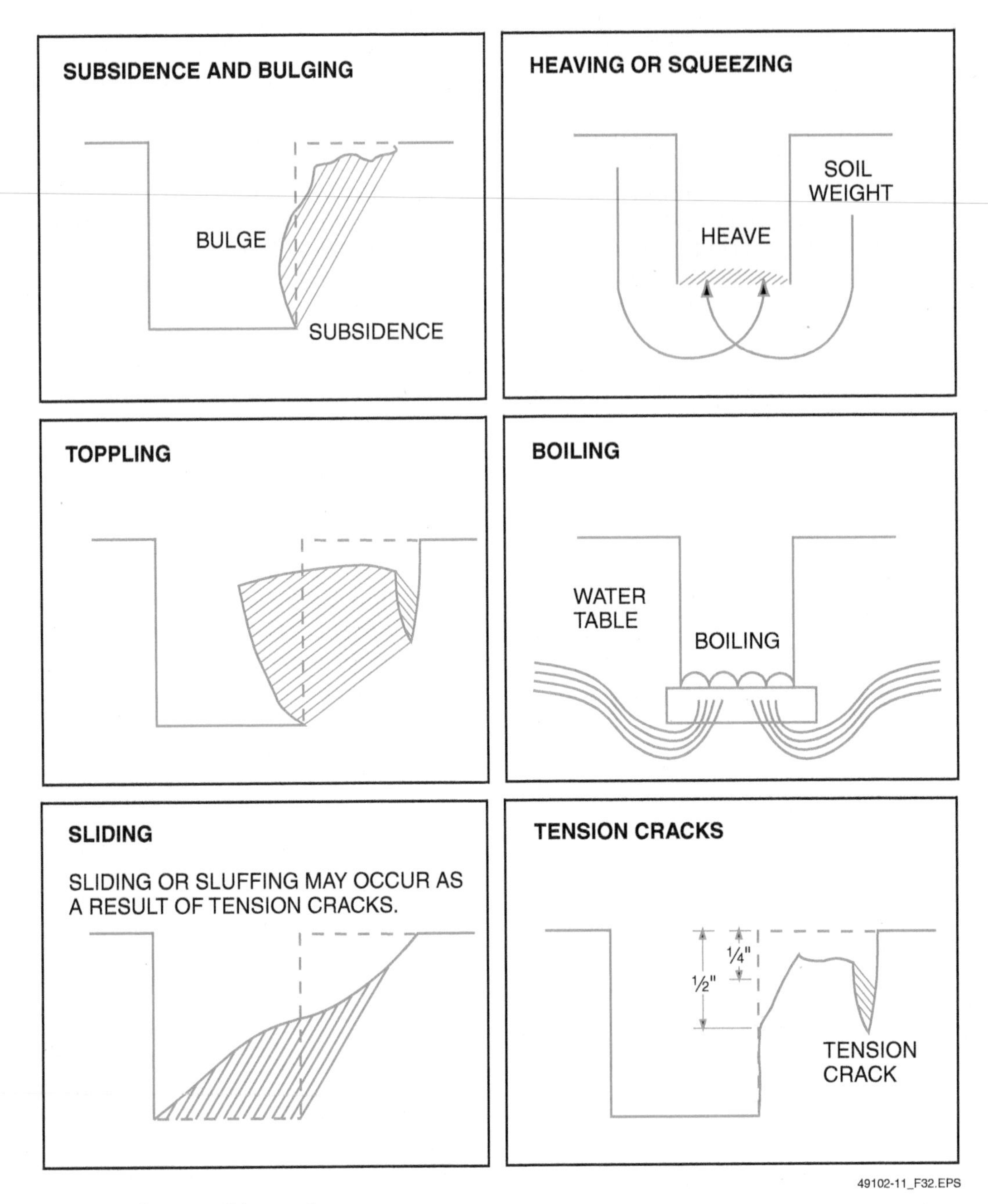

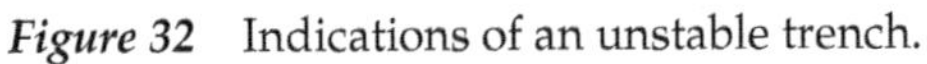

Figure 32 Indications of an unstable trench.

withstand the side weight of a collapsing trench wall to protect workers in the event of a cave-in.

Sloping a trench means cutting the walls back at an angle to its floor. This angle must be cut at least to the maximum allowable slope for the type of soil being used. The maximum allowable slope is the greatest angle above the horizontal plane at which a material will rest without sliding.

6.5.1 Ladders

There must be at least one method of entering and exiting all excavations over 4 feet deep. Ladders are generally used for this purpose. Ladders must be placed within 25 feet of each worker.

When ladders are used, there are a number of requirements that must be met:

- Ladder side rails must extend a minimum of 3 feet above the landing.
- Ladders must have nonconductive side rails if work will be performed near equipment or systems using electricity.
- Two or more ladders must be used where 25 or more workers are working in an excavation in which ladders serve as the primary means of entry and exit or where ladders are used for two-way traffic in and out of the trench.
- All ladders must be inspected before each use for signs of damage or defects.

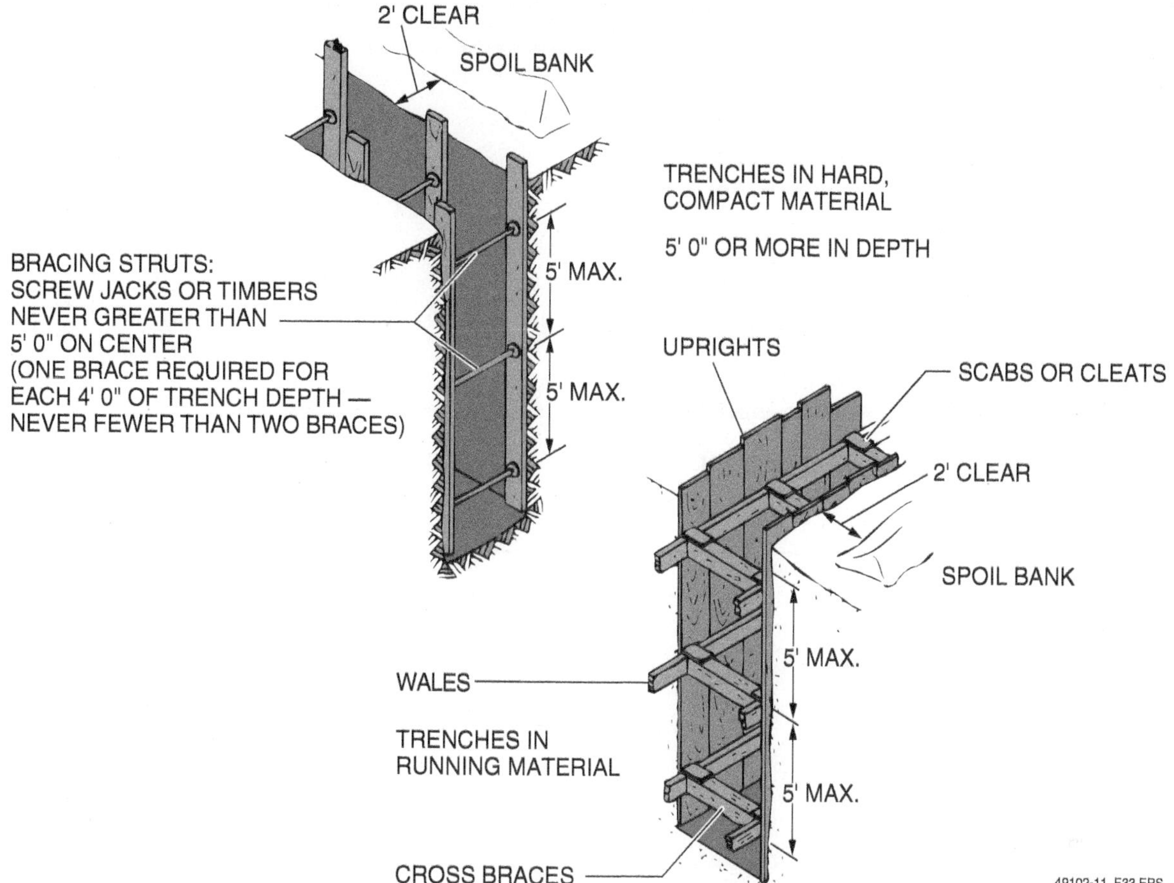

Figure 33 Trench box.

- Damaged ladders should be labeled Do Not Use and removed from service until repaired.
- Use ladders only on stable or level surfaces.
- Secure ladders when they are used in any location where they can be displaced by excavation activities or traffic.
- While on a ladder, do not carry any object or load that could cause you to lose your balance.
- Exercise caution whenever using a trench ladder.

6.6.0 Soil Hazards

Soil type is a major factor to consider in trenching operations. Only a company-assigned qualified person has the experience, training, and education to determine if the soil in and around a trench is safe and stable. However, it is still your responsibility to know the basics about soil and its associated hazards.

Soil is comprised of soil particles, air, and water in varying quantities. Surface soil often contains some amount of organic material, such as decaying plant matter. The soil that is found on most sites is a mixture of many mineral grains coming from several kinds of rocks. Average soils are usually a mixture of two or three materials, such as sand and silt or silt and clay. The type of mixture determines the soil characteristics.

Each of the various soil types, depending on the condition of the soil at the time of the excavation, behave differently. Sandy soil tends to collapse straight down. Wet clays and loams tend to slab off the side of the trench. Firm, dry clays and loams tend to crack. Wet sand and gravel tend to slide. These conditions are shown in *Figure 34*. You should be aware of the type of soil you are working in and know how it behaves.

6.6.1 Type A Soil

Type A soil refers to solid soil with a good strength. Examples of cohesive soils are clay, silty clay, sandy clay, and clay loam. Excavations in Type A soil can have a maximum allowable slope of 53 degrees, as shown in *Figure 35*.

No soil can be considered Type A if any of the following conditions exist:

- The soil is fissured. Fissured means a soil material has a tendency to break along definite planes of fracture with little resistance, or a material that has cracks, such as tension cracks, in an exposed surface.

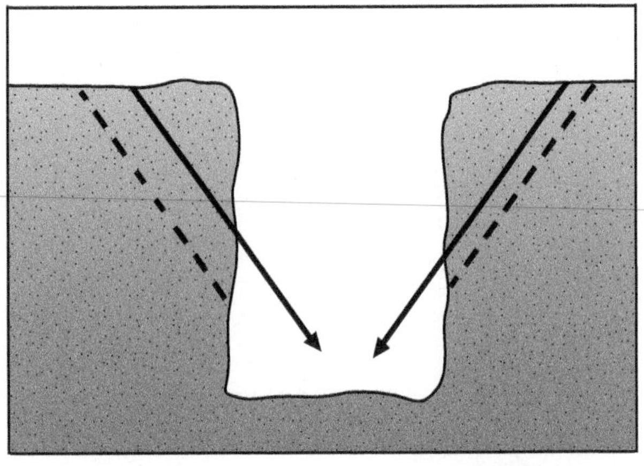

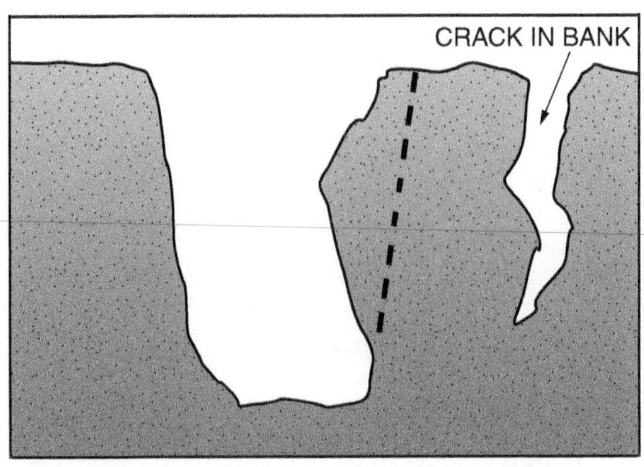

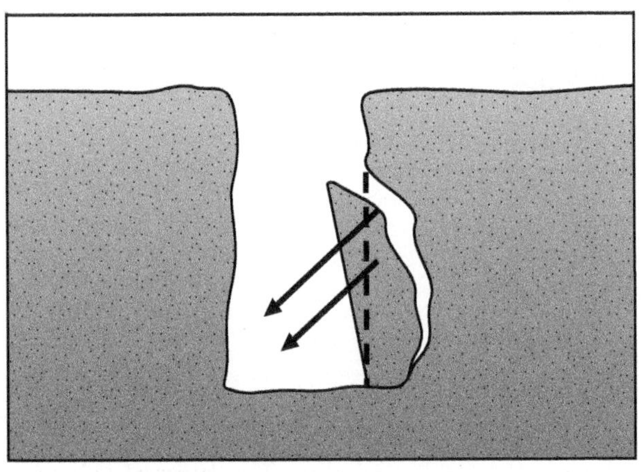

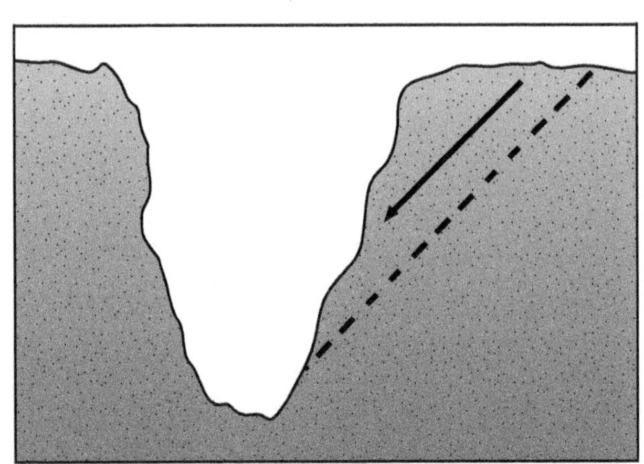

Figure 34 Soil behaviors.

- The soil can be affected by vibration from heavy traffic, pile driving, or other similar effects.
- The soil has been previously disturbed. An exception to this rule occurs when the excavation is 12 feet deep or less and will remain open 24 hours or less.

6.6.2 Type B Soil

Type B soil refers to cohesive soils with compression strength of greater than 0.5 ton per square foot but less than 1.5 tons per square foot. It also refers to granular soils, including angular gravel, which are similar to crushed rock, silt, sandy loam, unstable rock, and any unstable or fissured Type A soils. Type B soils also include previously disturbed soils, except those that would fall into the Type C classification. Excavations made in Type B soils have a maximum allowable slope of 45 degrees, as shown in *Figure 35*.

6.6.3 Type C Soil

Type C soil is the most unstable type. Type C refers to cohesive soil with compression strength of 0.5 ton per square foot or less. Gravel, loamy soil, sand, any submerged soil or soil from which water is freely seeping, and unstable submerged rock are considered Type C soils. Excavations made in Type C soils have a maximum allowable slope of 34 degrees, as shown in *Figure 35*.

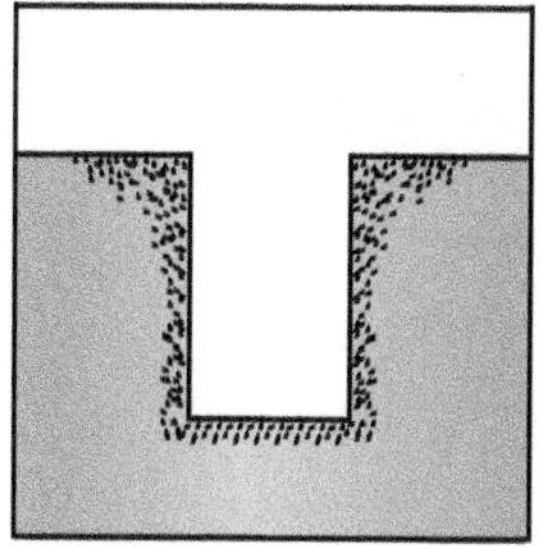

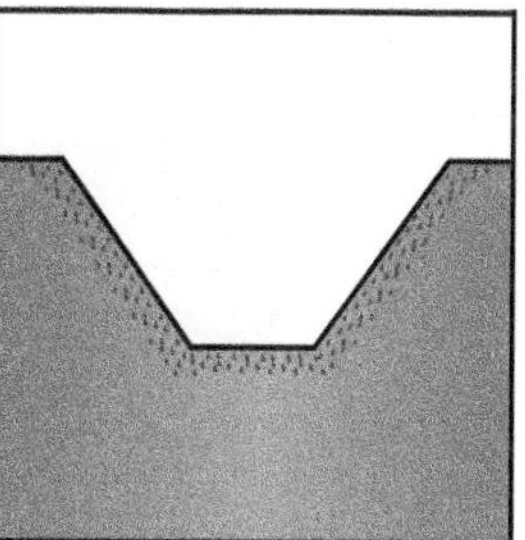

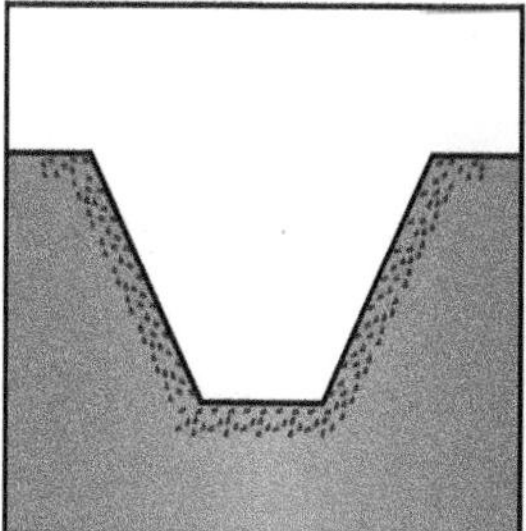

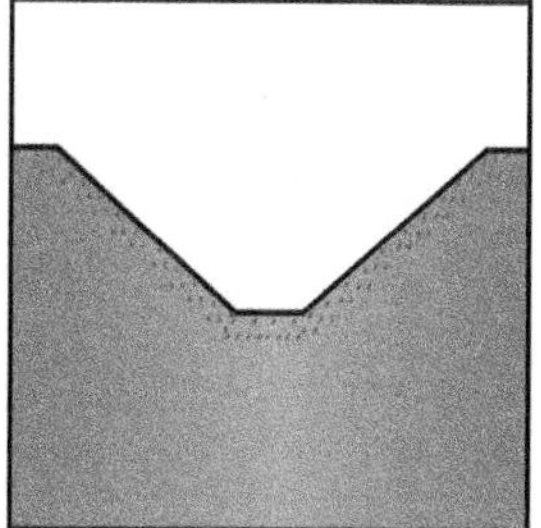

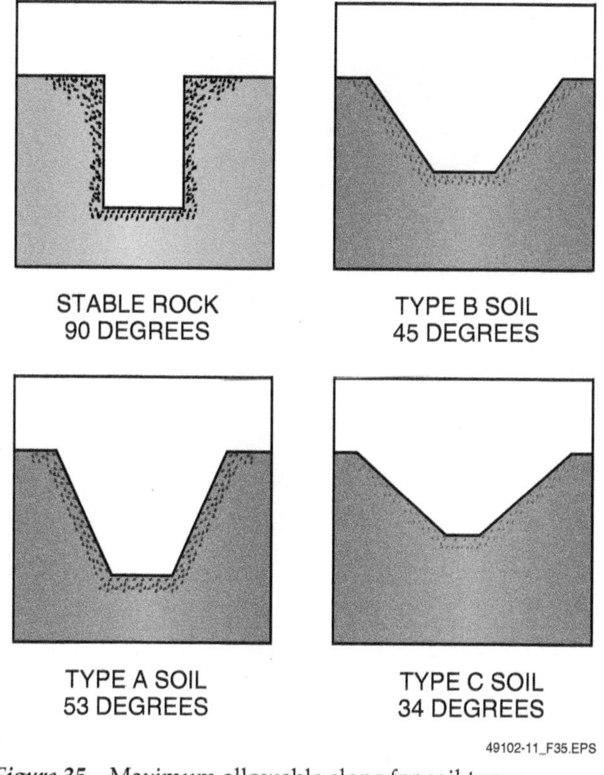

Figure 35 Maximum allowable slope for soil types.

SMALL RIG

Figure 36 HDD drill rigs.

7.0.0 Horizontal Directional Drilling

Horizontal directional drilling (HDD) allows workers to install cabling, conduit, and other types of piping under the ground without digging an open trench. An HDD project uses a boring machine (*Figure 36*) to rotate and push a rod or pipe through the earth. This creates an underground channel through which electrical and communications cabling can be run. This type of drilling is especially convenient for placing utilities in locations where trenching operations would dig up roads, sidewalks, and driveways. HDD is also used to place installations beneath small bodies of water.

There are several unique hazards associated with HDD operations. Anyone working near this equipment or assisting the boring machine operator in drilling operations must be aware of these hazards.

The entry and exit points of a bore hole are called the entry pit and the receiving pit (*Figure 37*). These pits are used to collect drilling fluids. Drilling fluids are drained from the pits at the end of the operation. These pits vary in size, depending on the job. For small operations, the pits are typically 1 to 3 feet deep, 1 to 2 feet wide, and 2 to 6 feet long. For larger operations, the entry and receiving pits may require shoring.

Case History

Drill Rod Dangers

Looks can be deceiving. A three-man crew was installing an underground telephone cable in a residential area. They had just completed a bore hole under a driveway using a horizontal boring machine. The bore hole rod had been removed from the hole. The operator straddled the rod and stooped over to pick it up. Unfortunately, the rod was still rotating and his trouser leg became entangled in the rotating rod and he was flipped over, striking tools and materials. He died from of his injuries.

Source: The Occupational Safety and Health Administration.

7.1.0 Setting Up the Drilling Site and Equipment

Before any work is started, the site must be surveyed to locate any buried utilities. This is generally done by notifying the local one-call agency in advance of the operation. They will send someone out to survey the site and mark utility locations. If this has not been done, or if there is any doubt,

49102-11_F37.EPS

Figure 37 Entry pit.

there are instruments designed to locate utilities (*Figure 38*).

After the area has been inspected, it is time to set up the drilling site and equipment. Before the boring machine can be unloaded from the trailer, make sure the following safeguards are in place:

- Properly placed barricades
- Visible warning signs
- Pedestrian and traffic controls

There are several hazards to recognize and avoid when unloading the boring machine from the trailer. They are as follows:

- *Being struck by moving equipment* – Workers are at risk of being struck by moving equipment when a rig is unloaded from the trailer. Rigs can overturn or slide rapidly down the trailer unloading ramps. This is dangerous because HDD rigs are heavy. Even a small rig can weigh up to 3,000 pounds.
- *Getting crushed by the drill rig if it flips or rolls off the trailer* – Because drill rigs are so heavy, there is an added risk of being crushed if a rig flips or rolls while it is being unloaded. To avoid being crushed and killed by a flipping or rolling rig, make sure that the unloading ramps are properly secured before moving the rig onto the ramps. Move the rig as slowly as possible and be aware of any quick shifts in weight.
- *Strains and sprains from moving loading ramps* – Loading ramps are designed to hold heavy equipment like HDD rigs, so they are often heavy themselves. Always get help when moving loading ramps.

Positioning and setting up the equipment has many hazards. Being aware of these hazards will help you to be more cautious when doing your job. The hazards involved in positioning and setting up the equipment are as follows:

49102-11_F38.EPS

Figure 38 Sub-surface utility locator.

- Strains and sprains when moving equipment
- Tripping and falling on tools or equipment
- Pinch points when handling the drill head, drill pipe, and hand tools
- Electrocution by exposed, energized utilities in the area
- Leaks and injection hazards from the hydraulic lines
- Bore and receiving pit cave-ins
- Contact with the rotating head and drill rod
- Tripping over equipment on the ground
- Falling into the bore or receiving pit

Follow these safeguards to avoid setup/positioning hazards:

- Make sure the ground strike system (*Figure 39*) has been set up and tested. This system alerts workers when an electrical line has been hit. It is generally included with the boring machine.
- Set up grounding mats for both the machine and the operator. Grounding mats are mesh mats that are placed underneath the drill rig. They protect workers from electrocution by providing a single electrical path to ground should the equipment come in contact with

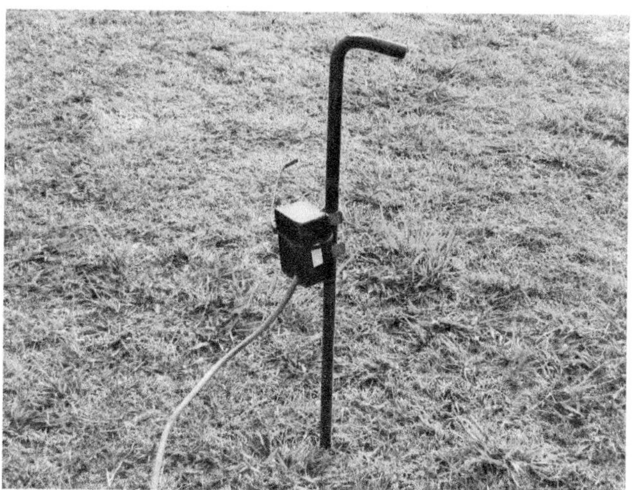

49102-11_F39.EPS

Figure 39 Ground strike detector.

an energized conductor during a utility strike. They can be considered part of the ground strike system.

- Stay away from hydraulic and pneumatic lines. Pressure can build up in these lines and cause leaks. The pressure in hydraulic lines can be strong enough to penetrate the skin.
- Pin all air hose connections to prevent injuries from a runaway hose.
- Test the operation of all controls to make sure they are working correctly. This includes the ground strike system and operator presence sensing system (seat switch) as well as the controls on the rig.
- Make sure the boring machine is turned off while workers are attaching the bore head. Workers can be struck by the bore head if the boring machine is powered up.

7.2.0 Boring Operations Safety

Boring operations can begin once the equipment is in position and set up. Potential hazards of the boring process include the following:

- Electric shock from a ground strike
- Rotating equipment that can strike workers, such as the drill string when it starts rotating
- Rotating equipment that workers or their clothes can get caught on
- Pinch points near the drilling rig
- Trips and falls near rotating drill rods

Following these safeguards can prevent you and others from getting hurt during the boring process:

- Wear nonconductive gloves and boots when operating the equipment.

Case History

Unsafe Conditions

Never work under unsafe conditions. A drill operator was injured while working on a 60-degree slope without adequate secure footing. The drill operator stood on the slope, which made reaching the controls awkward. He lost his footing, pinching his thumb between the stabilizer and the auger. The resulting cut required 10 stitches. No permanent damage to his thumb was anticipated, but the injury could have been considerably worse.

Source: U.S. Department of Energy.

- Keep all body parts clear of the drill rig while drilling is taking place. This will eliminate the possibility of being caught in or struck by the rotating drill string.
- Remove all tools that are used to loosen or tighten joints. If a wrench is left in place while the drill string is turning, it could hit the worker or fly off and hit other personnel.
- If maintenance needs to be performed on the equipment, be sure to properly de-energize it and then apply a lockout/tagout device to prevent unintended energizing. Disconnect the battery when servicing or repairing the electrical systems on the drill. This will ensure that equipment does not become energized and start operating.
- Turn the drill rig off when work is being done on the drill string. This helps to ensure that no one is caught or stuck on the rotating drill string.
- Make sure the operator communicates with all personnel before beginning to rotate the drill rod. It is important that everyone on the site knows what is going on at all times. Workers can easily be injured or killed if they approach equipment or an area of the site that is not safe because drilling is taking place.

If there is any risk of a utility strike, a pothole must be dug at or near the likely location of the utility to physically identify gas lines, electrical utilities, fiber optic lines, and communication lines. Potholing is also used to find out if the drill is where it is supposed to be according to locating equipment. Potholing can be done using mechanical equipment such as air blasters and excavation vacuums or digging by hand. The hazard of using mechanical equipment comes from contacting or cutting the utility line you are trying to locate. Digging the pothole by hand is a safe alternative.

If a ground strike occurs during boring operations, it is important to follow the correct safety procedures in order to avoid the possibility of electrocution. If you are on the ground when a strike occurs, the most important of these procedures is to do nothing until given clearance. Other procedures are as follows:

- Warn others that a strike has occurred.
- Make sure someone notifies the utility.
- Do not touch the drill string with hands or tools.
- Wait for the all-clear authorization before moving or resuming work.

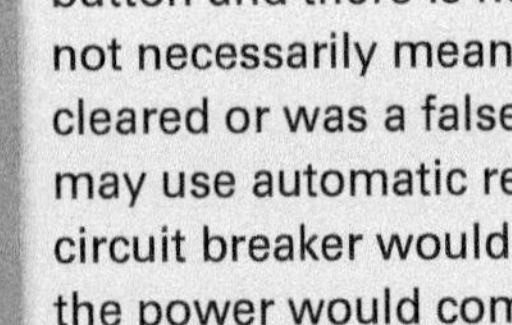

If the operator presses the strike system status button and there is no further alarm, it does not necessarily mean that the strike has been cleared or was a false alarm. The electric utility may use automatic re-closers. If so, a tripped circuit breaker would automatically reset and the power would come back on. The operator should wait about a minute and then check again for an alarm.

8.0.0 Confined Spaces

OSHA defines confined spaces as spaces, because of their configurations, that hinder the activities of employees who must enter into, work in, and exit from them. Employees who work in confined spaces may also face increased risk of exposure to serious physical injury from hazards, such as entrapment, engulfment, and hazardous atmospheric conditions. Confinement may pose entrapment hazards, and work in confined spaces may keep employees closer to hazards, such as an asphyxiating atmosphere, than they would be otherwise. For example, confinement, limited access, and restricted airflow can result in hazardous conditions that would not arise in an open workplace. A confined space is classified as a space that:

- Is large enough and configured so that an employee can bodily enter and perform assigned work
- Has a limited or restricted means of entry or exit
- Is not designed for continuous employee occupancy

These spaces may include, but are not limited to, tanks, storage bins, pits, vessels, sewers, underground vaults and manholes (*Figure 40*).

49102-11_F40.EPS

Figure 40 A manhole.

Most confined spaces have restricted entrances and exits. Workers can be injured as they enter or exit through small doors and hatches. It can also be difficult to move around in a confined space and workers can be struck by moving equipment. Escapes and rescues are much more difficult in confined spaces than they are elsewhere. Confined spaces are entered for inspection, equipment testing, repair, cleaning, or emergencies. They should only be entered for short periods of time.

A written confined-space entry program can protect you. It will identify the hazards and specify the equipment or support that is needed to avoid injury. All industrial sites have written confined space entry programs. You need to know and follow your company's policy.

8.1.0 Confined-Space Classification

Confined spaces must be inspected before work can begin. This helps to identify possible hazards. After an inspection by a company-authorized

Case History

Two Workers Injured in Underground Vault Explosion

On February 15, 2010, two electrical workers were preparing to replace a transformer in the underground vault when there was an explosion. Fortunately, the injuries both men received were not life threatening.

Source: StarBulletin.com.

person, the confined space is classified based on any hazards that are present. The two classifications are nonpermit-required and permit required.

WARNING!

No one may enter a confined space unless they have been properly trained and have a valid entry permit.

A nonpermit-required confined space is free of any mechanical, physical, electrical, and atmospheric hazards that can cause death or injury. After a space has been classified as a nonpermit-required space, workers can enter using the appropriate personal protective equipment for the type of work to be performed. Always check with your supervisor if it is unclear what personal protective equipment is required.

A permit-required confined space has real or possible hazards. These hazards can be atmospheric, physical, electrical, or mechanical. *OSHA CFR 1910.146* defines a permit-required confined space as a confined space that:

- Contains or has the potential to contain a hazardous atmosphere
- Contains a material that has the potential for engulfing an entrant
- Has an internal configuration such that an entrant could be trapped or asphyxiated by inwardly converging walls or by a floor that slopes downward and tapers to a small cross-section
- Contains any other recognized serious safety or health hazard

An entry permit must be issued and signed by the job-site supervisor before the confined space is entered. No one is allowed to enter a confined space unless there is a valid entry permit. The permit is to be kept at the confined space while work is being performed. Always check with your supervisor if it is unclear whether or not you need a permit to enter a confined space.

On Site

Manhole Safety

Manholes must be barricaded to prevent someone from falling in. In addition, air must be pumped into the manhole and a means of rescue must be ready to use.

49102-11_SA02.EPS

8.2.0 Entry Permits

Confined spaces can be extremely dangerous. Entry into the space begins when any part of your body passes the entrance or opening of a confined space. Before entering a permit-required confined space, you must have an entry permit (*Figure 41*). An entry permit is a job checklist that verifies that the space has been inspected. It also lets everyone on the site know about the hazards of the job. All entry permits must be filled out and signed by the authorized person before anyone enters the space. The permit must also be posted at the entrance to the site and be available for workers to review. Entry permits must include the following information:

- A description of the space and the type of work that will be done
- The date the permit is valid and how long it lasts
- Test results for all atmospheric testing including oxygen, toxin, and flammable material levels
- The name and signature of the person who did the tests
- The name and signature of the entry supervisor

Master Card No. ____________

1. Work Description

Area ____________ Equipment Location ____________

Work to be done:

2. Gas Test		Results	Recheck	Recheck
Required	☐ Instrument Check			
☐ Yes	☐ Oxygen % 20.8 Min.			
☐ No	☐ Combustible % LFL			
		Date/Time/Sig.	Date/Time/Sig.	Date/Time/Sig.

3. Special Instructions: ☐ Check issuer before beginning work ☐ None

4. Hazardous Materials: ☐ None What did the line / equipment last contain?

5. Special Protection Required: ☐ None ☐ Forced Air Ventilation

☐ Avoid Skin Contact ☐ Gloves ____________ ☐ Suit ____________

☐ Goggles or Face Shield ☐ Respirator ____________ ☐ Safety Harness

☐ Self-Contained Breathing Equipment ☐ Hoseline Breathing Equipment

☐ Other, specify ____________ ☐ Standby — Name: ____________

6. Fire Protection Required: ☐ None ☐ Portable Fire Extinguisher ☐ Fire Hose and Nozzle

☐ Fire Watch ☐ Other, specify ____________

7. Condition of Area and Equipment Required

THESE KEY POINTS MUST BE CHECKED

Yes	No	
		a. Lines disconnected & blinded or where disconnecting is not possible, blinds installed? (Includes drains, vents and instrument leads) and appropriate valves locked out?
		b. Equipment cleaned, washed, purged, ventilated?
		c. Low voltage or GFCI-protected electrical equipment provided?
		d. Explosion-proof electrical equipment provided?
		e. Life lines required to be attached to safety harnesses?

Comments

rev.6/10/10

SAMPLE

49102-11_F41A.EPS

Figure 41 Confined space entry permit (1 of 2).

8. Approval	Date	Time	Permit Authorization Area Supervisor	Permit Acceptance Maint./Contractor Supervisor/Engineer	Date	Time
Issued by						
Endorsed by						
Endorsed by						
Endorsed by						

9. Individual Review

I have been instructed on proper Safety Procedures and proper Confined Space Entry Procedures. I have signed in on the appropriate Master Card and have affixed personal locks on energy isolation devices as appropriate.

Signature of all personnel covered by this permit.

Forward to Production Superintendent 7 days after completion of work.

rev.6/10/10

49102-11_F41B.EPS

Figure 41 Confined space entry permit (2 of 2).

- A list of all workers, including supervisors, who are authorized to enter the site
- The means by which workers and supervisors will communicate with one another
- Special equipment and procedures that are to be used during the job
- Other permits needed for work done in the space, such as welding
- The contact information for the emergency response rescue team

> **WARNING!**
> All permit-required confined spaces must be identified, and the associated permit must be posted.

> **NOTE**
> Some confined spaces require both a permit to enter and a permit to work in the space.

8.3.0 Confined-Space Hazards

Confined spaces are dangerous. The main hazards include poor airflow and restricted movement. Poor ventilation can allow toxic gases such as hydrogen sulfide, which smells like rotten eggs, to build up. Physical hazards are more dangerous because escape is limited.

8.3.1 Atmospheric Hazards

Atmospheric hazards are the most common hazards in a confined space. In a hazardous atmosphere, the air can have either too little or too much oxygen, be explosive or flammable, or contain toxic gases. Special meters are used to detect these atmospheric hazards (*Figure 42*).

A confined space that does not have enough oxygen is called an oxygen-deficient atmosphere; a confined space that has too much oxygen is called an oxygen-enriched atmosphere. For safe working conditions, the oxygen level in a confined space must range between 19.5 percent and 23.5 percent by volume, with 21 percent being considered the normal level. Oxygen concentrations below 19.5 percent by volume are considered oxygen-deficient; those above 23.5 percent by volume are considered oxygen-enriched. Too much oxygen in a confined space is a fire hazard and can cause explosions. Some materials, such as

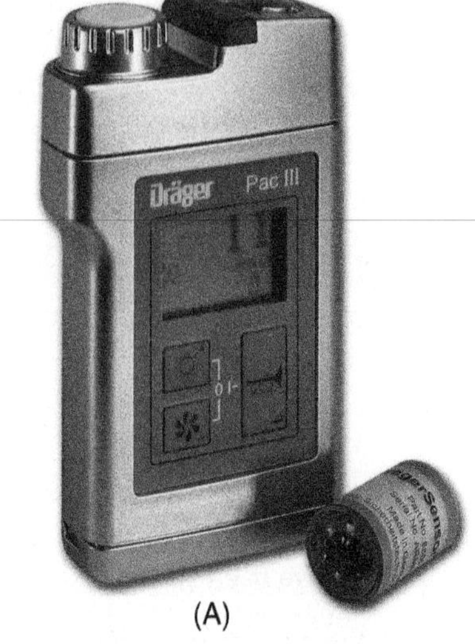

(A)

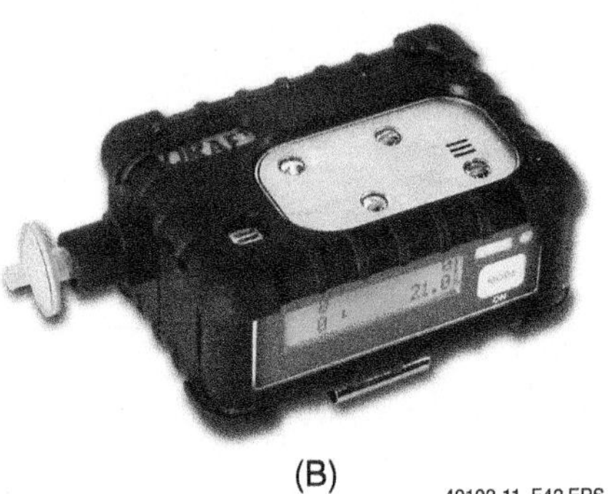

(B)

49102-11_F42.EPS

Figure 42 Detection meters.

clothing and hair, are highly flammable and burn rapidly in oxygen-enriched atmospheres. Fires can start easily in a confined space with oxygen-enriched air.

Air in a confined space becomes combustible when chemicals or gases reach a certain concentration. Flammable gases can be trapped in confined spaces. These include acetylene, butane, propane, methane, and others. Dust and work by-products from spray painting or welding can also form a combustible atmosphere. Some flammable gases are lighter than air and have a higher concentration at the top of a confined space. Vapors from fuels are generally heavier than air and form a greater concentration at the bottom of the space. A spark or flame will cause an explosion in a combustible atmosphere.

> **WARNING!**
> Never enter a confined space if you are not sure how safe it is.

Many of the processes that occur in a confined space use oxygen and may reduce the percentage of oxygen to an unsafe level. These processes include the following:

- Burning
- Rusting of metal
- Breaking down of plants or garbage
- Oxygen mixing with other gases

When the oxygen in a confined space is reduced, it becomes harder to breathe. Normal breathing in a confined space can create an oxygen deficiency. The symptoms of insufficient oxygen happen in this order:

1. Fast breathing and heartbeat
2. Impaired mental judgment
3. Extreme emotional reaction
4. Unusual fatigue
5. Nausea and vomiting
6. Inability to move your body freely
7. Loss of consciousness
8. Death

Toxic gases and vapors come from many sources. They can be deadly when they are inhaled or absorbed through the skin above certain concentration levels. In spaces with no ventilation, high concentrations can gather and quickly become toxic. Even in lower doses, some chemicals can seriously affect your breathing and brain functions.

WARNING!

Explosion-proof lights, motors, exhaust fans, and other equipment must be used to prevent fires and explosions in any combustible area.

8.3.2 Additional Hazards

In addition to atmospheric hazards, there are several other physical and environmental hazards in confined spaces. These hazards include the following:

- Electric shock
- Falling objects
- Poor visibility
- Extreme temperatures
- Noise
- Slick or wet surfaces
- Moving parts

Electric shock can occur when working on energized components. It is important to de-energize equipment when allowed and attach lockout/tagout devices and temporary grounds to prevent unintended re-energizing of equipment. When an arc hazard exists, it is important to wear the appropriate PPE. An arc flash in a confined space can be more dangerous than one in the open, because the confined space restricts the dissipation of energy, light, and noise.

Purging is needed when toxic, corrosive, or natural gases enter and mix with the air in a confined space. These gases can create an oxygen imbalance in the space that will suffocate workers almost immediately. To clear the gases, appropriate ventilation must be established and maintained to render the atmosphere safe. Air monitoring is necessary to verify air quality.

Materials or equipment can fall into confined spaces and strike workers. This usually happens when another worker enters or exits a confined space with a top-side opening, such as a manhole. Vibrations from equipment inside the space as well as those on the outside can also cause materials or tools to fall and strike workers.

Engulfment occurs when a worker is buried alive by a liquid or material that enters a confined space. Since many vaults are underground, they have sump pumps to keep out groundwater. When sumps fail, the vault can fill with water.

NOTE

The material safety data sheet (MSDS) attached to the entry permit will have information about any toxic substances you may encounter.

Confined spaces that are too hot or too cold can be hazardous to workers. Spaces that are too hot can cause heat stroke or heat exhaustion, while spaces that are too cold can cause **hypothermia**. Another temperature-related hazard in a confined space is steam. Steam is extremely hot and can cause serious or deadly burns.

Noise in a confined space can be very loud. This happens because the size of the space is small and sound bounces off the walls. Too much noise, or noise that is too loud, can permanently damage hearing. It can also prevent workers from communicating. If this happens, an evacuation warning could be missed.

Workers can get seriously hurt by slips and falls on slick or wet surfaces. Wet surfaces also add to the chance of electrocution from electrical circuits, equipment, and power tools.

Workers can get struck or trapped by moving parts of any equipment. This usually happens when a worker slips or falls.

8.4.0 Responsibilities and Duties

Everyone involved in confined space work must have special training. This includes entrants doing work in a confined space, attendants at the opening of the space, entry supervisors, and rescue workers.

8.4.1 Entrants

Entrants are people who enter a confined space to do the work. They must be aware of the dangers of the job and know how to protect themselves. The duties of an entrant are as follows:

- Obtain a valid entry permit.
- Know the atmospheric, fire, and toxic contamination hazards.
- Use the specified personal protective equipment, including face and eye protection, FR clothing, gloves, aprons, and coveralls as required.
- Stay in contact with the attendant to make sure they are being monitored and can be told to evacuate, if needed.
- Alert the attendant when warning signs or symptoms of exposure exist.
- Know how to escape when necessary.
- Exit the space immediately upon hearing an evacuation alarm or if an uncontrolled hazard is detected within the space.
- Keep track of tools, equipment, and personal items brought into the space and make sure it is all removed when the job is done.

8.4.2 Attendants

An attendant stays outside of the confined space and communicates directly with the entrant. The attendant has constant contact with the entrant through telephone, radio, visual, or other means. An attendant's basic responsibility is to protect the entrant. To do this, the attendant must do the following:

- Set up a station at the exit of the confined space.
- Keep count of the personnel inside.
- Know what work is allowed or disallowed in the space.
- Know how to monitor for safety.
- Test the atmosphere remotely using a probe and recognize safe and unsafe levels.
- Maintain contact with entrants and be able to recognize symptoms of physical distress.
- Order evacuation when problems occur.
- Know how to call rescuers and use the alarm system.
- Refuse entry to unauthorized personnel.
- Never enter the space for rescue attempts.

8.4.3 Supervisors

The entry supervisor is the person responsible for safe confined-space entry operations. This means that he or she is responsible for the lives of all workers involved in the operation. The entry supervisor must do the following:

- Authorize and oversee entry.
- Be well trained in entry procedures.
- Know the hazards of the confined space.
- Make sure entry permits are correct and complete.
- Make sure that proper lockout/tagout procedures are performed.
- Make sure that all equipment required for entry is available and understand how it works.
- Make sure workers understand the job.
- Set up rescue plans and participate in worker rescues.
- Be responsible for canceling entry authorization and terminating entry when unacceptable conditions are present or when the job is completed.
- Evaluate problems and recognize when reentry is safe.
- Ensure that explosion-proof equipment is used.
- Notify all workers in the area or facility that work is being done in a confined space.

8.5.0 Safeguards

It is important to understand how to protect yourself and your co-workers when working in a confined space. In order to do this, everyone must be aware of what is happening on the site and understand how to work safely. The following are the most common safeguards that should be used during confined-space operations:

- Monitoring and testing
- Ventilation
- Personal protective equipment
- Communications
- Training

8.5.1 Monitoring and Testing

The air in a confined space must be tested by the confined-space attendant before any workers enter. Testing must be done by a properly trained, qualified person. This person must be a company-approved or otherwise designated individual. The air is tested for oxygen content, explosive gases or vapors, and toxic chemicals. This can be done by inserting a wand attached to a gas meter into the confined space. It's important to remember that air being blown into or removed from the space doesn't mean the space is being ventilated.

Toxic gases can hide in confined spaces. Make sure the attendant has carefully tested the entire space before entering it.

The atmosphere in a confined space may need to be monitored during the entire job. This is done by attaching monitors to entrants or by using outside devices. When the atmosphere is monitored, workers can be assured that the air quality is good and that they will immediately know about changes in the atmosphere that would require them to leave. Always make sure the monitoring equipment is calibrated according to the equipment calibration schedule.

8.5.2 Ventilation

If the air in a confined space is hazardous or has the potential of becoming dangerous, the space must be ventilated immediately to remove toxic gases or vapors and replace lost oxygen. Ventilators blow clean air into the space. They must stay on as long as workers are in the space.

NOTE

Entrants and attendants have the right to witness or review the gas testing results and may request additional tests if deemed necessary.

8.5.3 Personal Protective Equipment

Every job requires some type of personal protective equipment. Standard personal protective equipment includes hard hats, safety goggles and glasses, boots, and gloves. On a confined-space job site, the following items may be needed in addition to the standard equipment:

- Full body harness
- Lifelines
- Air-purifying respirator
- Air-supplying respirator

NOTE

Always check the entry permit to make sure you have the personal protective equipment needed to enter a confined space.

8.6.0 Substation Entry

Safety requirements identified for confined space entry also apply in the case of substation entry. Specific hazards exist for substation entrants in the form of possible exposure to extreme high voltages, heated oil and other hazardous chemical components, hot gases, high temperature thermal exposure, and airborne shrapnel in the event of an arc flash. Only fully trained and authorized workers can enter a substation. Company and industry mandatory safety procedures must be followed by all personnel entering or working within a substation.

- All entrants must possess a properly issued entry permit to enter the substation. If work is to be performed, the entrant(s) must also possess a properly issued work permit. No work may be performed without a valid work permit. Know the difference and significance of both types of permits.
- All entrants and workers must don all company mandated safety apparel including fully inspected and tested FR clothing, a properly fitting hard hat and full face shield, fully inspected and tested insulated rubber gloves and protective leather outer gloves, and approved insulated work boots. When working with or around high voltage energized equipment, full FR coveralls and hood may also be required.
- The substation attendant or operator must verify all entry and work permits before entry. These permits must be posted at the entrance while the entrants are inside and the work is being performed.
- The attendant must remain at the entrance, monitor and be in constant contact with the workers. The attendant must also keep a current count of the number of workers within the substation. The attendant must be prepared to initiate and perform rescue procedures in the case of an accident.

Did You Know?

Digital Cameras for Safety

Workers can use digital cameras with a 20X zoom and flash to capture data on equipment located inside confined spaces. The camera can be used from outside the confined space to capture information such as data on tags, part numbers, and equipment configurations. In some cases, this eliminates the need for entry altogether, which in turn eliminates potential accidents and injuries.

Source: The Occupational Safety and Health Administration (OSHA).

- The attendant must inform the entrants of any special system conditions that may affect their safety. The attendant must also inform the entrants of the location of all energized equipment in or adjacent to the work area and the limits of any de-energized work area.
- All entrants and workers must be fully trained and knowledgeable about the substation equipment and the associated hazards.
- When work is performed on or around energized equipment of 600V or greater at least two workers must be present at all times.
- According to industry standard charts, minimum approach distances (MAD) must be calculated and barricades must be established around all energized equipment with tape, cones, or other company-approved markers. In the case of equipment containing extreme high voltages, it may be necessary for protective barriers to also be set up.
- All work performed on energized equipment must be performed with approved insulated tools, such as hot sticks or shotgun sticks.
- The attendant is responsible for canceling entry authorization and terminating entry when unacceptable conditions are present or when the job is completed.

9.0.0 Environmental Concerns

Just about everyone needs electricity—even those living in remote areas. To get electricity, the crew of line workers needs to clear a path, place poles, and string lines. After these lines are placed, line workers maintain them so consumers can enjoy uninterrupted electrical power. As a line worker, you will spend many of your working hours outside. You will work in congested urban areas, less populated suburban areas, and remote rural areas.

The work you do in each one of these areas has the potential to adversely affect the environment. Wherever you work, you are likely to encounter wildlife—sometimes you'll encounter endangered species in the oddest places. According to the Madison Gas and Electric web site (**www.mge.com**), in 2009 three peregrine falcons hatched at the Blount Generating Station in downtown Madison, Wisconsin. Peregrine falcons have been on the endangered species list in some areas since the 1970s.

Federal, state, and local governments all have laws governing the protection of endangered and migratory wildlife, as well as water and wetlands. It is important that you are aware of these laws and understand how to mitigate any adverse affects that performing your job can have on the environment.

GOING GREEN

You can help improve our environment by picking up roadside trash before you leave a work area. Some local governments sponsor this type of program and will even supply you with bags that can be left on the side of the road for pickup, or they may be dropped off at a designated point.

9.1.0 Clean Water Act

The Clean Water Act (CWA) is a federal law that protects all surface waters. The program includes lakes, rivers, estuaries, oceans, and wetlands. The CWA sets certain water quality standards and prohibits pollution. Any discharge into a body of water must be permitted. These permits are known as NPDES permits, for the National Pollutant Discharge Elimination System. The CWA also protects wetlands that provide habitat for many species of wildlife. Excavation or fill in wetlands or swamp areas is controlled by the CWA. Permits must be obtained for these activities.

Even when you are working far from a body of water, your actions can damage it. Oils and fuel spills from a roadway are washed into storm drains by rainwater and snow. Any water that is discharged unfiltered into a body of water can contaminate it. Further, road runoff in rural areas will most likely not go into a stormwater sewer but into the nearby earth where it can leech into the groundwater, contaminating it.

Line workers often need to work on unpaved surfaces. Simply driving over soft soil can produce ruts that allow a path for stormwater runoff. As stormwater runs down the ruts, it can pick up topsoil and contaminants that can be easily washed into nearby bodies. The soil particles that are washed away are called sediment. It is considered a pollutant when it gets into surface water. Contaminants and sediment can pollute water and destroy wildlife habitat. Further, the runoff of stormwater and sediment can lead to erosion of the earth's surface.

Because sediment is soil that has been moved from its original place, the first step to controlling sediment is to prevent erosion. During construction, some erosion is unavoidable, so most sites use various devices such as silt fences (*Figure 43*) to trap displaced soil. Workers can help to prevent sedimentation by avoiding travel across disturbed soil, washing soil from truck tires before leaving the site, and limiting vehicle operation to approved roads.

49102-11_F43.EPS

Figure 43 Silt fence.

49102-11_F44.EPS

Figure 44 Stormwater barrier.

One of the most effective ways to prevent erosion is to seed the area immediately after grading has been completed. Seeding an area is inexpensive and can show results in a few days. It can be difficult to seed a large area or slope. Hydromulch is a mixture of mulch, water, fertilizer, and grass seed, and it can be quickly sprayed long distances.

Stormwater barriers must be placed over storm drains (*Figure 44*) to prevent dislodged silt from entering the stormwater system.

9.2.0 Endangered Species Act

The Endangered Species Act (ESA) was passed into law in 1973. Its purpose is to protect and recover endangered species and their habitat. The ESA makes it illegal for a person to in any way harm or harass an endangered animal, including disrupting or destroying its habitat. This includes killing the animal or changing its behavior, including breeding, feeding, or sheltering. Simply moving a nest of or chasing an endangered animal is illegal. This act covers both public and private lands.

Often, endangered birds such as eagles build nests on the top of power poles. In fact, power companies set up power poles without power lines for the birds to nest. Never remove or destroy an endangered species habitat. When you encounter what you believe to be an endangered species, don't try to work around it. Leave the area and contact your supervisor for guidance.

9.3.0 Good Housekeeping Practices

The best way to preserve the environment is with good housekeeping practices. The first step in preventing such pollution is to think before you act. When you are working, be very careful not to spill chemicals, oil, grease, lubricants, and gasoline onto the ground. Whenever possible, place a covering on the ground to catch any spills. It's easier to clean a tarp or drip tray than it is to decontaminate soil.

GOING GREEN

Hazardous Waste

One easy way to reduce hazardous waste is to use a product that is not hazardous. Another way is to use as little of the product as you can to do the job. You may have no control over the selection of a product, but you do control how much you use.

When you create hazardous wastes, which can be as simple as a solvent-saturated rag that you used to clean a tool, be sure to dispose of it properly. Keep chemical or hazardous wastes separated from general debris. Even a small amount of a hazardous substance in a container of waste makes the whole container a hazardous waste. Costs to dispose of hazardous waste are much higher than those for disposing of non-hazardous trash.

In 1976, the Resource Conservation and Recovery Act (RCRA) was created to handle all aspects of hazardous waste management. Hazardous wastes are tracked from when they are created to when they are finally destroyed. RCRA covers hazardous waste generators, transporters, and treatment, storage, and disposal facilities. Hazardous wastes include gasoline, solvents, and some oils as well as many other products.

10.0.0 Job Safety Analyses and Task Safety Analyses

As you have already learned, OSHA guidelines require that employers provide employees with information and training about how to stay safe on the job. In addition, your employer must make it possible for workers to stay safe in the workplace. To fulfill these requirements, each workplace will develop a carefully thought-out safety program. Good safety programs help employers by reducing the number and severity of accidents that can injure workers and destroy equipment. Good safety programs help employees because they safeguard worker health and well-being. They also reduce costs for insurance coverage for on-the-job accidents, medical care for injured workers, and maintenance and repair of equipment damaged by an accident. In short, good safety programs are good for everyone.

Employers recognize that developing methods and controls to prevent accidents from occurring can dramatically increase the safety of a job or task. To do this, your employer will have specially trained workers conduct surveys and analyses at the workplace to identify hazards. The result of these analyses is safety policies and procedures that you will use on the job. The purpose of this section is to give you an understanding of how and why these policies and procedures were developed and to give you the basis for identifying hazards for any task that you perform.

You most likely perform safety analyses every day. When you're driving, you constantly scan the nearby traffic looking for hazards. When you spot one, you adjust your driving to prevent an accident. You don't even realize that you are doing it; you think that you're just having an uneventful drive to work. If you think back to when you were first learning to drive, it wasn't nearly so relaxing. How many near-misses did you have before you incorporated driving safety into your thinking?

Your employers use the same principle in the workplace. Employers have a formal safety program that carefully observes and breaks down activities in the workplace by conducting job safety analyses (JSAs) and task safety analyses (TSAs). A job safety analysis (*Figure 45*) is a careful study of a particular job to find all of the associated hazards. A task safety analysis (*Figure 46*) is a careful study of an individual task of the job to find all of the associated hazards with that task. *Figure 47* shows the relationship between a job and a task.

The terms *job safety analysis* and *job hazard analysis* are often used interchangeably. Each involves a methodical review of job steps identifying safety concerns. The JSA is more inclusive and covers all hazards, including safety, health, and environmental issues. It's important to note that a JSA is by far more common than a TSA, but the reasons and methods for performing both are similar.

JSAs can be time consuming, so not all jobs will be analyzed. Usually jobs that have one of more of the following characteristics will be analyzed:

- A history of accidents or injuries
- The potential for serious injuries
- The addition of a new job or equipment
- A change in procedure or routine
- A high turnover rate in personnel

The main benefit to conducting a JSA is that it provides information on potential hazards associated with a job or task. Besides identifying hazards, a JSA also helps to determine ways to avoid or minimize unsafe or unhealthy conditions. Identifying hazards and determining the appropriate steps, equipment, and controls needed to reduce those hazards will minimize accidents and injuries.

Hidden or underestimated hazards can be uncovered through the careful study and analysis involved in a JSA. As JSAs are conducted on individual jobs, a larger picture of the overall hazard level of a task will begin to develop. This information can be used to prevent accidents or illnesses and to reduce the level of risk at the workplace.

JSAs also provide a means for standardizing work practices. When conducting a JSA, a job is studied in great detail. Several workers may be observed, or even videotaped, while performing a job. This is the most desirable method of gathering information, but it can be distracting to workers. If you are ever involved in one of these studies, it is important that you perform your job exactly as you do normally.

Another way to gather information is to interview workers to determine all of the steps involved in performing the job. As a worker, it is important that you accurately and honestly describe the steps you take to perform any job or task.

This study produces a large volume of information that is analyzed to determine which practices work efficiently and safely and which do not. The final step in conducting a JSA is to develop practical methods of making the job safer. The goal is to prevent the occurrence of potential accidents and minimize the risk and severity of injury should an accident occur. The principal solutions are as follows:

- Find a safer way to do the job.
- Change the physical conditions that create the hazard.

JOB SAFETY ANALYSIS FORM

Job Title:	Date of Analysis:
Required PPE:	Conducted By:
Required Tools & Equipment:	Duration:
Materials Used:	

Task	Hazard	Quality Concern	Recommendation

49102-11_F45.EPS

Figure 45 Example of a JSA form.

Job:	Hazard	Severity (0 – 3)	Likelihood (0 – 3)
TASK 1			
TASK 2			
TASK 3			
TASK 4			
TASK 5			

49102-11_F46.EPS

Figure 46 Example of a TSA form.

JOB:	CHANGE BURNED OUT LIGHT BULB
TASK 1:	LOCATE BURNED OUT BULB.
TASK 2:	SET UP EQUIPMENT TO REACH BULB.
TASK 3:	DETERMINE BULB TYPE.
TASK 4:	REMOVE BURNED OUT BULB.
TASK 5:	INSTALL WORKING BULB.

49102-11_F47.EPS

Figure 47 The relationship between a job and a task.

- Change work methods or procedures to eliminate hazards.
- Try to reduce the necessity or frequency of doing the job or some of its tasks.

Developing solutions should be done with a team that includes safety experts, process engineers, and other experts such as ergonomists, fire experts, or medical doctors. Most changes will also require a commitment on the part of supervisory and management personnel. For that reason, it is a good idea to involve them at an appropriate point in the process.

11.0.0 Work Zone and Personal Safety

Your work can be hazardous, so you and your co-workers need to do everything you can to make it safer. You have two tasks to perform for safety's sake. The first is to accept responsibility for your own safety and act accordingly. The second is to keep yourself, your co-workers, and the public safe while performing your job.

11.1.0 Work Zone Safety

One complication to your job is that your work zone is always changing. In the morning, you could be working in an isolated rural area and in the afternoon on a busy roadway. Even when you work in the same area all day, the conditions can change from one hour to the next. You always need to be prepared to modify your safety strategy.

One of the ways accidents can be avoided is to effectively use various safety tools. These tools include signs, safety tags, barricades, and barriers, as well as other devices. When used properly, these devices can prevent accidents and assist in the smooth flow of work and permit you to efficiently complete your job.

Since some of your work will be located on or near public roads, you need to understand how to set up a safe work zone. Specific standard procedures must be used in these work zones. Using safety devices and putting rules and procedures in place to prevent accidents are only helpful if they are followed all of the time. It is up to you to work safely. Each person in the work crew has the personal responsibility to keep themselves and their co-workers safe.

When working in a public place, all vehicles should be positioned to minimize the disruption of the public's use of the area. If possible, vehicles should be parked as far off the road as is practical. TTC devices should be placed on the roadway to alert the public of an approaching work zone. The work zone should employ barricades and barriers to alert workers of the perimeter and to keep unauthorized persons out. Some common types of barricades and barriers are shown in *Figure 48.*

It is important to remember that the devices that you are using to barricade a work area serve only as a warning. They do not stop vehicles from crashing through them or people from going over or around them. You need to stay alert to make modifications and changes as needed.

Barricades establish the work zone for workers. Workers should not routinely step outside of the work zone. If you need to repeatedly leave the work zone to do your job, the work zone is not set up correctly and needs to be modified.

49102-11_F48.EPS

Figure 48 Typical barricades and barriers.

Before barricading a work zone, carefully examine the area that you intend to work in. Walk the area to ensure that there are no hidden hazards. This is especially important when there are obstructions, such as snow, mud, leaves, and other debris in the area. Never move a barricade—even in an emergency—unless you're sure that it is safe to do so.

Inside the work zone, there are varying degrees of safety. Some tasks are more dangerous than others, but as a trainee, you may not be able to distinguish the more hazardous tasks from the less hazardous ones. For you, the best defense is communication and to practice personal security. Use the following as a guide:

- Before work begins, ask your supervisor what you should do and where you should be doing it.
- If you are to observe, ask where to stand. Some tasks will be so hazardous that your supervisor will not want you in the vicinity and may ask you to sit in a vehicle. Try not to take this personally. Your supervisor is trying to protect you.
- Always wear your reflective vest.
- When performing dangerous work such as placing poles, hanging cable, and tensioning lines, stay alert for developing hazards. When observing, stay well away from the work area.
- Perform only tasks that you have been trained to do.
- Ask questions before work begins.
- When a hazardous situation develops, stay out of the way if you are not already involved, but be ready to help if asked.
- Wear your safety equipment even if you are not involved in a task. You may be called on to help.

Remember, PPE, barricades, and policies are just tools. They can't prevent accidents unless they are used correctly. The best tool for preventing accidents is to stay alert and aware of the work zone. Don't rely on someone else to keep you safe. Think a job through before starting it. Look around for potential hazards and stay alert for changes in the work area.

Most of the time you will use two-way communications to stay in contact with other workers. Sometimes voice communication is not practical, and it may be useful to communicate with hand signals. It is important for you to understand that there can be various meanings for the same hand signal. Always establish with your co-workers the hand signals that will be used and their meanings before starting the task. Hand signals can be safely used over long distances, but those using hand signals should never be out of each other's sight. You should always use slow and exaggerated gestures to be sure that you are understood. Some of the common hand signals and their meanings are described in *Table 6*.

Table 6 Typical Hand Signals

Signal	Meaning
Point index finger toward self	I/me
Point index finger toward object	It/them
Point index finger toward person	You/them
Circle index finger at group	We/us/all of us
Beckon by moving arm toward self	Come here
Extend arm and point hand to the right	Move right
Extend arm and point hand to the left	Move left
Point with thumb in a particular direction	Move this way/go this way
Bring index finger across throat	Quit
Slowly ease palm face down	Relax/slow down
Put palm over brow	Scout it out/check it out
Hold index finger up near head	Wait
Hands on top of head	I'm OK
Thumb up	Good/OK
Thumb down	Bad/not OK
Slap forehead	Bad idea
Palm down and rotated from side to side	Unsure/can't decide
Wave goodbye	Goodbye
Form a circle with thumb and index finger	OK/I understand/ agree
Military salute	I understand and will comply
Shake head from side to side	No/disagree
Shake head up and down	Yes/agree

11.2.0 Emergency Response

Victims of an electrical event are injured by any combination of electrical shock, arc flash, and arc blast. Most employers have emergency response procedures in place for each type of incident and practice them periodically. You must become familiar with these procedures before you need them. Often, emergency response professionals, such as fire, ambulance/EMTs, or air rescue, have a working relationship and plan of access for the facility, shaving minutes off a rescue effort. Communication may be the most important part of any emergency response plan. Take a moment at each job site and imagine how you would describe your location to a 911 operator in the event of an emergency.

11.2.1 Shock Victims

When a person gets an electrical shock, survival depends entirely on the path the voltage takes through their body. A surface path may only produce burns at the entry and exit points, while a path across the heart might result in death. Victims of electrical shock may not be able to let go of the object(s) shocking them. Anyone trying to help them must remember not to touch them, or the rescuer also becomes part of the circuit. If a co-worker is receiving a shock, turn off the power if possible, call 911, and administer CPR. If the source of the electricity cannot be turned off immediately, use an insulated rescue hook (*Figure 49*) to remove the person from the electrical source. After a victim has been removed, the area must be made safe for the rescuers so first aid can begin. A victim of electrical shock may go into physical shock. Anyone trying to help the victim must understand how to treat shock victims. This is why standby personnel must be trained in first aid and CPR.

11.2.2 Arc Flash Victims

The victim of an arc flash is often injured much more seriously than what is first apparent. Douse the victim with water and remove them to a safe location. Be aware that burns to the neck may interfere with the victim's ability to breathe. Arc flash victims often incur injuries so severe that they require treatment at a special burn center.

49102-11_F49.EPS

Figure 49 Insulated rescue hook.

11.2.3 Teamwork

When performing energized work, always have a trained person on standby to help if needed. *OSHA Standard 29, Section 1910.269* requires that CPR be available within four minutes where employees may be exposed to a shock hazard. Other industry standards identify training requirements for employees who may be exposed to shock hazards. Training includes first aid and emergency procedures, as well as methods of releasing shock victims from electrical contact.

Line workers may have to work in remote areas that are off-limits to other employees. These may include substations, roadways, or open country. Anyone working in these areas needs to make sure supervisors, managers, and co-workers know where they are and what they are doing.

11.2.4 Resuscitation

The heart of a shock victim may stop or lose its normal rhythm. This is known as fibrillation. If someone's heart has stopped, use CPR to maintain the victim's breathing and blood flow until help arrives. A heart in fibrillation can only be restored to normal rhythm through the use of a defibrillator. Automated external defibrillators (AEDs) are increasingly common in public areas, on airplanes, and in the workplace. An AED (*Figure 50*) can be used to check a victim's heart rhythm and indicate when a shock is required. These machines automatically sense, diagnose, and provide user guidance, requiring minimal training for effective use.

11.3.0 Personal Safety

Personal safety starts before you set foot on the job. Before you report to your workplace you must prepare yourself to work. Taking personal responsibility for your well-being helps protect you in hazardous situations. It also shows your employer that you are a professional who can be trusted.

Taking care of your health is your first step. Eating a healthy diet and getting enough exercise gives you the energy and stamina you need to perform your job. Getting adequate sleep helps you stay alert on the job.

Dressing appropriately can protect you from serious injury. Always wear clothing made of natural fibers, a shirt with long sleeves, and trousers made of durable fabric. Wear cotton undergarments and minimize any metal fasteners. Avoid wearing jewelry.

Dress for the weather. In cold weather, dress in layers so you can remove a layer as the day grows

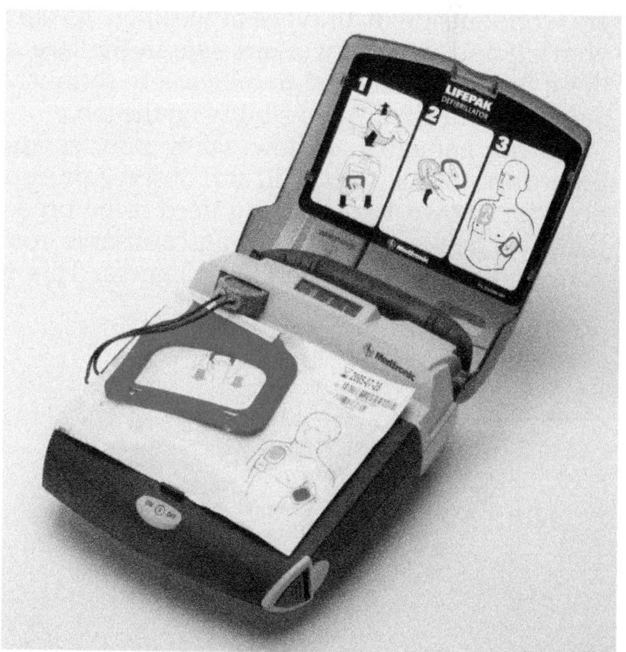

49102-11_F50.EPS

Figure 50 Example of an AED.

warmer. Always have warm gloves and a hat with you so you can wear them while not working. It is easier to stay warm in cold weather than it is to warm yourself when you get cold.

In hot weather, wear clothing that covers your skin to prevent sunburn. Apply sunscreen to your skin at least 20 minutes before you report to work and carry it with you so you can reapply it frequently. Apply insect repellent as needed. Always wash your hands thoroughly after applying sunscreen or repellent. Remember that the oils in these products can damage rubber insulating devices, so be sure to clean off your skin before donning PPE.

Carry water and high-protein snacks such as nuts with you. Drink adequate amounts of water in both warm and cold weather. Don't allow yourself to get too hungry. Dehydration and low blood sugar can cause you to become drowsy and can adversely affect your thinking.

Carry a first aid kit and keep your CPR skills up to date. Know how to use first aid equipment. Wear leather work gloves and keep shirt sleeves rolled down and secured to help prevent injuries. If you are injured, report it, and clean and dress the wound immediately. Keep your tetanus immunization up to date.

Look out for yourself, especially in rural areas. Look for holes and animal burrows as you move around. Stay alert for snakes and the nests of bees, hornets, and other stinging insects. If you are allergic to any insect stings or vegetation, keep your rescue kit handy and up to date. Be sure your co-workers know how to use the kit.

If you are digging and strike a utility, stop digging immediately and notify the entire crew of the accident. Turn off all vehicles and equipment if it is safe to do so. Notify the responsible utility company. For water, sewage, and gas lines, call 911 to notify the jurisdiction's emergency response service. When in doubt, call 911 for guidance. Never look into the cut ends of fiber optic cable; they can blind you.

Be proactive when it comes to safety. Learn required safety techniques and know how to use safety devices. Ask yourself "Is there a safer way to do this?" Don't do something just because everyone else does it. Think before you act.

SUMMARY

Power line workers are exposed to many unusual hazards in their work. These hazards include exposure to high voltages, confined space work, working in trenches, working underground, and working at heights. In addition, power line workers work outside in all types of weather. To keep yourself and your co-workers safe in the face of these hazards, you need to take safety seriously. Safety is your first responsibility on the job.

It's not enough to follow safety policies and procedures, wear your PPE, and observe precautions for known hazards. You need to be proactive. That is, you need to constantly observe your surroundings to stay alert for dangers. This is especially true when you are working in a public area where dangers can develop quickly. Being proactive means looking for and identifying job-related hazards and making sure they are corrected before you or your co-workers are exposed to those hazards.

To become a professional line worker you need to commit yourself to safety. To do this, you need to be prepared to learn, apply yourself to your studies, and communicate clearly to co-workers. Your employer is required to give you the tools you need to be safe, but only you can use those tools. That means that your safety is your own responsibility.

Review Questions

1. Step potential is a difference of potential between _____.
   a. a person's feet
   b. each 12-inch measure on a power pole
   c. a vehicle and the ground
   d. each 12-inch measure on a power line

2. Arcs occur because air is a good conductor of electricity.
   a. True
   b. False

3. To determine the PPE you should wear in a restricted area, you should _____.
   a. ask an experienced co-worker
   b. check the OSHA standard
   c. check the company safety policy
   d. look at the posted warning sign

4. Special safety procedures are required whenever the line-to-line voltage level exceeds _____.
   a. 50 volts
   b. 100 volts
   c. 500 volts
   d. 1000 volts

5. When voltages greater than 750 volts are present, unqualified workers should maintain a minimum distance of _____.
   a. 5 feet
   b. 10 feet
   c. 20 feet
   d. 50 feet

6. Personnel working in an arc flash protection boundary must wear _____.
   a. a respirator
   b. ballistic chaps
   c. reflective vests
   d. flame retardant clothing

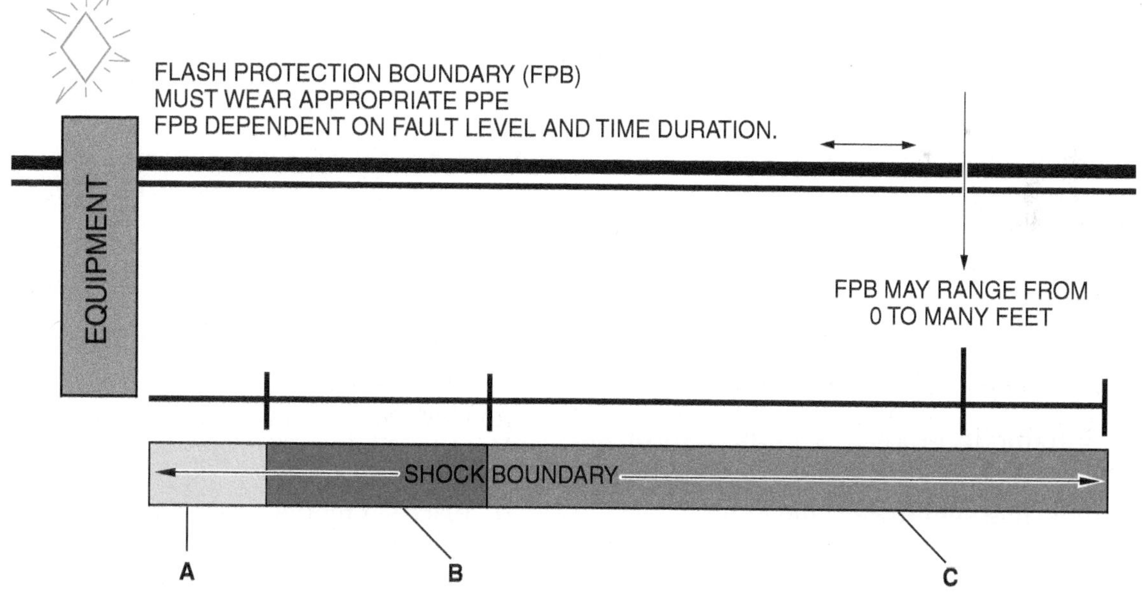

Figure 1

For Questions 7 through 9, match the letter on *Figure 1* to the boundary designation.

7. _____ Restricted shock boundary

8. _____ Limited shock boundary

9. _____ Prohibited shock boundary

10. Unqualified persons may cross a limited approach boundary _____.
    a. only in an emergency
    b. only to read warning signs
    c. when escorted by a qualified person
    d. only when escorted by a supervisor

11. A minimum approach distance (MAD) is a safety designation for _____.
    a. traffic control
    b. step potentials
    c. electrical equipment
    d. fall protection

12. The gloves that provide the highest level of protection against high voltages are in Class _____.
    a. 00
    b. 0
    c. 2
    d. 4

13. The common method used to check for holes in rubber insulated gloves is to use _____.
    a. glove dust and air
    b. light and powder
    c. water and air
    d. light and air

14. Flame-resistant clothing with the highest arc rating is rated Category _____.
    a. 00
    b. 4
    c. A
    d. F

15. The insulating device used to cover oddly shaped live connections is a _____.
    a. ballistic blanket
    b. rubber blanket
    c. dielectric boot
    d. rubber sleeve

16. To allow traffic to proceed, a traffic control flagger should _____.
    a. point the flag staff toward the ground
    b. hold the flag staff at shoulder level
    c. roll up the flag
    d. wave the flag up and down

17. One cubic yard of soil weighs about _____.
    a. 500 pounds
    b. 1,000 pounds
    c. 2,000 pounds
    d. 3,000 pounds

For Questions 18 through 20, match the correct protective system to its corresponding description.

18. _____ Shoring

19. _____ Trench shields

20. _____ Sloping

    a. Cutting the walls of the excavation back at an angle to its floor
    b. Supports the walls of the excavation and prevents their movement and collapse
    c. Placed in unshored excavations to protect personnel from excavation wall collapse
    d. Placing the spoil pile close to the edge of the trench

21. During an HDD operation, you need to dig a pothole when _____.
    a. the bore path curves
    b. there is the risk of a utility strike
    c. the bore path includes compacted soil
    d. there is an obstruction to the drill string

22. Entry into a confined space begins when _____.
    a. the permit is issued
    b. all of the entrants have been assigned a job
    c. any part of your body passes through the entrance
    d. the first person enters the confined space

23. An explosion in a combustible atmosphere can be caused by _____.
    a. flammable clothing or materials
    b. a source of ignition such as a flame or spark
    c. the presence of toxic vapors
    d. vibrations from tools outside the space

24. One of the most effective ways to prevent erosion is to _____.
    a. seed the area immediately after grading
    b. cover the area with straw after grading
    c. cover the area with plastic after grading
    d. place silt fences around it after grading

25. If a co-worker signals by slowly lowering his/her hand with the palm down, you can assume that he/she wants you to _____.
    a. take a break
    b. slow down
    c. stop traffic
    d. dig deeper

Trade Terms Quiz

Fill in the blank with the correct term that you learned from your study of this module.

1. A voltage between an energized object being touched and the ground is called ___________.
2. An approach limit at a distance from an exposed energized electrical conductor or circuit part within which work is considered the same as making contact with the energized conductor or part is called the ___________.
3. A dangerous condition caused by the enormous release of thermal energy in an electric arc is a(n) ___________.
4. A(n) ___________ is a dangerous condition associated with the release of energy caused by an electric arc.
5. A(n) ___________ is a potential danger in the air or a condition of poor air quality.
6. A method of protecting workers from cave-ins by excavating the sides of an excavation to form one or a series of horizontal levels or steps is called ___________.
7. ___________ means to drill a hole into the ground.
8. ___________ air or materials can explode and cause a fire.
9. A study investigating a worker's potential exposure to arc flash energy is called a(n) ___________.
10. A mesh mat that is placed underneath equipment to provide a single electrical path to ground should the equipment come in contact with an energized conductor during a utility strike is called a(n) ___________.
11. ___________ is a life-threatening condition caused by exposure to very cold temperatures.
12. An area in which wire mesh is embedded in concrete and bonded to metal structures is known as a(n) ___________.
13. ___________ materials resist combustion.
14. The incident energy limit that a flame-resistant material can withstand before it breaks down and loses its ability to protect the wearer that is expressed in joules/cm^2 or calories/cm^2 is ___________ or ___________ or ___________.
15. A(n) ___________ has the skills and knowledge related to the construction and operation of the electrical equipment and installations, and has received safety training to recognize and avoid the hazards involved.
16. A(n) ___________ is the approach limit at a distance from exposed energized electrical conductors or circuit parts within which a person could receive a second-degree burn if an electrical arc flash were to occur.
17. A device that is used to stop the HDD rig when the operator leaves the seat with the drill turned on is a(n) ___________.
18. ___________ is a high-energy discharge between two or more conductors.
19. A(n) ___________ atmosphere does not have enough oxygen.
20. The voltage between the feet of a person standing near an energized grounded object is ___________.
21. A(n) ___________ is an explosive expansion of air and metal in an arc path.
22. A(n) ___________ is an approach limit at a distance from an exposed energized electrical conductor or circuit part within which there is an increased risk of electrical shock.
23. ___________ can be permanent structures or can be designed to be portable and moved along in a trench as work progresses.
24. A(n) ___________ is caused by unbalanced stresses in the soil surrounding an excavation.
25. A shorting cable is an example of a(n) ___________.
26. A(n) ___________ occurs when the conductor or circuit part to be worked on or near has been disconnected from energized parts, locked/tagged in accordance with established standards, tested to ensure the absence of voltage, and grounded if necessary.
27. The greatest angle above the horizontal plane at which a material will adjust without sliding is the ___________.

28. An explosion similar to the detonation of dynamite that occurs during an arc flash incident is known as ________.

29. The ________ is a length of pipe that connects the boring machine to the drill head.

30. The distance from energized electrical conductors or circuit parts that a qualified person may approach without wearing rubber insulated PPE is called ________.

31. A ________ material has very little electrical conductivity.

32. The distance from an exposed energized electrical circuit within which an increased risk of electrical shock exists is referred to as ________.

33. The maximum incident energy resistance of a material is known as the ________.

34. The incident energy limit that a flame-resistant material can withstand is called its ________.

Trade Terms

Angle of repose
Arc blast
Arc fault
Arc flash
Arc rating
Arc thermal performance value (ATPV)
Atmospheric hazard
Benching
Blast hazard
Boring
Combustible
Dielectric
Drill string
Electrically safe work condition
Energy of break-open threshold (E_{BT})
Equipotential plane
Flame-resistant (FR)
Flash hazard
Flash hazard analysis
Flash protection boundary
Grounding mat
Hypothermia
Limited approach boundary
Maximum allowable slope
Minimum approach distance (MAD)
Operator presence sensing system (seat switch)
Oxygen-deficient
Prohibited approach boundary
Qualified worker
Restricted approach boundary
Shield
Step potential
Subsidence
Temporary grounding device (TGD)
Touch potential

Cornerstone of Craftsmanship

Gordon Johnson

Utility Worker, Instructor
BGE (formerly Baltimore Gas and Electric),
Baltimore, MD

Gordon Johnson started out climbing trees and switched to climbing utility poles. He had a family to support, and when the opportunity came up, he left the tree company where he worked and moved to Baltimore Gas and Electric (now BGE). With training and hard work, he went from utility worker to instructor, a role he enjoys.

How did you get started in the construction industry?
I was working as a foreman for a contract tree company, but I started out cutting and trimming, so I was already used to climbing and heights. Then an opportunity arose to join the Baltimore Gas and Electric team. I wanted a career and stability, so I eagerly made the move and became a lineman. It has proven to be a good choice.

Who inspired you to enter the industry?
I guess you could say necessity. I had a family to support, and the benefits in the utility industry were good. Besides, in my early days, I liked working outside and facing the elements. Every day brought a new challenge.

What do you enjoy most about your job?
It's satisfying. I liked working with my hands and climbing—all the activity. Now I'm finding just as much satisfaction in training the new hires. I want them to be safe and do a good job. It's very rewarding when you send people out, knowing they're well trained and can handle the job. There's a lot of camaraderie in this kind of work, where you depend on each other.

Do you think training and education are important in construction?
Essential. Training and education are very important. All this experience is gradually leaving the workforce in some form or other, and without training, who'll carry on? We need to keep up with new techniques and technology so the workforce stays strong.

How important do you think NCCER credentials will be in the future?
Very important. The courses were created by people who are experienced in their fields. They've agreed on the best and safest way to do things, and this sets a standard that employers will respect. When new workers have NCCER credentials, it will mean something in their industry.

How has training/construction impacted your life and your career?
I'll say one thing—training has kept me safe for a lot of years in a very hazardous job. Safety was impressed on me when I started, and now I try to impress it on others. This is not a career where you can walk in off the street and do it.

Would you suggest construction as a career to others?
Absolutely, if you enjoy being outdoors or working with your hands rather than sitting in an office behind a desk full time. I would offer one piece of advice: This job is unforgiving. It will not allow you to lose focus. Safety is very important.

How do you define craftsmanship?
It's having knowledge and superior skills, the ability to do a job safely and well.

Trade Terms Introduced in This Module

Angle of repose: The greatest angle above the horizontal plane at which a material will adjust without sliding.

Arc blast: An explosion similar to the detonation of dynamite that occurs during an arc flash incident.

Arc fault: A high-energy discharge between two or more conductors.

Arc flash: A dangerous condition caused by the enormous release of thermal energy in an electric arc, usually associated with electrical distribution equipment.

Arc rating: The maximum incident energy resistance demonstrated by a material (or a layered system of materials) prior to material breakdown, or at the onset of a second-degree skin burn. Expressed in joules/cm^2 or calories/cm^2.

Arc thermal performance value (ATPV): The incident energy limit that a flame-resistant material can withstand before it breaks down and loses its ability to protect the wearer. Expressed in joules/cm^2 or calories/cm^2.

Atmospheric hazard: A potential danger in the air or a condition of poor air quality.

Benching: A method of protecting workers from cave-ins by excavating the sides of an excavation to form one or a series of horizontal levels or steps, usually with vertical or near-vertical surfaces between levels.

Blast hazard: The explosive expansion of air and metal in an arc path. Arc blasts are characterized by the release of a high-pressure wave accompanied by shrapnel, molten metal, and deafening sound levels.

Boring: The process of drilling a hole into the ground.

Combustible: Air or materials that can explode and cause a fire.

Dielectric: A material with very little electrical conductivity.

Drill string: The length of pipe that connects the boring machine to the drill head.

Electrically safe work condition: A state in which the conductor or circuit part to be worked on or near has been disconnected from energized parts, locked/tagged in accordance with established standards, tested to ensure the absence of voltage, and grounded if necessary.

Energy of break-open threshold (E_{BT}): The incident energy limit that a flame-resistant material can withstand before the formation of one or more holes that would allow flames to penetrate the material. Expressed in joules/cm^2 or calories/cm^2.

Equipotential plane: An area where wire mesh or other conductive elements are embedded in or placed under concrete, bonded to all metal structures and fixed non-electrical equipment that may become energized, and connected to the electrical grounding system to prevent a difference in voltage from developing within the plane.

Flame-resistant (FR): The property of a material whereby combustion is prevented, terminated, or inhibited following the application of a flaming or non-flaming source of ignition, with or without subsequent removal of the ignition source.

Flash hazard: A dangerous condition associated with the release of energy caused by an electric arc.

Flash hazard analysis: A study investigating a worker's potential exposure to arc flash energy, conducted for the purpose of injury prevention and the determination of safe work practices and appropriate levels of PPE.

Flash protection boundary: An approach limit at a distance from exposed energized electrical conductors or circuit parts within which a person could receive a second-degree burn if an electrical arc flash were to occur.

Grounding mat: A mesh mat that is placed underneath equipment to provide a single electrical path to ground should the equipment come in contact with an energized conductor during a utility strike.

Hypothermia: A life-threatening condition caused by exposure to very cold temperatures.

Limited approach boundary: An approach limit at a distance from an exposed energized electrical conductor or circuit part within which a shock hazard exists.

Minimum approach distance (MAD): The distance from energized electrical conductors or circuit parts that a qualified person may approach without wearing rubber insulated PPE.

Operator presence sensing system (seat switch): A device that is used to stop the HDD rig when the operator leaves the seat with the drill turned on.

Oxygen-deficient: An atmosphere in which there is not enough oxygen. This is usually considered less than 19.5 percent oxygen by volume.

Prohibited approach boundary: An approach limit at a distance from an exposed energized electrical conductor or circuit part within which work is considered the same as making contact with the energized conductor or part.

Qualified worker: One who has the skills and knowledge related to the construction and operation of the electrical equipment and installations, and has received safety training to recognize and avoid the hazards involved.

Restricted approach boundary: An approach limit at a distance from an exposed energized electrical conductor or circuit part within which there is an increased risk of electrical shock.

Shield: A structure that is able to withstand the forces imposed on it by a cave-in and thereby protect employees within the structure. Shields can be permanent structures or can be designed to be portable and moved along as work progresses. Additionally, shields can be either premanufactured or job-built in accordance with *29 CFR 1926.652 (c)(3) or (c)(4).*

Step potential: The voltage between the feet (usually about 1m in length) of a person standing near an energized grounded object.

Subsidence: A depression in the earth that is caused by unbalanced stresses in the soil surrounding an excavation.

Temporary grounding device (TGD): A device, such as a shorting cable, that places ground potential on points of a de-energized circuit. It is used to protect personnel working on a de-energized circuit from electrical shock hazards should the circuit somehow become energized.

Touch potential: The voltage between the energized object being touched and the ground.

Additional Resources

This module presents thorough resources for task training. The following resource material is suggested for further study.

IEEE C2-2007, National Electrical Safety Code. Institute of Electrical and Electronics Engineers: New York, NY.

IEEE 1584-2002, IEEE Guide for Performing Arc-Flash Hazard Calculations. Institute of Electrical and Electronics Engineers: New York, NY.

NFPA 70B-2006, Recommended Practice for Electrical Equipment Maintenance. National Fire Protection Association: Quincy, MA.

NFPA 70E-2008, Standard for Electrical Safety in the Workplace. National Fire Protection Association: Quincy, MA.

Western Measures Step-Touch Potential. Gary Zevenbergen and Dennis Schurman, Western Area Power Administration. Transmission and Distribution World (http://tdworld.com/mag/power_western_measures_steptouch/index1.html)

Figure Credits

Salisbury Electrical Safety, Module opener, Figures 7–9, 13–20, SA01, 22–25, and 49

U.S. Department of Labor, Occupational Safety and Health Administration, Tables 1 and 3

U.S. Department of the Interior, Bureau of Reclamation, *Facilities Instructions, Standards, and Techniques Volume 5-1*, Figure 3

Jim Mitchem, Figure 6

Tables 2 and 4 reproduced with permission from *NFPA 70E®, Electrical Safety in the Workplace*, Copyright © 2008, National Fire Protection Association. This reprinted material is not the complete and official position of the NFPA on the referenced subject, which is represented only by the standard in its entirety.

Carolina Shoe Company, Figure 10

Elvex Corporation, Figure 11

Paulson Manufacturing, Figure 12

Panduit Corp., Figure 21

Hanco International, Figure 26

Manual on Uniform Traffic Control Devices (MUTCD), U.S. Department of Transportation, Federal Highway Association, Figures 27–30

© iStockphoto.com/gmcoop, Figure 31

Vermeer Manufacturing Company, Figure 36

Topaz Publications, Inc., Figures 37, 39, 43, 44, and 48

RIDGID®, Figure 38

Mike Powers, Figure 40, SA02

Draeger Safety, Inc., Figure 42

Professional Safety Associates, Inc., Figure 47

Physio-Control, Inc., Figure 50

CONTREN® LEARNING SERIES — USER UPDATE

NCCER makes every effort to keep its textbooks up-to-date and free of technical errors. We appreciate your help in this process. If you find an error, a typographical mistake, or an inaccuracy in NCCER's Contren® materials, please fill out this form (or a photocopy), or complete the online form at www.nccer.org/olf. Be sure to include the exact module number, page number, a detailed description, and your recommended correction. Your input will be brought to the attention of the Authoring Team. Thank you for your assistance.

Instructors – If you have an idea for improving this textbook, or have found that additional materials were necessary to teach this module effectively, please let us know so that we may present your suggestions to the Authoring Team.

NCCER Product Development and Revision

3600 NW 43rd Street, Building G, Gainesville, FL 32606

Fax: 352-334-0932
Email: curriculum@nccer.org
Online: www.nccer.org/olf

☐ Trainee Guide ☐ AIG ☐ Exam ☐ PowerPoints Other ____________

Craft / Level: Copyright Date:

Module Number / Title:

Section Number(s):

Description:

Recommended Correction:

Your Name:

Address:

Email: Phone:

Introduction to Electrical Circuits

49103-11

Trainees with successful module completions may be eligible for credentialing through NCCER's National Registry. To learn more, go to **www.nccer.org** or contact us at **1.888.622.3720.** Our website has information on the latest product releases and training, as well as online versions of our *Cornerstone* newsletter and Pearson's Contren® product catalog.

Your feedback is welcome. You may email your comments to **curriculum@nccer.org**, send general comments and inquiries to **info@nccer.org**, or use the User Update form at the back of this module.

V.1 7/11

49103-11

Introduction to Electrical Circuits

Objectives

When you have completed this module, you will be able to do the following:

1. Explain the difference between conductors and insulators.
2. Define voltage and identify the ways in which it can be produced.
3. Define the units of measurement that are used to measure the properties of electricity.
4. Explain the basic characteristics of series and parallel circuits.
5. Identify the meters used to measure voltage, current, and resistance.
6. Identify specialized test instruments used by power line workers.

Performance Tasks

This is a knowledge-based module; there are no performance tasks.

Trade Terms

Ammeter
Ampere (A)
Atom
Battery
Circuit
Conductor
Coulomb
Current
Electrical power
Electron
Insulator
Joule (J)
Kilo
Matter
Mega
Neutrons
Nucleus
Ohm (Ω)
Ohmmeter
Ohm's law
Power
Protons
Relay
Resistance
Resistor
Schematic
Series circuit
Solenoid
Transformer
Valence shell
Volt (V)
Voltage
Voltage drop
Voltmeter
Watt (W)

Required Trainee Materials

Calculator

Contents

Topics to be presented in this module include:

Figures and Tables

1.0.0 Introduction

Electricity is a form of energy that is used by electrical devices, such as motors, lights, TVs, and heaters, to perform work. Electricity is also used to control nonelectrical devices that perform work. For example, your car is driven by a gasoline engine, but you could not start it or turn it off without the electrical system. To work with electricity, you need to know how it is produced and how it acts in electrical circuits.

You will hear the term *circuit* throughout your training. An electrical circuit contains, at minimum, a voltage source, a load, and conductors (wires) to carry the electrical current (*Figure 1*). The circuit should also have a means to stop and start the current, such as a switch.

Electricity is concerned with cause and effect. The presence of voltage (volts) in a closed circuit causes current (amps) to flow. The more voltage you apply, the more current will flow. However, the amount of current flow is also determined by how much resistance, in ohms (Ω), the load offers to the flow of current. To convert electrical energy into work, the load consumes energy. The amount of energy a device consumes is called power, which is expressed in watts (W). Volts (V), amps, ohms, and watts are related in such a way that if any one of them changes, the others are proportionally affected. This relationship can be seen using basic math principles that are covered in this module. You will also learn how electricity is produced and how test instruments are used to measure electricity.

2.0.0 Atomic Theory

To understand electrical theory, you must first understand the basic concepts of atomic theory. Atomic theory explains the construction and behavior of atoms, including the transfer of electrons that results in current flow.

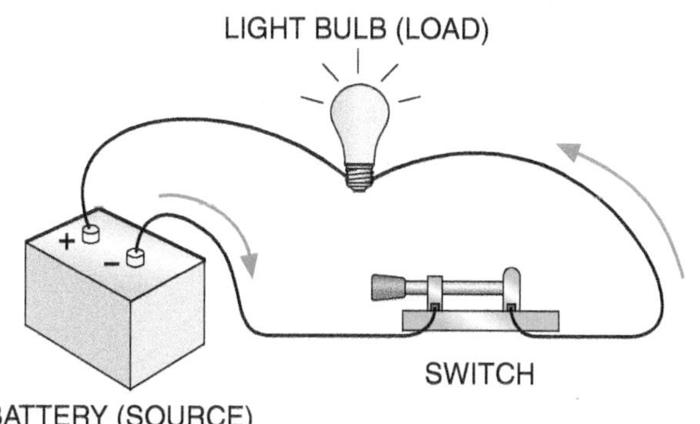

Figure 1 Basic electrical circuit.

On Site

Why Bother with Learning Theory?

Many trainees wonder why they need to bother learning the theory behind how things operate. They figure, why should I learn how it works as long as I know how to install it? The answer is, if you only know how to install something (run wire, connect switches, etc.), that is all you will ever be able to do. If you don't know how your car operates, how can you troubleshoot it? The answer is, you can't. You can only keep changing out the parts until you finally hit on what is causing the problem. (How many times have you seen people do this?) Remember, unless you understand not only how things work but why they work, you will only be a parts changer. With theory behind you, there is no limit to what you can do.

2.1.0 The Atom

The atom is the smallest part of an element that enters into a chemical change, and it does so in the form of a charged particle. The charged particles, called ions, are of two types—positive and negative. A positive ion is an atom that has become positively charged. A negative ion is an atom that has become negatively charged. One of the properties of charged ions is that ions of the same charge tend to repel one another, while ions of unlike charge attract one another. The term *charge* can mean a quantity of electricity that is either positive or negative.

The structure of an atom is best explained by observing the simplest of all atoms, that of the element hydrogen. The hydrogen atom (*Figure 2*) is composed of a nucleus containing one proton and one orbiting electron. As the electron revolves around the nucleus, it is held in orbit by two counteracting forces—centrifugal force and electrostatic force. Centrifugal force tends to cause the electron to fly outward as it travels around its circular orbit. Electrostatic force tends to pull the electron in toward the nucleus. Electrostatic force is provided by the mutual attraction between the positive nucleus and the negative electron. At some given radius, the two forces balance each other, providing a stable path for the electron.

- A proton (+) repels another proton (+).
- An electron (–) repels another electron (–).
- A proton (+) attracts an electron (–).

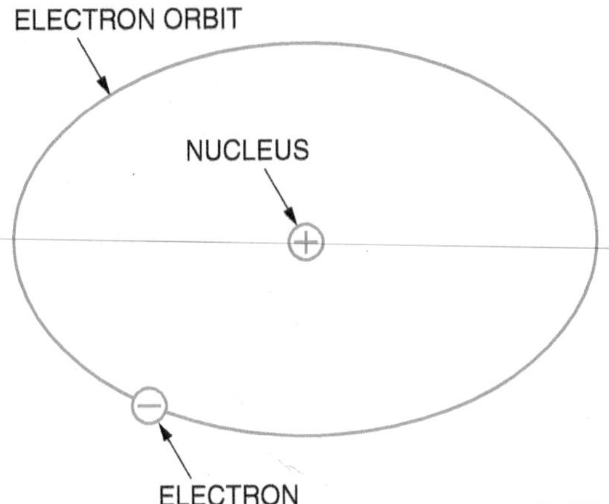

Figure 2 Hydrogen atom.

An atom contains three types of subatomic particles that are of interest in electricity: electrons, protons, and neutrons.

The protons and neutrons are located in the center, or nucleus, of the atom. The electrons travel around the nucleus in orbits.

Because protons are relatively heavy, the repulsive force they exert on one another in the nucleus of an atom has little effect.

The attracting and repelling forces on charged materials occur because electrostatic lines of force exist around the charged materials. In a negatively charged object, the lines of force of the excess electrons add up to produce an electrostatic field in which the lines of force are coming into the object from all directions. In a positively charged object, the lines of force of the excess protons add up to produce an electrostatic field in which the lines of force are going out from the object in all directions. The electrostatic fields either aid or oppose each other in their attracting or repelling.

2.1.1 The Nucleus

The nucleus is the central part of the atom. It is made up of heavy particles called protons and neutrons. The proton is a charged particle containing the smallest known unit of positive electricity. The neutron has no electrical charge. The number of protons in the nucleus determines how the atom of one element differs from the atom of another element.

Although a neutron is actually a particle by itself, it is generally thought of as an electron and proton combined. A neutron is electrically neutral. Because of this, neutrons are not considered relevant to the electrical nature of atoms.

Think About It

Electrical Charges

Think about the things you come in contact with every day. Where do you find examples of electrostatic attraction?

2.1.2 Electrical Charges

The negative charge of an electron is equal but opposite to the positive charge of a proton. The charges of an electron and a proton are called electrostatic charges. The lines of force associated with each particle produce electrostatic fields. Because of the way these fields act together, charged particles can attract or repel one another. The Law of Electrical Charges states that particles with like charges repel each other, and particles with unlike charges attract each other (*Figure 3*).

2.2.0 Conductors and Insulators

With respect to chemical activity and stability, the difference between atoms depends on the number and position of the electrons included within the atom. In general, the electrons reside in groups of orbits called shells. The shells are

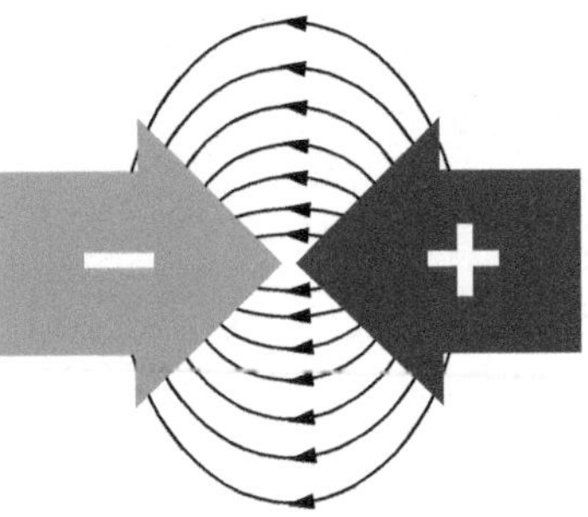

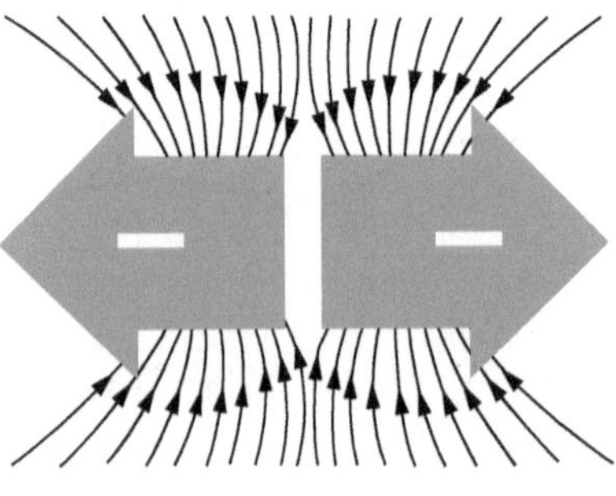

Figure 3 Law of electrical charges.

arranged in steps that correspond to fixed energy levels.

The outer shell of an atom is called the valence shell. The electrons contained in this shell are called valence electrons (*Figure 4*). The number of valence electrons determines an atom's ability to gain or lose an electron, which in turn determines the chemical and electrical properties of the atom. An atom that is lacking only one or two electrons from its outer shell will easily gain electrons to complete its shell. However, a large amount of energy is required to free any of its electrons. An atom having a relatively small number of electrons in its outer shell, compared to the number of electrons required to fill the shell, will easily lose its valence electrons.

The valence electrons are the main concern in electricity. These are the electrons that are easiest to break loose from the parent atom. A conductor normally has three or less valence electrons, an insulator has five or more, and semiconductors usually have four valence electrons.

All the elements of which matter is made fall into one of the following three categories: conductors, insulators, or semiconductors.

Conductors are elements such as copper and silver that readily conduct a flow of electricity. Because of their good conducting abilities, they are formed into wire and used to transfer electrical energy from one point to another.

Insulators do not conduct electricity to any great degree. Insulators are used to prevent the flow of electricity. Compounds such as porcelain and plastic are good insulators.

Materials such as germanium and silicon are not good conductors. However, they cannot be used as insulators because their electrical characteristics fall somewhere between those of conductors and insulators. These in-between materials are classified as semiconductors.

VALENCE (OUTER) SHELL

VALENCE ELECTRON

49103-11_F04.EPS

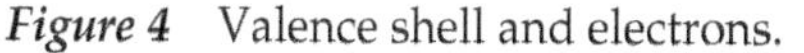

Figure 4 Valence shell and electrons.

2.3.0 Magnetism

The operation of many electrical components relies on the power of magnetism. Motors, relays, transformers, and solenoids all rely on magnetism to function. Magnetized iron generates a magnetic field made up of magnetic lines of force called magnetic flux lines (*Figure 5*). Magnetic objects within the field are attracted or repelled by the magnetic field. The more powerful the magnet, the more powerful the magnetic field around it is. Each magnet has a north pole and a south pole. Opposing poles attract each other, and like poles repel each other.

Electricity also produces magnetism. Current flowing through a conductor produces a small magnetic field around the conductor. If the conductor is coiled around an iron bar, the result is an electromagnet (*Figure 6*). Like iron magnets, electromagnets attract and repel other magnetic objects. This feature is the basis on which electric motors and other components operate.

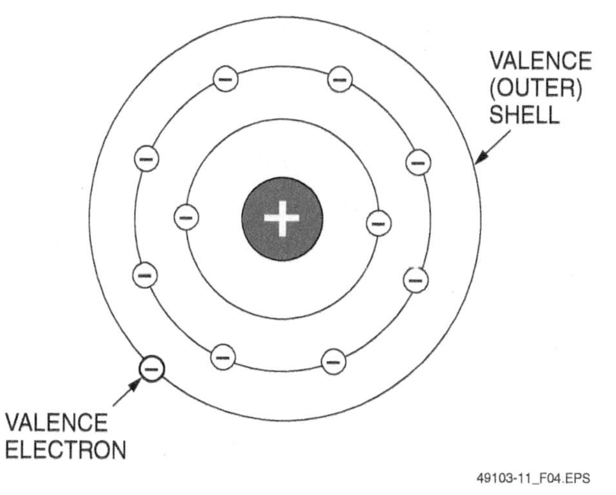

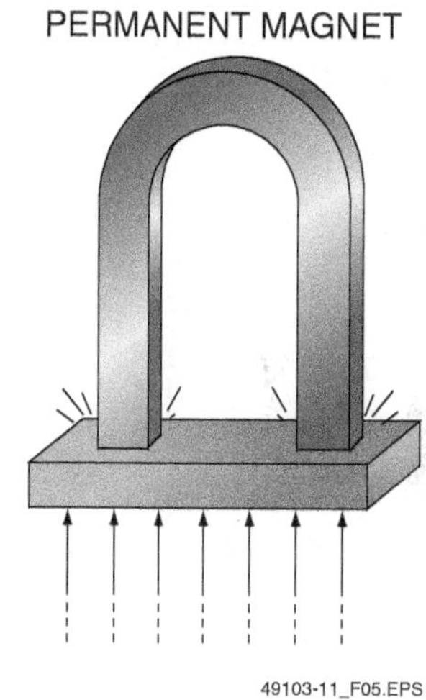

Figure 5 Magnetism.

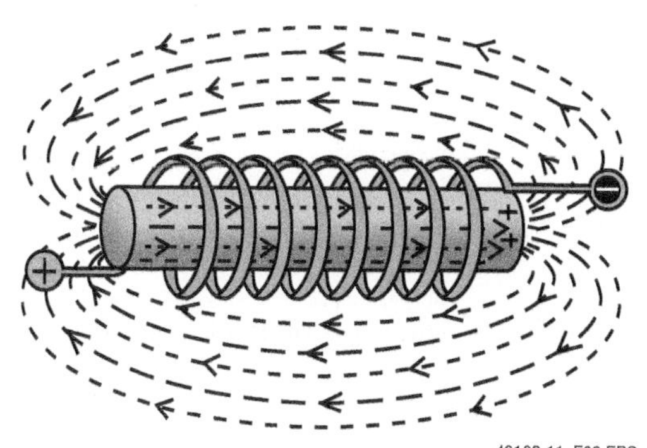

Figure 6 Electromagnet.

3.0.0 Electrical Power Generation and Distribution

Electricity comes from electrical generating plants (*Figure 7*) operated by utilities, such as local power companies. Steam from coal-burning or nuclear power plants is used to power huge generators called turbines. These turbines generate electricity. Another type of plant is the hydroelectric power plant, which uses water flowing through dams to drive the turbines.

The electrical power that travels through long-distance transmission lines may be as high as 750,000 volts (V). Transformers are used to step the voltage down to lower levels as it reaches electrical substations and eventually our homes, offices, and factories. The voltage received in a typical home is about 240V. At wall outlets where televisions and small appliances are plugged in, the voltage is about 120V (*Figure 8*). Electric stoves, clothes dryers, water heaters, and central air conditioning systems usually require the full 240V. Commercial buildings and factories may receive anywhere from 208V to 575V, depending on their power requirements.

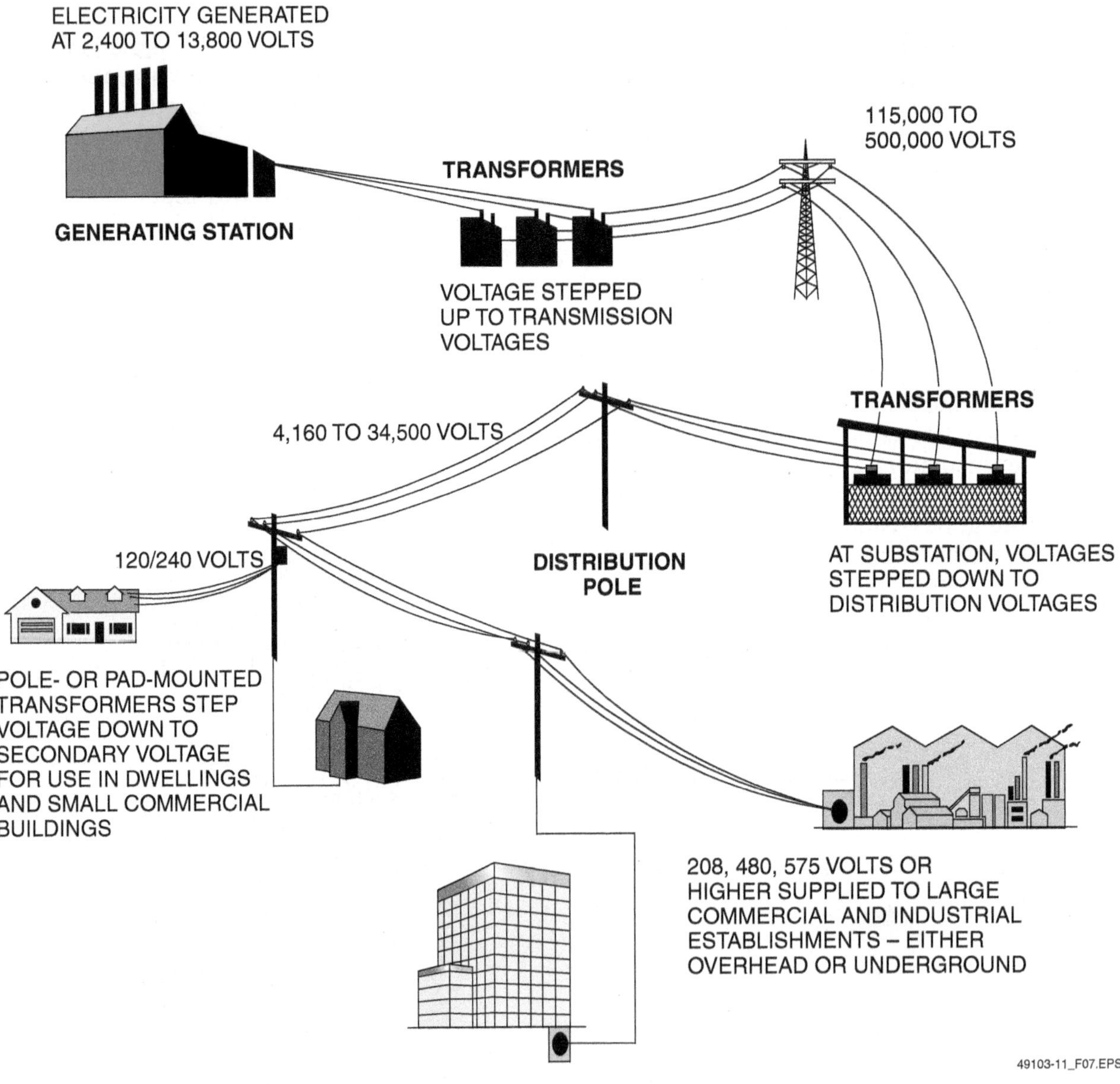

Figure 7 Electrical power distribution.

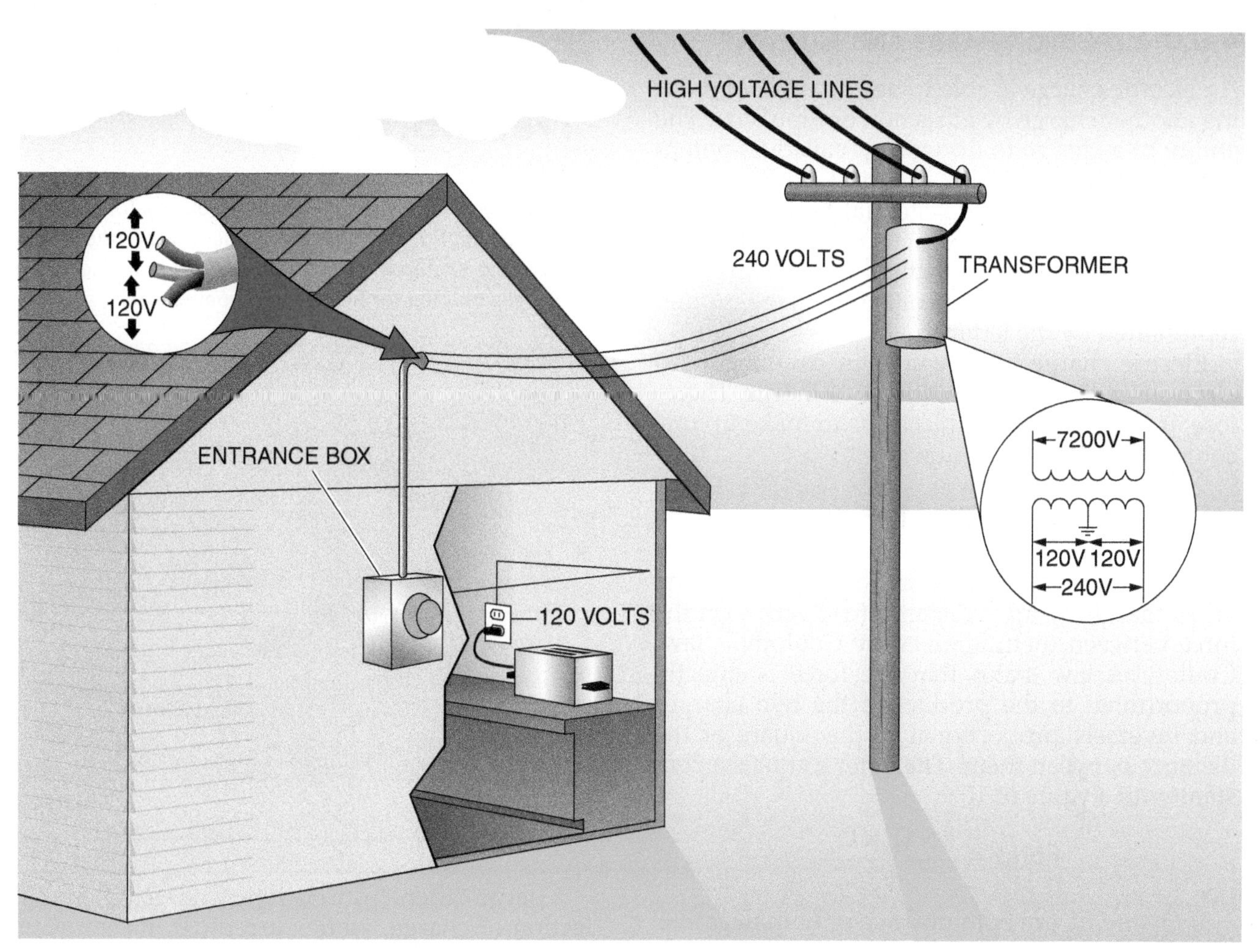

49103-11_F08.EPS

Figure 8 Internal power distribution.

GOING GREEN

Hydroelectric Plants

Hydroelectric plants use the power generated by water to drive turbines that produce electricity.

49103-11_SA01.EPS

4.0.0 Electric Charge and Current

An electric charge is able to do the work of moving another charge by attraction or repulsion. The ability of a charge to do work is called its potential. When one charge is different from another, there is a difference in potential between them. The sum of the differences in potential of all the charges in the electrostatic field is called the electromotive force (emf), or voltage. Voltage is often represented by the letter E.

Electric charge is measured in coulombs. An electron has 1.6×10^{-19} coulombs of charge. Therefore, it takes 6.25×10^{18} electrons to make up one coulomb of charge, as shown.

$$\frac{1}{1.6 \times 10^{-19}} = 6.25 \times 10^{18} \text{ electrons}$$

If two particles, one having charge Q_1 and the other charge Q_2, are a distance (d) apart, then the force between them is given by Coulomb's law. Coulomb's law states that the force is directly proportional to the product of the two charges and inversely proportional to the square of the distance between them. The letter *k* equals a constant with a value of 10^9.

$$\text{Force} = \frac{k \times Q_1 \times Q_2}{d^2}$$

If Q_1 and Q_2 are both positive or both negative, then the force will be positive and repulsive. If Q_1 and Q_2 are opposite charges, then the force will be negative and attractive.

4.1.0 Current Flow

The movement of the flow of electrons is called current. To produce current, the electrons must be moved by a potential difference. Current is represented by the letter *I*. The basic unit for measuring current is the ampere (A), or amp. As shown, the symbol for ampere is the letter *A*. One ampere of current is defined as the movement of one coulomb past any point of a conductor during one second of time. One coulomb is equal to 6.25×10^{18} electrons. Therefore, one ampere is equal to 6.25×10^{18} electrons moving past any point of a conductor during one second of time.

The definition of current can be expressed as an equation:

$$I = \frac{Q}{T}$$

Where:

I = current (amperes)
Q = charge (coulombs)
T = time (seconds)

On Site

Transformers

Large distribution transformers at power substations step down the power to the level required for local distribution. Pole transformers, such as the one shown here, step the power down further to the voltages needed for homes and businesses.

49103-11_SA02.EPS

Charge differs from current. Charge (Q) is a collection of charge, while current (I) measures the intensity of moving charges.

In a conductor, such as copper wire, the free electrons are charges that can be forced to move with relative ease by a potential difference. If a potential difference is connected across two ends of a copper wire (*Figure 9*), the applied voltage forces the free electrons to move. This current is a flow of electrons from the point of negative charge (–) at one end of the wire, moving through the wire to the positive charge (+) at the other end.

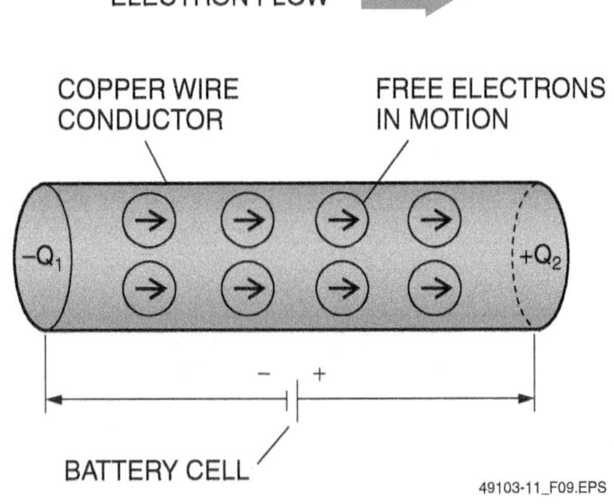

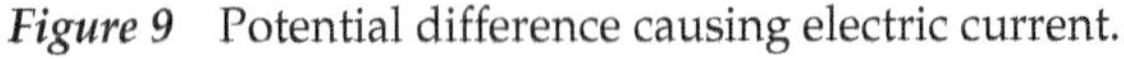

Figure 9 Potential difference causing electric current.

On Site

Units of Electricity and Volta

A disagreement with a fellow scientist over the twitching of a frog's leg eventually led 18th-century physicist Alessandro Volta to theorize that when certain objects and chemicals come into contact with each other, they produce an electric current. Believing that electricity came only from contact between metals, Volta coined the term *metallic electricity*. To demonstrate his theory, he placed two discs, one of silver and the other of zinc, into a weak acidic solution. When he linked the discs together with wire, electricity flowed through the wire. As a result, Volta introduced the world to the battery, also known as the Voltaic pile. Volta needed a term that would serve as a measure of the strength of the electric push or the flowing charge. The volt is that measure.

The direction of electron flow is from the negative side of the battery, through the wire, and back to the positive side of the battery. So the direction of current flow is from a point of negative potential to a point of positive potential.

4.2.0 Voltage

The force that causes electrons to move is called voltage, potential difference, or electromotive force (emf). One volt is the potential difference between two points for which one coulomb of electricity will do one joule (J) of work. A battery is one means of creating voltage. It chemically

On Site

Law of Electrical Force and de Coulomb

In the 18th century, a French physicist named Charles de Coulomb was studying how electric charges behaved. He watched the repelling forces electric charges exerted by measuring the twist in a wire. The weight of an object acted as a turning force to twist the wire. The amount of twist was proportional to the object's weight. After many experiments with opposing forces, de Coulomb proposed the Inverse Square Law, later known as the Law of Electrical Force.

Think About It

The Magic of Electricity

The effect of the flow of electrons occurs at a speed that is close to the speed of light, about 186,000 miles per second. How long does it take the light from the end of a flashlight to reach the floor? If you ran a light circuit from Maine to California and flipped the switch, how long would it take for the light to come on?

creates a large reserve of free electrons at the negative (–) terminal. The positive (+) terminal has electrons chemically removed, so it will accept electrons if an external path is provided from the negative (–) terminal. When a battery is no longer able to chemically deposit electrons at the negative (–) terminal, it is said to be dead, or in need of recharging. Batteries are normally rated in volts. Large batteries are also rated in ampere-hours, where one ampere-hour is a current of one amp supplied for one hour.

4.3.0 Resistance

Resistance is directly related to the ability of a material to conduct electricity. All conductors have very low resistance. Insulators have very high resistance.

4.3.1 Characteristics of Resistance

Resistance is the opposition to current flow. To add resistance to a circuit, electrical components called resistors are used. A resistor is a device whose resistance to current flow is a known, specified value. Resistance is measured in ohms. The symbol for an ohm is Ω. In equations, an ohm is represented by the letter *R*. One ohm is defined as the amount of resistance that will limit the current in a conductor to one ampere when the voltage applied to the conductor is one volt.

The resistance of a wire is proportional to the length of the wire and inversely proportional to its cross-sectional area. In addition, resistance is

Think About It

Current Flow

Why do you need two wires to use electrical devices? Why can't current simply move to a lamp and be released as light energy?

On Site

Joule's Law

While other scientists of the 19th century were experimenting with batteries, cells, and circuits, James Joule was studying the relationship between heat and energy. He discovered that work did not just move heat from one place to another—work generated heat. He also demonstrated that over time, a relationship existed between the temperature of water and electric current. These ideas formed the basis for the concept of energy. In his honor, the modern unit of energy was named the joule.

Table 1 Conductor Properties

Metal	Specific Resistance (Resistance of 1 CM/ft in ohms)	
	32°F or 0°C	**75°F or 23.8°C**
Silver, pure annealed	8.831	9.674
Copper, pure annealed	9.390	10.351
Copper, annealed	9.590	10.505
Copper, hard-drawn	9.810	10.745
Gold	13.216	14.404
Aluminum	15.219	16.758
Zinc	34.595	37.957
Iron	54.529	62.643

49103-11_T01.EPS

dependent upon the type of material of which the wire is made. The relationship for finding the resistance of a wire is as follows:

$$R = \rho\frac{L}{A}$$

Where:

R = resistance (ohms)
L = length of wire (feet)
A = area of wire (circular mils, CM, or cm^2)
ρ = specific resistance (ohm-CM/ft or microhm-CM)

A mil equals 0.001 inch. A circular mil is the cross-sectional area of a wire one mil in diameter.

The specific resistance is a constant that depends on the material of which the wire is made. *Table 1* lists the properties of various wire conductors.

Table 1 shows that at 75°F, a one-mil diameter, one-foot long pure annealed copper wire has a resistance of 10.351 ohms. At the same temperature, a one-mil diameter, one-foot-long aluminum wire has a resistance of 16.758 ohms. Temperature is important in determining the resistance of a wire. The hotter a wire, the greater its resistance is.

5.0.0 Ohm's Law

Ohm's law defines the relationship between current, voltage, and resistance. There are three ways to express Ohm's law mathematically.

- The current in a circuit is equal to the voltage applied to the circuit divided by the resistance of the circuit:

$$I = \frac{E}{R}$$

On Site

The Visual Language of Electricity

Learning to read circuit diagrams is like learning to read the printed word—first you learn to recognize the letters, then you learn to read the words, and before long you are reading without paying attention to the individual letters anymore. With circuit diagrams, you may struggle at first with the individual pieces, but before long you will be reading a circuit without having to really think about it. Study the table provided. It will help you to understand the language of electricity.

What Is Measured	Unit of Measurement	Symbol	Ohm's Law Symbol
Amount of current	Amp	A	I
Electrical power	Watt	W	P
Force of current	Volt	V	E
Resistance to current	Ohm	Ω	R

- The resistance of a circuit is equal to the voltage applied to the circuit divided by the current in the circuit:

$$R = \frac{I}{E}$$

- The applied voltage to a circuit is equal to the product of the current and the resistance of the circuit:

$$E = I \times R = IR$$

Where:

I = current (amperes)
R = resistance (ohms)
E = voltage or emf (volts)

If any two of the quantities E, I, or R are known, the third can be calculated.

The Ohm's law equations can be memorized and practiced using an Ohm's law circle (*Figure 10*). To find the equation for E, I, or R when two quantities are known, cover the unknown third quantity. The other two quantities in the circle will indicate how the covered quantity may be found.

Example 1:

Find I when E = 120V and R = 30Ω.

$$I = \frac{E}{R}$$

$$I = \frac{120V}{30\Omega}$$

$$I = 4A$$

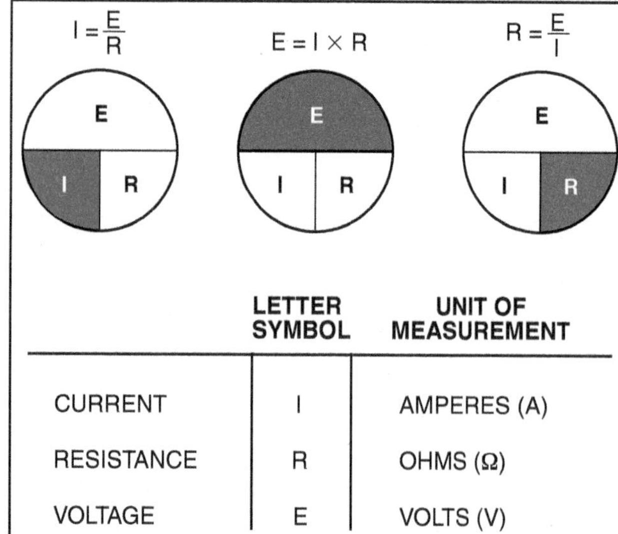

	LETTER SYMBOL	UNIT OF MEASUREMENT
CURRENT	I	AMPERES (A)
RESISTANCE	R	OHMS (Ω)
VOLTAGE	E	VOLTS (V)

Figure 10 Ohm's law circle.

Think About It

Voltage Matters

Standard household voltage is different the world over. It ranges from 100V in Japan to 600V in Bombay, India. Many countries have no standard voltage. In France, voltage varies from 110V to 360V. If you were to plug in your 120V hair dryer in England, where household voltage is 240V, you would burn out the dryer. Use basic electric theory to explain exactly what would happen to destroy the hair dryer.

This formula shows that in a DC circuit, current (I) is directly proportional to voltage (E) and inversely proportional to resistance (R).

Example 2:

Find R when E = 240V and I = 20A.

$$R = \frac{E}{I}$$

$$R = \frac{240V}{20A}$$

$$R = 12\Omega$$

Example 3:

Find E when I = 15A and R = 8Ω.

$$E = I \times R$$

$$E = 15A \times 8\Omega$$

$$E = 120V$$

6.0.0 Schematic Representation of Circuit Elements

The simple electrical circuit shown earlier in this module is presented in both pictorial and schematic forms in *Figure 11*. The schematic diagram is a shorthand way to draw an electric circuit. Circuits are usually shown in this way. Besides the connecting wire, three components are shown using symbols: the battery, the switch, and the lamp. Note the positive (+) and negative (–) markings in both the pictorial and schematic views of the battery. The schematic diagram shows the pictorial components in a simplified form. A schematic diagram uses graphic symbols to show the electrical connections and functions of the different parts of a circuit.

The standard graphic symbols for common electrical and electronic components are shown in *Figure 12*.

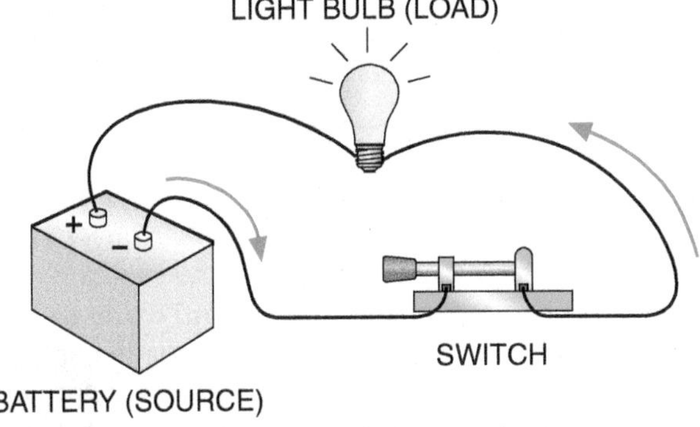

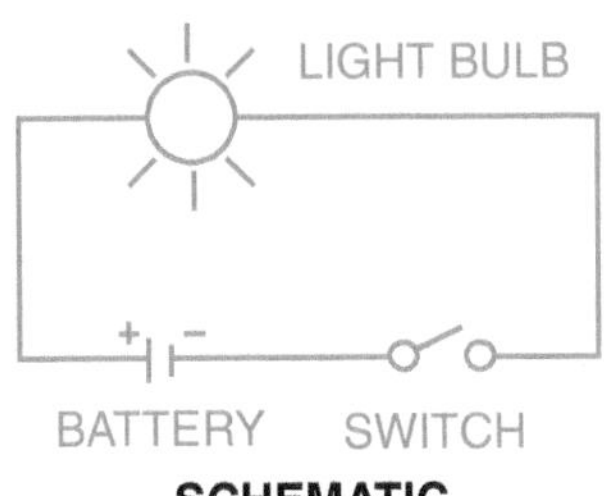

49103-11_F11.EPS

Figure 11 Electrical circuit.

On Site

Using Your Intuition

Learning the meanings of various electrical symbols may seem challenging. But if you take a moment to study *Figure 12*, you will see that most of the symbols are shaped (in a symbolic way) to represent the actual object. For example, the battery shows plus (+) and minus (–) signs, similar to an actual battery. The motor has two arms, suggesting a spinning rotor. The transformer shows two coils. The resistor has a jagged edge to indicate pulling or resistance. Connected wires have a black dot that looks similar to solder. Unconnected wires simply cross. The fuse stretches out in both directions, as if to provide extra slack in the line. The circuit breaker shows a line with a break in it. The capacitor shows a gap. The variable resistor has an arrow like a swinging compass needle. As you learn to read schematics, take the time to make mental connections between the symbols and the objects they represent.

Symbol	Symbol
Ammeter — (A)	Motor (DC) — (B)
Battery — + –	Resistor (fixed)
Capacitor (fixed)	Resistor (variable)
Capacitor (variable)	Rheostat
Circuit breaker	Switch
Crystal	Semiconductor diode
Fuse	Transformer (general)
Generator (AC)	Transformer (iron-core)
Generator (DC) — (G)	Transistor (NPN) — B C E
Ground — or	Transistor (PNP) — B C E
Inductor (air-core)	Voltmeter — (V)
Inductor (iron-core)	Wattmeter — (W)
Inductor (tapped)	Wires (connected)
Lamp	Wires (unconnected)
Motor (AC)	Zener diode

49103-11_F12.EPS

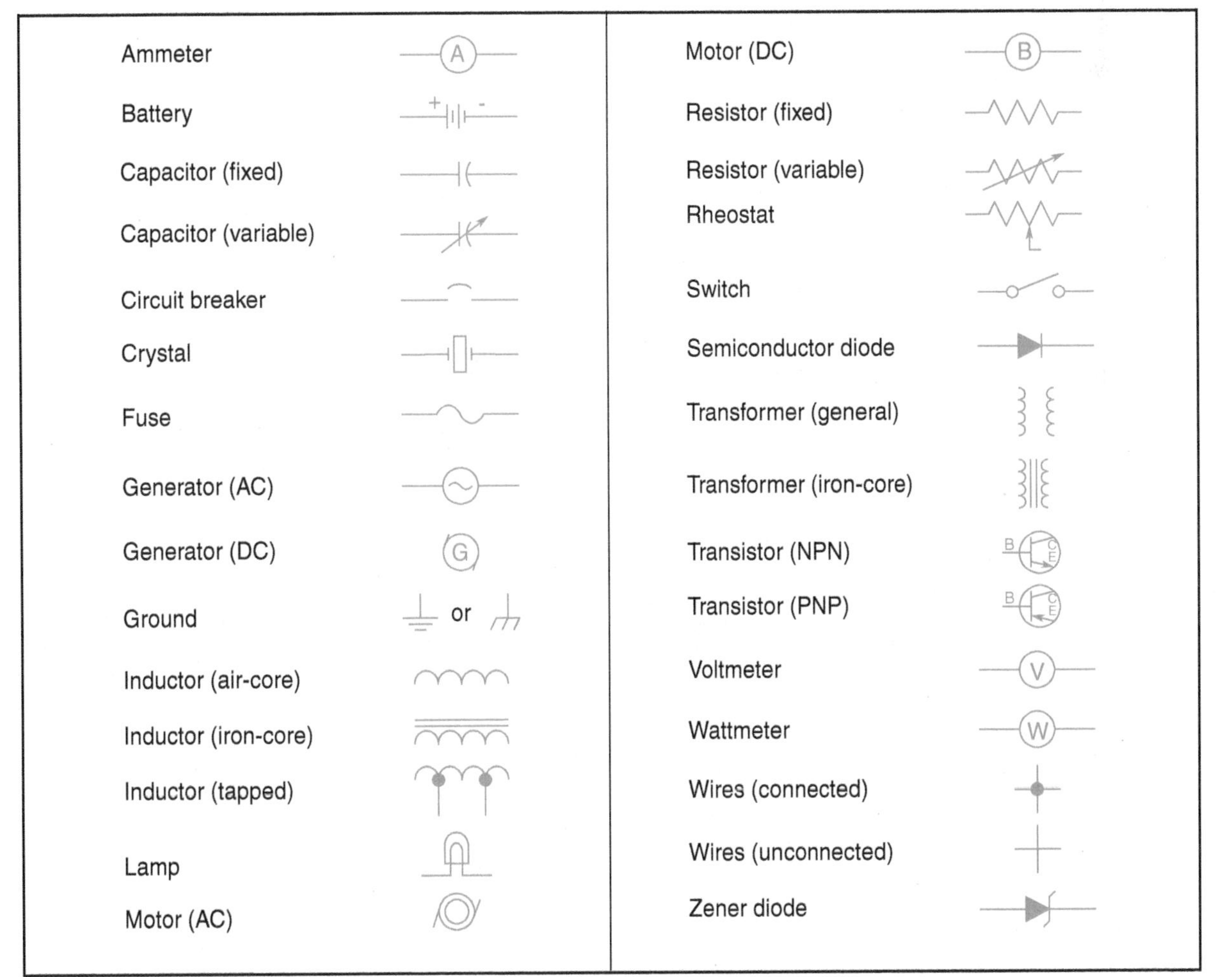

Figure 12 Standard schematic symbols.

Think About It

Drawing a Schematic

Draw a schematic diagram showing a voltage source, switch, motor, and fuse.

7.0.0 Resistors

The function of a resistor is to resist current flow. For a given current and known resistance, the change in voltage, or voltage drop, across the component can be predicted using Ohm's law. Voltage drop refers to a specific amount of voltage used, or developed, by that component. An example is a very basic circuit of a 10V battery and a single resistor in a series circuit. The voltage drop across that resistor is 10V because it is the only component in the circuit. All voltage must be dropped across that resistor. For a given applied voltage, the current that flows may be predetermined by selection of the resistor value. The wattage dictates the type and size of a resistor.

The two most common electronic resistors are the wire wound and carbon composition types. A typical wire wound resistor consists of a length of nickel wire wound on a ceramic tube and covered with porcelain. Low-resistance connecting wires are provided. The resistance value is usually printed on the side of the component. *Figure 13* shows the construction of typical resistors. Carbon composition resistors are made by molding mixtures of powdered carbon and insulating materials into a cylindrical shape. An outer

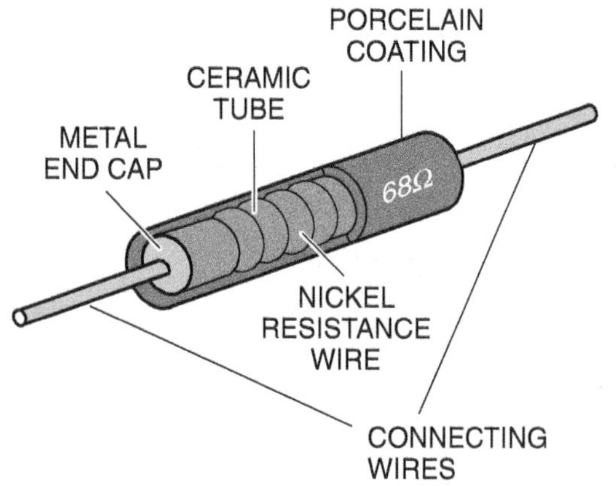

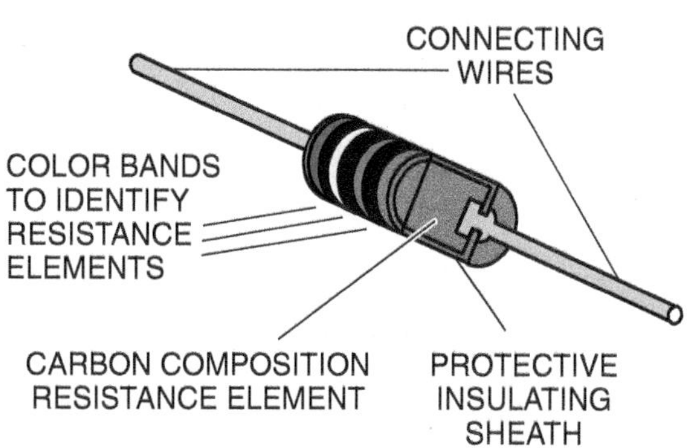

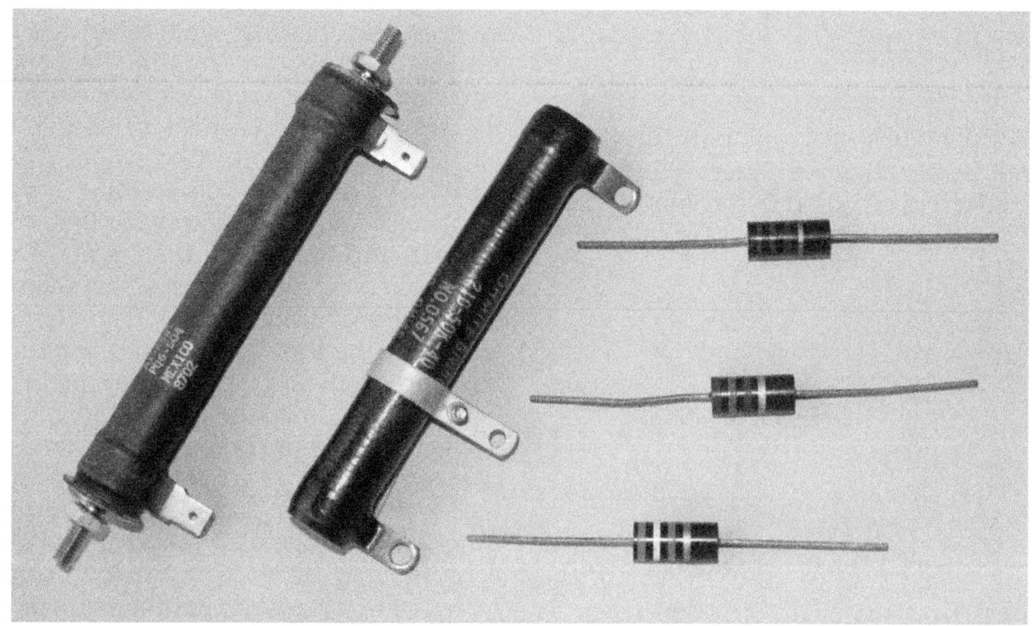

49103-11_F13.EPS

Figure 13 Common resistors.

sheath of insulating material provides mechanical and electrical protection. Copper connecting wires are provided at each end. Carbon composition resistors are smaller and less expensive than the wire wound type. However, the wire wound resistor is the more rugged of the two. It is also able to survive much larger power dissipations than the carbon composition type.

Most resistors have standard fixed values and are called fixed resistors. Variable, or adjustable, resistors are used a great deal in electronics. Two common symbols for a variable resistor are shown in *Figure 14*.

A variable resistor consists of a coil of closely wound, insulated resistance wire formed into a partial circle. The coil has a low-resistance terminal at each end. A third terminal is connected to a movable contact with a shaft adjustment facility. The movable contact can be set to any point on a connecting track that extends over one (uninsulated) edge of the coil.

Using the adjustable contact, the resistance from the terminal at either end to the center terminal may be adjusted from zero to the maximum coil resistance.

Another type of variable resistor is the decade resistance box. This is a laboratory component that contains precise values of switched series-connected resistors.

8.0.0 Electrical Circuits

The terms *series circuit* and *parallel circuit* are commonly used when discussing electrical circuits. These terms refer to the way loads are connected in the circuit.

8.1.0 Series Circuits

A series circuit provides only one path for current flow and is a voltage divider. The total resistance of the circuit is equal to the sum of the individual resistances. The 12V series circuit in *Figure 15* has two 30Ω loads. The total resistance is 60Ω. The amount of current flowing in the circuit is 0.2A.

$$I = \frac{E}{R} = \frac{12V}{60\Omega} = 0.2A$$

If there were five 30Ω loads, the total resistance would be 150Ω. The current flow would be the same through all the loads. The voltage measured across any load (voltage drop) depends on the resistance of that load. The sum of the voltage drops equals the total voltage applied to the circuit. Circuits containing loads in series are uncommon. An important trait of a series circuit is that if the circuit is open at any point, no current will flow. For example, if five light bulbs are connected in series and one of them blows, all five lights will go out.

8.2.0 Parallel Circuits

In a parallel circuit, each load is connected directly to the voltage source. As a result, the voltage drop is the same through all loads, and the current is divided between the loads. The source sees the circuit as two or more individual circuits containing one load each. In the parallel circuit shown in *Figure 15*, the source sees three circuits, each containing a 30Ω load. The current flow through any load is determined by the resistance of that load. Thus, the total current drawn by the circuit is the sum of the individual currents. The total resistance of a parallel circuit is calculated differently from that of a series circuit. In a parallel circuit, the total resistance is less than the smallest of the individual resistances. This can seem illogical compared to what we just learned about total resistance being the sum of all resistances in simple series circuits. It can be helpful to consider electron flow (current) to resolve this seeming contradiction. Consider a parallel circuit with three resistors of 2Ω, 4Ω, and 8Ω. When the circuit is energized, the applied current moves across each resistor in each of the circuit branches. The volume of electrons flowing through each of the parallel circuit branches depends on the amount of resistance encountered, since the applied force (voltage) is the same for all the branches. The largest volume of electrons moves across the 2Ω resistor among the three resistors. At the same time smaller volumes of electrons also move across the 4Ω and 8Ω resistors. The electrons moving across the 2Ω resistor are joined on the other side of the circuit by electrons that crossed the 4Ω and 8Ω resistor

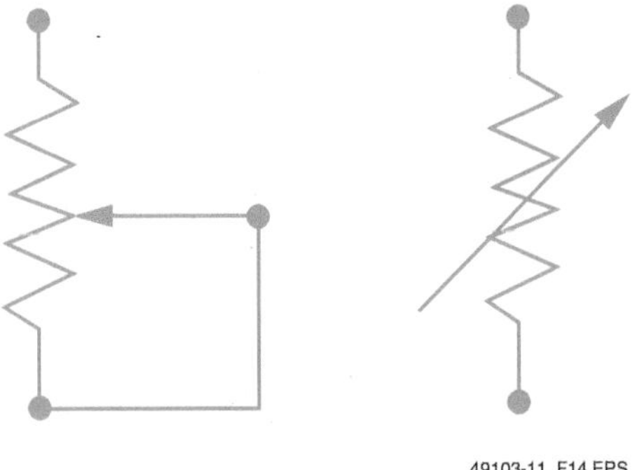

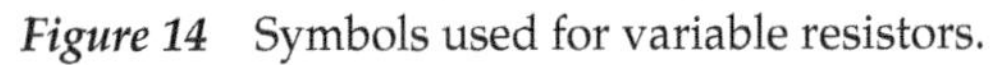

Figure 14 Symbols used for variable resistors.

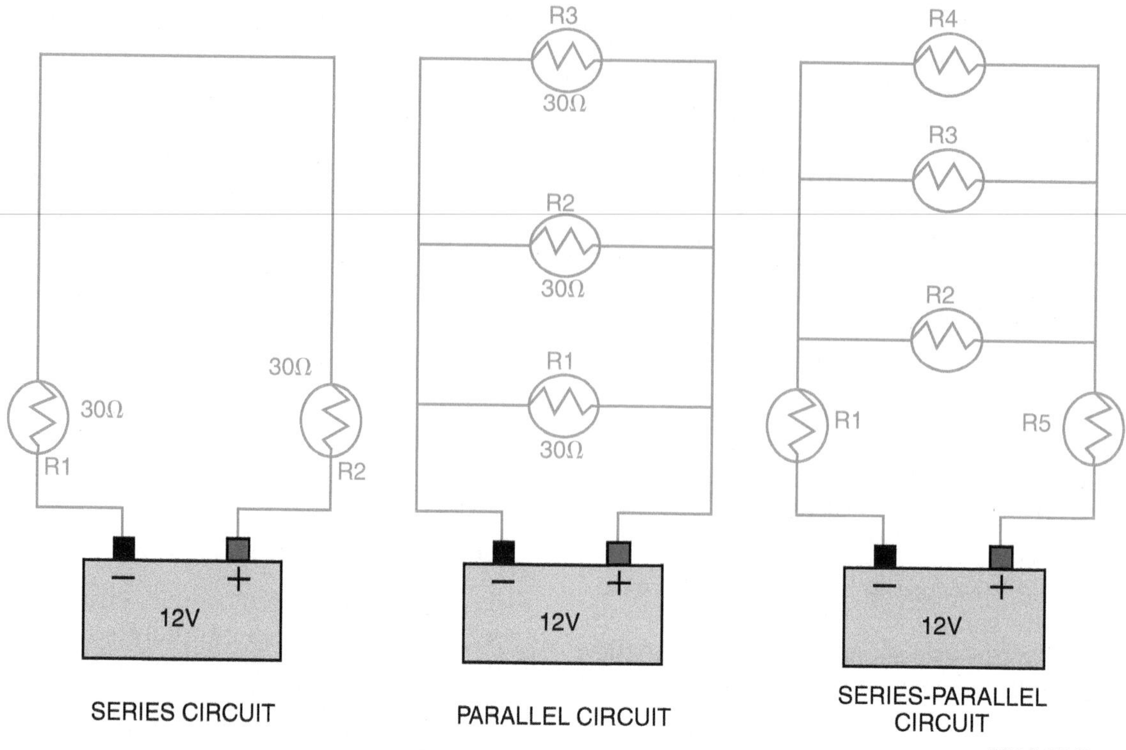

Figure 15 Types of circuits.

On Site

Is It a Series Circuit?

The term *series circuit* refers to the way the loads are connected. The same is true for parallel and series-parallel circuits. Loads connected in series or in a series-parallel arrangement are rare. The simple circuit shown here illustrates this point. It may appear at first that this is a series-parallel circuit. However, closer examination shows that there are only two loads—the relay and the contactor—and they are connected in parallel. So it is a parallel circuit. The control devices are wired in series with the loads, but only the loads are considered in determining the type of circuit.

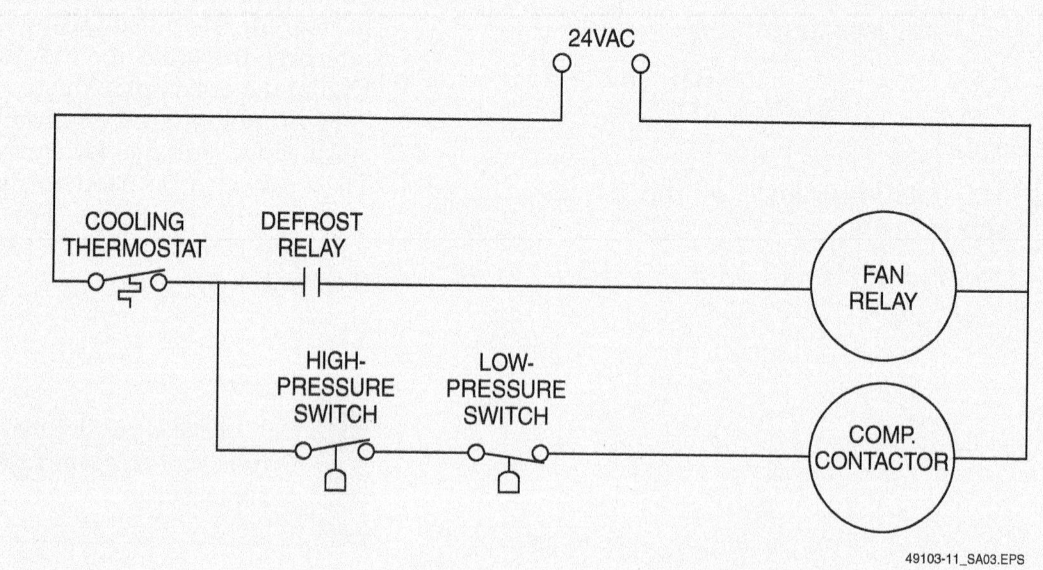

branches. Overall, a larger volume of electrons crosses the various parallel branches than could cross the 2Ω resistor branch alone. So, overall the total resistance of the circuit is lower than that of the single 2Ω branch.

In *Figure 15*, each of the 30Ω loads draws 0.4A at 12V; therefore, the total current is 1.2A:

$$I = \frac{E}{R} = \frac{12V}{30\Omega} = 0.4A \text{ per circuit}$$

$$0.4A \text{ per circuit} \times \text{three circuits} = 1.2A$$

Now, Ohm's law can be used again to calculate the total resistance:

$$R = \frac{E}{I} = \frac{12V}{1.2A} = 10\Omega$$

This example was simple because all the resistances were the same value. The process is the same when the resistances are different, but the current calculation must be done for each load. The individual currents are added to get the total current.

Unlike series circuits, parallel circuits continue working even if one circuit opens. Household circuits are wired in parallel. Almost all the load circuits you will encounter will be parallel circuits.

Either of the following formulas can be used to convert parallel resistances to a single resistance value. The first formula is used when there are two resistances in parallel. The second is used when there are three or more.

$$\text{Total resistance} = \frac{R1 \times R2}{R1 + R2}$$

$$\text{Total resistance} = \frac{1}{\frac{1}{R1} + \frac{1}{R2} + \frac{1}{R3}}$$

Example:

1. The total resistance of the parallel circuit shown is 6V.

10Ω

15Ω

$$\frac{R1 \times R2}{R1 + R2} = \frac{10 \times 15 = 150}{10 + 15 = 25} = 6$$

49103-11_UA01.EPS

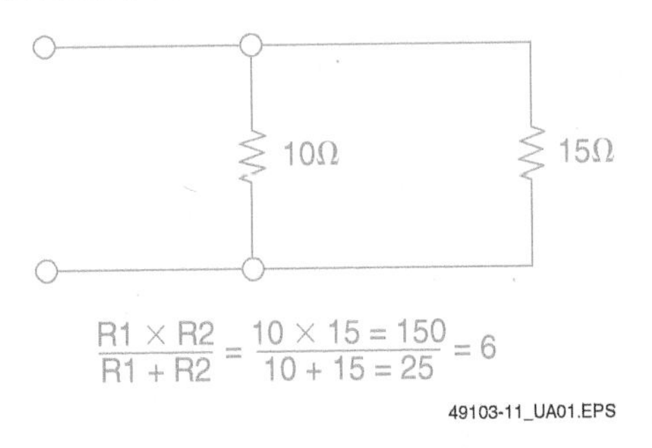

2. The total resistance of the parallel circuit shown is 4.76V.

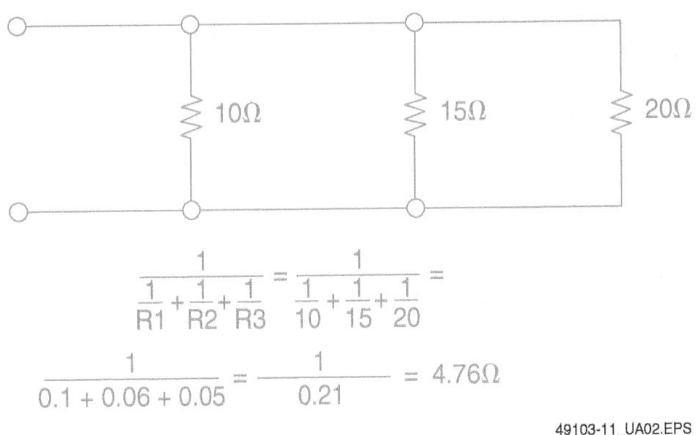

49103-11_UA02.EPS

8.3.0 Series-Parallel Circuits

Electronic circuits often contain a hybrid arrangement known as a series-parallel circuit (*Figure 15*). It is unlikely that you will ever have to determine the electrical characteristics of one of these circuits. If this becomes necessary, the parallel loads must be converted to their equivalent series resistance. Then the load resistances are added to determine total circuit resistance.

9.0.0 Electrical Measuring Instruments

Test meters are used to measure voltage, current, and resistance. The most common test meter is the volt-ohm-milliammeter (VOM), also called the multimeter. *Figure 16* shows both digital and analog multimeters. In the analog meter, the pointer moves in proportion to the value being measured. The person using the meter must then interpret the scale to find the measured value. Digital meters display the result on the screen.

Multimeters are commonly used to measure AC and DC voltage, DC current, and resistance. They can also be used to measure AC current in the milliamp range. For larger AC current values, a clamp-on ammeter is usually needed (*Figure 17*).

NOTE

Only qualified persons may use these meters. Consult your company's safety policy for applicable rules.

9.1.0 Measuring Current

A clamp-on ammeter (*Figure 18*) is used to measure current. The jaws of the ammeter are placed around a single conductor. Current flowing through the wire creates a magnetic field, which induces a proportional current in the ammeter's jaws. This current is read by the meter movement and appears as either a direct readout or, on an analog meter, as a deflection of the meter needle.

In-line ammeters (*Figure 19*) are less common. This type of meter must be connected in series with the circuit, which means that the circuit must be opened.

The following are good safety practices to keep in mind when measuring current:

- If the ammeter jaws are dirty or misaligned, the meter will not read correctly.
- To avoid damaging an analog meter, always start at a high range and work down.
- Do not clamp the meter jaws around two different conductors at the same time. This will cause an inaccurate reading.

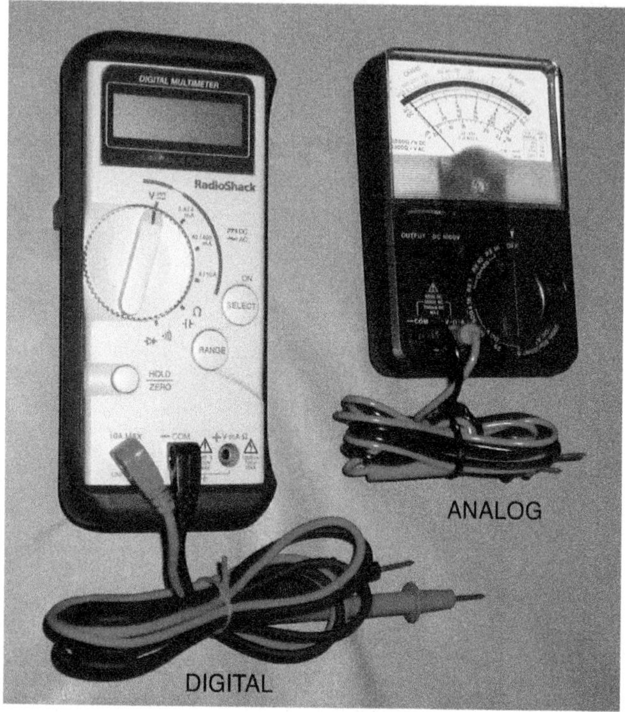

49103-11_F16.EPS

Figure 16 Digital and analog meters.

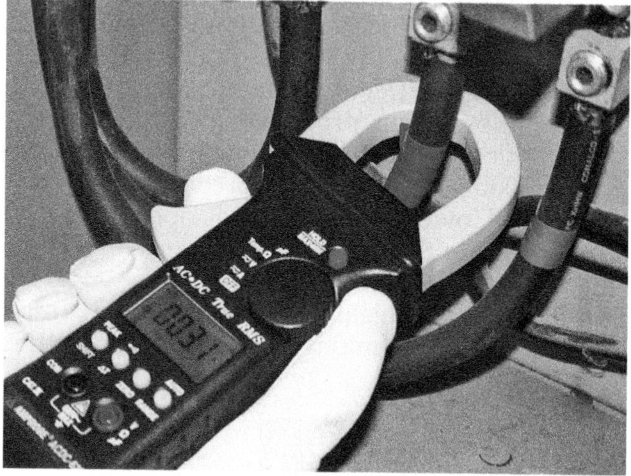

49103-11_F18.EPS

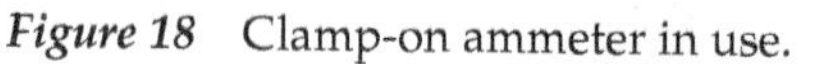

Figure 18 Clamp-on ammeter in use.

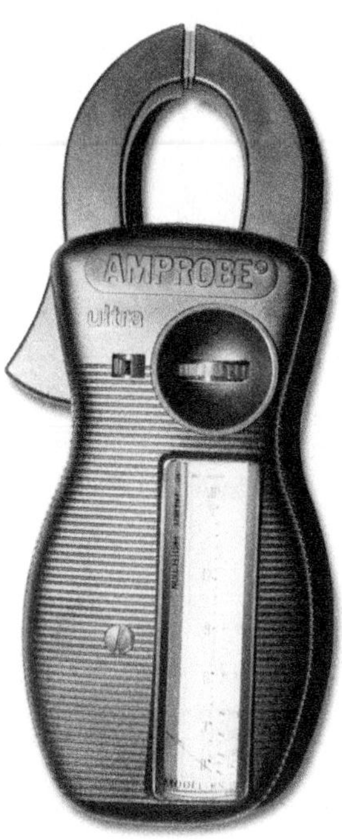

49103-11_F17.EPS

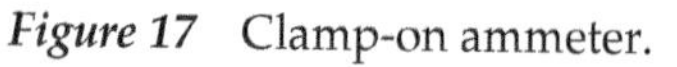

Figure 17 Clamp-on ammeter.

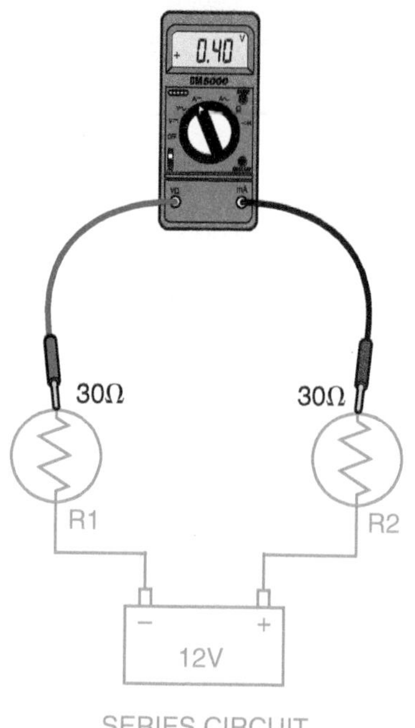

49103-11_F19.EPS

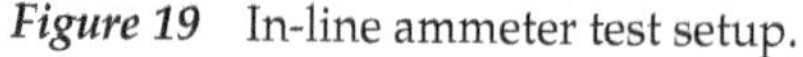

Figure 19 In-line ammeter test setup.

9.2.0 Measuring Voltage

A voltmeter must be connected in parallel with (across) the component or circuit to be tested (*Figure 20*). If a circuit function is not working, the voltmeter can determine if the correct voltage is available to the circuit. Voltage must be checked with the power applied.

9.3.0 Measuring Resistance

An ohmmeter contains an internal battery that acts as a voltage source. Resistance measurements are always made with the system power shut off. Sometimes an ohmmeter is used to measure resistance in a load. Motor windings are a good example of this type of use. An ohmmeter is often used to check continuity in a circuit. A wire or closed switch offers very little resistance. With the ohmmeter connected (*Figure 21*) and the three switches closed, the current produced by the ohmmeter battery flows unopposed, and the meter shows zero resistance. The circuit has continuity (is continuous). However, if a switch is open, there will be no path for the current. The meter will see infinite resistance, or a lack of continuity.

A continuity tester (*Figure 22*) is a simple device. It consists mainly of a battery and either an audible or visual indicator. A continuity tester can be used in place of an ohmmeter to test the continuity of a wire and to identify individual wires contained in a conduit or other raceway. To test the continuity of a wire, strip the insulation off the end of the wire to be tested at one

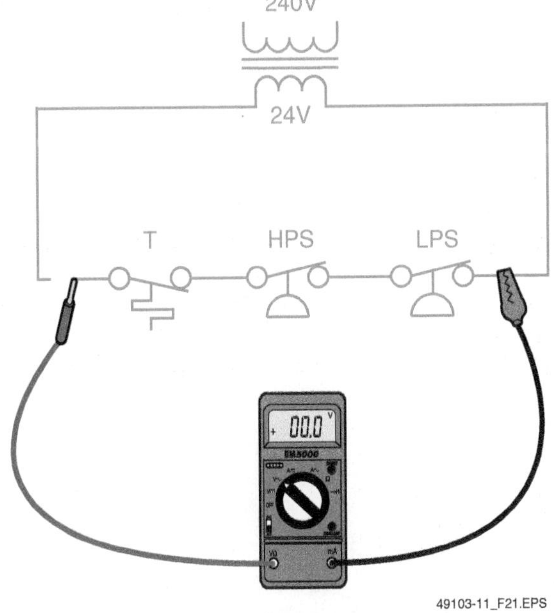

Figure 21 Ohmmeter connection for continuity testing.

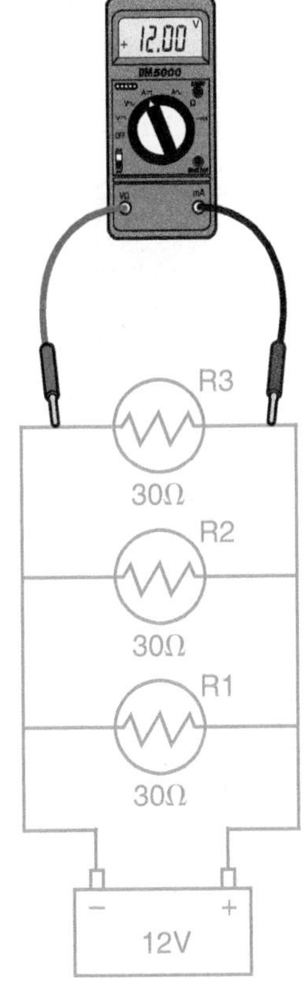

Figure 20 Voltmeter connection.

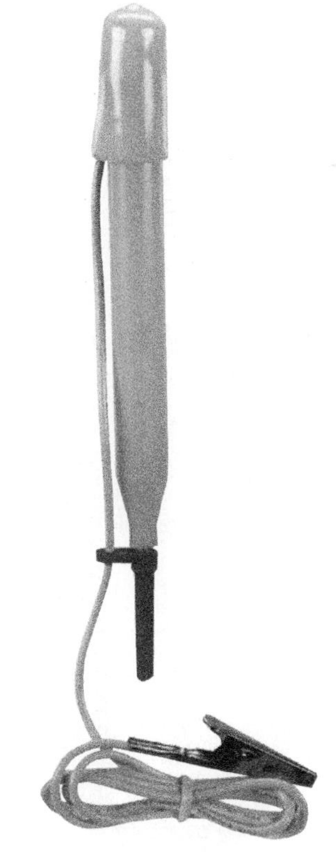

Figure 22 Continuity tester.

On Site

Meter Applications

Different electrical systems may have different levels of available current and different voltages. The meter must be matched to the application. For example, a meter that has enough protective features to be used safely on interior branch circuits may not be safe to use on the service feeders.

end of the conduit run. Next, connect (short) the wire to the metal conduit. At the other end of the conduit run, clip the alligator clip lead of the tester to the conduit, and touch the probe to the end of the wire under test. If the tester audible alarm sounds or the indicator light comes on, there is continuity. Note that this only shows that there is continuity between the two points being tested. It does not reveal the actual value of the resistance. If there is no indicator (alarm or light), the wire is open.

To identify individual wires in a conduit run, touch the tester probe to the wires in the conduit one at a time until the tester audible alarm sounds or the indicator lights up. Then, put matching identification tags on both ends of the wire. Continue this procedure until all the wires have been identified.

9.4.0 Voltage Testers

Figure 23 shows one of the many different devices used to check for the presence of voltage. A voltage tester can be used as a troubleshooting tool. It can also serve as a safety device to make sure that the voltage is turned off before touching any terminals or conductors. When the probes are touched to the circuit, the light on the instrument turns on if a voltage is present. This type of instrument is available in several voltage ranges. As a

On Site

Test Instruments—Old and New

Early electricians used individual meters to test circuits. Those instruments now seem primitive. Today, all-purpose instruments are available, such as the combination multimeter and clamp-on ammeter with its direct digital readout.

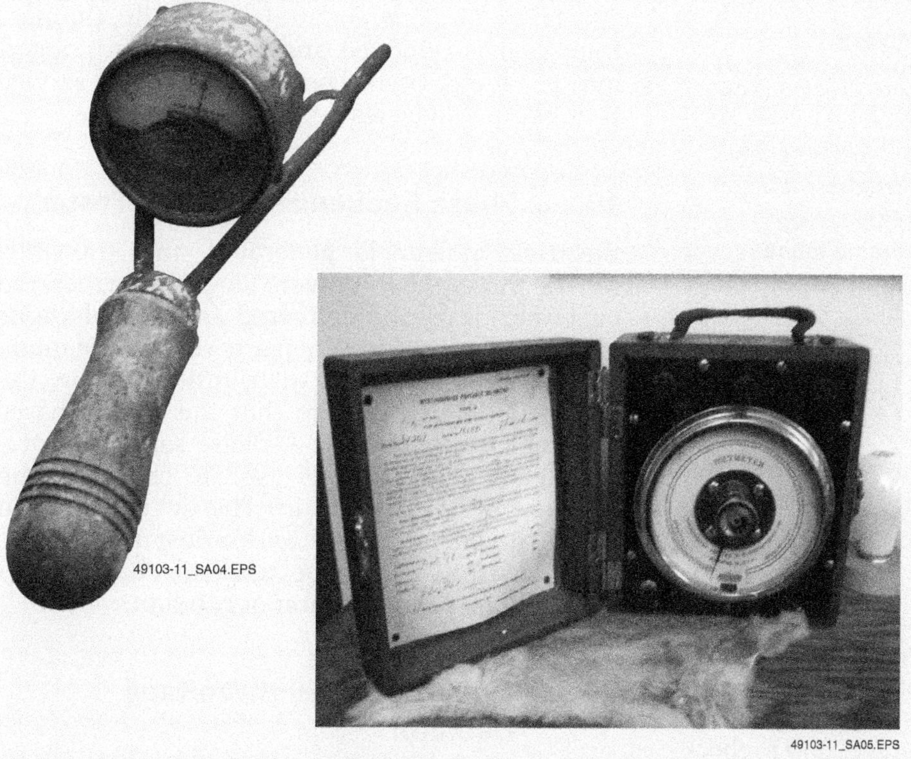

49103-11_SA04.EPS

49103-11_SA05.EPS

49103-11_SA06.EPS

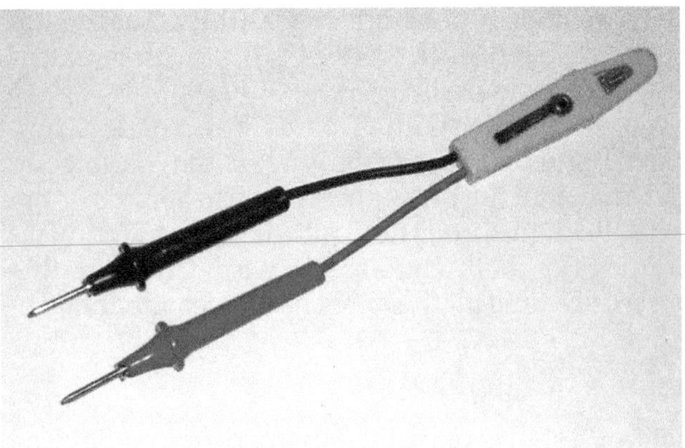

49103-11_F23.EPS

Figure 23 Voltage tester.

On Site

Using an Ohmmeter

An ohmmeter has its own battery to test the resistance or continuity of a circuit. The circuit must be de-energized. This is because the ohmmeter is calibrated for its own power source.

On Site

Ohmmeter Batteries

When using an ohmmeter to measure resistance or test continuity, the power to the circuit under test must be turned off. The internal multimeter battery provides a small DC current that is used to measure resistance. To ensure accurate readings, always replace or recharge the battery according to the meter manufacturer's instructions.

result, it is important to know something about the circuit being checked.

10.0.0 Specialized Power Transmission and Distribution System Test Instruments

Power transmission and distribution system technicians encounter problems and conditions that are beyond the capability of the test instruments already mentioned. Special test instruments are needed to measure very high voltages and to check the special electrical system components found in power transmission and distribution systems. These instruments are introduced in this section.

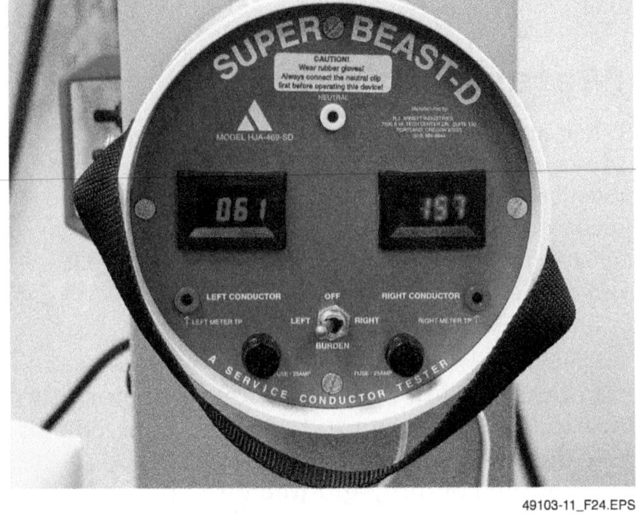

49103-11_F24.EPS

Figure 24 Secondary service conductor tester.

10.1.0 Secondary Service Conductor Tester

Secondary service conductors are used to link utility customers' electric meters to the power pole belonging to the utility. The conductors can be underground or overhead. Secondary service conductor testers (*Figure 24*) are often used to identify problems in the secondary conductors. Problems may include open circuits, unwanted resistance, and open neutrals. Typically, the electric meter is removed and the tester is plugged into the meter base. Then the secondary service conductors can be tested.

10.2.0 Phase Sequence Rotation Tester

Electrical power is generated and transmitted as three-phase power. Once it reaches the customer, it can be delivered as three-phase or single-phase power. When any repairs are made to a three-phase power distribution system, the technician must be sure that the three phases are in proper sequence. A phase sequence rotation tester (*Figure 25*) is used to check for the correct phase relationship. The device can also detect reversed phases and a lost phase. The same instrument can be used to check three-phase motors for direction of rotation.

10.3.0 Non-Contact High-Voltage Detector

Voltages in a power distribution system are often well over 1,000V. The technician must be very careful when checking these voltages to avoid an

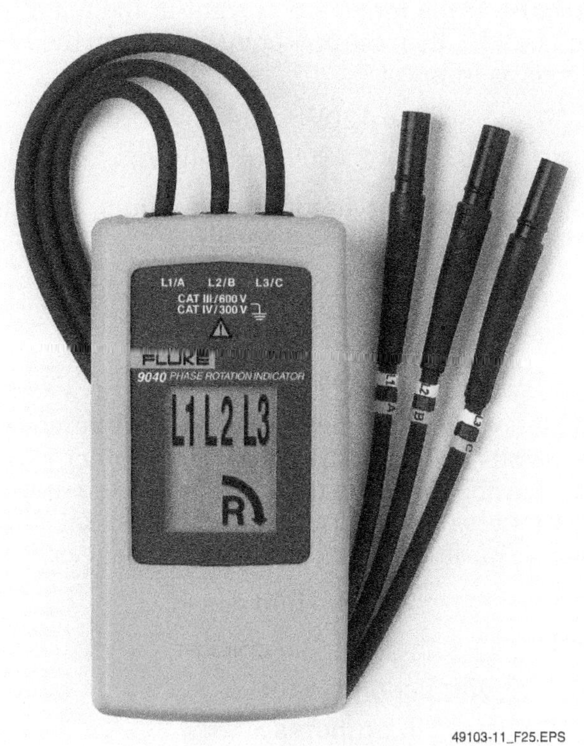

49103-11_F25.EPS

Figure 25 Phase rotation meter.

49103-11_F26.EPS

Figure 26 Non-contact high-voltage tester.

electric shock. The voltmeters discussed earlier are not suitable for safely measuring very high voltages. A non-contact high-voltage detector (*Figure 26*) is a type of voltmeter that can safely measure high voltages. The voltage detector is usually placed on the end of an insulated pole, called a hot stick. The detector is positioned near (not touching) the conductor being tested. The presence of high voltage is shown by a light and/or buzzer.

10.4.0 Transformer Tester

Transformers are widely used in power distribution systems to step up and step down voltage. A transformer can fail due to a short circuit or open circuit in its windings. A transformer tester (*Figure 27*) can check for opens and shorts in power distribution transformers. The tester is connected across the transformer windings terminals.

49103-11_F27.EPS

Figure 27 Transformer tester.

11.0.0 Electrical Power

Power is defined as the rate of doing work. This is the rate at which energy is used or dissipated. Electrons passing through a resistance dissipate energy in the form of heat. In electrical circuits, power is measured in units called watts (W). The power in watts equals the rate of energy conversion. One watt of power equals the work done in one second by one volt of potential difference in moving one coulomb of charge. One coulomb per second is an ampere. Therefore, power in watts equals the product of amperes times volts.

The work done in an electrical circuit can be useful or wasted. In both cases, the rate at which the work is done is still measured in power. The turning of an electric motor is useful work. The heating of wires or resistors in a circuit is wasted work, since no useful function is performed by the heat.

The unit of electrical work is the joule. A joule is the amount of work done by one coulomb flowing through a potential difference of one volt. For example, if five coulombs flow through a potential difference of one volt, five joules of work are done. The time it takes the coulombs to flow through the potential difference has no bearing on the amount of work done.

When working with circuits, it is more convenient to think of amperes of current rather than coulombs. As previously discussed, one ampere equals one coulomb passing a point in one second. Using amperes, one joule of work is done in one second when one ampere moves through a potential difference of one volt. This rate of one joule of work in one second is the basic unit of power, the watt. A watt is the power used when one ampere of current flows through a potential difference of one volt (*Figure 28*).

Mechanical power is usually measured in units of horsepower (hp). To convert from horsepower to watts, multiply the number of horsepower by 746. To convert from watts to horsepower, divide the number of watts by 746. Conversions for common units of power are given in *Table 2*.

The kilowatt-hour (kWh) is commonly used for large amounts of electrical work or energy. The prefix kilo means one thousand. To find the number of kilowatt-hours, multiply the power in kilowatts by the time in hours that the power is used. If a light bulb uses 300W or 0.3kW for 4 hours, the amount of energy is 0.3×4, or 1.2kWh.

Very large amounts of electrical work or energy are measured in megawatts (MW). The prefix mega means one million.

Electricity usage is figured in kilowatt-hours of energy. The power line voltage is fairly constant

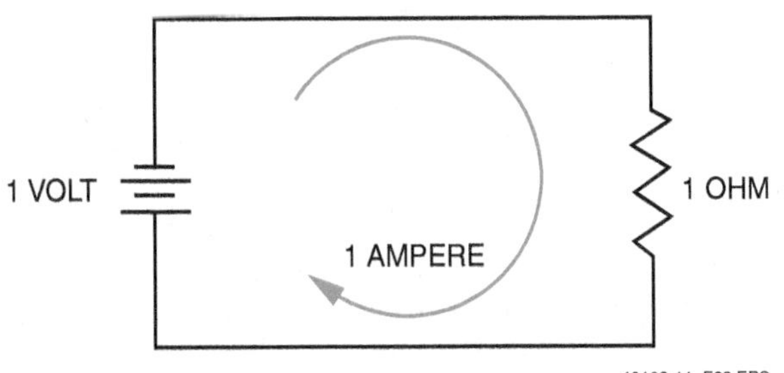

Figure 28 One watt.

Table 2 Conversion Table

1,000 watts (W)	=	1 kilowatt (kW)
1,000,000 watts (W)	=	1 megawatt (MW)
1,000 kilowatts (kW)	=	1 megawatt (MW)
1 watt (W)	=	0.00134 horsepower (hp)
1 horsepower (hp)	=	746 watts (W)

49103-11_T02.EPS

at 120V. If the total load current in the main line equals 20A, then the power in watts from the 120V line is as follows:

$$P = 120V \times 20A$$
$$P = 2{,}400W \text{ or } 2.4kW$$

If the power is used for five hours, then the energy of work supplied equals:

$$2.4kW \times 5 = 12kWh$$

11.1.0 Power Equation

When one ampere flows through a difference of two volts, two watts must be used. In other words, the number of watts used is equal to the number of amperes of current times the potential difference. This is expressed in equation form as:

$$P = I \times E \text{ or } P = IE$$

Where:

P = power used in watts
I = current in amperes
E = potential difference in volts

The equation is sometimes called Ohm's law for power because it is similar to Ohm's law. This equation is used to find the power consumed in a circuit or load when the values of current and voltage are known. The second form of the equation is used to find the voltage when the power and current are known:

$$E = \frac{P}{I}$$

The third form of the equation is used to find the current when the power and voltage are known:

$$I = \frac{P}{E}$$

Using these three equations, the power, voltage, or current in a circuit can be calculated when any two of the values are already known.

Think About It

Power

We take electrical power for granted, never stopping to think how amazing it is that a flow of submicroscopic electrons can pump thousands of gallons of water or light up a skyscraper. Our lives constantly rely on the electron to do work. Think about your day up to this point. How has electrical power shaped your experience?

Example 1:

Calculate the power in a circuit where the source of 100V produces 2A in a 50Ω resistance.

$$P = IE$$
$$P = 2A \times 100V$$
$$P = 200W$$

This means the source generates 200W of power while the resistance dissipates 200W in the form of heat.

Example 2:

Calculate the source voltage in a circuit that consumes 1,200W at a current of 5A.

$$E = \frac{P}{I}$$
$$E = \frac{1{,}200W}{5A}$$
$$E = 240V$$

Example 3:

Calculate the current in a circuit that consumes 600W with a source voltage of 120V.

$$I = \frac{P}{E}$$
$$I = \frac{600W}{120V}$$
$$I = 5A$$

Components that use the power dissipated in their resistance are usually rated in terms of power. The power is rated at normal operating voltage, which is usually 120V. For example, an appliance that draws 5A at 120V would dissipate 600W. The rating for the appliance would then be 600W/120V.

To calculate I or R for components rated in terms of power at a specified voltage, the power formula may be used in different forms. There are three basic power formulas. Each of the three can be rearranged into two other forms for a total of nine combinations:

$$P = IE \qquad P = I^2R \qquad P = \frac{E^2}{R}$$
$$I = \frac{P}{E} \qquad R = \frac{P}{I^2} \qquad R = \frac{E^2}{P}$$
$$E = \frac{P}{I} \qquad I = \sqrt{\frac{P}{R}} \qquad E = \sqrt{PR}$$

Note that all of these formulas are based on Ohm's law (E = IR) and the power formula (P = I × E). *Figure 29* shows all of the applicable power, voltage, resistance, and current equations.

11.2.0 Power Rating of Resistors

If too much current flows through a resistor, the heat produced will damage or destroy the resistor. This heat is caused by I^2R heating, which is power loss in watts. Every resistor is given a wattage, or power rating. This rating shows how much I^2R heating the resistor can withstand before it burns out. A resistor with a power rating of one watt will burn out if it is used in a circuit in which the current causes it to dissipate heat at a rate greater than one watt.

If the power rating of a resistor is known, the maximum current it can carry is found by using an equation derived from $P = I^2R$:

$$P = I^2R$$
$$I^2 = P/R$$
$$I = \sqrt{P/R}$$

Using this equation, find the maximum current that can be carried by a 1Ω resistor with a power rating of 4W:

$$I = \sqrt{P/R} = \sqrt{4W/1\Omega} = 2A$$

If the 1Ω resistor conducts more than 2 amperes, it will dissipate more than its rated power and burn out.

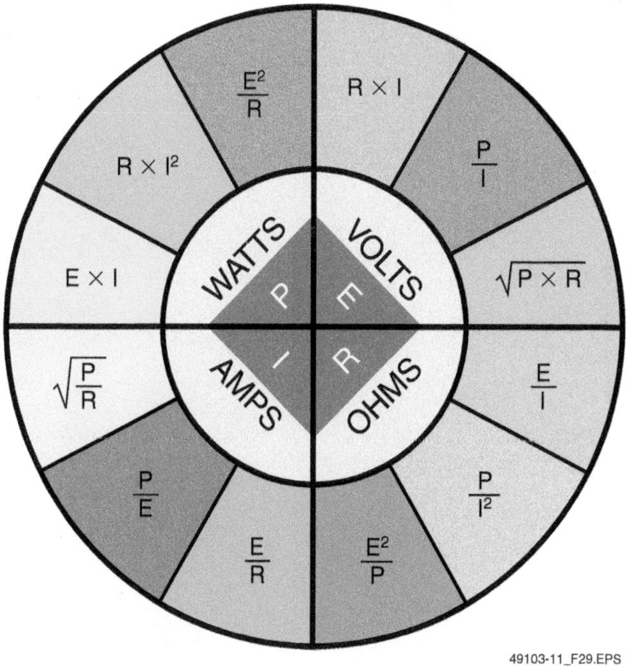

Figure 29 Expanded Ohm's law circle.

Power ratings from resistor manufacturers usually assume that the resistors are mounted in an open location where there is free air circulation and the temperature is no higher than 104°F (40°C). If a resistor is mounted in a small, enclosed space or where the temperature is above 104°F, it may burn out before its power rating is exceeded. Also, some resistors are designed to be attached to a chassis or frame that will carry away the heat.

Summary

Electrical technicians are often called upon to test and troubleshoot electrical circuits. To do this work safely and more effectively requires an understanding of the theory of electricity. This means becoming familiar with the interrelationships between voltage, current, resistance, and power. The basic tool for understanding these relationships is Ohm's law.

Testing and troubleshooting electrical circuits involves the use of test instruments. One typical device is the multimeter, or VOM. The multimeter combines the voltmeter, ammeter, and ohmmeter in a single instrument. There are also a number of specialized instruments available for doing high-voltage work.

In an analog meter, a pointer moves across a scale in proportion to the current flowing though the meter. In a digital meter, the measured value is displayed directly on a screen.

Review Questions

1. Silicon is a material that acts as a(n) _____.
   a. insulator
   b. conductor
   c. semiconductor
   d. voltage source

2. Which material acts as the best conductor of electricity?
   a. Silver
   b. Germanium
   c. Porcelain
   d. Silicon

3. What is the number of electrons typically found in the valence shell of a conductor?
   a. Three or less
   b. Five or more
   c. Seven
   d. Nine

4. The ability of an electric charge to do work is called _____.
   a. kinetic energy
   b. positive force
   c. motivation
   d. potential

5. Which of the following terms is another way of defining voltage?
   a. A flow of electrons
   b. Electromotive force
   c. Opposition
   d. Negative charge

6. A device that creates voltage by chemical means is called a(n) _____.
   a. alternator
   b. generator
   c. reactor
   d. battery

7. Resistance in a circuit is measured in _____.
   a. amperes
   b. watts
   c. ohms
   d. coulombs

8. When current is multiplied by resistance, the resulting value is stated in _____.
   a. ohms
   b. volts
   c. amps
   d. watts

9. To find current in a circuit, voltage applied to the circuit is divided by the circuit's _____.
   a. resistance
   b. power factor
   c. wire size
   d. amperage

10. The formula E = I × R is used to find _____.
    a. power
    b. resistance
    c. current
    d. voltage

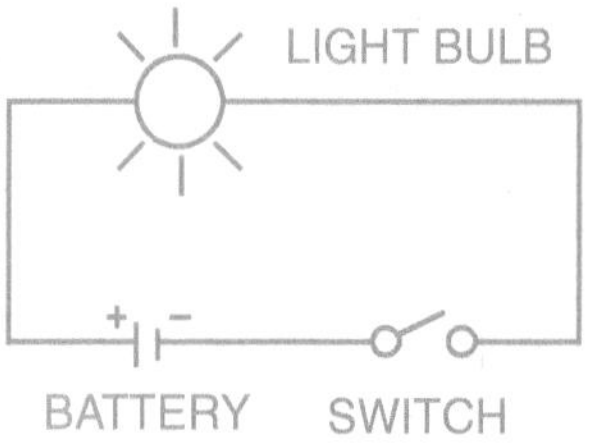

LIGHT BULB
BATTERY
SWITCH
49103-11_RQ01.EPS

Figure 1

11. What type of circuit is shown in *Figure 1*?
    a. Series
    b. Parallel
    c. Series-parallel
    d. Short

12. In a 12V series circuit containing two resistors, 1A flows through one resistor. What is the current flowing through the other resistor?
    a. 0.25A
    b. 0.5A
    c. 1.0A
    d. 2.0A

13. In a circuit containing five 50-ohm resistors, total circuit resistance is calculated at 250 ohms. What type of circuit is this?
    a. Parallel
    b. Series
    c. Series-parallel
    d. Open

14. Three loads of equal resistance are connected in parallel across a 120V power supply. What is the voltage drop across each load?
    a. 30V
    b. 40V
    c. 120V
    d. 360V

15. When 50-ohm, 40-ohm, and 12-ohm resistors are connected in parallel, the total circuit resistance will always be _____.
    a. greater than 50 ohms
    b. greater than 102 ohms
    c. less than 102 ohms
    d. less than 12 ohms

16. Placing the jaws of a clamp-on ammeter around two conductors will result in _____.
    a. an inaccurate reading
    b. a more accurate reading
    c. double the current reading
    d. an oscillating reading

17. How must a voltmeter be connected to measure the voltage applied to a component?
    a. In series with the component
    b. Across the component
    c. With the component disconnected from the circuit
    d. With the circuit de-energized

18. A continuity tester operates similar to a(n) _____.
    a. ammeter
    b. voltmeter
    c. ohmmeter
    d. clamp-on ammeter

19. To check the direction of rotation of a three-phase motor, use a _____.
    a. secondary service conductor tester
    b. clamp-on ammeter
    c. multimeter
    d. phase sequence rotation tester

20. To check for the presence of 275,000 volts on a power line, use a _____.
    a. multimeter
    b. secondary service conductor tester
    c. non-contact high-voltage detector
    d. clamp-on ammeter

Trade Terms Quiz

Fill in the blank with the correct term that you learned from your study of this module.

1. An instrument for measuring electrical current is a(n) ___________.
2. Measured in amperes, ___________ is the flow of electrons in a circuit.
3. The force required to produce a current of one ampere through a resistance of one ohm is a(n) ___________.
4. Voltage is measured with a(n) ___________.
5. One volt applied across one ohm of resistance causes a current flow of one ___________.
6. One volt is the potential difference between two points for which one coulomb of electricity will do one ___________ of work.
7. The common unit used for specifying the size of a given charge is a(n) ___________.
8. The driving force that makes current flow in a circuit is ___________.
9. The basic unit of measurement for electrical power is the ___________.
10. The smallest particle of an element that will still retain the properties of that element is the ___________.
11. The ___________ is at the center of an atom.
12. Found in the nuclei of atoms are electrically positive particles called ___________ and electrically neutral particles known as ___________.
13. The outermost ring of electrons orbiting the nucleus of an atom is known as the ___________.
14. A negatively charged particle that orbits the nucleus of an atom is a(n) ___________.
15. Any substance that has mass and occupies space is known as ___________.
16. The prefix used to indicate one thousand is ___________.
17. The prefix used to indicate one million is ___________.
18. A DC voltage source consisting of two or more cells that convert chemical energy into electrical energy is a(n) ___________.
19. A complete path for current flow is a(n) ___________.
20. A material through which it is relatively easy to maintain an electric current is called a(n) ___________.
21. A material through which it is difficult to conduct an electric current is a(n) ___________.
22. The basic unit of measurement for resistance is the ___________.
23. The instrument that is used to measure resistance is called a(n) ___________.
24. A statement of the relationship between current, voltage, and resistance in an electrical circuit is known as ___________.
25. The rate of doing work or the rate at which energy is used or dissipated is called ___________.
26. The electrical property that opposes the flow of current through a circuit and is measured in ohms is ___________.
27. A component that normally opposes current flow in a DC circuit is a(n) ___________.
28. A type of drawing in which symbols are used to represent the components in a system is called a(n)___________.
29. A circuit with only one path for current flow is a(n) ___________.
30. The change in voltage across a component caused by the current flowing through it and the amount of resistance opposing it is called ___________.
31. An electromechanical switching device consisting of a coil and one or more sets of contacts is a(n) ___________.
32. A device consisting of one or more coils of wire wrapped around a common core that is used to step voltage up or down is a(n) ___________.
33. An electromagnetic coil used to control a mechanical device, such as a valve, is a(n) ___________.
34. The rate of doing electrical work, which is measured in watts, is ___________.

Trade Terms

Ammeter
Ampere (A)
Atom
Battery
Circuit
Conductor
Coulomb
Current
Electrical power
Electron
Insulator
Joule (J)
Kilo
Matter
Mega
Neutrons
Nucleus
Ohm (Ω)
Ohmmeter
Ohm's law
Power
Protons
Relay
Resistance
Resistor
Schematic
Series circuit
Solenoid
Transformer
Valence shell
Volt (V)
Voltage
Voltage drop
Voltmeter
Watt (W)

Trade Terms Introduced in This Module

Ammeter: An instrument for measuring electrical current.

Ampere (A): A unit of electrical current. For example, one volt across one ohm of resistance causes a current flow of one ampere.

Atom: The smallest particle into which an element may be divided and still retain the properties of that element.

Battery: A DC voltage source consisting of two or more cells that convert chemical energy into electrical energy.

Circuit: A complete path for current flow.

Conductor: A material through which it is relatively easy to maintain an electric current.

Coulomb: An electrical charge equal to 6.25×10^{18} electrons or 6,250,000,000,000,000,000 electrons. A coulomb is the common unit of quantity used for specifying the size of a given charge.

Current: The movement, or flow, of electrons in a circuit. Current (I) is measured in amperes.

Electrical power: The rate of doing electrical work. Electrical power is measured in watts (W).

Electron: A negatively charged particle that orbits the nucleus of an atom.

Insulator: A material through which it is difficult to conduct an electric current.

Joule (J): A unit of measurement that represents one newton-meter (Nm), which is a unit of measure for doing work.

Kilo: A prefix used to indicate one thousand. For example, one kilowatt is equal to one thousand watts.

Matter: Any substance that has mass and occupies space.

Mega: A prefix used to indicate one million. For example, one megawatt is equal to one million watts.

Neutrons: Electrically neutral particles (neither positive nor negative). Neutrons have the same mass as a proton and are found in the nucleus of an atom.

Nucleus: The center of an atom, which contains protons and neutrons.

Ohm (Ω): The basic unit of measurement for resistance.

Ohmmeter: An instrument used for measuring resistance.

Ohm's law: A statement of the relationships among current, voltage, and resistance in an electrical circuit. Current (I) equals voltage (E) divided by resistance (R). This is generally expressed as a mathematical formula: $I = E/R$.

Power: The rate of doing work, or the rate at which energy is used or dissipated.

Protons: The smallest positively charged particles of an atom, found in the nucleus.

Relay: An electromechanical device consisting of a coil and one or more sets of contacts. Used as a switching device.

Resistance: An electrical property that opposes the flow of current through a circuit. Resistance (R) is measured in ohms.

Resistor: Any device in a circuit that resists the flow of electrons.

Schematic: A type of drawing in which symbols are used to represent the components in a system.

Series circuit: A circuit with only one path for current flow.

Solenoid: An electromagnetic coil used to control a mechanical device, such as a valve.

Transformer: A device consisting of one or more coils of wire wrapped around a common core. Transformers are commonly used to step voltage up or down.

Valence shell: The outermost ring of electrons that orbit about the nucleus of an atom.

Volt (V): The unit of measurement for voltage, also known as electromotive force (emf) or difference of potential. One volt is the force required to produce a current of one ampere through a resistance of one ohm. In addition, one volt is the potential difference between two points for which one coulomb of electricity will do one joule (J) of work.

Voltage: The driving force that makes current flow in a circuit. Voltage (E) is also referred to as electromotive force (emf) or difference of potential.

Voltage drop: The change in voltage across a component caused by the current flowing through it and the amount of resistance opposing it.

Voltmeter: An instrument for measuring voltage. The resistance of a voltmeter is fixed. When the voltmeter is connected to a circuit, the current passing through it is directly proportional to the voltage at the connection points.

Watt (W): The basic unit of measurement for electrical power.

Additional Resources

This module presents thorough resources for task training. The following resource material is suggested for further study.

Electronics Fundamentals: Circuits, Devices, and Applications. 8th ed. Thomas L. Floyd. Upper Saddle River, NJ: Prentice Hall, 2009.

Principles of Electric Circuits: Electron Flow Version. 8th ed. Thomas L. Floyd. Upper Saddle River, NJ: Prentice Hall, 2006.

Figure Credits

H. J. Arnett Industries, Module opener, Figure 24

U.S. Army Corps of Engineers, SA01

Topaz Publications, Inc., SA02, Figures 13 (photo), 16, 17, 23, SA04, and SA05

Tim Dean, Figure 18

Amprobe Instruments, Figure 22

Extech Instruments, SA06

Fluke Corporation, Reproduced with permission, Figure 25

Photograph provided by AEMC® Instruments, Figure 26

HD Electric Co., Figure 27

CONTREN® LEARNING SERIES — USER UPDATE

NCCER makes every effort to keep its textbooks up-to-date and free of technical errors. We appreciate your help in this process. If you find an error, a typographical mistake, or an inaccuracy in NCCER's Contren® materials, please fill out this form (or a photocopy), or complete the online form at www.nccer.org/olf. Be sure to include the exact module number, page number, a detailed description, and your recommended correction. Your input will be brought to the attention of the Authoring Team. Thank you for your assistance.

Instructors – If you have an idea for improving this textbook, or have found that additional materials were necessary to teach this module effectively, please let us know so that we may present your suggestions to the Authoring Team.

NCCER Product Development and Revision

3600 NW 43rd Street, Building G, Gainesville, FL 32606

Fax: 352-334-0932
Email: curriculum@nccer.org
Online: www.nccer.org/olf

☐ Trainee Guide ☐ AIG ☐ Exam ☐ PowerPoints Other ______________

Craft / Level: Copyright Date:

Module Number / Title:

Section Number(s):

Description:

Recommended Correction:

Your Name:

Address:

Email: Phone:

Introduction to Electrical Theory

49104-11

Trainees with successful module completions may be eligible for credentialing through NCCER's National Registry. To learn more, go to **www.nccer.org** or contact us at **1.888.622.3720.** Our website has information on the latest product releases and training, as well as online versions of our *Cornerstone* newsletter and Pearson's Contren® product catalog.

Your feedback is welcome. You may email your comments to **curriculum@nccer.org**, send general comments and inquiries to **info@nccer.org**, or use the User Update form at the back of this module.

V.1 7/11

49104-11

Introduction to Electrical Theory

Objectives

When you have completed this module, you will be able to do the following:

1. Explain the basic characteristics of series, parallel, and combination circuits.
2. Using Ohm's law, find the unknown values in series, parallel, and series-parallel circuits.
3. Explain the purpose of grounding and bonding.

Performance Tasks

This is a knowledge-based module; there are no performance tasks.

Trade Terms

Bonding
Grounding
Kirchhoff's current law
Kirchhoff's voltage law
Parallel circuits
Series-parallel circuits

Required Trainee Materials

Calculator

Contents

Topics to be presented in this module include:

Figures and Tables

1.0.0 INTRODUCTION

The fundamental concept of Ohm's law will now be used to analyze more complex series circuits, parallel circuits, and combination series-parallel circuits. This module explains how to calculate resistance, current, and voltage in these complex circuits. Ohm's law will be used to develop a new law for voltage and current determination. This law, called Kirchhoff's law, will become the new foundation for analyzing circuits.

2.0.0 RESISTIVE CIRCUITS

Resistance is calculated in different ways, depending on whether the circuit is series or parallel. Resistance is measured and calculated in ohms (Ω).

2.1.0 Resistances in Series

In a series circuit, there is only one path for current flow. Resistance is measured in ohms. In the series circuit shown in *Figure 1*, the current (I) is the same in all parts of the circuit. This means that the current flowing through R_1 is the same as the current flowing through R_2 and R_3. It is also the same as the current supplied by the battery.

When resistances are connected in series, the total resistance in the circuit is the sum of the resistances of all the parts of the circuit:

$$R_T = R_1 + R_2 + R_3$$

Where:

R_T = total resistance

$R_1 + R_2 + R_3$ = resistances in series

Example 1:

The circuit shown in *Figure 2(A)* has 50Ω, 75Ω, and 100Ω resistors in series. Find the total resistance of the circuit.

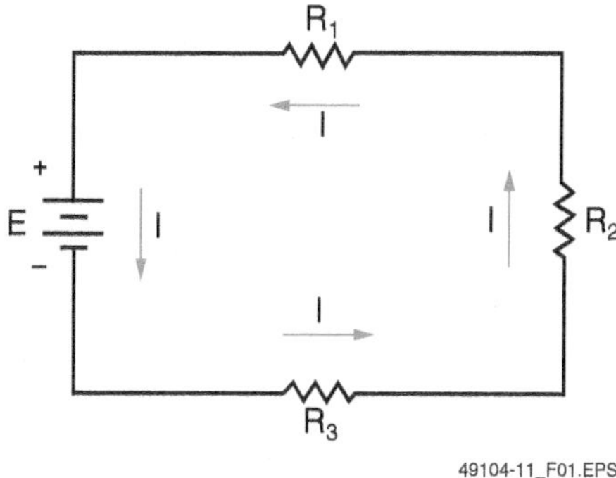

Figure 1 Series circuit.

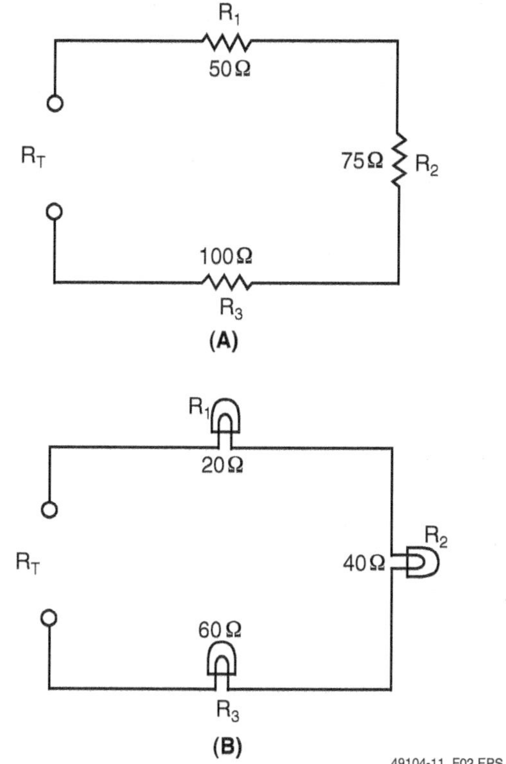

Figure 2 Total resistance.

Add the values of the three resistors in series:

$$R_T = R_1 + R_2 + R_3 = 50\Omega + 75\Omega + 100\Omega = 225\Omega$$

Example 2:

The circuit in *Figure 2(B)* has three lamps connected in series with the resistances shown. Find the total resistance of the circuit. Add the values of the three lamp resistances in series:

$$R_T = R_1 + R_2 + R_3 = 20\Omega + 40\Omega + 60\Omega = 120\Omega$$

Think About It

Series Circuits

Simple series circuits are seldom seen in practical wiring. One example of a simple series circuit is an older strand of Christmas lights, in which the entire string went dead when one bulb burned out. Think about what the wiring of a series circuit would look like in household receptacles. How would the circuit physically be wired? What kind of lighting would result from wiring household receptacles in series and plugging half a dozen lamps into those receptacles?

2.2.0 Resistances in Parallel

The total resistance in a parallel resistive circuit is given in the following formula, known as the reciprocal formula:

$$R_T = \frac{1}{\frac{1}{R_1} + \frac{1}{R_2} + \frac{1}{R_3} + \frac{1}{R_n}}$$

Where:

R_T = total resistance in parallel

R_1, R_2, R_3, and R_n = branch resistances

Example 1:

Find the total resistance of the 2Ω, 4Ω, and 8Ω resistors in parallel shown in *Figure 3*.

Write the formula for the three resistances in parallel:

$$R_T = \frac{1}{\frac{1}{R_1} + \frac{1}{R_2} + \frac{1}{R_3}}$$

Substitute the resistance values:

$$R_T = \frac{1}{\frac{1}{2} + \frac{1}{4} + \frac{1}{8}}$$

$$R_T = \frac{1}{0.5 + 0.25 + 0.125}$$

$$R_T = \frac{1}{0.875}$$

$$R_T = 1.14\Omega$$

Note that when resistances are connected in parallel, the total resistance is always less than the resistance of any single branch.

In this case:

$$R_T = 1.14\Omega < R_1 = 2\Omega, R_2 = 4\Omega, \text{ and } R_3 = 8\Omega$$

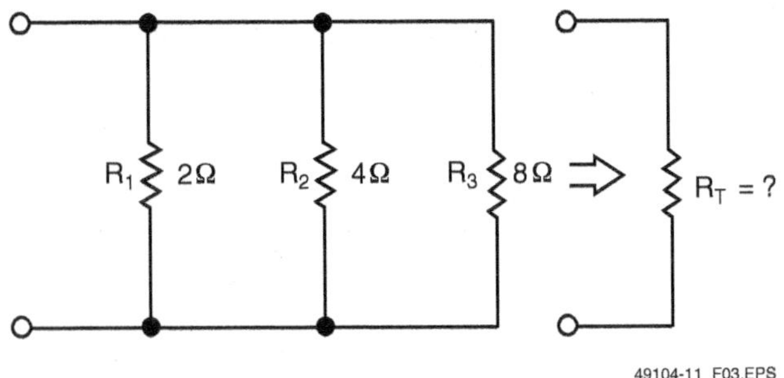

Figure 3 Parallel branch.

Example 2:

Add a fourth parallel resistor of 2Ω to the circuit in *Figure 3*. What is the new total resistance? What is the net effect of adding another resistance in parallel?

Write the formula for four resistances in parallel:

$$R_T = \frac{1}{\frac{1}{R_1} + \frac{1}{R_2} + \frac{1}{R_3} + \frac{1}{R_4}}$$

Substitute values:

$$R_T = \frac{1}{\frac{1}{2} + \frac{1}{4} + \frac{1}{8} + \frac{1}{2}}$$

$$R_T = \frac{1}{0.5 + 0.25 + 0.125 + 0.5}$$

$$R_T = \frac{1}{1.375}$$

$$R_T = 0.73\Omega$$

The net effect of adding another resistance in parallel is a reduction of the total resistance from 1.14Ω to 0.73Ω.

2.2.1 Simplified Formulas

The total resistance of equal resistors in parallel is equal to the resistance of one resistor divided by the number of resistors:

$$R_T = \frac{R}{N}$$

Where:

R_T = total resistance of equal resistors in parallel

R = resistance of one of the equal resistors

N = number of equal resistors

Think About It

Parallel Circuits

An interesting fact about circuits is the drop in total resistance in a parallel circuit as more resistors are added. But this does not mean that an endless number of devices, such as lamps, can be added in a parallel circuit. Why not?

On Site

Parallel Circuits

Most practical circuits are wired in parallel, like the pole lamps shown here.

49104-11_SA01.EPS

For example, if two resistors with the same resistance are connected in parallel, the equivalent resistance is half of that value (*Figure 4*).

The two 200Ω resistors in parallel are the equivalent of one 100Ω resistor. The two 100Ω resistors are the equivalent of one 50Ω resistor. And the two 50Ω resistors are the equivalent of one 25Ω resistor.

When any two unequal resistors are in parallel, it is often easier to calculate the total resistance by multiplying the two resistances and then dividing the product by the sum of the resistances. This is known as the product over sum formula.

$$R_T = \frac{R_1 \times R_2}{R_1 + R_2}$$

Where:

R_T = total resistance of unequal resistors in parallel

R_1, R_2 = two unequal resistors in parallel

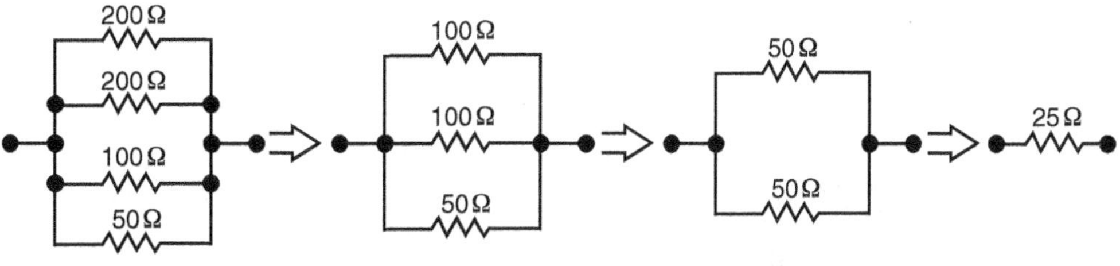

49104-11_F04.EPS

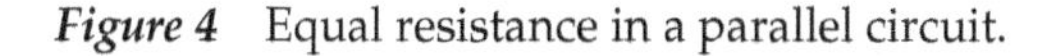

Figure 4 Equal resistance in a parallel circuit.

Example 1:

Find the total resistance of a 6Ω (R_1) resistor and an 18Ω (R_2) resistor in parallel:

$$R_T = \frac{R_1 \times R_2}{R_1 + R_2} = \frac{6 \times 18}{6 + 18} = \frac{108}{24} = 4.5\Omega$$

Example 2:

Find the total resistance of a 100Ω (R_1) resistor and a 150Ω (R_2) resistor in parallel:

$$R_T = \frac{R_1 \times R_2}{R_1 + R_2} = \frac{100 \times 150}{100 + 150} = \frac{15{,}000}{250} = 60\Omega$$

2.3.0 Series-Parallel Circuits

Finding current, voltage, and resistance in series and parallel circuits is fairly easy. When working with either type, use only the rules that apply to that type. In a series-parallel circuit, some parts of the circuit are series-connected and other parts are parallel-connected. In some parts of the circuit, the rules for series circuits apply. In other parts, the rules for parallel circuits apply. Solving a problem involving a series-parallel circuit requires recognizing which parts of the circuit are series-connected and which parts are parallel-connected. This is obvious if the circuit is simple. Many times, however, the circuit must be redrawn and put into a form that is easier to recognize.

In a series DC circuit, the current is the same at all points. In a parallel circuit, there are one or more points where the current divides and flows in separate branches. In a series-parallel circuit, there are both separate branches and series loads. The easiest way to find out if a circuit is a series, parallel, or series-parallel circuit is to start at the negative terminal of the power source. From there, trace the path of current through the circuit back to the positive terminal of the power source. If the current does not divide anywhere, it is a series circuit. If the current divides into separate branches, but there are no series loads (resistances), it is a parallel circuit. If the current divides into separate branches and there are also series loads, it is a series-parallel circuit. *Figure 5* shows electric lamps connected in series, parallel, and series-parallel circuits.

After determining that a circuit is series-parallel, redraw the circuit so that the branches and the series loads are more easily recognized. This is especially helpful when computing the total resistance of the circuit. *Figure 6* shows resistors connected in a series-parallel circuit and the equivalent circuit redrawn to simplify it.

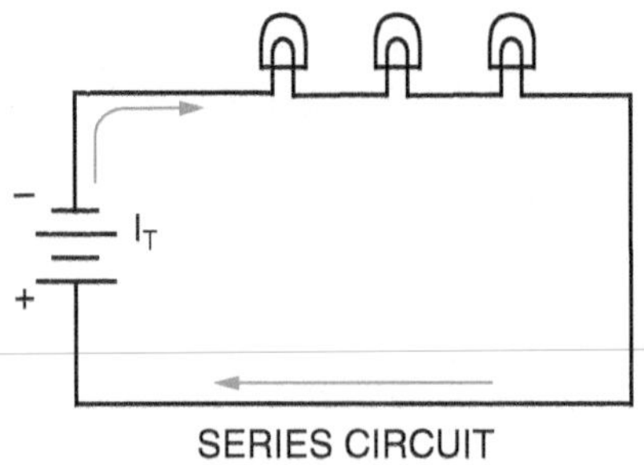

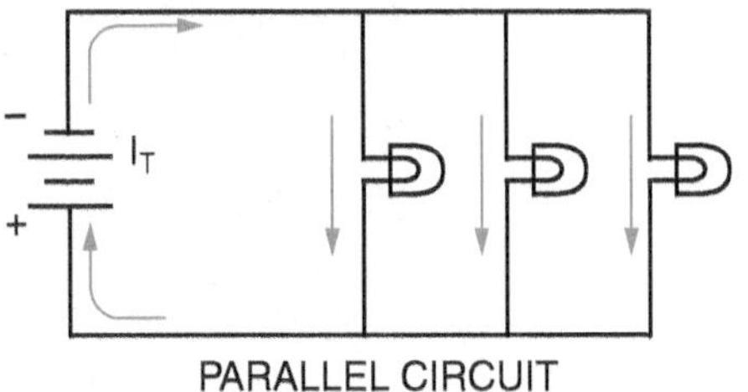

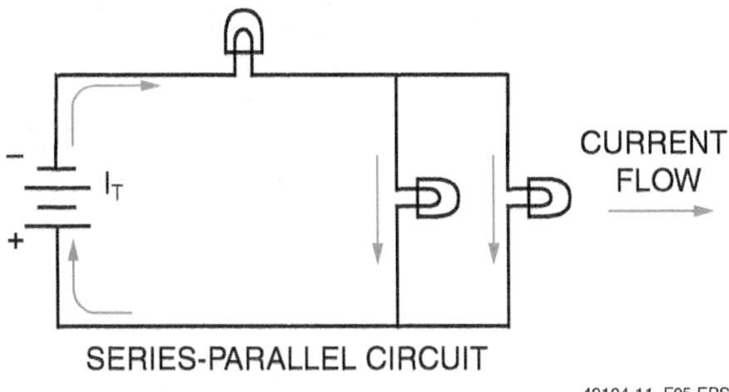

Figure 5 Series, parallel, and series-parallel circuits.

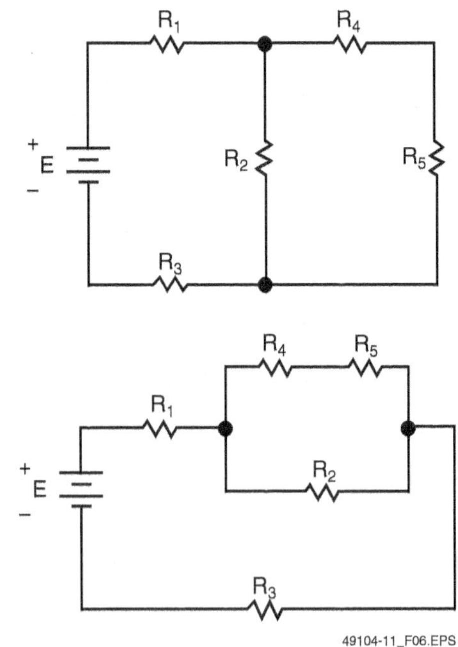

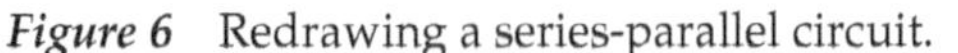

Figure 6 Redrawing a series-parallel circuit.

Think About It

Series-Parallel Circuits

Explain *Figure 6.* Which resistors are in series and which are in parallel?

2.3.1 Reducing Series-Parallel Circuits

Often all that is known about a series-parallel circuit is the applied voltage and the values of the individual resistances. To find the voltage drop across any of the loads or the current in any of the branches, the total circuit current must usually be known. But to find the total current, the total resistance of the circuit must be known. To find the total resistance, reduce the circuit to its simplest form, which is usually one resistance that forms a series circuit with the voltage source. This simple series circuit has the equivalent resistance of the series-parallel circuit from which it was derived. It also has the same total current. There are four basic steps in reducing a series-parallel circuit:

- If necessary, redraw the circuit so that all parallel combinations of resistances and series resistances are easily recognized.
- For each parallel combination of resistances, calculate the effective resistance.
- Using the resistance formulas, replace each of the parallel combinations with one resistance whose value is equal to the effective resistance of that combination. This provides a circuit with all series loads.
- Find the total resistance of the circuit by adding the resistances of all the series loads.

Examine the series-parallel circuit shown in *Figure 7*, and reduce it to an equivalent series circuit.

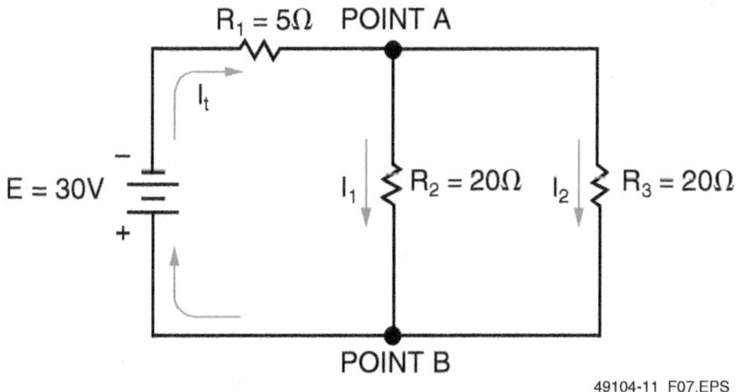

Figure 7 Reducing a series-parallel circuit.

In this circuit, resistors R_2 and R_3 are connected in parallel, but resistor R_1 is in series with both the battery and the parallel combination of R_2 and R_3. The current I_T leaving the negative terminal of the voltage source travels through resistor R_1 before it is divided at the junction of resistors R_1, R_2, and R_3 (Point A) to go through the two branches formed by resistors R_2 and R_3.

Given the information in *Figure 7*, calculate the resistance of R_2 and R_3 in parallel and the total resistance of the circuit, R_T.

The total resistance of the circuit is the sum of R_1 and the equivalent resistance of R_2 and R_3 in parallel. To find R_T, first find the resistance of R_2 and R_3 in parallel. Because the two resistances have the same value of 20Ω, the resulting equivalent resistance is 10Ω. Therefore, the total resistance (R_T) is 15Ω (5Ω + 10Ω).

2.4.0 Applying Ohm's Law

In resistive circuits, unknown circuit values can be found by using Ohm's law and the techniques for determining equivalent resistance.

2.4.1 Voltage and Current in Series Circuits

Ohm's law may be applied to an entire series circuit or to the individual parts of the circuit. When it is used on a particular part of a circuit, the voltage across that part is equal to the current in that part multiplied by the resistance of that part.

For example, given the information in *Figure 8*, calculate the total resistance (R_T) and the total current (I_T).

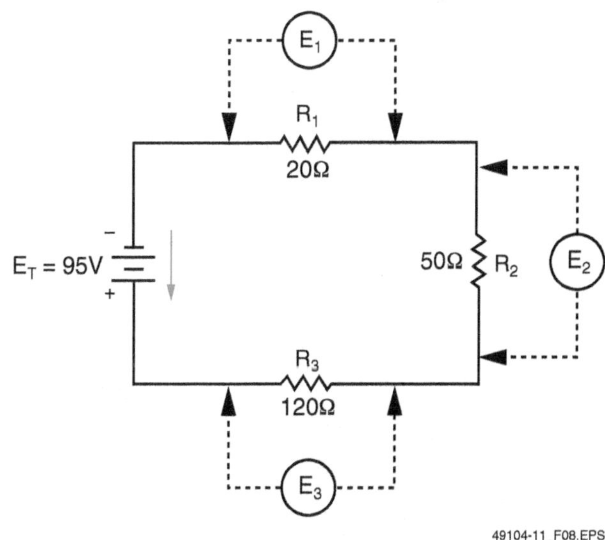

Figure 8 Calculating voltage drops.

To find R_T:

$$R_T = R_1 + R_2 + R_3$$
$$R_T = 20 + 50 + 120$$
$$R_T = 190\Omega$$

To find I_T using Ohm's law:

$$I_T = \frac{E_T}{R_T}$$
$$I_T = \frac{95V}{190\Omega}$$
$$I_T = 0.5A$$

Find the voltage across each resistor. In a series circuit, the current is the same; that is, I = 0.5A through each resistor:

$$E_1 = IR_1 = 0.5A(20\Omega) = 10V$$
$$E_2 = IR_2 = 0.5A(50\Omega) = 25V$$
$$E_3 = IR_3 = 0.5A(120\Omega) = 60V$$

The voltages E_1, E_2, and E_3 found for *Figure 8* are known as voltage drops or IR drops. Their effect is to reduce the voltage that is available to be applied across the rest of the components in the circuit. The sum of the voltage drops in any series circuit is always equal to the voltage that is applied to the circuit. The total voltage (E_T) is the same as the applied voltage. This can be verified in the example ($E_T = 10 + 25 + 60 = 95V$).

2.4.2 Voltage and Current in Parallel Circuits

A parallel circuit is a circuit in which two or more components are connected across the same voltage source (*Figure 9*). The resistors R_1, R_2, and R_3 are in parallel with each other and with the battery. Each parallel path is then a branch with its own individual current. When the total current I_T leaves the voltage source E, part I_1 of the current I_T will flow through R_1, part I_2 will flow through R_2, and the remainder I_3 will flow through R_3. The branch currents I_1, I_2, and I_3 can be different.

Think About It

Voltage Drops

Calculating voltage drops is not just a classroom exercise. It is important to know the voltage drop when sizing circuit components. What would happen if you sized a component without accounting for a substantial voltage drop in the circuit?

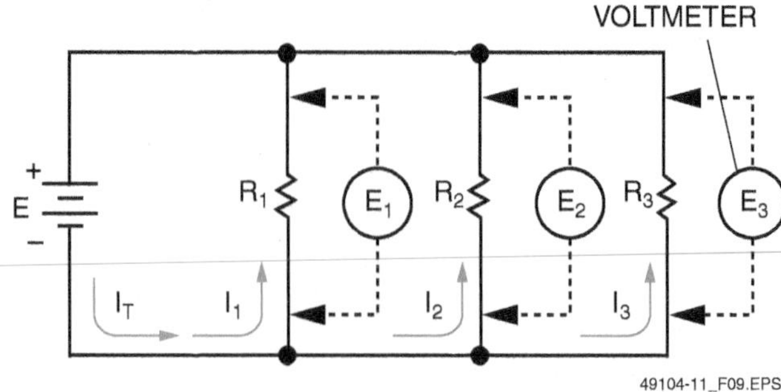

Figure 9 Parallel circuit.

However, if a voltmeter is connected across R_1, R_2, and R_3, the respective voltages E_1, E_2, and E_3 will be equal to the source voltage E.

The total current I_T is equal to the sum of all branch currents. This formula applies for any number of parallel branches whether the resistances are equal or unequal.

Using Ohm's law, each branch current equals the applied voltage divided by the resistance between the two points where the voltage is applied. Hence, the following equations apply for each branch in *Figure 9*:

$$\text{Branch 1: } I_1 = \frac{E_1}{R_1} = \frac{E}{R_1}$$
$$\text{Branch 2: } I_2 = \frac{E_2}{R_2} = \frac{E}{R_2}$$
$$\text{Branch 3: } I_3 = \frac{E_3}{R_3} = \frac{E}{R_3}$$

With the same applied voltage, any branch that has less resistance allows more current to flow through it than a branch with higher resistance.

Example 1:

In *Figure 10(A)*, the two branches R_1 and R_2 across a 110V power line draw a total line current of 20A. Branch R_1 takes 12A. What is the current I_2 in branch R_2?

Transpose to find I_2 and then substitute given values:

$$I_T = I_1 + I_2$$
$$I_2 = I_T - I_1$$
$$I_2 = 20 - 12 = 8A$$

Example 2:

As shown in *Figure 10(B)*, the two branches R_1 and R_2 across a 240V power line draw a total line current of 35A. Branch R_2 takes 20A. What is the current I_1 in branch R_1?

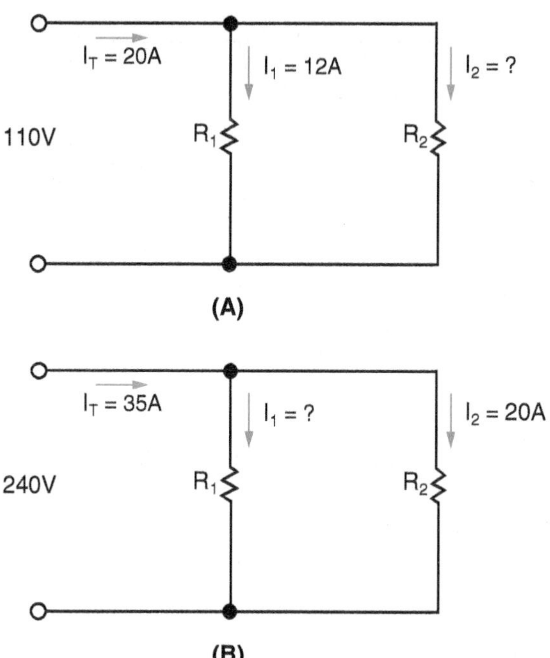

Figure 10 Solving for an unknown current.

Transpose to find I_1 and then substitute given values:

$$I_T = I_1 + I_2$$
$$I_1 = I_T - I_2$$
$$I_1 = 35 - 20 = 15A$$

2.4.3 Voltage and Current in Series-Parallel Circuits

Series-parallel circuits combine the elements and characteristics of both the series and parallel circuits. By properly applying the equations and methods already discussed, the values of individual components of the circuit can be found. *Figure 11* shows a simple series-parallel circuit with a 1.5V battery.

To find the current and voltage of each component, first simplify the circuit to find the total current. Then work across the individual components.

This circuit can be broken into two components: the series resistances R_1 and R_2 and the parallel resistances R_3 and R_4.

R_1 and R_2 can be added together to form the equivalent series resistance R_{1+2}:

$$R_{1+2} = R_1 + R_2$$
$$R_{1+2} = 0.5k\Omega + 0.5k\Omega$$
$$R_{1+2} = 1k\Omega$$

R_3 and R_4 can be totaled using either the general reciprocal formula or, since there are two

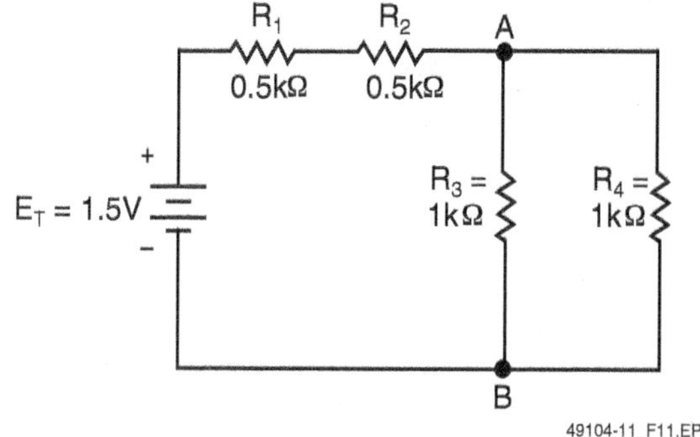

Figure 11 Series-parallel circuit.

resistances in parallel, the product over sum method. Both methods are shown here.

$$R_{3+4} = \frac{1}{\frac{1}{R_3} + \frac{1}{R_4}} = \frac{1}{\frac{1}{1k\Omega} + \frac{1}{1k\Omega}}$$
$$= \frac{1}{\frac{2}{1{,}000\Omega}} = \frac{1}{0.002} = 0.5k\Omega$$

$$R_{3+4} = \frac{R_3 \times R_4}{R_3 + R_4} = \frac{1k\Omega \times 1k\Omega}{1k\Omega + 1k\Omega}$$
$$= \frac{1{,}000{,}000\Omega}{2{,}000\Omega} = 0.5k\Omega$$

The equivalent circuit containing the R_{1+2} resistance of 1kΩ and the R_{3+4} resistance of 0.5kΩ is shown in *Figure 12*.

Using the Ohm's law relationship that total current equals voltage divided by circuit resistance, the circuit current can be determined. First, total circuit resistance must be found. Since the simplified circuit consists of two resistances in series, they are added together to obtain total resistance.

$$R_T = R_{1+2} + R_{3+4}$$
$$R_T = 1k\Omega + 0.5k\Omega$$
$$R_T = 1.5k\Omega$$

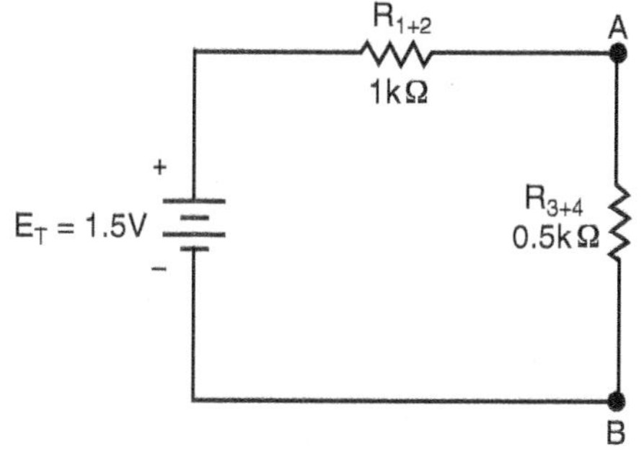

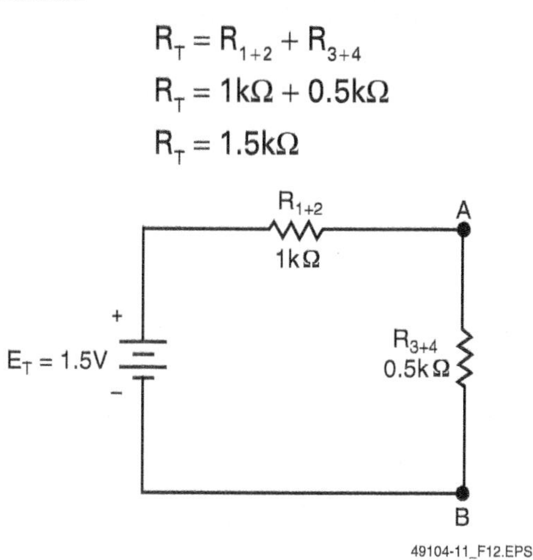

Figure 12 Simplified series-parallel circuit.

Applying this to the current/voltage equation:

$$I_T = \frac{E_T}{R_T}$$

$$I_T = \frac{1.5V}{1.5k\Omega}$$

$$I_T = 1mA \text{ or } 0.001A$$

Now that the total current is known, voltage drops across individual components can be determined:

$$E_{R1} = I_T R_1 = 1mA \times 0.5k\Omega = 0.5V$$

$$E_{R2} = I_T R_2 = 1mA \times 0.5k\Omega = 0.5V$$

Since the total voltage equals the sum of all voltage drops, the voltage drop from A to B can be determined by subtraction:

$$E_T = E_{R1} + E_{R2} + E_{A+B}$$

$$E_T - E_{R1} - E_{R2} = E_{A+B}$$

$$1.5V - 0.5V - 0.5V = E_{A+B} = 0.5V$$

Since R_3 and R_4 are in parallel, some of the total current must pass through each resistor. R_3 and R_4 are equal, so the same current should flow through each branch. Using the relationship:

$$I = \frac{E}{R}$$

$$I_{R3} = \frac{E_{R3}}{R_3} \qquad I_{R4} = \frac{E_{R4}}{R_4}$$

$$I_{R3} = \frac{0.5V}{1k\Omega} \qquad I_{R4} = \frac{0.5V}{1k\Omega}$$

$$I_{R3} = 0.5mA \qquad I_{R4} = 0.5mA$$

$$0.5mA + 0.5mA = 1mA$$

The total current for the circuit passes through R_1 and R_2 and is evenly divided between R_3 and R_4.

3.0.0 Kirchhoff's Laws

Kirchhoff's laws provide a simple, practical method of solving for unknown values in a circuit.

In its general form, Kirchhoff's current law states that at any point in a circuit, the total current entering that point must equal the total current leaving that point. For parallel circuits, this implies that the current in a parallel circuit is equal to the sum of the currents in each branch.

Kirchhoff's current law is the basis for the practical rule in parallel circuits that the total line current must equal the sum of the branch currents.

Did You Know?

Electrical Laws

Ohm's law and Kirchhoff's laws were named after real people. George Ohm first published his theory on the mathematical relationships of current, voltage, and resistance in 1826. Gustav Kirchhoff followed 19 years later with his work on the calculation of currents, voltages, and resistances in electrical networks. As a result of their breakthrough research, both men went on to head mathematics departments at major German universities.

Kirchhoff's voltage law states that the sum of the voltages around any closed path is zero.

4.0.0 Grounding and Bonding

The grounding system is a major part of an electrical system. The purpose of a grounding system is to protect life and equipment against the various electrical faults that can occur. Sometimes higher-than-normal voltages may appear at certain points in an electrical system or in the electrical equipment connected to the system. Proper grounding ensures that the electrical charges that cause these high voltages are channeled to the earth or ground before they can harm people or damage equipment. A circuit is grounded to limit the voltage on the circuit. Grounding also improves overall operation of the electrical system and continuity of service.

In electrical work, the term *ground* means being connected to the earth. A conductor is said to be grounded when it is connected to the earth or to some conducting body that extends the ground connection, such as a driven ground rod (electrode) or cold-water pipe. A grounding system is permanently installed. *Figure 13* shows a few of the many ways that grounding can be achieved.

Utility poles are grounded to provide a path to ground for lightning strikes (*Figure 14*). A molded cover over the ground wire keeps it from being damaged and protects people in case the ground wire becomes energized. The ground wire is usually a No. 6 or No. 4 AWG copper conductor. A common method of grounding is to drive a ground rod into the earth and connect the ground wire to the rod using a bronze clamp.

Bonding is related to grounding. Bonding is a means of interconnecting conductive metal objects that are not intended to carry current in order to protect people from electric shock. This causes all the objects to be at the same potential.

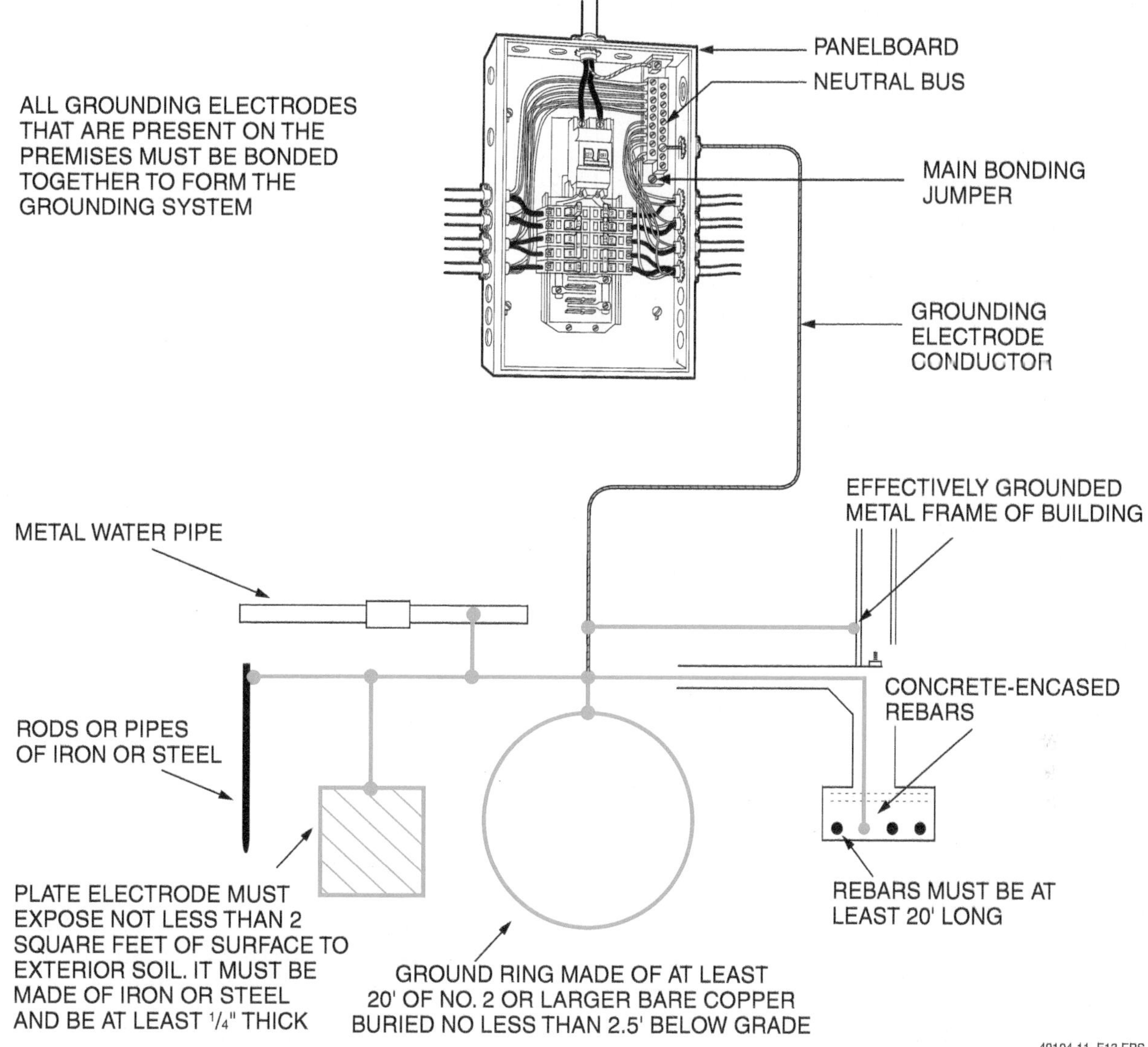

Figure 13 Common methods used to ground electrical systems at the terminal end.

If one metal object in a bonded system comes into contact with a live conductor, a person touching that metal object and another one would not experience a difference in potential or the resultant shock.

4.1.0 Protective Grounding

Power transmission and distribution system technicians must often work on de-energized power lines. Even though a power line may be de-energized, there are conditions under which it could become accidentally energized. In other situations, stray voltages that are hazardous to humans may be present on de-energized power lines. Under certain weather conditions, wind blowing across a power line can cause a static charge to build up on the power line. If a de-energized power line is located near an energized power line, the energized line can induce a voltage in the de-energized line. To protect workers from these hazardous voltages, cables of the correct size, equipped with special clamps, are temporarily attached to the de-energized power lines (*Figure 15*). These personal protective grounding cables are then attached to an adequate earth ground.

In temporary safety grounding, all the conductors in the work area are temporarily bonded and connected to a common ground (*Figure 16*). Temporary safety grounding ensures that everything the workers contact is at the same potential. This protects them from hazards, such as feedback within the utility system or from a generator.

The safety and bonding connections are temporary. After the repair has been made, the personal protective grounding cables and other equipment are removed before the power line is re-energized.

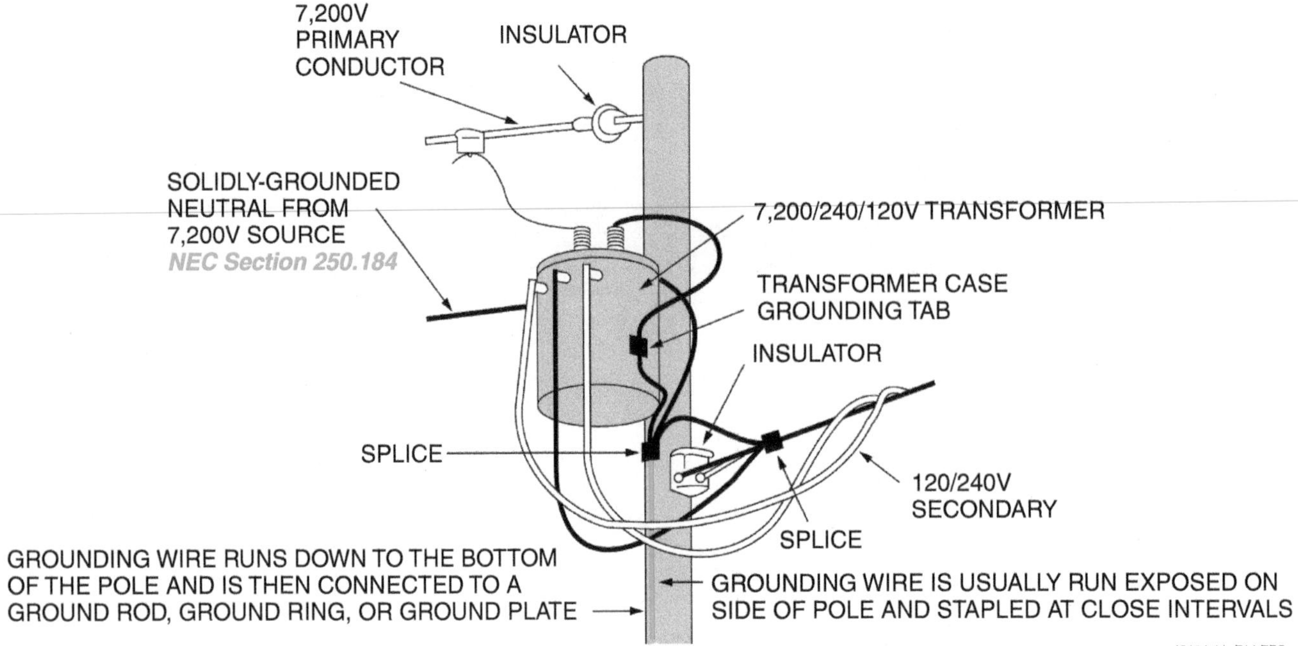

Figure 14 Utility pole grounding.

Figure 15 Grounding devices on de-energized power lines.

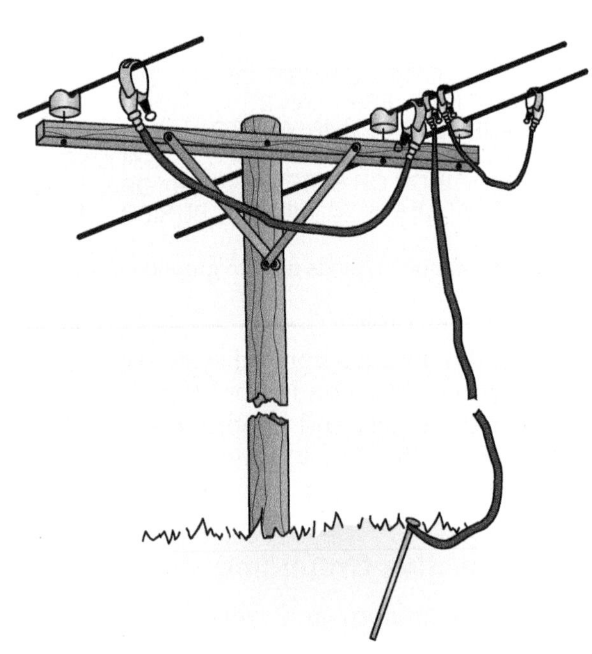

Figure 16 Single-point grounding.

SUMMARY

The relationships among current, voltage, resistance, and power in Ohm's law are the same for both DC series and DC parallel circuits. The ability to understand and apply these concepts is necessary for effective circuit analysis and troubleshooting. DC series-parallel circuits also have these fundamental relationships. DC series-parallel circuits are a combination of simple series and parallel circuits. Calculating I, E, R, and P for series-parallel circuits is no more difficult than calculating these values for simple series or parallel circuits. However, for series-parallel circuits, the calculations require more careful circuit analysis in order to apply Ohm's law correctly.

Review Questions

1. The formula for calculating the total resistance in a series circuit with three resistors is _____.
   a. $R_T = R_1 + R_2 + R_3$
   b. $R_T = R_1 - R_2 - R_3$
   c. $R_T = R_1 \times R_2 \times R_3$
   d. $R_T = \dfrac{1}{\frac{1}{R_1} + \frac{1}{R_2} + \frac{1}{R_3}}$

2. Find the total resistance in a series circuit with three resistances of 10Ω, 20Ω, and 30Ω.
   a. 1Ω
   b. 15Ω
   c. 20Ω
   d. 60Ω

3. The formula for calculating the total resistance in a parallel circuit with three resistors is _____.
   a. $R_T = R_1 + R_2 + R_3$
   b. $R_T = R_1 - R_2 - R_3$
   c. $R_T = R_1 \times R_2 \times R_3$
   d. $R_T = \dfrac{1}{\frac{1}{R_1} + \frac{1}{R_2} + \frac{1}{R_3}}$

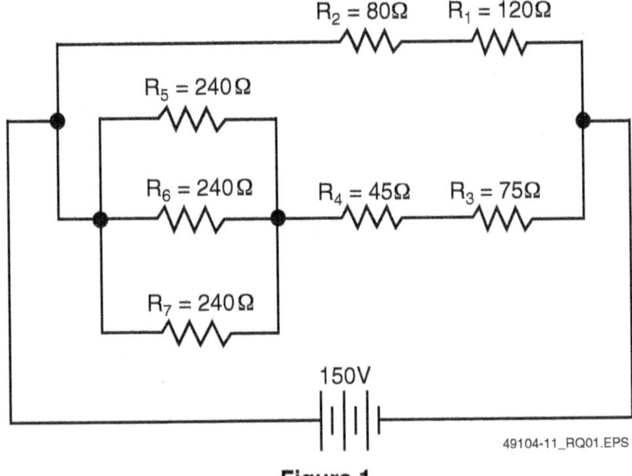

Figure 1

4. The total resistance in *Figure 1* is _____.
   a. 100Ω
   b. 129Ω
   c. 157Ω
   d. 1,040Ω

5. The voltage drop across the R_1/R_2 combination in *Figure 1* is _____.
   a. 75V
   b. 92.5V
   c. 110V
   d. 150V

6. In a parallel circuit, the largest amount of current will flow through the smallest resistor.
   a. True
   b. False

7. Which of the following is a correct statement about parallel circuits?
   a. Current flow is the same through all branches.
   b. The voltage drop is the same through all branches.
   c. The largest resistor will drop the most voltage.
   d. Total resistance equals the sum of the parallel resistors.

8. In a parallel circuit, the voltage across each path is equal to the _____.
   a. total circuit resistance times path current
   b. source voltage minus path voltage
   c. path resistance times total current
   d. applied voltage

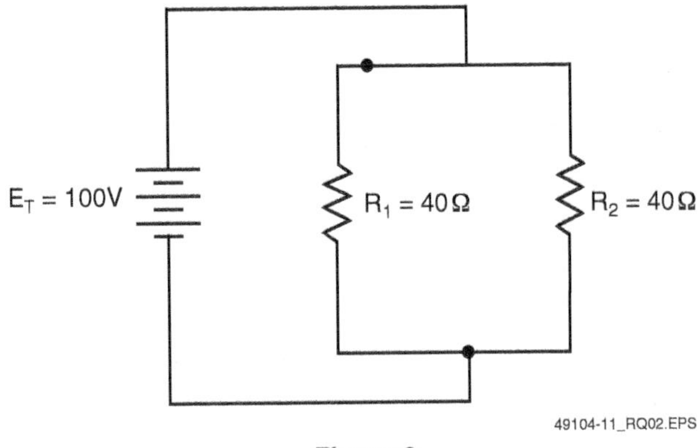

Figure 2

9. The value for total current in *Figure 2* is _____.
   a. 1.25A
   b. 2.50A
   c. 5A
   d. 10A

10. A resistor of 32Ω is in parallel with a resistor of 36Ω, and a 54Ω resistor is in series with the pair. When 350V is applied to the combination, the current through the 54Ω resistor is _____.

    a. 2.87A
    b. 3.26A
    c. 4.93A
    d. 5.86A

11. A 242Ω resistor is in parallel with a 180Ω resistor, and a 420Ω resistor is in series with the combination in a 27V circuit. A current of 22mA flows through the 242Ω resistor. The current through the 180Ω resistor is _____.

    a. 60mA
    b. 45mA
    c. 37mA
    d. 30mA

12. Two 24Ω resistors are in parallel, and a 42Ω resistor is in series with the combination. When 78V is applied to the three resistors, the voltage drop across the 42Ω resistor is about _____.

    a. 49.8V
    b. 55.8V
    c. 60.5V
    d. 65.3V

13. In a series circuit, the actual voltage dropped across all the loads is equal to _____.

    a. zero
    b. the applied voltage
    c. half the applied voltage
    d. a negative value

14. A grounded system is connected to _____.

    a. copper
    b. earth
    c. any metal
    d. water pipes

15. The safety measure in which conductive metal objects that are not intended to carry current are interconnected to prevent electrical shock is known as _____.

    a. temporary safety grounding
    b. personal protective grounding
    c. de-energizing
    d. bonding

Trade Terms Quiz

Fill in the blank with the correct term that you learned from your study of this module.

1. The statement that the total amount of current flowing through a parallel circuit is equal to the sum of the amounts of current flowing through each current path is known as ___________.
2. The statement that the sum of all the voltage drops in a circuit is equal to the source voltage of the circuit is known as ___________.
3. Circuits that contain both series and parallel current paths are called ___________.
4. Circuits that contain two or more parallel paths through which current can flow are known as ___________.
5. A means of interconnecting conductive metal objects that are not intended to carry current in order to protect people from electric shock is called ___________.
6. Connecting equipment to ground or to a conductive body that extends the ground connection is termed ___________.

Trade Terms

Bonding
Grounding
Kirchhoff's current law
Kirchhoff's voltage law
Parallel circuits
Series-parallel circuits

Mark Lagasse

Transmission and Distribution Superintendent
Cianbro Corporation,
Pittsfield, ME

When Mark Lagasse was about 13 years old, his parents built a new home for their family. It could have been a different story, but the electrician, a nice guy, welcomed Mark's enthusiastic (if unskilled) help. That's when Mark decided on his career.

How did you get started in the construction industry?
I began in the residential/commercial wiring world. After high school I went to a vocational school and then took my journeyman electrician's test and passed. My job included everything from wiring new homes to renovating commercial buildings such as supermarkets. I always had an eye on the local utility company and their line crews, and decided I wanted to do that some day. So in 1985, I signed on with the local utility for an internally run line school. After seven years of line work, I went into the planning and engineering departments and eventually landed in a line superintendent's position.

Who inspired you to enter the industry?
A nice guy who took the time to answer an eager kid's questions. When I was about thirteen years old, my parents decided to build a new home for us. I have to say it was when I assisted—at least that's how I saw it—the electrician while he wired our new home. I decided that was the field I would enjoy as a career.

What do you enjoy most about your job?
It's especially rewarding to assist with storm restoration and have customers thank you for getting them back in power. And I like tutoring young linemen who are excited to learn the trade and who enjoy working outside in all the elements. I hope I can pass that sense of value on to the linemen I help train.

Do you think training and education are important in construction?
This kind of work can be dangerous. Without the proper training, safety would be more of an issue and many more workers could be exposed to situations that could cause harm to them or others. Also, it's important that we keep current with any new OSHA laws and in new technology and safe work practices.

How important are NCCER credentials to your career?
Being a part of the NCCER team brings me a very important credential. It's a nationally recognized institution with a wealth of experienced members and is well thought of in the construction industry. NCCER credentials would assist me in numerous companies around the U.S. should I ever relocate to another part of the country.

How has training/construction impacted your life and your career?
In my career, I've gained significant knowledge from training in many aspects of utility work. I now have an excellent understanding of how electricity is generated, transmitted, and utilized. I also know what it takes to get it safely to its destination and what equipment is necessary to do so.

I had the opportunity early in my career to be part of the entire process, from wiring homes to hooking up homes to the "grid," and then providing or producing the electricity to be consumed. This broad perspective had a huge impact on my understanding of the cycle.

A well-trained workforce is vital in today's complex industries, whether electrical or any other area of the construction industry.

Would you suggest construction as a career to others?
Very much so. There's so much opportunity in the construction industry that I feel it is a great career to be in. To keep this country moving forward, we must continue to attract young, intelligent workers so we will maintain our global lead in growth and infrastructure.

How do you define craftsmanship?
Professionally performed, high-quality, skilled work using hands, tools, and materials.

Trade Terms Introduced in This Module

Bonding: A means of interconnecting conductive metal objects that are not intended to carry current in order to protect people from electric shock. Bonding causes all the objects to be at the same potential.

Grounding: Connecting to ground or to a conductive body that extends the ground connection.

Kirchhoff's current law: The statement that the total amount of current flowing through a parallel circuit is equal to the sum of the amounts of current flowing through each current path.

Kirchhoff's voltage law: The statement that the sum of all the voltage drops in a circuit is equal to the source voltage of the circuit.

Parallel circuits: Circuits containing two or more parallel paths through which current can flow.

Series-parallel circuits: Circuits that contain both series and parallel current paths.

Additional Resources

This module presents thorough resources for task training. The following resource material is suggested for further study.

Electronics Fundamentals: Circuits, Devices, and Applications. 8th ed. Thomas L. Floyd. Upper Saddle River, NJ: Prentice Hall, 2009.

Principles of Electric Circuits: Electron Flow Version. 8th ed. Thomas L. Floyd. Upper Saddle River, NJ: Prentice Hall, 2006.

Figure Credits

Tony Vazquez, Module opener

Topaz Publications, Inc., SA01

Salisbury Electrical Safety, Figures 15 and 16

CONTREN® LEARNING SERIES — USER UPDATE

NCCER makes every effort to keep its textbooks up-to-date and free of technical errors. We appreciate your help in this process. If you find an error, a typographical mistake, or an inaccuracy in NCCER's Contren® materials, please fill out this form (or a photocopy), or complete the online form at www.nccer.org/olf. Be sure to include the exact module number, page number, a detailed description, and your recommended correction. Your input will be brought to the attention of the Authoring Team. Thank you for your assistance.

Instructors – If you have an idea for improving this textbook, or have found that additional materials were necessary to teach this module effectively, please let us know so that we may present your suggestions to the Authoring Team.

NCCER Product Development and Revision

3600 NW 43rd Street, Building G, Gainesville, FL 32606

Fax: 352-334-0932
Email: curriculum@nccer.org
Online: www.nccer.org/olf

☐ Trainee Guide ☐ AIG ☐ Exam ☐ PowerPoints Other ____________

Craft / Level: Copyright Date:

Module Number / Title:

Section Number(s):

Description:

Recommended Correction:

Your Name:

Address:

Email: Phone:

Climbing Wooden Poles

49105-11

Trainees with successful module completions may be eligible for credentialing through NCCER's National Registry. To learn more, go to **www.nccer.org** or contact us at **1.888.622.3720.** Our website has information on the latest product releases and training, as well as online versions of our *Cornerstone* newsletter and Pearson's Contren® product catalog.

Your feedback is welcome. You may email your comments to curriculum@nccer.org, send general comments and inquiries to **info@nccer.org,** or use the User Update form at the back of this module.

V.1 7/11

49105-11

Climbing Wooden Poles

Objectives

When you have completed this module, you will be able to do the following:

1. Identify all required and recommended safety equipment.
2. Demonstrate the knowledge and proper use of required climbing equipment.
3. Demonstrate the ability to inspect climbing equipment prior to climbing.
4. Identify the hazards associated with climbing wooden poles.
5. Demonstrate the ability to inspect a wooden pole for defects and hazards prior to climbing.
6. Identify and demonstrate proper climbing ascent, descent, and lateral positioning techniques.
7. Demonstrate the ability to safely climb over obstructions.
8. Demonstrate the ability to withstand working at heights above 32 feet.
9. Demonstrate the ability to perform pole-top rescue with and without the presence of a cross arm.

Performance Tasks

Under the supervision of the instructor, you should be able to do the following:

1. Demonstrate the ability to inspect climbing equipment prior to climbing.
2. Demonstrate the ability to inspect a wooden pole for defects and hazards prior to climbing.
3. Demonstrate proper climbing ascent, descent, and lateral positioning techniques.
4. Demonstrate the ability to safely climb over obstructions.
5. Demonstrate the ability to withstand working at heights above 32 feet.
6. Demonstrate the ability to perform pole-top rescue with and without the presence of a cross arm.

Trade Terms

Body belt
Carabiner
Cutout
D-ring
D-size
Fall arrest belt
Fall restraint belt
Gaff gauge
Gaff guard
Gaff protectors
Hitchhiking
Pole climbing gaffs
Pole-top rescue
Positioning lanyard
Rescue line
Snap ring
V-shaped stance

Contents

Figures and Tables

1.0.0 INTRODUCTION

If you enjoy being outdoors in all types of environments and conditions and are not afraid of heights, then working as a professional line or utility worker may be the right career for you. It can be rewarding and satisfying work, even exhilarating at times. As you read this module, you will become familiar with the equipment and techniques needed to safely climb wooden poles and perform the work of a professional line or utility worker (*Figure 1*).

> **WARNING!**
>
> Practice climbs should always be supervised and monitored by a skilled and experienced climbing instructor.

Utility workers are often required to climb tall poles and install, perform maintenance on, or repair components or the poles themselves. Many opportunities are available to those who have the ability to climb and work at significant heights. The constant demands placed on distribution systems require that they be continually maintained and expanded. Weather-related damage to poles and power lines also plays a significant role in the need for experienced climbers.

This module focuses on the equipment, procedures, and safe working habits involved in climbing. It also tests your mental and physical ability to withstand the stress of climbing and working in an elevated environment. Some of the most important material covered will concentrate on techniques intended to maximize your personal safety. The smallest lapse in judgment or attention can place you or others in a dangerous situation. For aspiring linemen and utility workers, there is no substitute for the proper training and preparation.

49105-11_F01.EPS

Figure 1 Apprentice training.

No job should be considered successfully completed unless it is done free of injury to the worker. Although significant risks are involved in climbing and working at heights, many professionals believe that the benefits far outweigh the risks.

2.0.0 SAFETY EQUIPMENT

OSHA is charged with the responsibility of ensuring that employers protect workers from workplace hazards. OSHA standards must be considered as the minimum requirements. OSHA provides a number of regulations that apply to power line workers. These regulations are primarily related to fall protection and electrical shock hazards. OSHA regulations associated with fall protection require that specific precautions be taken to protect employees who work at heights and are exposed to the potential of falling 4 feet or more. In such cases, the use of fall protection systems is required. The fall protection requirement related to climbing wooden poles can be satisfied through the use of either fall restraint or fall arrest equipment. Fall restraint equipment helps to prevent a fall from occurring but does not stop or arrest a fall in progress. Fall arrest equipment is much safer and adds increased protection by stopping and taking control of a fall in progress. Employers can impose additional conditions and policies, but OSHA regulations are the minimum requirements. Land or equipment owners may also create their own set of guidelines when work is being conducted on their property.

2.1.0 Additional Safety Gear

Besides needing fall protection equipment, climbers require other safety gear. This additional personal protective equipment, which must be used when required, includes the following:

- *First aid kits* – These must be readily available, although they are not carried by the climber in most cases. Vehicles on site should be equipped with at least minimal first aid gear, and workers should be properly trained in its use.
- *Head protection* – This is required by OSHA. A number of different styles are acceptable, and many are better suited to one or more applications or crew positions. Hard hats that are well-fitted and secure on the head are necessary for climbers. Many hard hats developed for rock climbing have been adapted to meet construction standards. Verify that your hard hat meets the required OSHA standards and has been tested to ANSI standards for the intended use.

- *Footwear* – Footwear must comply with the OSHA regulations and ANSI standards for steel or fiberglass safety toes. It must also have a full steel shank, or at minimum, a half-shank with non-slip soles. Footwear must cover the ankle and have a well-defined heel. An excellent fit is crucial because of the consistent stress placed on the feet during climbing. Bootlaces must be tucked in to avoid possible entanglement with equipment.
- *Hand protection* – Gloves are a necessity for climbing and working on wooden poles. Pole climbing presents many ways to experience a hand injury. Gloves should be made of leather, fit snugly, and allow the freedom of movement needed to do the job. A good tight fit helps to minimize any potential for the glove itself to be a safety hazard. Gloves with built-in knuckle protection are also available.
- *Eye protection* – Eye protection is as important in pole work as it is in any other trade work. An eye injury can be especially hazardous for a climber, who may then be unable to safely descend the pole. Eyewear must meet *ANSI Standard Z87.1*.
- *Long pants and long sleeves* – This type of clothing should always be worn, regardless of outdoor conditions. Long pants and long sleeves help to prevent skin irritations and other injuries from contact with the pole, belts, or ropes. They also help to protect workers from prolonged exposure to the sun and assist in minimizing hazards from insects. A good fit is essential to ensure that pants and shirts do not restrict movement. However, clothing should not be so loose as to become entangled in climbing gear.
- *Wrist straps* – These are recommended for use on all tools. Wrist straps help to lessen the inconvenience of dropping tools. They also prevent possible injuries to workers on the ground and damage to equipment.
- *Personal grooming* – Good grooming habits are essential for safety and comfort while working. Long hair should be pulled back and contained to avoid its being caught in the climbing equipment. Jewelry, necklaces, and rings can create safety hazards and should not be worn while working.

2.2.0 Additional Safety Concerns

Other safety concerns when climbing and working on poles include weather conditions, birds, animals, and insects. Climbers must be aware of these hazards and be prepared to effectively deal with them.

2.2.1 Weather

Weather conditions and the possible dangers they present must be considered before every climb. A clear day can quickly turn stormy. Temperature and wind conditions at pole-top can be different from those on the ground.

- *Rain and dew* – Moisture from rain and dew presents hazards to the climber from slippery surfaces. Wet surfaces cannot always be avoided because repair work often follows weather damage. Always make sure that the correct fall protection equipment is used for the existing conditions.
- *Snow and ice* – Whenever snow or ice is present, the work should not be considered safe. Keep in mind that when the temperature drops, a wet pole can quickly become icy as work is being performed. Always use the appropriate gripper-style fall protection equipment when a climb is unavoidable under these weather conditions. When heavy ice is present, watch for falling ice.
- *Wind* – Wind can create noise, frustration, and a sense of urgency, simply by its presence. Strong winds can create movement or swaying of the pole, which can be unsettling and distracting. Sudden strong winds can even separate a climber from the pole. Be aware that winds at climbing height can be much more severe than those at ground level. Good judgment based on experience must be used before the work is performed. Under windy conditions, it may be possible to perform simple tasks, while installation or securing of components may not be possible.
- *Lightning* – Lightning presents a deadly hazard to an unprotected climber at an elevated height. Approaching storms and lightning may be visible at a great distance before any thunder is heard. If thunder is heard, the climber should immediately descend the pole and take shelter. Lightning can strike up to 10 miles away from the storm. Be sure the storm is well past before attempting to resume any work on the pole.
- *Sun* – Exposure to the sun can cause discomfort and can even damage the skin. Glare from the sun can result in impaired vision or eyestrain. Climbers should always be equipped with good quality, impact-resistant sunglasses or dark safety glasses. Climbers must also protect themselves as much as possible from the sun's burning rays by using sunscreen and wearing the proper clothing. On hot days, climbers must have access to an ample supply of water to prevent dehydration and fatigue. Dehydration can cause a loss of agility and strength, which can be extremely dangerous to a climber.

2.2.2 Birds and Animals

Birds can be a general nuisance and distraction to climbers. If the pole is located in an area where birds gather or roost, heavy bird droppings may collect. Bird droppings are smelly and can create slippery surfaces for climbers. They can also carry a variety of infectious bacteria. Contact with droppings or other bird discharges should always be avoided. Special attention is needed if the birds fall into a protected migratory bird class.

Some animals, such as raccoons and bobcats, can and do climb poles, especially in secluded areas. Never attempt to climb the pole and remove a wild animal, no matter how small or cute it may appear to be from the ground. Contact a supervisor and always follow company guidelines for handling such situations.

Snakes can also present a hazard to the ground crew, especially in secluded, overgrown areas. While most snakes do not climb poles, birds of prey sometimes drop live snakes onto poles. The danger in the rare occurrence of encountering a snake on the pole is in the overreaction by the climber.

2.2.3 Insects

Insects, especially stinging insects such as wasps and bees, can present problems for climbers. Poles and installed equipment are common sites for the nests of paper wasps. Avoid the nests, if at all possible, or follow company guidelines for handling the situation. In the warmer months, always be on the alert for wasps and bees. Ticks are also a problem when working in brush or tall grass.

Climbers can also encounter spiders, although poisonous species are rarely found at heights. The danger lies in being startled or overreacting to the sight of the spider. The overreaction is usually far more serious than any damage the spider can do.

2.2.4 Nighttime Work

In the event of a power failure, crews may have to work after dark to restore power. Such work requires special care to avoid electrocution and other hazards. Your truck should be equipped with lighting implements that can be used by one of the crew while the other is up on the pole.

3.0.0 Climbing Equipment

Specialized equipment is required for climbing wooden poles. All equipment must be correctly sized to the climber and must be maintained and thoroughly inspected before each climb. Climbing equipment includes the following items:

- Correctly sized climbing gaffs and pads
- Gaff guards
- Gaff gauge
- Gaff protectors
- Correctly sized body belt
- Safety belts
- Positioning lanyards

3.1.0 Pole Climbing Gaffs

Pole climbing gaffs, also called climbers or hooks, are sturdy, narrow metal bars that extend down the lower leg, flare out slightly around the ankle, and form a stirrup that curves under the work boot, just in front of the heel (*Figure 2*).

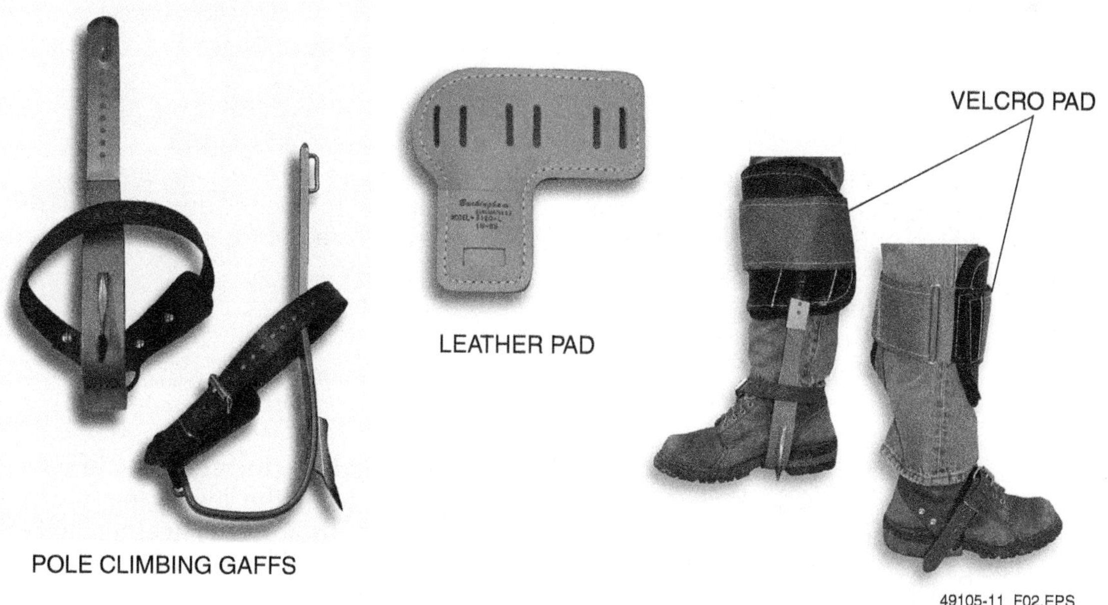

Figure 2 Pole climbing gaffs.

On Site

Climbing Irons

In an effort to design a more ergonomic climbing tool, climbing irons, such as those shown, have been created and are in use in some European countries. Climbing irons consist of a set of curved rods that have climbing teeth on the front and rear inside surfaces. Foot mounts allow the climber to climb and stand in a more natural, flat-footed stance. Ascending the pole is much like climbing stairs. Once the climber is at working height, the flat standing position is designed to be less tiring and stressful on the legs. Because the climber is in a more upright position, the fall restraint belt is shorter and is worn around the torso, rather than the hips. Climbing irons have not been fully assessed for practical use or for the versatility and safety required for common tasks, such as climbing over obstructions.

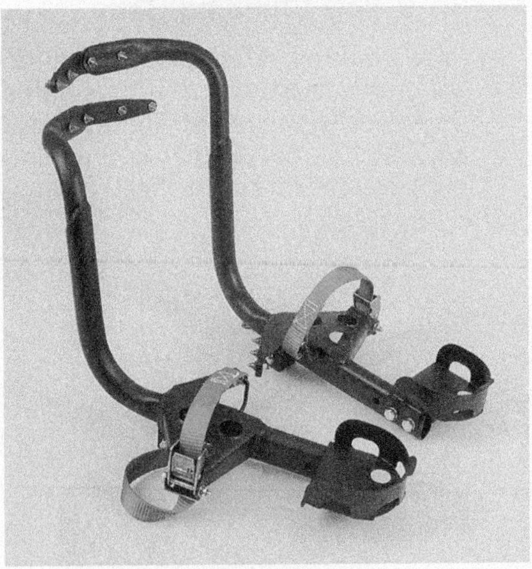

49105-11_SA01.EPS

Most climber shanks are approximately 1" wide and 3⁄8" thick. A strong, durable, V-shaped gaff (spike) is securely attached at the lower curve of the climber. The gaffs are configured into a sharp chisel point and are about 1¼" in length. Pads are attached to the top end of the climbers to protect the wearer's legs during use. Climbers are worn on the inside of the lower leg and are secured to the leg with a sturdy leather belt-style strap. Climbers are also secured at the lower stirrup end by a similar strap, which wraps around the boot and is securely fastened.

Correctly sized climbers extend from just below the knee downward and fit snugly under the boot. Climbers should always be securely strapped onto the legs and tightly strapped around the boots.

3.1.1 Gaff Guards

As the name implies, gaff guards (*Figure 3*) are used to safeguard the gaffs and to preserve their integrity and sharpness.

Whenever the pole climbing gaffs are not in use, the gaffs should be covered by gaff guards. This includes times when the climber is in transit, walking to or between poles. Gaff guards should only be removed when the climber is ready to mount the pole.

Gaff guards are available in various styles, from hard-cover clip-on types to leather guards with Velcro closures.

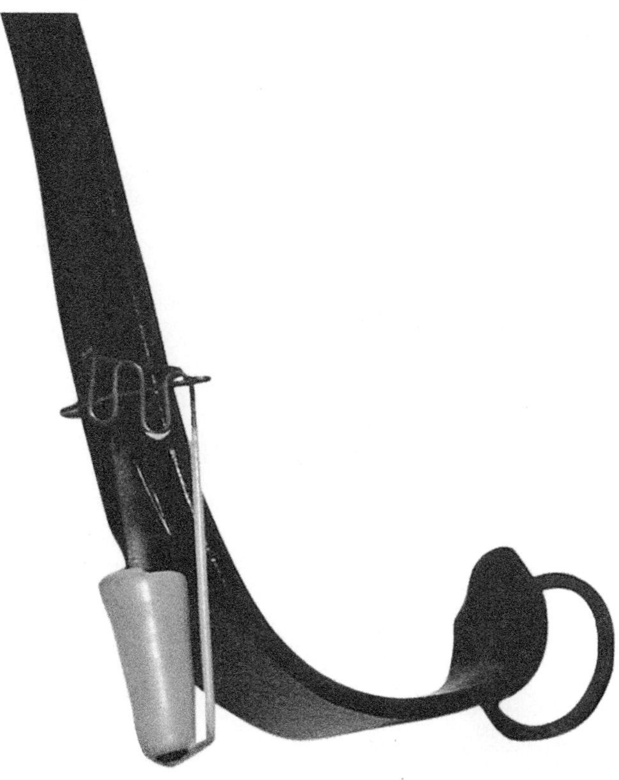

49105-11_F03.EPS

Figure 3 Gaff guards.

3.1.2 Gaff Gauge

The gaff gauge is a manufacturer-specific measuring tool for inspecting the condition of the climbers and gaffs (*Figure 4*).

The gaff gauge should be used at each pre-climb inspection to check for wear and damage to the climbers and gaffs. Several aspects of the gaffs and climbers must be checked, including the following:

- Gaff length
- Gaff thickness
- Gaff body, point shape, and angle
- Climber thickness
- Climber angle and stress twist

Gaffs must be sharpened and/or reshaped after a period of normal use. Some manufacturers recommend using a fine-tooth file, while others recommend using only a honing stone. In either case, always file or hone the gaffs lengthwise, toward the point.

CAUTION

Never file or hone the gaffs cross-laterally. This can create small grooves in the gaffs, forming the beginning of weak points that can later become stress fractures.

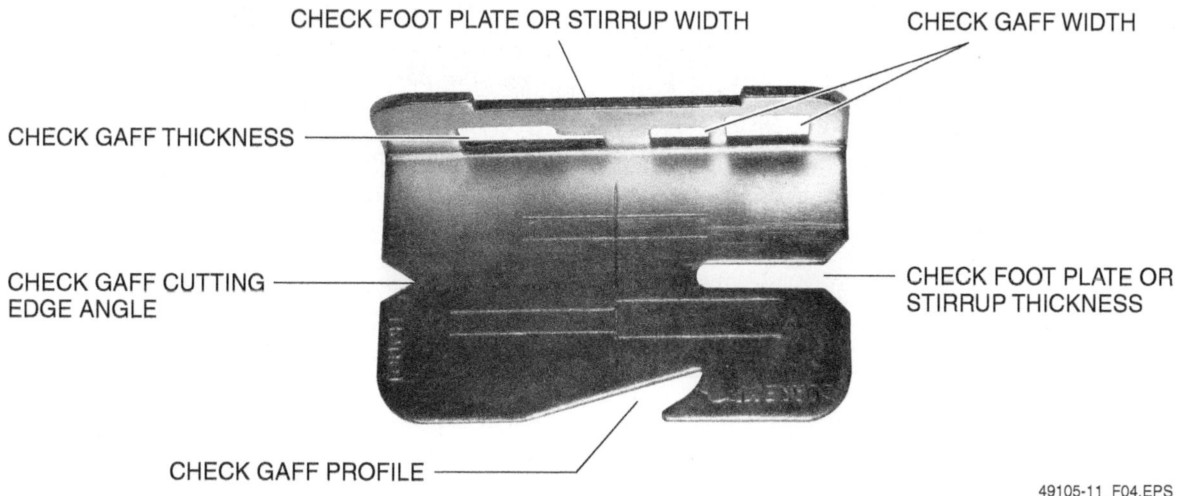

49105-11_F04.EPS

Figure 4 Gaff gauge.

3.1.3 Gaff Protectors

Gaff protectors provide protection from the sharp gaffs (*Figure 5*). They are attached to the lower climbers behind the gaffs. The gaff protectors cover the insides of the ankles and lower legs. They help to prevent the gaffs from causing painful punctures to those areas. It is highly recommended that all apprentice climbers use gaff protectors.

3.2.0 Body Belt

The body belt has two functions. It serves as both the anchor point for the fall restraint safety belt and as a utility belt for carrying all necessary tools. Body belts are available in various styles. Belts can be fixed, semi-floating, or full-floating. Padded and back-support belts are also available (*Figure 6*). A fixed belt is a solid leather belt. A semi-floating belt is made in two parts, which allows some lateral movement in the belt. A full-floating belt provides the most lateral movement for the wearer. It also helps to reduce pole-to-belt friction wear, which extends the life of the fall restraint belt.

3.2.1 Body Belt Sizing

Correctly sizing the body belt is critical, no matter which style of belt is selected. The belt is worn about 4 inches below the waist and across the hips. Always follow the manufacturer's specific instructions for correctly sizing the belt. The critical measurement is the distance between the heels of the D-rings, which is identified as the D-size (*Figure 7*).

The D-rings must be positioned just in front of the hip bones (approximately 1 inch). D-rings must never be placed directly on the hip bones. Some manufacturers instruct users to measure across the backside from hip bone to hip bone, and then add 2 inches for a standard belt or 4 inches for a padded belt. Whatever the case, always follow the manufacturer's specific sizing instructions. In addition, always wear your normal work clothes when measuring for the belt.

49105-11_F06.EPS

Figure 6 Body belt.

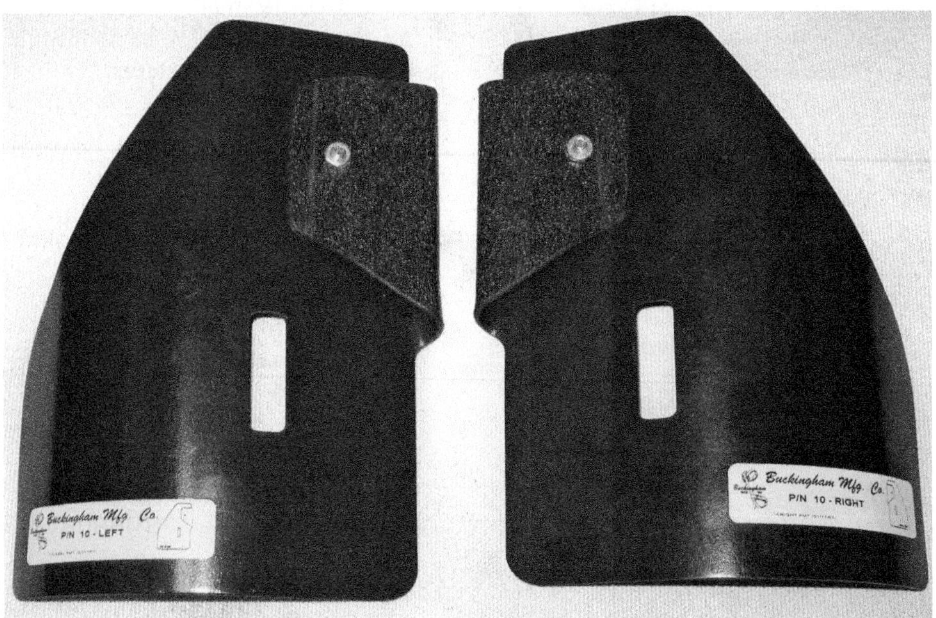

49105-11_F05.EPS

Figure 5 Gaff protectors.

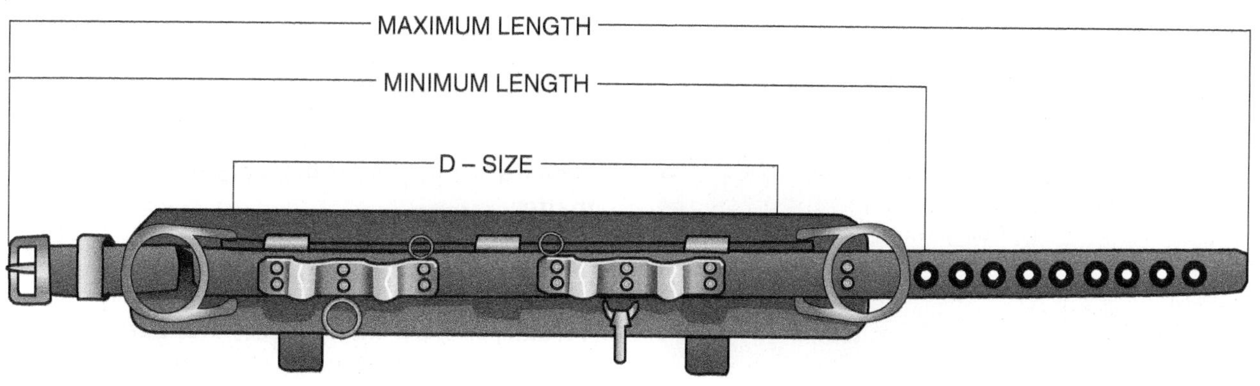

TYPICAL BELT SIZING CHART

D-SIZE	D18	D19	D20	D21	D22	D23	D24	D25	D26	D27	D28	D29	D30
MIN	32"	33"	34"	35"	36"	38"	40"	42"	44"	46"	47	48"	49"
MAX.	42"	43"	44"	45"	46"	48"	50"	52"	54"	56"	57"	58"	59"

49105-11_F07.EPS

Figure 7 Body belt sizing.

3.3.0 Safety Belts

Safety belts are classified as either fall restraint belts or fall arrest belts. Both types wrap around the pole and provide the necessary hands-free support for the climber performing tasks on the pole.

3.3.1 Fall Restraint Belts

The fall restraint belt functions only as a positioning support for the climber. It does not stop or arrest a fall in progress. Fall restraint belts are adjustable, wide leather straps that have snap rings attached at either end. The belt is wrapped around the pole, and the snap rings are connected to the D-rings of the body belt (*Figure 8*). The belt length is adjusted to maintain the climber at a working angle of approximately 30 degrees from the pole.

3.3.2 Fall Arrest Belts

Fall arrest belts are generally designed as two-piece, adjustable, wide nylon web belts with snap rings attached at either end (*Figure 9*). An outer belt wraps around the outside of the pole and an inner belt, attached to the outer belt, covers the inside surface of the pole. In this way, the pole is sandwiched between the two belts. Unlike the fall restraint belt, the fall arrest belt can stop a fall in progress and is the safest to use. When correctly worn and adjusted, the fall arrest belt tightly cinches around the pole when weight is suddenly applied during a fall. This cinching action stops the fall within a few inches.

49105-11_F08.EPS

Figure 8 Fall restraint belt.

Figure 9 Pole cinching fall arrest belt.

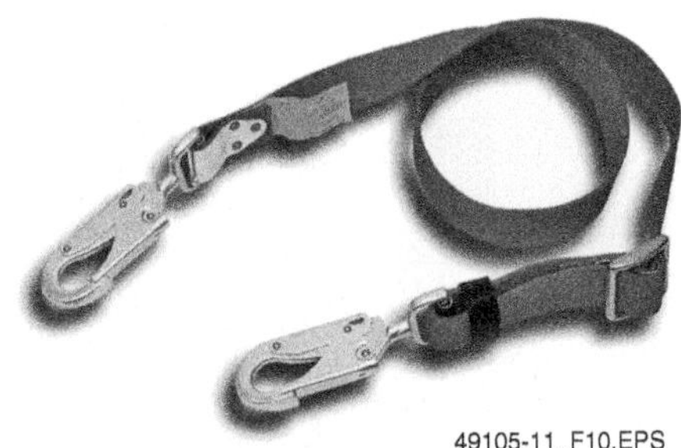

Figure 10 Positioning lanyards.

3.4.0 Positioning Lanyards

Positioning lanyards are used as secondary safety belts when climbing over obstructions (*Figure 10*). The lanyard provides support for the climber when the fall restraint or fall arrest belt is temporarily disconnected for repositioning above the obstruction. The positioning lanyards most commonly used are properly rated adjustable nylon ropes that have snap rings attached at either end. Web-style lanyards are also available. Their use is covered later in this module.

4.0.0 Pre-Climb Equipment Inspection

All personal climbing equipment must be thoroughly inspected for damage and wear before each climb. This includes climbing gaffs and all belts and lanyards. Do not climb with damaged or questionable equipment.

4.1.0 Climbing Gaffs

Climbing gaffs provide the foothold for the climber. The importance of their overall condition cannot be overstated. Always consider the pre-climb equipment inspection as a very serious matter.

4.1.1 Climbers and Gaffs

Visually inspect the climbers and gaffs (*Figure 11*) for any obvious signs of damage or wear. Thoroughly inspect all aspects of the climbers and gaffs with the gaff gauge. If any repairable wear is discovered, such as, dull, misshapen, or damaged gaffs, correctly sharpen or replace the damaged or out-of-tolerance gaff before climbing. It is a good practice to keep spare gaffs on hand for a quick change-out when necessary.

4.1.2 Leather Mounting Straps and Buckles

Carefully inspect the leather mounting straps, including the belt holes or grommets, for signs of cracking, wear, or deterioration. Check for any frayed or broken stitching in the leather straps. No additional holes may be punched in straps.

Check the leather strap's buckles for cracks, distortions, or rough edges. Buckle tongues must not be bent or distorted, must close and engage the buckle properly, and should move freely with no binding. If grommets are present, check for any looseness or stretching.

4.2.0 Belts and Lanyards

It is very important to thoroughly inspect all belts and lanyards before use. Check them for any signs of cracking, cuts, broken fibers or stitching, or other signs of wear or deterioration (*Figure 12*).

4.2.1 Web Belts or Lanyards

To inspect the webbing, hold the belt with your hands 6 inches apart, and bend the section of belt upwards. Check for damaged fibers or cuts in the bend. Using this method, inspect the full length of the belt on both sides. Also examine for frayed edges, broken or pulled stitching, a general lack of elasticity/flexibility, or other deterioration. No additional holes may be punched in any belt.

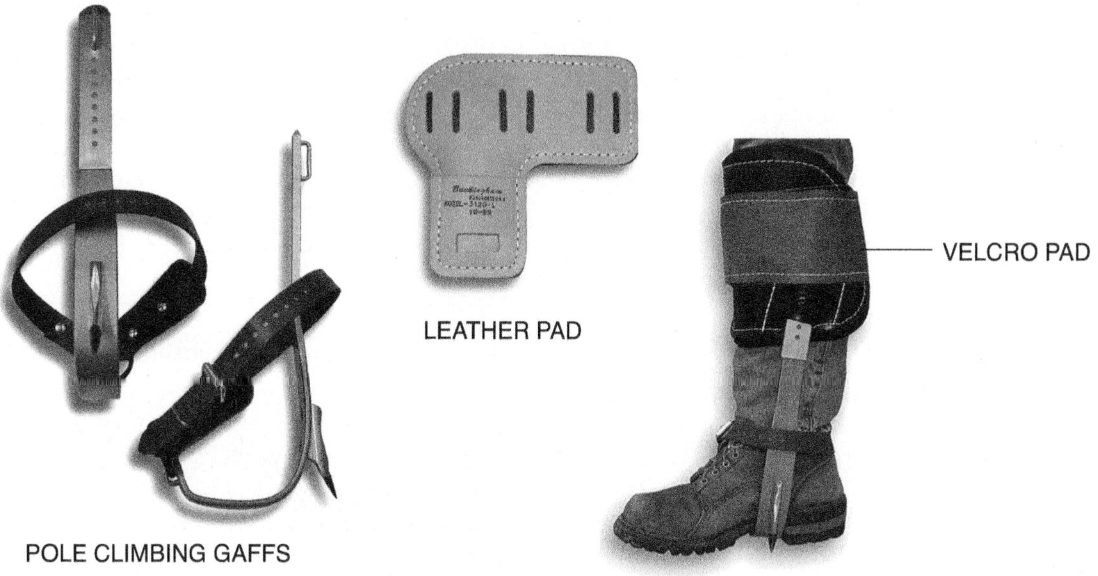

Figure 11 Pole climbing gaffs.

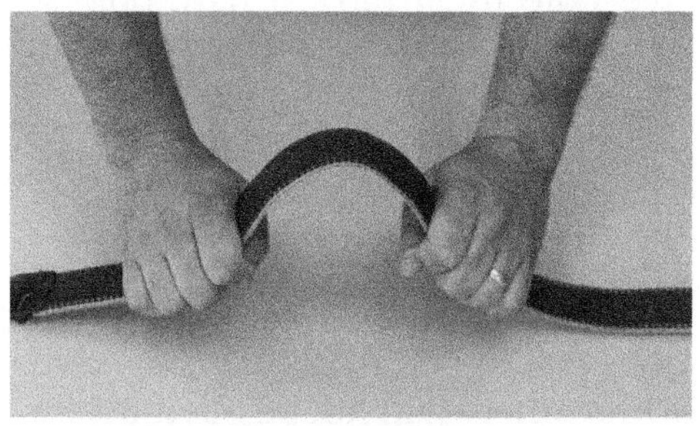

CHECK WEBBING

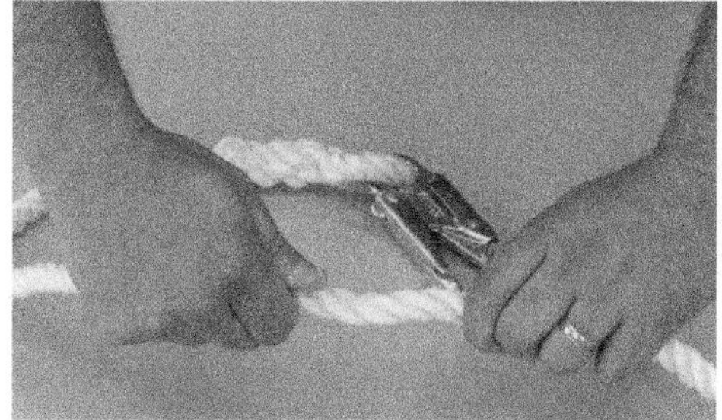

CHECK ALL ROPES

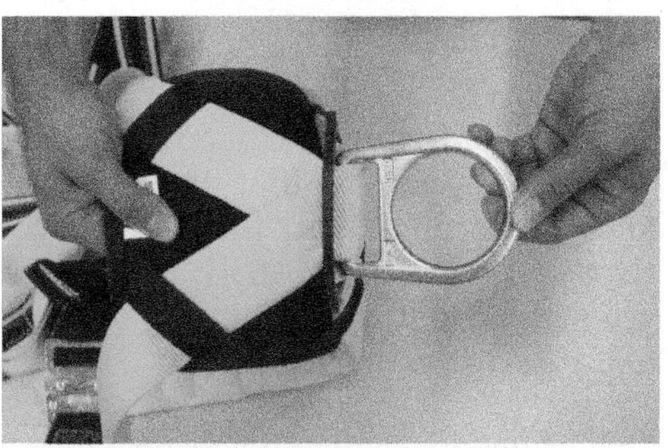

CHECK D-RINGS AND O-RINGS

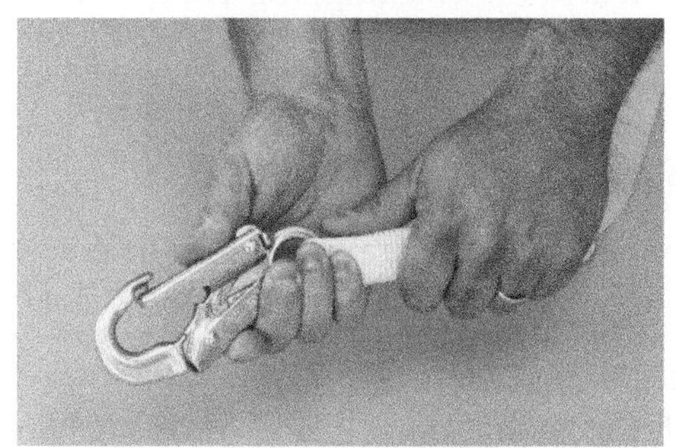

CHECK SNAP RINGS/CARABINERS

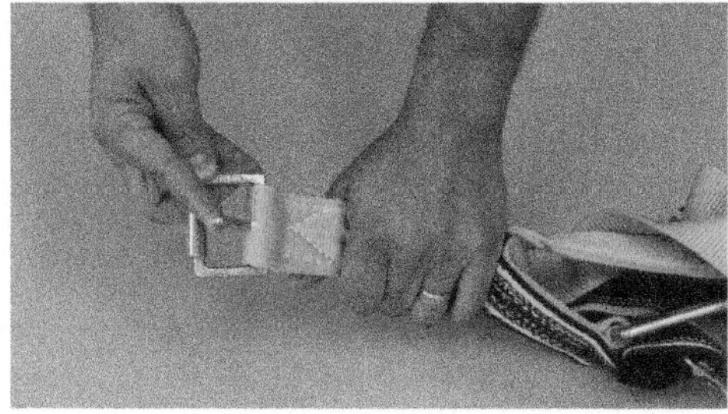

CHECK ALL BUCKLES

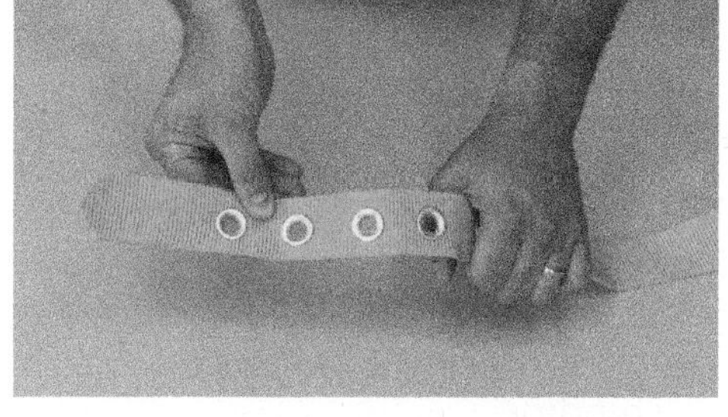

CHECK ALL BELTS AND GROMMETS

49105-11_F12.EPS

Figure 12 Belts and lanyards inspection.

WARNING! Always keep oil and petroleum products away from belts and lanyards. These substances cause deterioration. If aluminum hardware is used, note that a petroleum-based inhibitor is used to prevent oxidation. This substance must be removed before the belt or lanyard is used, as it causes deterioration.

4.2.2 Leather Fall Restraint Belts

Inspect the leather belt using the same method used with the webbing. Bend each section of belt along its full length on both sides. Check for cracks, friction wear, and broken or pulled stitching. Examine the belt's stitching and hardware at the snap ring ends for damage, wear, or looseness.

4.2.3 Rope Lanyards

Bend and rotate the length of the rope checking for damaged or cut fibers. Inspect the diameter along the full length of the rope. Any areas of reduced diameter could indicate stress or weakness due to excess loading. Check the rope's stitching and hardware at the snap ring ends for damage, wear, or looseness.

4.2.4 D-Rings and O-Rings

Check the D-rings or O-rings for any cracks, stretching, warping, twisting, or rough edges. Make sure that the D-rings pivot freely within the body belt.

4.2.5 Snap Rings and Carabiners

Inspect the snap rings or carabiners for any cracks, stretching, warping, twisting, or rough edges. The latch should seat without any binding or distortion. The latch spring must keep the latch firmly closed. Refer to the manufacturer's warnings and instructions for wear and tolerance limits. Do not try to repair damaged rings or carabiners. Replace them.

4.2.6 Buckles

Examine all belt and strap buckles for any cracks, distortions, or rough edges. Buckle tongues must not be bent or distorted and must close and engage the buckle properly. Buckle tongues should move freely with no binding.

5.0.0 Checking the Pole Condition Before Climbing

Before climbing and performing any work on a wooden pole, the physical condition and soundness of the pole must be determined. Defects in or deterioration of the pole can cause it to fail during stresses applied to it during climbing, repairs, or component installation.

WARNING! If a pole is deemed unsafe after inspecting, do not climb the pole. Also, never assume that someone such as a supervisor or engineer has inspected the pole properly. Do it yourself to be sure.

In cases of necessity, a moderately defective pole can sometimes be secured temporarily by a boom truck or additional guys or ropes while the work is being performed. A severely defective pole must have a new pole set and lashed alongside the defective one; all work should be performed from the new pole.

5.1.0 Inspecting the Pole

Testing the condition of the pole is done by both visual and physical inspection. Certain external defects will be obvious during visual inspection, while others will not be obvious.

5.1.1 Visual Inspection

Visually check the pole for any signs of damage, cracks, leaning, bending, or splintering. These signs can indicate that the pole is rotted, broken, or damaged. In the case of vertical cracks, they may not necessarily mean that the pole is unsound (*Figure 13*).

NOTE Although vertical cracks do not necessarily indicate that the pole is unsound, they do present a hazard for gaffs. Always avoid the cracks when climbing and setting the gaffs.

During the visual inspection, check the pole and surrounding area for the following signs:

- Inspect for hollow areas or holes, such as those made by woodpeckers. Any open holes may indicate hollow spots and areas of weakness.

49105-11_F13.EPS

Figure 13 Vertical crack.

- Check for any areas of external rotting or deterioration. These can indicate a defective internal condition of the pole.
- Examine the pole for areas of knots. If these are present, look for cracking or deterioration around the knots, as this can indicate a weak point.
- Look for the birthmark that is often stamped or branded on the pole at the time of manufacture. The birthmark is normally found at about eye level on a set pole. If the mark is much higher, and if signs indicate that the previous ground line was substantially higher than the current ground level, the pole may no longer be buried deep enough for a safe climb.
- Check for extremely soft or wet soil resulting from weather or other causes. These conditions may temporarily make a pole unstable and unsafe to climb.
- Look for burn marks from brush fires around installed components or at the base of the pole. They can indicate a weakness in the pole. Thoroughly check the soundness of the pole before climbing.
- Be on the alert for any bent or leaning poles, which may still be considered safe to climb after inspection. If this is the case, only climb the pole on the high side. Severely leaning poles may require bracing prior to climbing.
- Check for signs that a pole inspection service was used to inspect the pole. If such a service was employed, date tags will be affixed to the pole. If the pole is deemed unsafe, Do Not Climb or No Climb tags will be attached. Familiarize yourself with these tags so that they will be easily recognizable.

5.1.2 Physical Inspection

Several common methods are used to physically check the condition of the pole. These procedures should always be followed before climbing any pole:

- Use about a three-pound hammer to sharply rap the pole from the ground line up to head height all the way around the pole. Solid wood will produce a sharp, clear sound, and the hammer will quickly rebound or bounce back. Areas of internal decay will produce a dull sound, and the hammer will not bounce back as quickly.
- If decay or rot is seen at ground line, probe the pole with a screwdriver that is at least 6 inches long to determine the depth of the decay. If the soundness of the pole's base is suspect, dig down around the pole to a depth of about a foot, and then probe the pole with the screwdriver and a hammer. When substantial decay is present, consider the pole unsafe to climb. Contact a supervisor about the pole's condition.
- If the stability of a pole is suspect, attempt to rock it back and forth in line with the power lines. This can be done by hand, by pushing with a pry pole, or by pulling the pole with a rope. Never rock a pole laterally to the power lines. This can cause the lines to sway and possibly contact each other.

6.0.0 Climbing the Pole

After all climbing equipment has been inspected and deemed safe for climbing and the pole has been checked for soundness, the climber is ready to climb the pole. The steps to take when climbing the pole are as follows:

- Put on the climbing gaffs
- Ascend the pole
- Maneuver laterally on the pole
- Climb over obstructions
- Descend the pole

6.1.0 Putting on the Climbing Gaffs

Position the climbers on the inside of the lower legs with the stirrup under the boot, just in front of the heel. Place the top of the climbers two fingers below the knee at the top of the shin bone. Fasten the top strap and pad around the leg as tightly as possible. The straps will loosen and stretch to some extent while climbing, especially on new climbers. Wrap the lower strap around the back of the boot, over the top of the Achilles tendon, and fasten the buckle as tightly as possible. Before moving to the pole, make sure that the gaff guards are attached and cover the gaffs. Remove the gaff guards when you have approached the pole and are ready to climb.

> **CAUTION**
>
> Gaff guards should always be attached when walking to the pole, walking between poles, and anytime the climbers are not in use. They should only be removed before beginning the climb.

6.2.0 Ascending the Pole

After a thorough inspection of all climbing equipment and the condition of the pole, the climber is ready to ascend the pole. Before mounting the pole, the climber should identify the best climbing route to avoid obstructions, if possible, and any hazards in reaching the climbing destination.

Hitchhiking is the safest way to climb the pole and is the method recommended or required by most companies. Hitchhiking consists of connecting the fall restraint/fall arrest belt around the pole before mounting the pole. The steps used in the hitchhiking method are as follows:

Step 1 Pass the fall restraint/fall arrest belt around the pole to the right hand. With the right hand, connect the snap ring to the D-ring on the right side of the body belt. Visually and physically verify that the snap ring is fully connected by looking at it and tugging on it.

Step 2 Position the fall restraint/fall arrest belt at chest height. Firmly grasp both sides of the belt with gloved hands close to the pole. The grasp on the belt produces a choking action around the pole and provides the necessary support. As a result, it is not necessary to grasp the pole with the hands.

Step 3 While grasping the belt, raise the right foot 6 to 8 inches, and smoothly step down placing the gaff into the pole while maintaining a wide, V-shaped stance with the feet. In a V-shaped stance, one foot will be positioned to one side of the pole and the opposite foot on the opposite side of the pole to form the shape of a V.

> **NOTE**
>
> Keep in mind that pressure-treated (green) poles may be hard to penetrate with the gaffs. It may also be harder to remove the gaffs.

Step 4 Lift the weight onto the right leg, extend the leg straight, lock the knee, and keep the heel down and the toe up. Do not lift yourself with your arms or they will quickly tire. Use your legs to lift your weight. The arms are used mainly for balance.

Step 5 Slightly release the tension on the belt, and position it again at chest height. Firmly grasp both sides of the belt with gloved hands close to the pole. While grasping the belt, raise the left foot 6 to 8 inches, and smoothly step down placing the gaff into the pole while maintaining a wide, V-shaped stance.

Step 6 Lift the weight onto the left leg, extend the leg straight, lock the knee, and keep the heel down and the toe up. This position must be maintained to avoid a dangerous condition called a cutout (*Figure 14*).

> **WARNING!**
>
> Because the steel shank boot is rigid and does not flex, the center of the boot, or gaff area, becomes a pivot point. For this reason, if the toe rotates down, the heel and gaff will automatically rotate up, dislodging the gaff from the pole or causing the gaff to unexpectedly split out of the pole. This is known as a cutout. Cutouts are the number one cause of fall accidents and injuries. Climbers must always maintain the heel down, toe up, knee locked position on the weight-bearing leg.

Step 7 Continue the alternating right leg/left leg climb using the fall restraint belt for support. Always look up ahead to verify your climbing route as you ascend the pole. Position and maintain the belt at chest height and continue the climb until the working height has been reached. Once at the working height, set both feet at the same level in the V-shaped position with both knees locked.

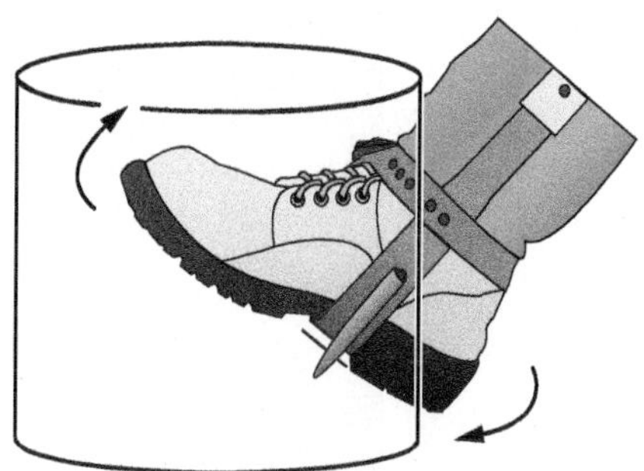

Figure 14 Toes down, you go down!

The fall restraint belt provides the necessary three points of contact and leaves the hands free to perform work. The belt's length must always support the climber at about a 30-degree angle to the pole (*Figure 15*).

6.3.0 Maneuvering Laterally on the Pole

While performing tasks on the pole, the climber must have the ability to safely move laterally, left or right. Adequate practice under the supervision of a skilled instructor is the best way to gain the confidence and skill necessary for such maneuvers. All initial practice by apprentice climbers should be conducted only a few feet from the ground.

Once correctly in position on the pole, the climber moves to the right using the following steps:

Step 1 Place the weight onto the left leg, lock the knee, position the foot in the V-shaped stance, keep the heel down, and shift the hips to the right.

Step 2 While grasping the pole with the right hand, lift the right foot about 4 inches. Extend the foot to the right about 6 to 8 inches, and step down placing the right gaff into the pole, always maintaining the heel down, V-shaped stance.

Figure 15 Correct working angle.

Step 3 Move the fall restraint belt to around chest height, then lift up and straighten the right leg into the locked position.

Step 4 Lift the left foot, removing the gaff from the pole. Position the left foot a few inches from the right heel, and step down placing the left gaff back into the pole. This should move you about one-fourth of the distance around the pole. Veteran climbers can usually get around the pole in about three such moves.

Step 5 After reaching the desired location, step back down the pole a few inches with the right foot. Place the right gaff into the pole and lock the knee, maintaining the V-shaped stance.

Step 6 Lift the left foot and step down even with the right foot, keeping heels about 6 inches apart. Place the left gaff into the pole and lock the knee, maintaining the V-shaped stance, with both heels down and toes up. Position the fall restraint belt at about chest height. The belt and feet maintain the necessary three points of contact, freeing the hands to perform the task.

Step 7 A move to the left is performed in the same manner with these exceptions: the hips are shifted to the left, the pole is grasped with the left hand, and the left gaff is moved and resettled first.

6.4.0 Climbing Over Obstructions

Climbers must be able to safely climb above installed equipment or other obstructions in the climbing path. A properly rated positioning lanyard is used as a temporary fall restraint belt for this task, which is described as follows:

- When an obstruction is encountered, ascend as close as possible to it using the fall restraint belt.
- With the left hand, release the snap ring at one end of the positioning lanyard. Lean forward at the waist, and then pass the lanyard around the pole and above the obstruction, into the right hand (*Figure 16*).
- With the right hand, connect the snap ring to the D-ring on the right side of the body belt. Verify the connection visually and physically. Adjust the length of the lanyard as needed to maintain the proper 30-degree angle.
- Using the lanyard as a temporary fall restraint device, disconnect the fall restraint belt's snap ring at the right side of the body belt (*Figure 17*).
- Pass the fall restraint belt around the pole to the left hand, and connect the snap ring to the D-ring at the left side of the body belt.

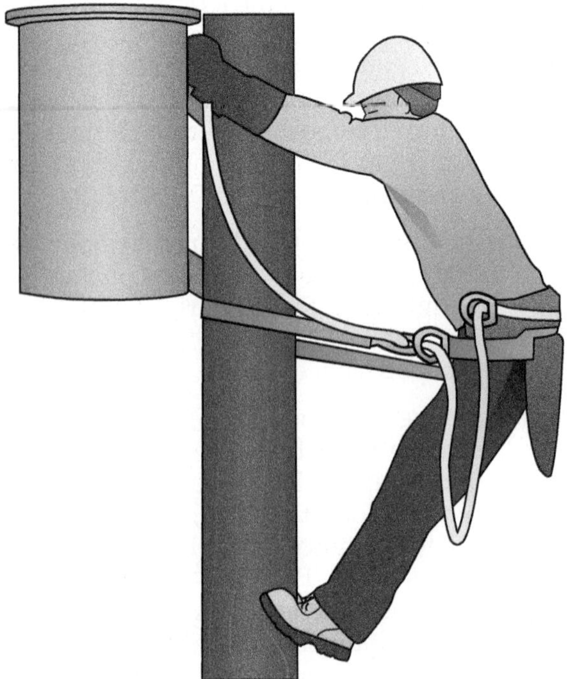

Figure 16 Connecting the lanyard.

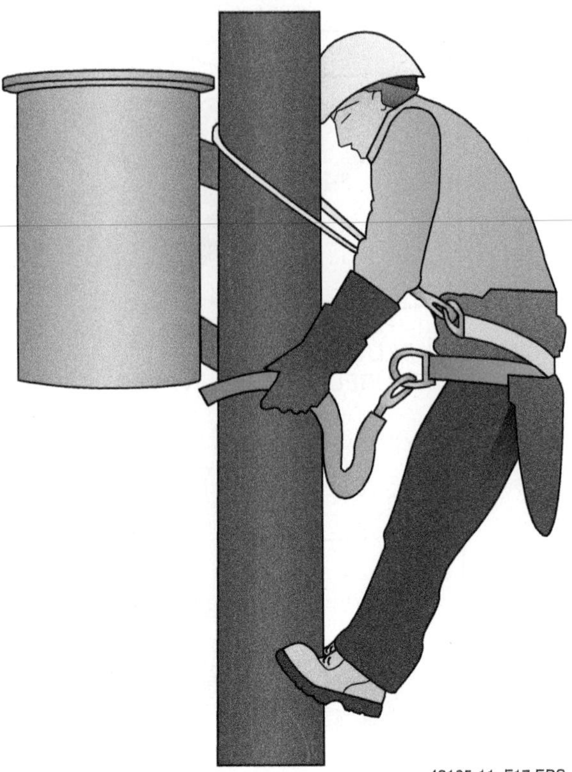

Figure 17 Releasing the fall restraint belt.

- Climb high enough to allow access above the obstruction. With the left hand, disconnect the front fall restraint belt's snap ring (the one closest to the body belt buckle) from the D-ring at the left side of the body belt. Pass it around the pole, above the positioning lanyard, to the right hand.
- Reconnect the fall restraint belt's snap ring to the D-ring on the right side of the body belt. Visually and physically verify the connection (*Figure 18*).
- With the right hand, disconnect the positioning lanyard's snap ring at the right side of the body belt.
- Pass the lanyard around the pole to the left hand, and connect the snap ring back into the D-ring on the left side of the body belt. Both lanyard snap rings are now connected to the same D-ring on the left side of the body belt.
- Continue the climb as usual.

6.5.0 Descending the Pole

The descent should be accomplished with a smooth, steady rhythm, similar to the ascent. The following are guidelines to follow when descending the pole:

- Avoid descending the pole too quickly. Moving too fast can cause the attached tools and belts to swing wildly back and forth, which can disrupt the climber's balance.

49105-11_F18.EPS

Figure 18 Reconnecting the fall restraint belt.

- Look down during the descent, and check the placement of the gaffs as you step down. Be aware of and avoid any possible cutout hazards on the pole as you place the gaffs.
- When descending the pole, release the gaff, straighten the leg with the knee locked, and drop the gaff into the pole about one foot down. You do not have to jam the gaff into the pole. Your body weight will sufficiently set the gaff on the straightened leg. After some practice and experience, you will be able to take a longer stride on descent.
- Always take your time in descending until you are comfortable with performing the moves. Keep in mind that no job is successfully completed until you are safely back on the ground again.

6.6.0 Pole Steps

While not typically used on publicly accessible poles, pole steps may be installed on poles in enclosed or guarded areas. Metal-rod steps are most common, although flat-surfaced, removable steps are also used. Climbing pole steps is similar to climbing a ladder. Even though climbing gaffs are not worn, a fall restraint belt is still required for climbing pole steps. Climbers must exercise caution and remain alert for wet or slippery step conditions. They must check the soundness of each step before applying full weight as a hand hold or foot hold. Steps are not always installed at even distances from each other. The climber must take time in ascending, and especially in descending the pole, in order to verify that the foot is in firm contact with each step before applying full body weight.

7.0.0 Pole-Top Rescue

All climbers must be trained in and have the ability to perform the pole-top rescue of an injured worker. Should this situation arise, an injured, incapacitated, or unconscious worker must be secured and safely lowered to the ground. The climber performing the rescue must also be trained in and able to administer mouth-to-mouth resuscitation and CPR. Training is conducted using a life-size mannequin that approximates the size and weight of an unconscious worker. Students learning this procedure often practice with a dummy (*Figure 19*).

7.1.0 Rescue Equipment

All line trucks should be equipped with a dedicated rescue kit. The contents of the kit can vary by company, but the universal item is the rescue line, which is a properly rated rope at least ½ inch in diameter. The rope must be twice the length of the climbing height plus 10 feet. It is used to secure the victim and lower the individual to the ground.

49105-11_F19.EPS

Figure 19 Pole-top rescue.

7.1.1 Performing a Rescue When a Cross Arm or Substantial Anchor Is Available

Most poles that are climbed and worked on have cross arms or other substantial anchor points available. In the case of a pole-top rescue, such a structure is used as an anchor point for the rescue rope. The following is a basic procedure for a pole-top rescue:

Step 1 Call to the victim. If there is no response from the victim, follow your company's rescue procedures. Suit-up and ascend the pole.

Step 2 If a rescue line (hand line) is not on the pole, carry one with you on the ascent. Remove the line out of the block and throw the short end over the arm twice. Be sure that the line is not overlapped. Pull enough slack, 5 to 6 feet below the victim's chest, to tie around the victim's chest.

Step 3 Route the short end of the rope around the victim's back and under the armpits. Tie two half hitch knots and slide the knots to the victim's armpit nearest to the rescuer. Snug the knot in the pectoral area under the victim's armpits and around the chest.

Step 4 Using one hand, remove the excess slack from the fall line. Do not wrap the line around the hand.

Step 5 Place your hip against the victim to prevent him or her from swinging.

Step 6 While gripping tightly on the fall line, cut the victim's safety belt. After cutting the belt, smoothly control drop the victim. It may be necessary to use a hand-over-hand technique.

Step 7 Once the victim is safely on the ground, descend the pole and start CPR or first aid as needed until medical personnel arrive.

You must be able to secure and lower the victim to the ground in five minutes or less. The average rescue time for experienced, veteran line workers is about three minutes.

7.1.2 Performing a Rescue When No Cross Arm Is Available

When there is no cross arm or other anchor point on the pole, a temporary anchor point must be established to secure the rescue rope and victim. Though rescue kits can vary by company, it is common for a small pulley to be included. The pulley is driven into the pole above the victim and serves as the anchor point for the rescue rope. For this type of rescue, follow the previous steps using the pulley rather than the cross arm as the anchor point. The use of a pulley is a typical example; tools and kits vary by company. Always follow company directives, and use the tools/kits provided.

Step 1 Call to the victim. If there is no response from the victim, follow your company's hurt man rescue procedures. Suit-up and ascend the pole.

Step 2 Throw the rope over any substantial hardware (such as a bolt or clevis) that will not cut the rope. If no hardware is present, drive an 8- to 10-inch shank screwdriver into the pole at a 45 degree angle to use as an anchor point.

Step 3 Follow Steps 3 through 7 as previously described for a cross arm rescue.

Summary

Power line workers are often called out to replace damaged utility poles or perform repairs such as replacement of conductors, transformers, and components on utility poles. Therefore, the ability to safely climb a pole and to work safely and comfortably while on a pole are important skills. Safety gear and safe climbing practices are a primary concern when working at heights and in proximity to active power lines and transformers. The climber must also be aware of hazards associated with weather conditions, as well as birds and animals.

Climbing gear, including gaffs, safety belts, and positioning lanyards, must be used. This gear must be properly maintained and must be inspected before each use. Poles can be damaged or can deteriorate, so it is important to perform a thorough inspection of every pole before climbing it.

Proper climbing technique is learned through training and practice. The climber must learn how to properly ascend and descend a pole. An additional skill that every line worker must learn to perform is a pole-top rescue of a worker who has been injured or has fallen ill while working on the pole.

Review Questions

1. Footwear and head protection worn by climbers must meet _____.
   a. company requirements
   b. ANSI standards
   c. the manufacturer's recommendations
   d. the climber's personal preference

2. The tips of pole climbing gaffs are shaped as _____.
   a. spear points
   b. barbs
   c. sharp chisel points
   d. conical spikes

3. The straps that secure the climbing gaffs to the legs are normally made of _____.
   a. nylon
   b. nylon/rayon web
   c. leather
   d. Velcro

4. Gaff guards protect the _____.
   a. gaffs
   b. other climbers
   c. boots
   d. poles

5. Filing or honing gaffs cross-laterally can cause the formation of dangerous _____.
   a. curves
   b. pits
   c. grooves
   d. rough edges

6. Gaff protectors protect the climber's legs from _____.
   a. scrapes
   b. blisters
   c. puncture wounds
   d. sun and insects

7. Correctly sizing the body belt is only critical for sizing certain styles of body belts.
   a. True
   b. False

8. Body belt D-rings must never be placed _____.
   a. above the waist
   b. below the waist
   c. directly over the spine
   d. directly over the hip bone

9. A fall arrest belt functions as a _____.
   a. harness
   b. shock absorbing lanyard
   c. pole cinching system
   d. rappelling system

10. In some circumstances, it is permissible to climb with damaged or questionable equipment.
    a. True
    b. False

11. The condition of a pole's core can be checked with a _____.
    a. ¾-inch drill bit
    b. 3-pound hammer
    c. pry bar
    d. hydrostatic probe

12. When hitchhiking up the pole, the arms are used mainly for _____.
    a. lifting your weight
    b. resting
    c. balance
    d. carrying tools

13. When ascending a pole, the fall restraint/fall arrest belt should be positioned _____.
    a. at eye level
    b. at chest height
    c. at waist level
    d. approximately 2 inches below waist level

14. When ascending a pole and attempting to move your weight up, how should the heel and toe be positioned?
    a. Heel up and toe down
    b. Heel down and toe up
    c. Heel up and toe up
    d. Heel down and toe down

15. When moving laterally left on the pole, what is moved first?
    a. Your hips
    b. Your left hand
    c. Your right hand
    d. Your left gaff

Trade Terms Quiz

Fill in the blank with the correct term that you learned from your study of this module.

1. A manufacturer-specific measuring tool used for inspecting the condition of the climbers and gaffs is a(n) ____________.
2. A belt that functions as the anchor point for the fall restraint safety belt and as a utility belt for carrying all necessary tools is a(n) ____________.
3. A connecting ring for belts and straps, which was first designed and used for mountain climbing, is a(n) ____________.
4. A device that safeguards the gaffs and preserves their integrity and sharpness is a(n) ____________.
5. The term commonly taught to apprentice climbers in which their heels are close together and their toes turned outwards is the ____________.
6. An unexpected and unintentional dislodging of a gaff from the pole is ____________.
7. A belt that only functions as a support for the climber and will not stop or arrest a fall in progress is a(n) ____________.
8. A device used as a secondary safety belt when climbing over obstructions is a(n) ____________.
9. Sturdy, narrow metal bars that extend down the lower leg, flare out slightly around the ankle, and form a stirrup that curves under the work boot just in front of the heel are ____________.
10. The metal connecting ring used as the anchor point for straps and lanyards is the ____________.
11. The rescue of an injured, incapacitated, or unconscious worker, who must be secured and safely lowered to the ground, is a(n) ____________.
12. A connecting ring used to secure belts and lanyards is a(n) ____________.
13. The distance between the heels of the D-rings on a body belt is the ____________.
14. A properly rated rope of at least ½ inch in diameter that is used to secure and lower an injured worker during a pole-top rescue is a(n) ____________.
15. Equipment that protects the climber's ankles and lower legs from sharp gaffs is ____________.
16. A belt that provides support for the climber and that will stop a fall in progress is a(n) ____________.
17. The term for ascending the pole with the fall restraint belt connected is ____________.

Trade Terms

Body belt
Carabiner
Cutout
D-ring
D-size
Fall arrest belt
Fall restraint belt
Gaff gauge
Gaff guard
Gaff protectors
Hitchhiking
Pole climbing gaffs
Pole-top rescue
Positioning lanyard
Rescue line
Snap ring
V-shaped stance

Cornerstone of Craftsmanship

James (Shane) Smith

Instructor
Oneonta Job Corps

Shane Smith has been in the electrical industry for over 35 years. He started out as an electrician's helper and took advantage of the training and opportunities available and is now a lineman instructor. He takes pride in helping young people grow and learn the trade.

How did you get started in the construction industry?
I got my first job 35 years ago as a laborer on an electrical job by bugging the foreman until he gave me a job. There is something about building things; for me, it's much more than a paycheck. You finish a job and walk away and people use it. You can point to it and say I built that 15 years ago. There is personal pride and craftsmanship. At the end of the day, you have something to show for your work.

I started as a laborer and worked my way up. About 10 years ago, Job Corps wanted to cross-train electricians on overhead work. I trained with New York State Electric and Gas and then came here and developed our training program.

What do you enjoy most about your job?
I enjoy watching young people "get it" as they develop and grow in their skills. This is a fairly physical and technical trade. I see people struggling at the beginning of the program. By the end of the program, they can do anything you ask them to. I take pride in watching them grow.

Do you think training and education are important in construction?
Oh, yes, continuing education is critical. The technology, the materials, safety requirements, and procedures change frequently. It is a constant learning process to keep current in the trade. If you are not taking the course, you are teaching the course.

How important do you think NCCER credentials will be in the future?
It is becoming recognized as a standard in the industry. It puts you on par with other people who have been trained and ahead of those who haven't. The hiring person knows exactly what level of training you have. You become a known quantity, while they may be unsure of the other person's skill level.

How has training/construction impacted your life and your career?
It is my career. I've been at it for 35 years. I started as a laborer and became a residential electrician, and then a residential and commercial electrician. I started teaching residential and commercial electricians and now my classroom is 45 feet in the air. It's the same field but different procedures. That's part of learning and growing.

Would you suggest construction as a career to others?
Definitely, especially as an electrician. No matter what happens in the economy, people will always need electricians. As we become more and more dependent on our computers and other electronic tools, we need a dependable, consistent supply of electricity.

How do you define craftsmanship?
Pride in a job well done. At the bottom of every contract it says "All work will be performed in a workmanlike manner, in accordance with standard procedures, time being of the essence." That pretty much sums it up. You show up on time, you work 8 hours. As a tradesman, you do the work as it should be done, the way a professional does it. I get paid to do what I know how to do.

Linemen follow a strict set of safety protocols and procedures. The company procedures may vary, but the standards remain the same. You don't come off the pole till the lights are on.

Trade Terms Introduced in This Module

Body belt: A belt that functions as the anchor point for the fall restraint safety belt and as a utility belt for carrying all necessary tools.

Carabiner: A connecting ring for belts and straps, which was first designed and used for mountain climbing.

Cutout: An unexpected and unintentional dislodging of a gaff from the pole.

D-ring: The metal connecting ring used as the anchor point for straps and lanyards. On the body belt, the fall restraint strap is connected to the D-rings.

D-size: The distance between the heels of the D-rings on a body belt. D-size is the critical measurement for correctly sizing body belts.

Fall arrest belt: A belt that provides support for the climber and that will stop a fall in progress.

Fall restraint belt: A belt that only functions as a support for the climber; it will not stop or arrest a fall in progress.

Gaff gauge: A manufacturer-specific measuring tool used for inspecting the condition of the climbers and gaffs.

Gaff guard: A device that safeguards the gaffs and preserves their integrity and sharpness.

Gaff protectors: Equipment that protects the climber's ankles and lower legs from sharp gaffs.

Hitchhiking: The term for ascending the pole with the fall restraint belt connected.

Pole climbing gaffs: Pole climbing gaffs, also called climbers, are sturdy, narrow metal bars that extend down the lower leg, flare out slightly around the ankle, and form a stirrup that curves under the work boot just in front of the heel. A strong and durable V-shaped gaff (spike) that is securely attached at the lower curve of the device serves as the foot hold for the climber.

Pole-top rescue: The rescue of an injured, incapacitated, or unconscious worker, who must be secured and safely lowered to the ground. All climbers must be properly trained and able to perform the pole-top rescue of an injured worker.

Positioning lanyard: A device used as a secondary safety belt when climbing over obstructions.

Rescue line: A properly rated rope of at least ½ inch in diameter that is used to secure and lower an injured worker during a pole-top rescue. The line, which is an important part of the rescue kit, must be twice the length of the climbing height plus 10 feet.

Snap ring: A connecting ring used to secure belts and lanyards.

V-shaped stance: The term for the stance commonly taught to apprentice climbers in which their heels are close together and their toes turned outwards. This stance correctly orients the gaff angle to the pole.

Additional Resources

This module presents thorough resources for task training. The following resource material is suggested for further study.

OSHA Regulation 1926, Subpart M, Fall Protection, Latest edition. Occupational Health and Safety Administration. www.osha.gov.

Figure Credits

Tony Vazquez, Module opener, Figures 1, 8, and 19

Photos provided courtesy of Buckingham Manufacturing Company, Inc., Figures 2–5 and 9–11

Courtesy of Sahlins, SA01

Fall protection materials provided courtesy of Miller Fall Protection, Franklin, PA, Figures 6, 7, and 12

Samuel Kenyon, Figure 13

CONTREN® LEARNING SERIES — USER UPDATE

NCCER makes every effort to keep its textbooks up-to-date and free of technical errors. We appreciate your help in this process. If you find an error, a typographical mistake, or an inaccuracy in NCCER's Contren® materials, please fill out this form (or a photocopy), or complete the online form at www.nccer.org/olf. Be sure to include the exact module number, page number, a detailed description, and your recommended correction. Your input will be brought to the attention of the Authoring Team. Thank you for your assistance.

Instructors – If you have an idea for improving this textbook, or have found that additional materials were necessary to teach this module effectively, please let us know so that we may present your suggestions to the Authoring Team.

NCCER Product Development and Revision

3600 NW 43rd Street, Building G, Gainesville, FL 32606

Fax: 352-334-0932
Email: curriculum@nccer.org
Online: www.nccer.org/olf

☐ Trainee Guide ☐ AIG ☐ Exam ☐ PowerPoints Other ______________

Craft / Level: Copyright Date:

Module Number / Title:

Section Number(s):

Description:

Recommended Correction:

Your Name:

Address:

Email: Phone:

Climbing Structures Other Than Wood

49106-11

Trainees with successful module completions may be eligible for credentialing through NCCER's National Registry. To learn more, go to **www.nccer.org** or contact us at **1.888.622.3720.** Our website has information on the latest product releases and training, as well as online versions of our *Cornerstone* newsletter and Pearson's Contren® product catalog.

Your feedback is welcome. You may email your comments to **curriculum@nccer.org**, send general comments and inquiries to **info@nccer.org**, or use the User Update form at the back of this module.

V.1 7/11

Climbing Structures Other Than Wood

Objectives

When you have completed this module, you will be able to do the following:

1. Identify the required safety equipment for proper climbing.
2. Demonstrate the ability to inspect required safety equipment before use.
3. Identify the various environmental hazards requiring consistent attention from the climber.
4. Conduct a proper pre-climb inspection of steel poles and towers and the surrounding area.
5. Identify the appropriate climbing routes of various structures.
6. State the practices for safely ascending and descending steel poles and towers.
7. Demonstrate the physical and mental ability to endure the unique stresses of working at high elevations.
8. Safely ascend and descend a steel tower.

Performance Tasks

Under the supervision of the instructor, you should be able to do the following:

1. Demonstrate the ability to inspect required safety equipment before use.
2. Identify the various environmental hazards requiring consistent attention from the climber.
3. Conduct a proper pre-climb inspection of steel poles and towers and the surrounding area.
4. Identify the appropriate climbing routes of various structures.
5. Demonstrate the physical and mental ability to endure the unique stresses of working at high elevations.
6. Safely ascend and descend a steel tower.

Trade Terms

Anchor point
Body harness
Cable grabs
Carabiner
Connecting devices
Fall arrest
Fall restraint
Newtons
Point of daylight
Rappel
Swing zone

Contents

Topics to be presented in this module include:

Figures and Tables

1.0.0 INTRODUCTION

Power line workers are often required to climb tall towers (*Figure 1*) in order to install conductors and components. Maintenance and repair of the pole or tower itself is also a common chore. However, even if the required task is completed, the job should never be considered successful unless it is done free of injury to the worker. Great risks are involved in climbing steel towers, but when all proper attention is given to the details and safe execution of a climb, great rewards will result.

Many opportunities are available to those who have the ability to climb and work at extreme heights. Power transmission and distribution systems are spread across the country like a giant web. The constant demands placed upon these systems require continual maintenance and expansion. Weather also plays a significant role in the need for experienced climbers by causing damage to power lines and towers. As power generation continues to evolve, and wind-generated power (*Figure 2*) plays an increasing role, the working environment of the climber and opportunities for climbers will evolve and grow as well.

In addition to challenging the technician to learn and understand the safe working habits and rules involved with climbing, this module also provides you with your first opportunity to test your mental and physical ability to withstand the stress of climbing and working in the elevated environment. The vast majority of this module focuses on personal safety, and it is essential that

49106-11_F01.EPS

Figure 1 Power transmission tower.

49106-11_F02.EPS

Figure 2 Wind tower.

these lessons be learned. The smallest lapse in judgment or attention can prove disastrous. There is no replacement for proper training and preparation for a climbing assignment.

2.0.0 Safety Equipment

The list of safe working policies and regulations that could be imposed may seem endless. A number of organizations, both government and private, have attempted to create a widely accepted version of comprehensive tower and pole climbing regulations or guidelines. However, there is still considerable difference between the creators of such publications. OSHA provides a number of regulations that apply to the trade, primarily consisting of directives for fall protection. Remember that OSHA is charged with the responsibility of ensuring that employers protect workers from hazards in the workplace. Their requirements must be considered the minimum. Beyond that, individual organizations and employers impose other requirements and policies for their internal use. Other entities, such as land or equipment owners, may also invoke their own set of guidelines when work is being conducted on their property.

Although many guidelines internal to an organization address other issues, fall prevention and protection is truly the focus. There is certainly plenty of incentive to do so because more than 100,000 injuries or deaths occur each year from falls in the workplace. OSHA regulations require that specific precautions be taken to protect employees who work at heights. When workers are exposed to the potential for falling 6 feet or more, OSHA requires installation of one or more of these three primary fall prevention systems: a guardrail system, safety net system, or a personal fall protection system. Guardrail systems and safety nets are not likely to be found on towers. That places the focus on personal fall protection.

The fall protection requirement can be satisfied through the use of either fall restraint equipment or fall arrest equipment. Fall restraint equipment helps prevent any significant fall from occurring, while fall arrest systems stop or take control of a fall in progress. The climber must take responsibility for using fall arrest and/or fall restraint equipment on each and every climb, 100 percent of the time.

2.1.0 Personal Fall Arrest Systems

Personal fall arrest systems (PFAS) combine several pieces of equipment together into a complete fall protection system. Many different types and styles exist, with each one designed for specific uses. In addition, the personal aspect of a PFAS must also be emphasized. One size does not fit all, and proper fit is essential to avoid injury to the worker as a fall is arrested. *OSHA Regulation 1926.502, Subpart M, Section (d)*, lists requirements that must be met, such as the test strength and use of D-rings and snap hooks, as well as the specifications for anchor points on the structure.

OSHA defines several important characteristics for a PFAS:

- It must limit the maximum arresting force imparted to the body to 1,800 pounds with a full body harness.
- The free fall distance must be limited to 6 feet and be rigged to prevent contact with anything below.
- It must bring the body to a stop within an additional 3½ feet.
- The system must be strong enough to withstand twice the possible force of a body falling from a distance of 6 feet.

It should be somewhat obvious that any PFAS is only as good as the weakest component incorporated into it. They should never be used for any other task, such as lifting tools or materials to the work area. The specifications and usage instructions for components such as the full body harness (*Figure 3*) identify the working load limit, and this limit should be strictly followed. Climbers must use common sense and not overload themselves

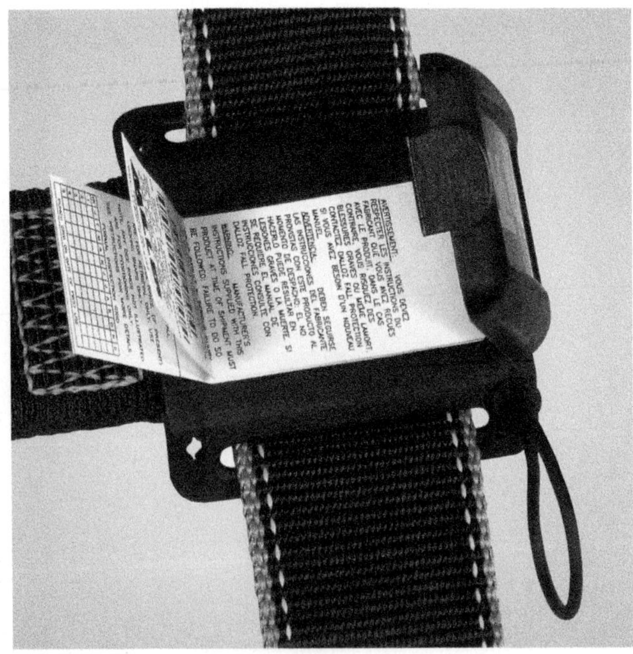

49106-11_F03.EPS

Figure 3 Body harness labeling.

with equipment, even if the weight does fall within the limits of the PFAS specifications.

A complete PFAS (*Figure 4*) is made up of three primary components. Anchor points are related to the structure. The type and availability of anchor points help determine what other equipment should be chosen. The body harness is the system of belts, rings, and hooks worn by the climber. Connecting devices, or connectors, are used to maintain attachment between anchor points and the PFAS or positioning belts. They include lanyards and various pieces of hardware.

2.1.1 Anchor Points

OSHA requires that an anchor point (*Figure 5*) be rated at 5,000 pounds breaking strength per person attached to it. The anchor point cannot be

On Site

ANSI

The American National Standards Institute membership is made up of government agencies, trade and professional associations, companies, manufacturers, academic bodies, and individuals. It is a non-profit organization, not a government agency. ANSI's stated purpose is to "oversee the creation, promulgation and use of thousands of norms and guidelines that directly impact businesses in nearly every sector: from acoustical devices to construction equipment, from dairy and livestock production to energy distribution, and many more."

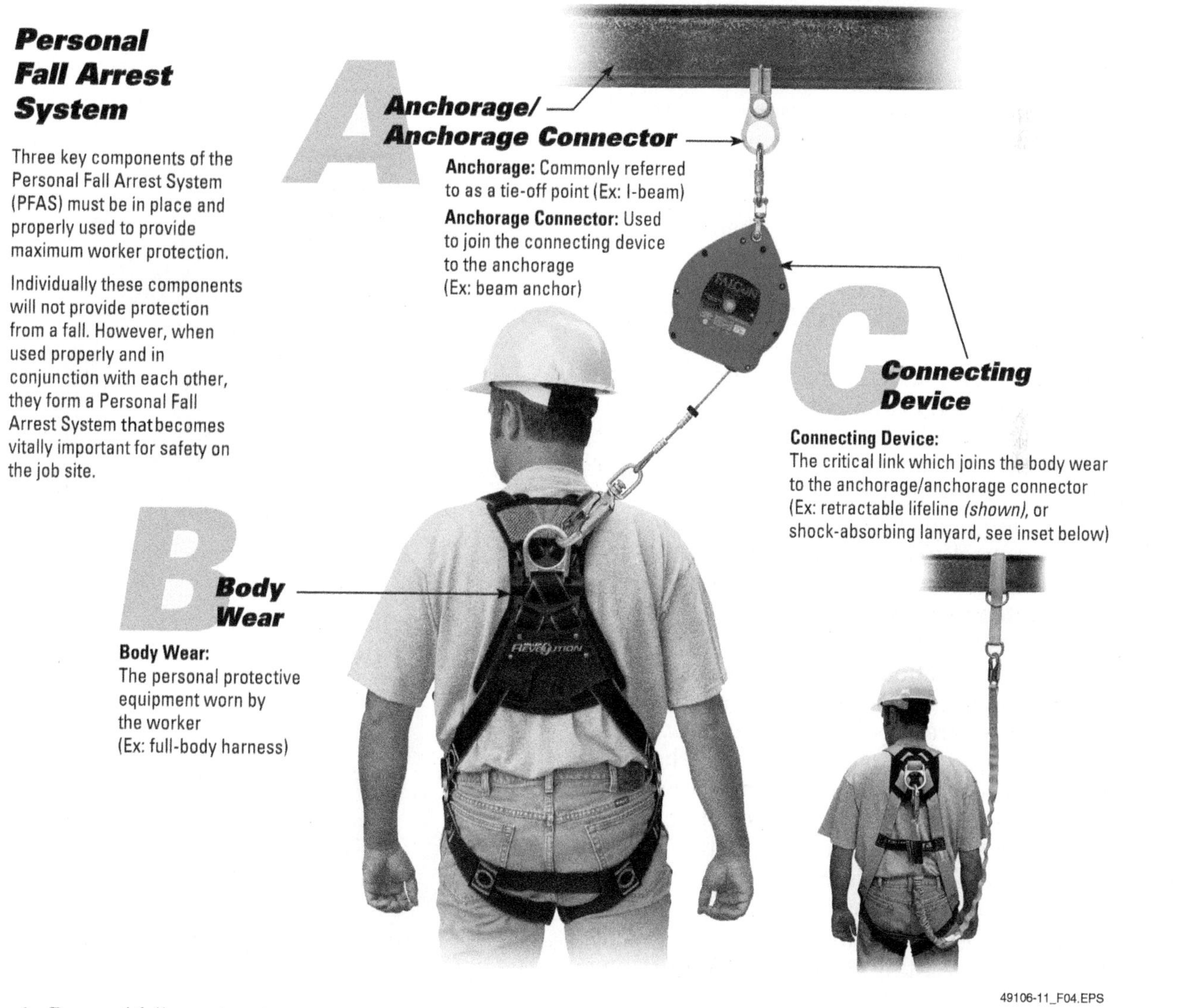

Personal Fall Arrest System

Three key components of the Personal Fall Arrest System (PFAS) must be in place and properly used to provide maximum worker protection.

Individually these components will not provide protection from a fall. However, when used properly and in conjunction with each other, they form a Personal Fall Arrest System that becomes vitally important for safety on the job site.

A Anchorage/ Anchorage Connector

Anchorage: Commonly referred to as a tie-off point (Ex: I-beam)

Anchorage Connector: Used to join the connecting device to the anchorage (Ex: beam anchor)

B Body Wear

Body Wear: The personal protective equipment worn by the worker (Ex: full-body harness)

C Connecting Device

Connecting Device: The critical link which joins the body wear to the anchorage/anchorage connector (Ex: retractable lifeline *(shown)*, or shock-absorbing lanyard, see inset below)

49106-11_F04.EPS

Figure 4 Personal fall arrest system.

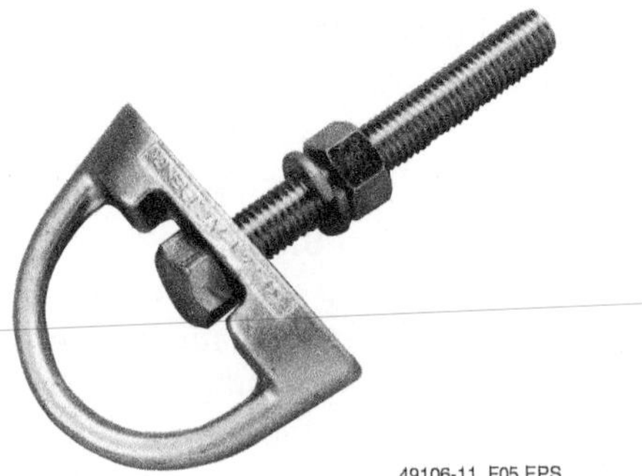

49106-11_F05.EPS

Figure 5 Typical anchor point.

part of the support used for tower platforms. This can be a challenging requirement for the climber, as some towers may not have any structures that conform and no global standard presently exists for clearly marking or installing them during construction. Final selection of the anchor point during a climb is left to the professional judgment of the climber. When no such 5,000 pound point exists or can be identified, the next standard the climber must meet is to calculate the PFAS anchor point to provide a safety factor of 2. This means that the anchor point must be capable of supporting twice as much weight as the maximum that could be applied to it during a fall arrest situation. In other words, if a chosen anchor point is expected to support 2,500 pounds, then you must ensure that you will only apply 1,250 pounds of force in a worst-case scenario of fall arrest.

The anchor point shown in *Figure 5* is one that can be either temporarily or permanently installed. It is only one example of the many different styles you may encounter. Properly tested anchor point hardware installed on the tower specifically for use by climbers is uncommon. There are no tower construction standards that dictate the installation or location of anchors. The tower erectors or owners may specify them, however.

Anchor points should always be above the back D-ring, selected to minimize any swing zone hazards as well as the possible free fall distance. The swing zone is a function of the amount of slack in the lanyard. Swing zones are minimized when the anchor point is directly above the climber. Horizontal climbing situations can be far more hazardous in terms of the swing zone, as an anchor point directly above may not exist. It should be noted that the potential injury and damage that occurs as a mass (such as the human body) strikes an immoveable object while swinging as a pendulum can cause serious injury. Although the PFAS may work fine by preventing the climber from falling a great distance, serious injury or death can still occur from striking an object in the swing zone.

Anchor points are also often needed to secure the position of the climber, leaving the hands free to accomplish the task at hand. Ideally, climbers should select anchor points, and connect as needed, to maintain a potential fall distance of 2 feet. Positioning connections are a factor in fall arrest, as the connection to a positioning strap will likely modify the swing zone and/or distance of the fall during an accident. Positioning anchor points are not considered the primary fall arrest anchor. For that reason they are required by OSHA to be rated at a 3,000 pound strength instead of the 5,000 pound rating for primary fall arrest. Keep in mind that positioning lanyards, connected to D-rings on the harness or belt other than the back D-ring, are fall restraints, not fall arrests.

2.1.2 Belts and Harnesses

The full body harness (*Figure 6*) is the center of the PFAS. It should be worn any time work is performed more than 6 feet above the ground. Although a variety of rings may be part of the full body harness assembly, the back D-ring (*Figure 7*)

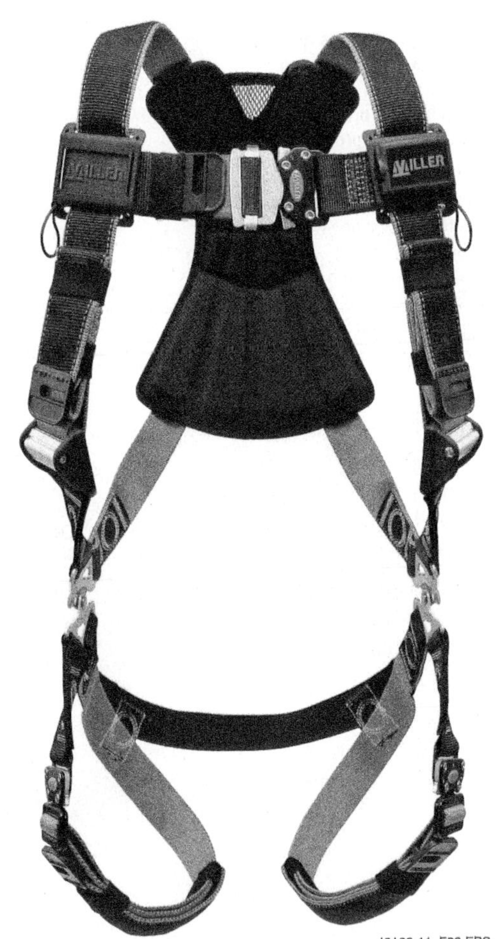

49106-11_F06.EPS

Figure 6 Full body harness.

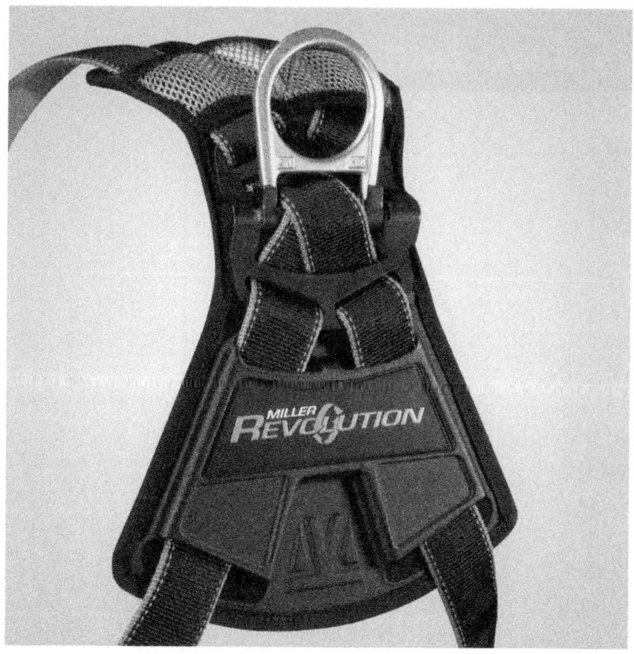

49106-11_F07.EPS

Figure 7 Back D-ring.

is the only one used to connect the harness to the anchor point for primary fall arrest purposes. Saddle D-rings may be installed for use in suspended work, while hip D-rings are generally used for positioning connections. D-rings mounted to shoulders are often used for rescue situations. All can be used for fall restraint, but the back D-ring is the primary connection for fall arrest.

The body harness must fit properly to ensure that it protects the climber. Under no circumstances should additional holes be made in the harness components. No modifications should be attempted. Installation, maintenance, and inspection instructions are provided for every harness, and it is the responsibility of the climber to read and understand these instructions. Also remember that the harness straps are generally designed with some stretch to help absorb some of the potential force. This means good, taut installation on the body is essential so that any slack, plus stretch, does not allow the climber to fall out of the harness.

Figure 8 shows the proper method for putting on one particular full body harness. The items that follow should be considered during the fitting and wearing of a full body harness. Please note, however, that these guidelines are general in nature and the instructions provided for specific equipment by the equipment manufacturer always take precedence. It is the climber's responsibility to be familiar with the purpose of each ring and belt of the system.

The back D-ring location is vital to proper fall arrest. This ring should be positioned between the shoulder blades. If it is too low, it will tend to cause the body to hang in a more horizontal (prone) position, increasing pressure on the diaphragm and affecting breathing. If it is positioned too high or with too much slack, the D-ring may strike the back of the climber's head in the event of a fall. The shoulder straps may be pulled too tightly into the neck and restrict blood flow. The force of impact at the base of the fall arrest can be dramatic. This force must be spread all around the body to prevent injury to any one portion.

Chest straps generally form either an "H" pattern or "X" pattern. "H" pattern straps should generally be adjusted to land between the bottom of the breastbone and the navel. This helps ensure the horizontal portion of the "H" does not contact the throat during a fall, choking the climber. Some harness designs may not allow for this adjustment, with the final position of the "H" based solely on proper fit. The position of the chest straps is also crucial for systems using the "X" pattern. The "X" should be positioned at or just below the breastbone.

Groin straps are an essential part of the harness. They must be adjusted for a good, snug fit. Too much slack will cause extreme discomfort in a fall, when the impact snatches them up tight and you are left suspended this way. Some harnesses have extra padding both in the groin and in the shoulder straps to avoid this discomfort. When done properly and without any interference to the working parts of the harness, this practice would not be considered a modification.

A separate positioning or tool belt, while not considered a true fall arrest component of the system, must be fitted properly as well. Its use for positioning the worker is best employed with the D-rings precisely located at the hip sides, rather than in the front or rear. It should not be adjusted in a way that could apply pressure to the kidneys or lower back. If the waist belt is not an integral part of the harness, it must not be worn on the outside of the harness. Put the harness on first then put the belt over it. This is also true of added tool belts.

Saddles, like the waist belt, are also not considered a true part of the PFAS. They are optional and are often detachable. They are generally used by the worker to allow a seated, suspended position and changes in weight distribution when the task may require long periods in the same location.

2.1.3 Connecting Devices

Connecting devices connect the PFAS or positioning belts to anchor points. They include several different types of hardware, as well as the

6 Easy Steps That Could Save Your Life

How To Don A Harness

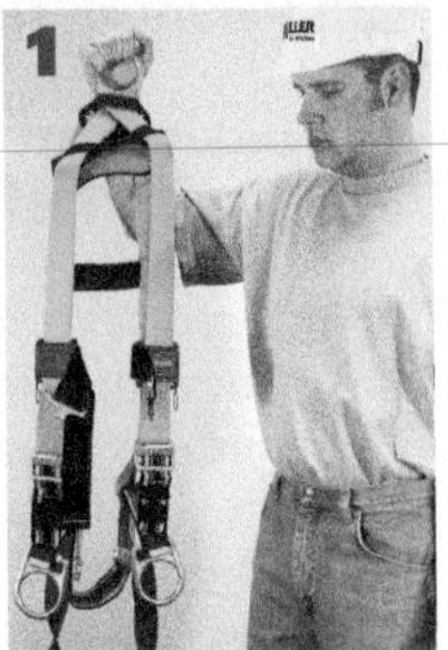

Hold harness by back D-ring. Shake harness to allow all straps to fall in place.

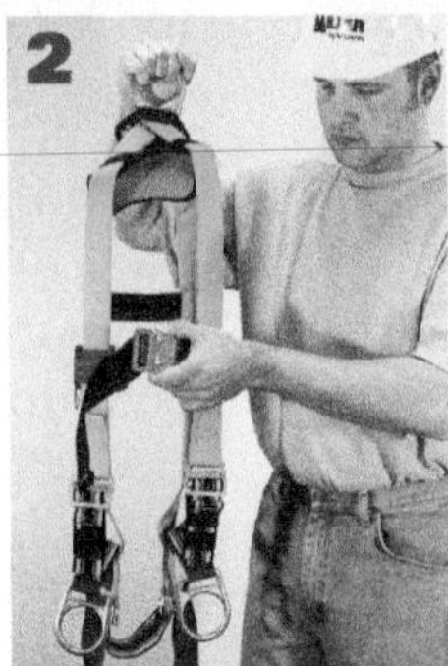

If chest, leg and/or waist straps are buckled, release straps and unbuckle at this time.

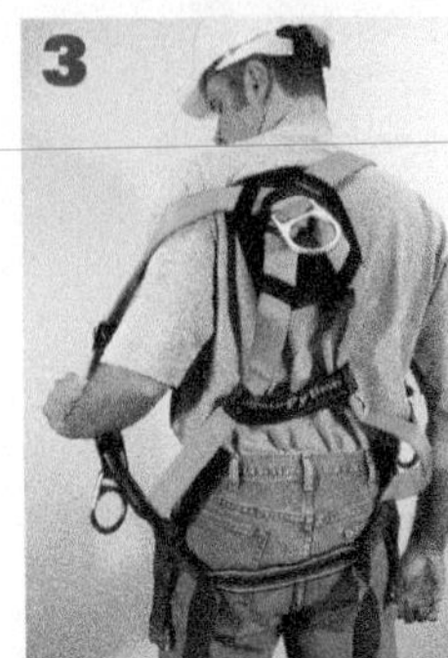

Slip straps over shoulders so **D-ring is located in middle of back between shoulder blades.**

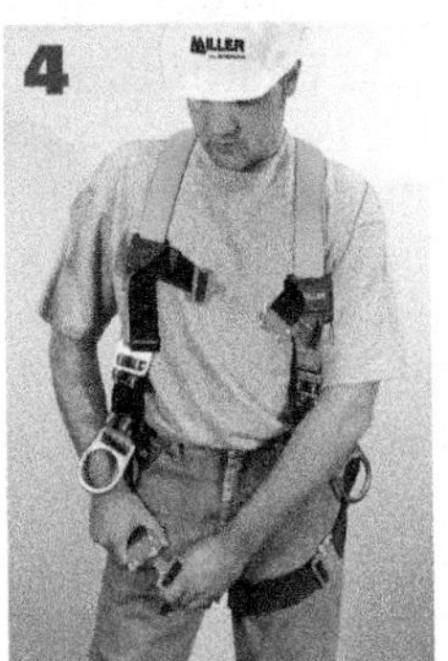

Pull leg strap between legs and connect to opposite end. Repeat with second leg strap. If belted harness, connect waist strap after leg straps.

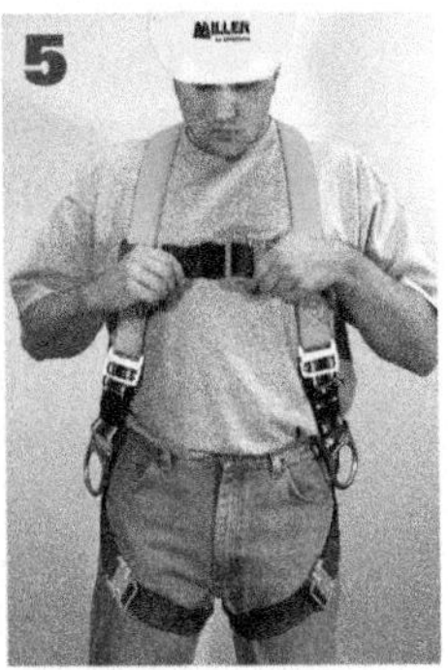

Connect chest strap and position in midchest area. Tighten to keep shoulder straps taut.

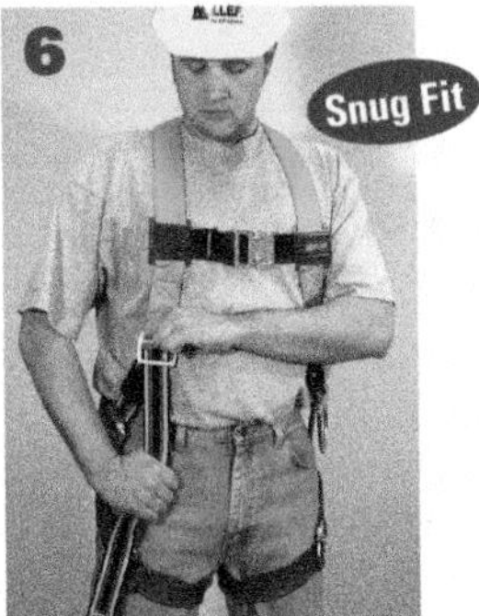

After all straps have been buckled, **tighten all buckles so that harness fits snug but allows full range of movement.** Pass excess strap through loop keepers.

49106-11_F08.EPS

Figure 8 Installing the body harness.

> **On Site**
>
> ### Adjusting the Harness
>
> It may be necessary to adjust the harness to allow for extra clothing on cold days. In addition, the use of additional PPE such as chaps for welding and cutting work may also require readjustment of the harness.

lanyards. Lanyards are discussed in detail in the next section. Some of the hardware available is to be used only for attaching tools and equipment to you or to structures. Hardware used as connecting devices as part of your PFAS must be drop-forged steel and have a corrosion-resistant finish that will withstand salt spray per ANSI standards. Any type of hook or carabiner must be equipped with safety gates or keepers to prevent the hooked object from being disconnected accidentally. In most cases, these safety gates must be two-step (also called double action) devices. Designs for these features vary, but those designed such that both movements required to open the gate can be done with a single hand are generally better. Using both hands to manipulate a single connector can be a hazard in itself.

A carabiner (*Figure 9*) is like an elongated ring, with some type of safety gate. It is shaped much like a chain link. The safety gate on some may be screw-type, requiring that the gate be screwed closed. Carabiners generally have no sharp edges and are used to attach lines. They are tested by impact to determine their strength and are rated in Newtons of force.

Hooks (*Figure 10*) are usually curved and also have an opening to allow connection to a line that can then be securely closed. They are usually not as consistent in appearance as a carabiner, and come in different shapes. The type most often used is referred to as pelican style due to the shape of the hook and the closure. When used as part of your PFAS, the security closure should be automatic. Hooks are tested under direct load rather than the force applied, and are then rated in pounds as the minimum breaking strength. Hooks are designed to connect to D-rings primarily, not to each other.

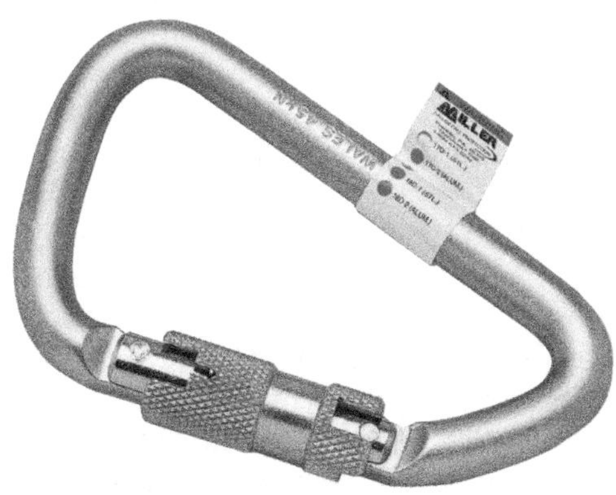

49106-11_F09.EPS

Figure 9 Example of a carabiner.

49106-11_F10.EPS

Figure 10 Example of a hook.

On Site

What Is a Newton?

A Newton is a metric measurement of force combining mass, energy, and time. In the meter-kilogram-second system, it is the amount of force required to accelerate a mass of one kilogram one meter per second, per second. In the English system of measurement, it is expressed in feet-pounds-seconds. One Newton equals a force of about 0.2240 lb in the English system. It is named for Sir Isaac Newton, whose second law of motion in the late 1600s describes the changes that force can produce in the movement of a body. Considered by some to be the father of modern science, he may be best known as the man who made sense out of gravity.

2.2.0 Lanyards

Lanyards generally consist of the primary material of construction (rope, webbing, aircraft cable, etc.) with a connecting device integrated into the ends. Lanyards come in a great variety of lengths from as small as 1½ feet up to 30 feet, and should never be connected together to increase their length. Depending on the use, adequate padding may be integrated at the point of manufacture or added in the field to protect against sharp steel edges. Climbers should remember never to tie a knot in a lanyard, as knots can severely reduce the load limit. Lanyards should never be wrapped around a structure then choked by placing a pelican hook around the lanyard itself, unless they are of a design specifically allowing this use.

There are two main categories of lanyards: shock absorbing (*Figure 11*) and non-shock absorbing (*Figure 12*). Lanyards for fall arrest should always be shock absorbing. Non-shock absorbing

49106-11_F11.EPS

Figure 11 Shock absorbing lanyard.

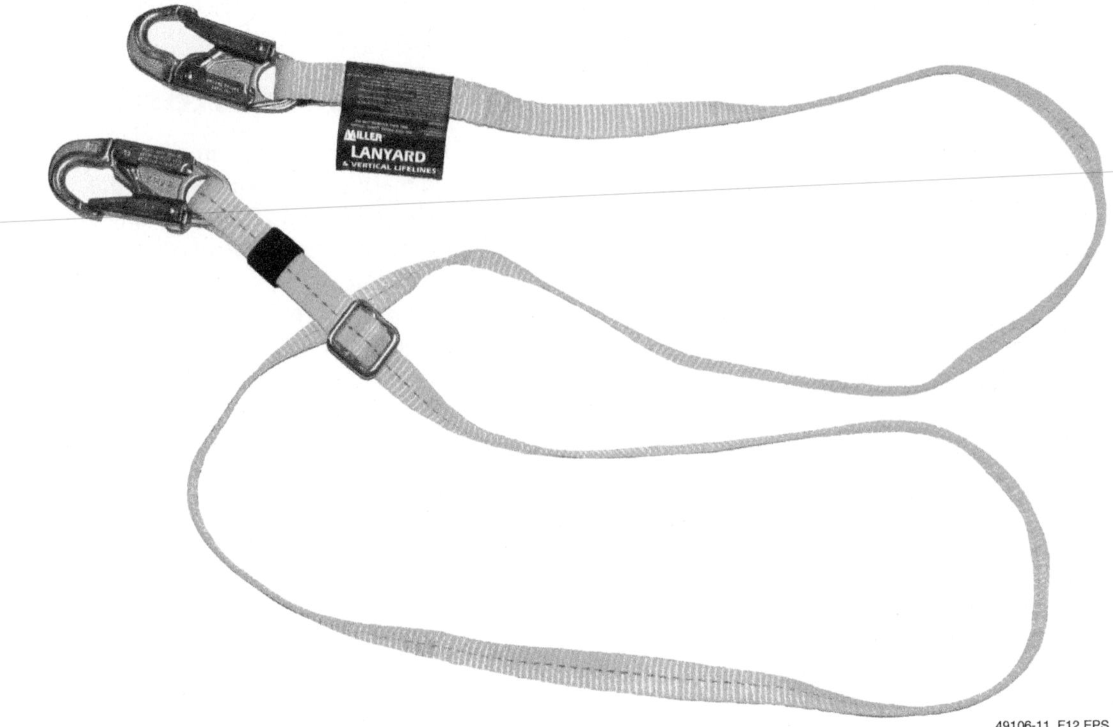

Figure 12 Non-shock absorbing lanyard.

lanyards are used for positioning and fall restraint. Lanyards used for positioning yourself for work are not considered part of the fall arrest system. They are fall restraints and should be attached to D-rings on the harness other than the back D-ring. Remember that fall restraint is all about preventing a fall from happening. For this reason, lanyards used for positioning should not allow workers to fall or move more than 2 feet. That's a pretty short distance, and it means that a positioning lanyard may have to be wrapped several times around the anchor point to get the needed length. As previously mentioned, do not choke the lanyard with the end hook in an effort to adjust the lanyard to the desired length. Some lanyards with special D-rings are designed for this purpose.

When used for fall restraint, fall arrest, or positioning, the line of support should always be at eye level or above. Lanyards may have to be wrapped around an anchor point to allow the correct allowable working distance. When the structure has sharp edges that might cut the lanyard, use a wire or aircraft cable lanyard or a lanyard with proper padding to guard against it.

2.2.1 Shock Absorbing Lanyards

Shock absorbing, or deceleration, lanyards have shock absorbing properties built in and are designed for fall arrest. There are also some shock absorber devices that are designed to provide the needed protection to non-shock absorbing lanyards. This type of lanyard is capable of reducing the fall arrest force on the body by as much as 80 percent.

A section of these lanyards is built to extend under severe stress, such as that encountered in a fall. The section or end that contains the energy-absorbing feature should generally be connected to the climber's back D-ring rather than the anchor point. Because of their duty, deceleration lanyards should be no more than 6 feet in usable length and allow no more than an additional 3½ feet of extension when the shock-absorbing feature is activated. These lengths must be considered when they are employed so that climbers know what their total distance of fall will be. As mentioned earlier, the primary anchor point should be straight above the body whenever possible to minimize the potential fall distance. The projected fall distance must be less than any structure below that you could fall into.

The D-ring used on your PFAS for fall arrest is only allowed to be connected to one live connection at a time. This can be challenging when trying to move from one point to another, especially horizontally. The climber must be able to reach back and disconnect a lanyard, while maintaining one connection at all times. There are some special Y-configured lanyards (*Figure 13*) used for this purpose, where a single point of attachment at the D-ring is used to accommodate two lanyards.

2.2.2 Self-Retracting Lanyards

Self-retracting lanyards (SRLs) arrest falls as they happen by their reaction to tension. Like deceleration lanyards, climbers must consider the operating principles as they decide on anchor points and calculate fall distance and swing zones. SRLs (*Figure 14*) generally restrict free fall to 2 feet. Deceleration of the weight may take up to an additional 1½ feet. The total fall distance then would be 3½ feet. SRLs allow for far greater mobility for the climber, both horizontally and vertically.

One very simple type of SRL with an effective length of up to roughly 10 feet is often used. These shorter styles are often built without true braking components and are more like deceleration lanyards. Other units allow for much greater freedom of movement for the climber, and are equipped with braking systems. This more sophisticated style of SRL can be found in lengths up to nearly 200 feet. The line moves in and out slowly as the climber moves about, until the line begins paying out at high speed. Braking operation then begins to slow and stop the fall. It is important that too much slack does not develop in the line connection to the D-ring. If slack or loose line is present, it will add to the distance of the fall before the braking mechanism takes over.

Because of the freedom allowed by long SRLs, the climber must be very careful to avoid hazardous situations. It is all too easy to wrap the lanyard in and out of structural members as you move, or to move much too far from the anchor point vertically or horizontally.

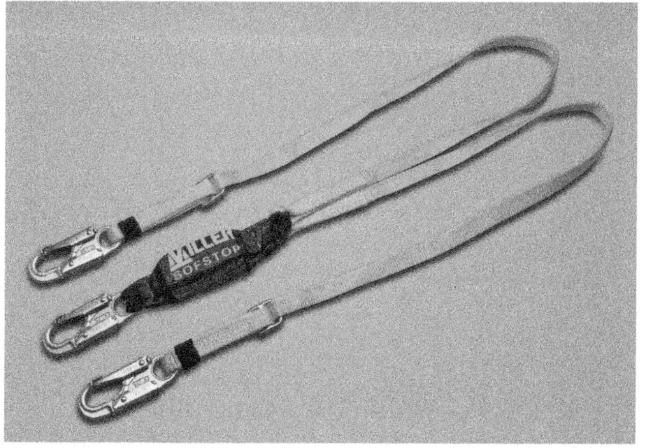

49106-11_F13.EPS

Figure 13 Y-configured shock absorbing lanyard.

49106-11_F14.EPS

Figure 14 Self-retracting lanyard.

2.3.0 Safe Climbing Assistance Devices

There are several different systems and devices that allow the climber to climb more confidently and safely. These systems eliminate the situation where the climber must connect to an anchor, climb, connect to another anchor, disconnect the first anchor, climb further, etc. Safe climbing devices are generally classified as permanent attachments to the structure, although this section also covers the use of temporary vertical lifelines.

Permanently installed systems vary in design. Some incorporate a rail with a sliding puck to which the climber connects. The puck locks up when pulled in the opposite direction or when tension is not maintained in an upward direction. However, climbers more often encounter setups that use cable routed up the preferred climbing path (*Figure 15*). Cable grabs, also called shuttles or pucks, are mounted on the cable for the climber to use as a connection point. The cable grab locks down on the cable to help restrain a fall when movement is too rapid or aggressive. It is important for the climber to differentiate between such systems installed with the intent of fall restraint or fall arrest. When designed as a fall restraint device, as most are, then a lanyard should not be used between the climber and the cable grip—a direct connection is made to the body harness.

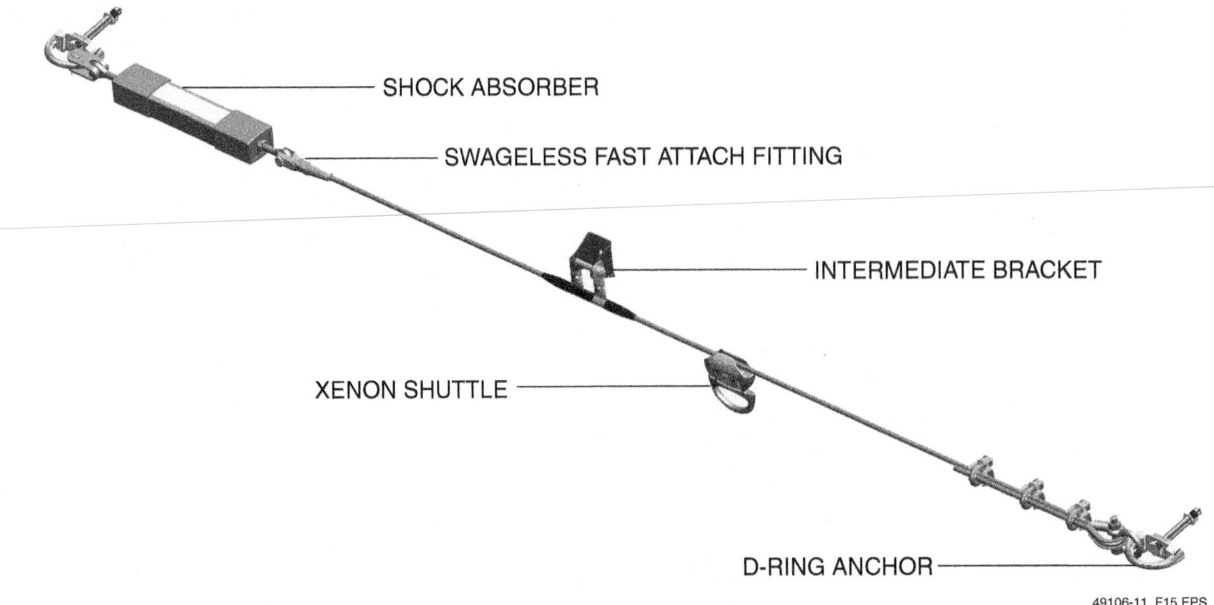

Figure 15 Permanent safe climb lifeline system.

Temporary vertical lines, using a safety rope, are also popular (*Figure 16*). They are rigged at the time of need and are installed with fall arrest in mind in regards to their anchorage. Rope grabs (*Figure 17*) are much like cable grabs but cannot withstand the same stresses as cable. A deceleration lanyard should be used and connected to the back D-ring for proper fall arrest protection. As the climber ascends, the connection to the rope will trail below and slightly behind the climber, as far as the connecting lanyard will allow. This adds to the potential fall distance of the climber before arrest begins. For this reason, and since this is a fall arrest application, the lanyard used should be no more than 3 feet in length to avoid a fall that is too long and imparts a force to the rope beyond its ratings.

2.4.0 Other Safety Gear

In addition to the PFAS and related equipment, other common safety gear is needed by the climber. First aid kits must be readily available, but are not carried by the climber in most cases. Tower work sites and/or vehicles on site

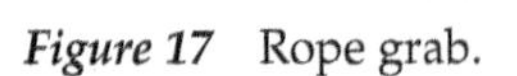

Figure 16 Temporary safe climb rope system.

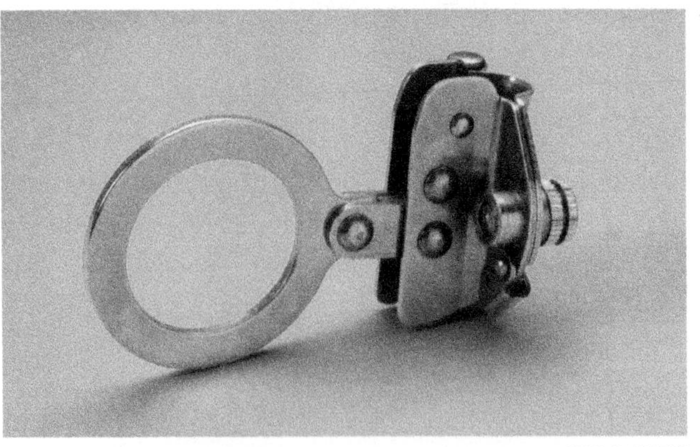

Figure 17 Rope grab.

should be equipped with minimal first aid gear, and workers should be properly trained in its use. Other types of equipment that should be considered and used when appropriate include the following:

- *Head protection (required by OSHA)* – A number of different styles are acceptable and many are better suited to one or more applications or crew positions. Hard hats that are well-fitted and secured to the head are necessary for climbers. Many of these hard hats were originally developed for the recreational rock climber and have been adapted to meet construction standards. Check your chosen hard hat to ensure that it meets the required OSHA standards and has been tested to ANSI standards for the projected use.
- *Hand protection* – Towers are generally made of hot-dip galvanized material and can be rough on the hands. The galvanized coating often results in sharp burrs that easily penetrate hands. Therefore, leather gloves should be worn. The gloves should fit snugly and allow the freedom of movement needed to do the job. A good tight fit also helps minimize any potential for the glove itself to be a safety hazard. Gloves with built-in knuckle protection may also be in order.
- *Footwear* – ANSI standards and OSHA require footwear that incorporates steel or fiberglass safety toes and reinforced soles. They should cover the ankle, have a well-defined heel, and a non-slip sole. One style of shoes may not be a good choice for every climbing situation. Due to the consistent and unique stress that is placed on the foot by climbers, an excellent fit is crucial. Soft-soled shoes without reinforcement will quickly tire or injure a climber, since the stress will not be spread across the foot. Also keep in mind that bootlaces can become entangled with ropes, lanyards, and hardware. Laces should be tucked in, secured in some other way, or avoided entirely.
- *Eye protection* – Eye protection is as important for tower work as it is for any other trade work. An eye injury can become far more complex when the worker is in the air and is unable to see well enough to descend safely.
- *Ear muffs or ear plugs* – Ear muffs or ear plugs can be used as needed for hearing protection.
- *Knee pads* – These are optional, but can greatly increase comfort for certain jobs.

It is suggested that long pants and long sleeve shirts with collars always be used regardless of outdoor conditions. This helps prevent minor skin injuries and irritations that come from the tower surface, cables, ropes, and webbing. It also helps prevent unnecessary exposure to the sun and minimizes access to the skin for insects. Regardless of your choice, good fit is essential to ensure that clothes do not restrict movement but also do not become entangled in your PFAS, lanyards, or ropes.

Climbers should also carefully consider other personal grooming choices. Long hair that is not properly controlled, and necklaces and rings can create safety hazards. Jewelry should not be worn, and hair should be contained to prevent it from being caught up in the PFAS webbing.

2.5.0 Equipment Inspection

All personal safety equipment must be inspected by the user before each climb. Any equipment found to have defects must be taken out of service. It is also important to note that any such equipment that has been involved in a fall arrest situation and subjected to the loads imposed by a fall must be taken out of service and replaced. Do not reuse harnesses, lanyards, ropes, and other devices that have played a role in a fall arrest situation.

The guidelines that follow are general in nature are not intended to replace the recommendations for use, inspection, or maintenance provided by the manufacturer. A proper inspection of safety equipment must always be done in accordance with the manufacturer's information and any additional directives that exceed those recommendations.

2.5.1 Belts and Harnesses

Inspect belts and harness as follows:

- Make sure the manufacturer's label is attached and legible.
- Check all webbing material for evidence of fraying or stress.
- Inspect all sewing and lacing to ensure it is intact.
- Check any leather components for signs of deterioration.
- Verify that there are no additional holes or rings that were not installed at the point of manufacture. Check grommets for signs of stress and make sure they are secure.
- Ensure that all D-rings, buckles, and other metal parts are in good condition and show no signs of corrosion or distortion (*Figure 18*).
- Ensure that the belt or harness design provides a proper fit for the individual user.

2.5.2 Ropes and Lanyards

Ropes and lanyards must also be inspected. The following checks should be performed before any rope or lanyard is used:

- Ensure that lanyards have the manufacturer's label attached and that it is legible.
- Check lanyard webbing to ensure it is relatively clean, and is not frayed or damaged. Grab the webbing material between your two hands about a foot apart, and bend the lanyard up toward you, forming an upside-down U (*Figure 19*). Inspecting the webbing while it is bent in this manner will help broken fibers or strands to stand out.
- Inspect ropes to ensure there is no fraying or tearing anywhere along the length.
- Ensure that all termination connections are sound and show no sign of damage or failure.

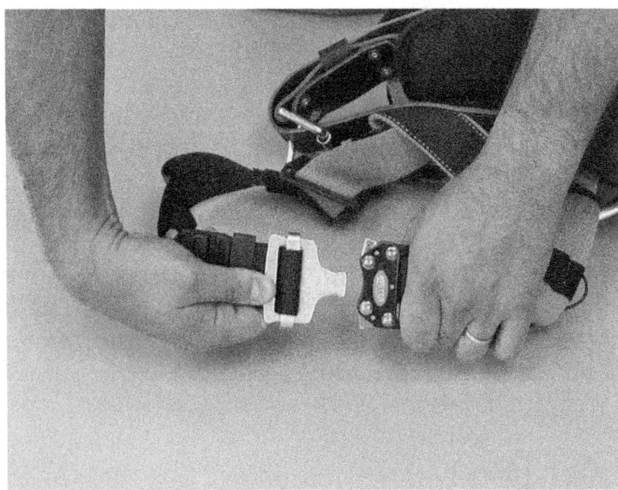

49106-11_F18.EPS

Figure 18 Buckle inspection.

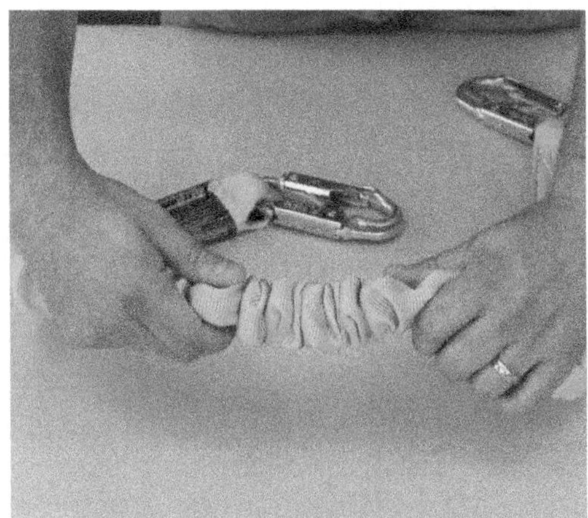

49106-11_F19.EPS

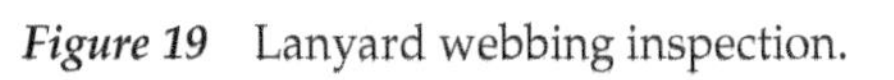

Figure 19 Lanyard webbing inspection.

- Check any D-rings attached for signs of corrosion or deformation (*Figure 20*).
- Verify that deceleration and/or shock absorbing lanyards show no sign of extension or decompression from previous use. A warning flag will sometimes be exposed when the lanyard shock absorbing feature has been deployed.
- Check wire and cable lanyards for signs of corrosion or broken wires.

2.5.3 Self-Retracting Lanyards

There are specific checks for self-retracting lanyards. These checks include the following:

- Check the SRL housing or casing (*Figure 21*) for any sign of physical damage and verify that all assembly hardware is in place.
- Inspect the line itself regularly for any sign of concealed damage.
- Pull the line out slowly several feet, then hold it lightly as it retracts back into the housing.
- Test the braking mechanism by grabbing the line above the impact indicator and pulling it sharply downward (*Figure 22*). This should engage the brakes. Once you begin to release tension, the brakes should disengage and the line should retract.

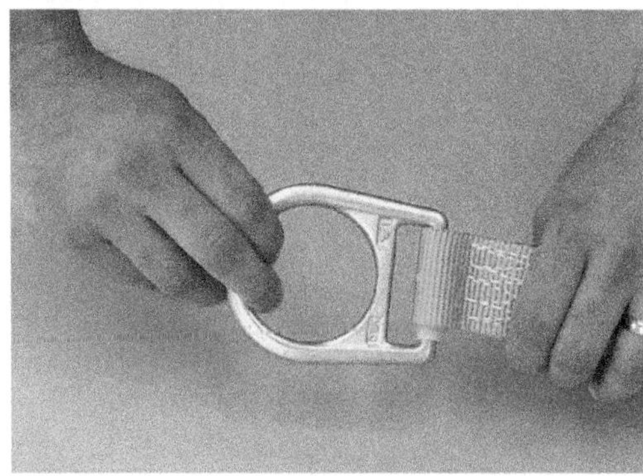

49106-11_F20.EPS

Figure 20 D-ring inspection.

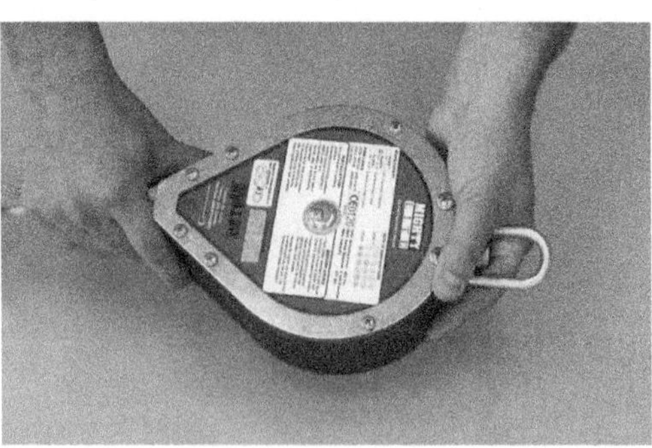

49106-11_F21.EPS

Figure 21 SRL casing inspection.

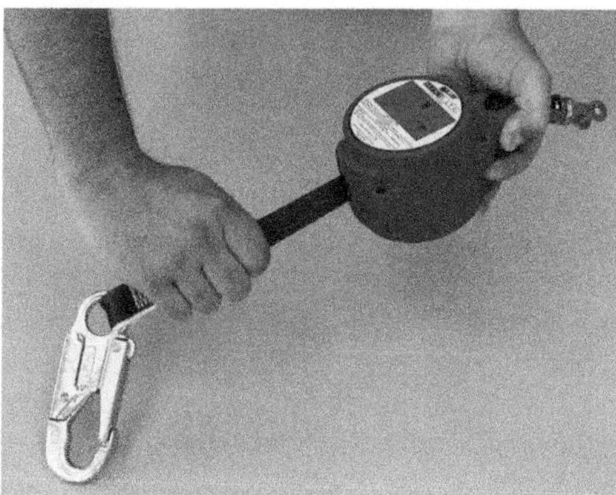

49106-11_F22.EPS

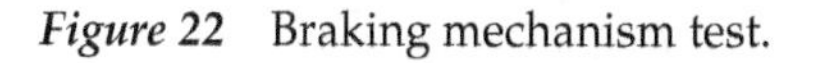

Figure 22 Braking mechanism test.

2.5.4 Hooks, Carabiners, and Cable/Rope Grabs

Hooks, carabiners, and other hardware must also be inspected. Inspect your hardware as follows:

- Check hooks to be used as part of the PFAS for proper load capacity information stamped into the metal. The unit of measure will be in pounds. Those used for fall arrest connections should be rated at 5,000 pounds.
- Inspect carabiners for the presence of appropriate force capacity information stamped into the metal. The unit of measure will be in thousands of Newtons, or kN. Carabiners and hooks should be rated at a minimum of 22kN.
- Check hooks and carabiners for any sign of corrosion, sharp edges, or shape distortion.
- If provided, check the load indicator on hooks for an indication that the hook has been heavily loaded (*Figure 23*). The connecting pin between the hook body and swivel connection will change position as shown.

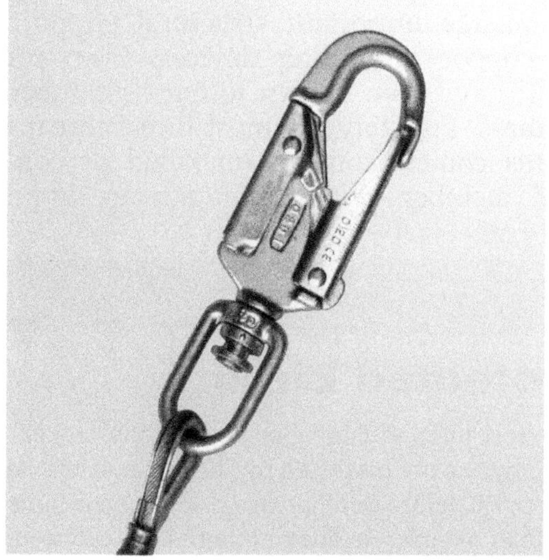

49106-11_F23.EPS

Figure 23 Hook load indicator.

- Ensure that the safety gate or closure on both hooks and carabiners operate freely, lock, and are undamaged in any way. Make sure the double safety works.
- Cable and rope grabs should be checked for any sign of stress, corrosion, or missing parts. It is also essential to ensure that the grabs used are specifically designed for the type and diameter of the rope or cable to which they apply.

Each climber must take responsibility for his or her own safety equipment. When not in use, safety equipment and harnesses should be properly stored in a dry location and away from sunlight, as well as paints, oils, and corrosives that could cause deterioration.

3.0.0 Hazards of the Environment

There are hazards other than falls that must be considered. The most obvious of these hazards is the potential for electrical shock associated with proximity to high-voltage power lines. Other types of hazards are covered in the sections that follow.

3.1.0 Noise and Communication

The potential noise that any work on the tower will create should be considered before leaving the ground, and proper hearing protection should be on hand. Noise can also interfere with communications. Portable radios are a common communication tool, but excessive noise, as well as radio frequency interference, may prevent or inhibit their use. A second or even third means of communication, such as hand signals or the use of a code based on whistle blows, should be considered to ensure that some means of communicating with the ground crew is available.

3.2.0 Living Dangers

Birds, animals, insects, and even humans can be potentially hazardous to the climber. It is important to note that the hazards discussed here may

> **On Site**
>
> **Extra Gear**
>
> It is always a good idea to have one extra set of fall protection equipment for a crew. That way, if equipment is damaged, there will be a spare on hand.

not be extremely dangerous directly, but any external factor that causes the climber to be distracted while working or climbing can prove to be disastrous. Trainees will likely spend a great deal of time as members of a ground crew while experienced technicians perform tasks that require climbing. Many of these same threats are just as dangerous to non-climbing crew members.

3.2.1 Birds

Birds can be a nuisance as well as direct hazard. Although the average sparrow can inflict very little damage and has little incentive to do so, birds of prey are another matter. Towers often provide a unique habitat for some birds, such as the osprey (*Figure 24*) and other predatory birds known as raptors. The raptor classification includes eagles, ospreys, hawks, and vultures. In an attempt to protect the nest, these birds can be very territorial. The beak and claws of these birds can inflict damage and injury. Although a tower that is free of nests represents a much smaller potential for bird attacks, they can and do happen anyway.

Bird droppings can be both smelly and slippery. Some birds of prey, like the vulture, have highly corrosive digestive juices which enable them to digest putrid meat that would be lethal to man and other animals. When threatened, the contents of the stomach are used as a means of defense through projectile vomiting. This is not the way you want to start your day.

Bird droppings can also carry a variety of diseases. Contact with droppings and other bird discharges should always be avoided.

Figure 24 Osprey, or fish eagle.

3.2.2 Insects

Towers and poles are common sites for the nests of stinging insects, and they should be avoided. Although some experienced climbers may have developed their own solutions to the problem, avoidance is always best. Paper wasps seem particularly drawn to communications and power transmission towers, although the reason is not clear. Bee stings can produce a serious allergic reaction in some people. If not immediately treated, people with such allergies can die from these stings.

Spiders are also a concern to the climber. It is very rare though, that a poisonous species would be found on the higher structures. Some people, however, tend to overreact to the mere sight of a spider and other insects. For most insects, the danger from your own reaction is far more serious than the damage the insect itself is likely to inflict.

3.2.3 Snakes

Snakes are generally considered to be a hazard that is limited to the ground crew. Most snakes that would climb a short distance up a tower are non-poisonous as well. But hawks and eagles prey on snakes, and it is not unheard of to find live snakes that have been dropped by these birds onto a tower. In general though, it is the ground crew members that must keep an eye out for snakes, especially in remote areas.

3.2.4 Humans

One would not generally think of humans as being a potential hazard for the climber. But, in fact, humans can represent a significant danger. Agitated homeowners and others who may disapprove of the tall steel structures have been known to target climbers. Vandals and thieves can damage important structural supports or steal copper grounding devices. Even private aircraft have been known to harass climbers on structures. Probably the most direct threat from humans comes from an unskilled or compromised member of a climbing team. Improper

> **On Site**
>
> **Protected Birds**
>
> Migratory birds, including eagles, herons, cranes, and ospreys, are protected by Federal law. It is illegal to trap, kill, injure or otherwise disturb these birds and their habitat. Violation of these laws can result in fines and jail time.

handling of tools and equipment by both climbers and ground crew members can be extremely dangerous.

3.3.0 Weather

Weather conditions and the potential dangers they represent must be considered well beyond the immediate situation. A clear day can quickly become a stormy one. There are a number of ways to deal with these issues while safely on the ground, but the climber does not share that same advantage. It is also important to remember that temperatures and wind conditions can be very different at a higher elevation than on the ground.

In all matters related to weather, it is essential the climber be familiar with the guidelines and regulations of their employer, and follow them without deviation.

3.3.1 Sunshine

The sun can cause discomfort in a variety of ways. Climbers should always be equipped with sunglasses to protect their eyes. Part of the day, the sun's glare may not be a factor, but the glare may increase as the sun shifts. Climbers must also protect themselves from burning rays with proper clothing and sunscreen. On warmer days, water should either be carried by the climber or provided by the ground crew to prevent the possibility of dehydration. Climbers must remain properly hydrated to prevent a loss of strength or agility that can affect the descent after many hours in position.

3.3.2 Rain, Snow, and Ice

Water makes almost any surface more slippery, and steel towers and poles are no exception. It is very difficult to avoid wet surfaces completely, as a climber will often be called upon to climb while dew or the remnants of recent rain are still present. The greater concern will come from the other forms water often takes—ice and snow (*Figure 25*).

Slipping and sliding are the most obvious hazards of ice. It can be difficult to negotiate an icy surface with both feet on the ground. Imagine the challenge at 100 feet in the air on a steel structure. As mentioned earlier, temperatures and conditions are different at a higher elevation. Water at ground level may become ice as you ascend. Under the right conditions, a tower can be wet from top to bottom during the climb, then freeze while work is being conducted and your attention is diverted from the weather. Ice also imposes structural loads on a tower due to its weight. It can form deadly projectiles as pieces break off and fall to earth. Any time ice is present or is expected to form during a climb, the work should not be considered safe. However, an experienced climber may accept this challenge in an emergency.

49106-11_F25.EPS

Figure 25 Ice and snow can interfere with climbing and needed repairs.

The hazards of snow can vary due to differences in its consistency. Dry snow may present an acceptable hazard, while wet snow may not. Experienced climbers, especially those in areas where snow is a constant consideration, are generally better suited to evaluate the level of concern it presents.

3.3.3 Lightning

A steel structure high in the air is an obvious target for a lightning strike (Figure 26). Protecting the structure from lightning damage is of the utmost importance in its design and construction.

The view from an elevated position is an advantage. From this perspective, storms may be visible at a great distance, allowing sufficient time to react. Because light travels much faster than sound, you will see lightning before you hear the thunder. If you can see lightning but cannot hear thunder, the lightning probably hit within 10 miles. The sound of thunder travels at about 5 seconds per mile, so if you hear thunder 10 seconds after you see lightning, the lightning is about two miles away. Under no circumstances should a climber ascend a tower, or remain on a tower at work, if thunder is audible. Keep in mind that it can take 30 minutes to descend, and plan accordingly.

49106-11_F26.EPS

Figure 26 Lightning.

3.3.4 Wind

Wind is an enemy of the climber. At a minimum it creates noise, frustration, and a sense of urgency by its presence. At its worst, it can separate the climber from the structure.

Wind has a profound effect on the tower itself, often creating movement that is both disconcerting and distracting. Since poles and towers are often well above any type of wind break that protects you at ground level, wind conditions that are acceptable near the ground may be severe farther up. The towers and poles themselves are designed to withstand the flexing and other stresses, but the climber may not enjoy the experience. Occasionally, especially when other weaknesses are present in the structure, the tower does fail.

What can be identified as an acceptable amount of wind varies depending on the nature and height of the work to be performed. Simple tasks at a lower altitude may be accomplished in a 15-knot wind, while the installation and securing of a new component may not be feasible in a wind of that magnitude.

4.0.0 Climb Preparations

There are a few things that need to be done in preparation for any climb. The work site and the structure itself need to be assessed. In addition, crewmembers should be aware of the location and availability of medical kits and the nearest medical facility. Everyone needs to know what tasks are to be done on the tower itself. These are all points of discussion in a proper pre-climb meeting.

Pre-climb meetings should take place on a daily basis, even on the same jobsite, and they must be taken seriously. All information exchanged and issues discussed should be documented as well.

4.1.0 Site and Tower Assessment

A site and tower assessment must be performed before a pre-climb meeting so that the results can be presented to the group. A thorough assessment of the site includes observing the skies as well as the ground. You have already reviewed hazards associated with the environment, including weather issues. During the site assessment, you put this learning to the test. Look for the various types of wildlife and insects that may be a distraction and deal with them if possible. For insects, be prepared to distribute repellents. Take a walk around the perimeter of the work area and structure, looking for hazards as you go. Make sure that a recent weather forecast has been monitored, and look for weather issues yourself. If it is a sunny day, workers need sunscreen and eye protection. Record all this information for discussion during the pre-climb meeting.

The tower must also be examined carefully, with your known task and work location on the tower in mind. First, think of the big picture in your preliminary analysis of the tower. Ensure that the tower is standing up straight and required guy wires are in place. Look for any evidence of twists or turns in the structure that may indicate a problem. Then begin a more detailed, up-close inspection, checking for loose or missing pieces of the structure, rust and corrosion, missing hardware, or any signs that vandals have damaged the structure. Pay particular attention to tower ground cables and their connections, as they are a favorite target of thieves.

If the tower does have guy wires, there will likely be several different styles of attachment and adjustment that you will encounter. Here are some general guidelines for examining them:

- Inspect the anchors where the guy wire attaches to the ground, especially at the point where the anchor shaft meets the ground or the concrete footing. Sometimes known as the **point of daylight**, this is the location that will first show signs of deterioration and chemical reactions.
- If turnbuckles are in place for adjustment of the guy wire, examine them for loose or missing hardware or damage. Both male threaded shafts should be completely through the adjusting nut, and they should be roughly equal in their length through.

- Inspect other guy wire connecting components as well. Ensure that all connections, thimbles, and terminations appear to be in sound condition and show no evidence of slippage or movement.
- All guy wires should be inspected to make sure they are not loose. They should always be taut. A loose guy wire is an indication that something has changed. Determine the reason for a loose guy wire and correct it before climbing. Consider that when a loose guy wire is noticed, a guy wire on the opposite side will likely be too tight. This indicates a tower that is no longer tuned properly and should be evaluated further before climbing.

After examining the overall condition of the structure, evaluate the climbing route. Inspect whatever step bolts or other climbing devices are permanently installed on the structure that you can see. Those out of sight must be evaluated during the climb. Take a look at any safe climbing assist systems and make sure the proper size and style of rope or cable grab is available. Think about and evaluate where you need to work, plan the climb, and make adjustments as you progress. Know how you will deal with any obstacles in your path, or plan to avoid them altogether. Imagine yourself actively climbing and what you will likely use as the next anchor point on the route.

4.2.0 The Pre-Climb Meeting

The pre-climb meeting should occur on site daily and be well-documented. One important part of the pre-climb meeting is the pre-job briefing. A form is used to document the discussion and important data that should be readily available to all site workers. OSHA requires that a pre-job briefing be conducted prior to the start of each job, and the following five topics should be covered:

- Hazards of the job
- Specific work procedures
- Special precautions required
- Required PPE
- Energy source controls

The hazards of electrical energy on the tower must be discussed during this meeting. The minimum approach distance (MAD) for the nominal voltage on the tower must not be violated by workers, nor by any conductive object in their possession. Any energized lines or equipment must be tested for voltage with an approved voltage tester, proved to be absent of voltage and properly grounded before an employee enters the MAD. Induced voltage from nearby energized circuits can be hazardous to the point of being lethal. Properly ground the conductors on the tower to be climbed and follow approved work practices to avoid the hazards of induced voltage.

A thorough discussion of the task to be performed is needed. Required tools and materials and their locations should be identified, and the order in which tasks will be undertaken should be reviewed so that all climbers and ground crew members know their respective responsibilities. The following are some general topics that should be covered in addition to the specifics of the work to be performed:

- *First aid* – OSHA requires employers to provide adequate access and availability to medical care. If a hospital or emergency room is nearby, you still need to provide reasonable first aid supplies. Because the term *nearby* is subjective, it is best to follow the alternate ruling of OSHA, which requires that a person adequately trained in first aid be available on the site. Each crew member should know location of first aid supplies and everyone should know who is properly trained and who isn't.
- *Emergency medical facilities* – The phone number and location of the nearest emergency medical facility should be clearly written on the pre-job briefing form and made available to all crew members. Workers should have knowledge of the site location in order to guide medical responders to the scene if necessary. GPS coordinates and the physical address of the site should be recorded on the pre-job briefing form. If any hazardous materials are involved with the job, the related MSDS, material handling procedures, and emergency response plans should be reviewed.
- *Rescue planning* – A rescue plan must be in place in the event that a climber becomes incapacitated. This is an OSHA requirement and a good common sense rule. All parties should know who is designated as the rescue climber if one or more climbers needs help, and what form of rescue equipment is going to be used. Climbers should receive tower rescue training every 12 months.
- *Site and tower assessment* – The results of the site and tower assessment completed before the meeting should be reviewed. Make sure the team knows which hazards are present and how to deal with them. Crew members must also be aware of any weather conditions that should be monitored.

- *Equipment inspection* – This may be the chosen time for climbers to inspect their personal gear. If this is done as a group, it is more likely that problems will be noticed.

If you take responsibility for the pre-climb meeting or any aspect of it, take it seriously and always be on the lookout for changes in the environment or working conditions. Once the pre-climb meeting is complete, it is time to put the job in motion.

5.0.0 Basic Climbing Skills

It would be virtually impossible to list every possible guideline or tip that could be associated with climbing, due to the extreme variety of situations and climbing tasks. As a climber, you are subject to the regulations and policies of your employer, federal and state agencies, and possibly those of the equipment owner. There is no shortage of rules to follow as you engage in this line of work. But learning the rules is only a part of the process. In order to learn how to climb, you must practice. The physical process of climbing can be explained to some degree, but much of the learning must come from ascending a structure in the presence of a skilled and experienced climbing instructor.

Climbing safety is not a subject you learn once, take a test, and perform forever based on that single experience. It is a continual learning process that requires climbers to pay close attention to every successful climb, and closer attention to those that did not go well or resulted in an injury. Not only must you be vigilant about every aspect of the climb, but it is quite likely you will be working on or near live, high voltage electrical systems. The distractions created by the electrical hazards and the work at hand cannot be allowed to interfere with your attention to climbing safety. Your full attention must be given to personal safety and fall avoidance, then to the work to be done. The task cannot be done if there is a failure to focus on personal safety first.

5.1.0 Preliminary Considerations

Safe work habits for the individual climber come into play before arrival at the job site. The following basic advice should be considered by each individual:

- Recognize that not everyone was born to climb and work at heights. Some take to it naturally; others work their way into proficiency; while still others are never comfortable with climbing. If you are unable to conquer the fear of climbing to significant heights and working in a relaxed and focused manner, you should not be a climber.
- Understand that the most important safety factor is you, and assume that your safety and the successful completion of the job lies solely with you. Although team members must be able to rely on each other, the best possible outcome is when *all* crew members and climbers are thinking the same thing—that safe execution of the climb relies solely on them.

Think About It

The FACE Program

The National Institute for Occupational Safety and Health (NOISH) publishes a report as part of a program known as Fatality Assessment and Control Evaluation (FACE). The intent of the program is to provide interested users with access to hundreds of fatality investigation reports from a variety of environments. Detailed reports on climbing fatalities are a part of the program. Access to these reports is available at **www.cdc.gov/niosh/face/**. A history and complete description of the program, as well as relevant links to other sites of interest, can also be found there.

It is important to note that the vast majority of deaths due to falls have nothing to do with tower climbers. A review of past reports shows that, of those deaths that are in fact related to tower climbers, they are overwhelmingly related to climbers in the telecommunications industry rather than power transmission and distribution workers. A special report created in 2001 entitled *NIOSH Publication 2001-156, Preventing Injuries and Deaths from Falls During Construction and Maintenance of Telecommunication Towers* is an excellent review of cases and the conclusions or recommendations that were derived from their study. The fact that the report focused on telecommunications towers is an indication of the disparity in fall incidence rates.

Why do you think this difference exists between the two industries in fall incidence? Discuss as a class if possible.

49106-11_SA02.EPS

- Actively participate in all training programs offered to you. Consider safety a lifestyle rather than something you practice only at the job site.
- Follow the policies and procedures outlined by your employer and/or client. If you are not in agreement with them, feel free to discuss such issues with supervisors before the climb. You will not generally be released from following the guidelines and policies simply because you disagree, but changes and improvements in procedures are usually the result of discussion and mutual understanding.
- Climbing is a profession with significant physical and mental demands. As a professional, you must practice proper maintenance of yourself, as well as your equipment. Try to maintain good physical condition, eat properly, drink plenty of fluids, and allow plenty of time for proper sleep.
- Maintain a drug-free lifestyle. Do not climb while under the influence of alcohol or drugs. Note that some over-the-counter medicines, such as cold and allergy medicine, can impair performance. Avoid climbing if you are taking such medicines.
- Do not allow yourself to climb if you feel your health or condition is impaired in any way.

5.2.0 Ascent

Climbing is a very physical process and should be done in a relaxed, unrushed, and confident manner. There is no award for speed, except possibly in a rescue attempt. Climbers often forget the physical exertion involved with climbing and positioning and find themselves suddenly tired. If you tire, stop and attach a short positioning lanyard for a brief rest.

- Inspect your equipment before every climb. Don't assume anything.
- Crewmembers should not be standing directly under the climbing path or the work area at any time. If a fall occurs or a tool is dropped, crewmembers in this area are likely to be injured.
- Climbing is to be done with the legs, shifting weight from one leg to another smoothly and gradually. Do not depend on your arms to pull you up – allow your legs to do the lifting work. Move the legs up first, then reposition each hand higher to prepare for the next step up.
- Connect your back D-ring to the safe climbing assist cable system or the vertical lifeline if installed. Never trust the clicking sound of hardware as a means of verifying connection. Regardless of what you hear or expect to hear, and regardless of how well you think you know your personal equipment, you must visually ensure that every connection made is secure. Sound may help your confidence, but it is no guarantee that a connection was made.
- If safe climb systems are not in place, you should plan to use other structures for anchor points as you ascend. Step bolts or any similar protrusion do not qualify as anchor points because they do not generally meet the load requirements. In addition, lanyards or hooks can slip off them too easily. They are convenient, but are not valid points of connection. You can alternate between fall restraint use (very short effective lanyard length) and fall arrest on your choice of anchor points, or alternate between fall restraint connections alone.
- Look up as you climb, not down. You should be assessing the route ahead of you, looking for any structural issues or obstructions, and planning your series of anchor points. If you are connected to a fall arrest system, climb with any slack in the strap collecting behind you.
- As you climb, you must maintain three points of contact at all times. Hopping or jumping from point-to-point is not the way to go. Move one foot or hand at a time, always remembering to keep three points of the body in contact with the structure. If you must provide your own fall arrest or restraint on the way up (no safe climb system installed), then disconnect and reconnect to anchor points while continuing to maintain three-point contact.
- Step bolts or steps that are provided may not always be at perfectly consistent distances nor perfectly oriented to each other. Think about your foot position as you climb and evaluate each new step on its own.
- Climb at a speed and rhythm that is smooth and free of exaggerated movements to help prevent lanyards, harnesses, and equipment from swinging wildly back and forth.

5.3.0 Maneuvering and Positioning

- Once you arrive at the work location, secure yourself with a positioning strap. You should never disconnect from any safe climbing system until you have attached to another anchor point. Evaluate your situation and determine what else needs to be done to secure your safety in the needed work position.
- Remember that fall restraint lanyards, when used for positioning, should leave no more than 2 feet of movement and must be attached to a 3,000 pound anchor point. Your fall arrest

lanyards should be attached to a 5,000-pound anchor point and should be higher than your shoulders whenever possible to minimize the fall distance and the swing zone. For each climber attached to a single point, the load capacity is increased another 5,000 pounds. In other words, for you and two other workers to share a fall arrest anchor point, it must be able to handle a 15,000-pound load. Take a good look at it and consider if you would suspend three or four pickup trucks from the anchor and expect it to hold.

- Towers generally have many diagonals in their structure. Always think twice about the result when connecting lanyards to diagonals. Remember that there will always be a tendency for the strap to slide down the diagonal to the next structural intersection. Visualize your connections here and consider what to do to avoid a problem that either causes a fall or creates a very challenging disconnect situation.
- When attaching a PFAS system and using deceleration lanyards, remember that there is a 6-foot fall and roughly 3-foot deceleration distance, at a minimum. When using self-retracting lanyards that are of significant length, never forget that the convenience has a price. It may save a number of connects and disconnects, but you must always be mindful of, and respect, your swing zone and potential fall distance when using them. Consider the SRL anchoring position carefully, and it is quite likely that you will need to climb to your chosen anchor point using fall restraint only until you have it securely in place and connected to your back D-ring.

5.4.0 Descent

- If you are using a safe climbing system, return to it to begin your descent. Do not disconnect fall restraint lanyards until you are connected to the rope or cable grabs.
- Make a final check to ensure tools and equipment will not interfere with your descent and you are well balanced.
- During the descent, the hands are moved down first, then the legs are moved. The hands and arms should still not be supporting your weight in any way, nor should your grip be too tight. The legs continue to do the heavy work, but they are not pushed down—simply allow the weight of your body and the leg itself to lower it into position.
- Don't skip step bolts or try to descend too rapidly.
- Do not be tempted to rappel or slide down the tower. This is forbidden.
- Be aware of your rhythm, just as you were during the ascent. Most climbers tend to descend a bit quicker and with more rhythm in the step, which causes all of your equipment to swing. Any displays of excitement about finally being down from a difficult all-day aerial position should be saved for ground contact. Keep your mind on the descent and your fall arrest and restraint system use.
- Once your climb is complete, look over all of your equipment and stow it away in an organized manner. Look for any damage that might have occurred. You know better than anyone what happened up there, and where a lanyard might have taken some abuse.

SUMMARY

For those that have the skill and aptitude for climbing, opportunities in the electrical industry will continue to expand. Safe climbing begins with a thorough understanding of the equipment that is your connection to security. You must inspect and care for your equipment with the knowledge that, at any moment, it may save your life.

Climbers are faced with a number of challenges unique to their environment. Some of these can, and should, prevent a climbing event altogether, while others present a simple distraction that can lead to an accident. Preparation for a climb in the form of a site and tower inspection helps reveal hazards and challenges in the general area and on the tower itself. Pre-climb meetings and job briefings should precede each climb, and include a variety of pertinent subjects.

Consistent, quality training is essential to every climber and their continued safety. Although learning the basics of moving up and around a tower from a textbook is an appropriate start, climbing is a physical skill that must be learned and practiced with hands on steel.

Review Questions

1. All of the following items are considered by OSHA to be acceptable fall prevention systems for working at heights above 6 feet *except* for _____.
   a. a guardrail system
   b. personal fall protection
   c. the two-climber buddy system
   d. a safety net system

2. Fall arrest systems _____.
   a. stop or take control of a fall in progress
   b. prevent falls from ever occurring
   c. ensure that the climber does not swing from side-to-side during a fall
   d. ensure that every fall is reported to the proper authorities

3. OSHA requires that a PFAS limits the maximum force imparted to the body during fall arrest to _____.
   a. 500 pounds
   b. 1,200 pounds
   c. 1,800 pounds
   d. 2,500 pounds

4. The free fall distance allowed by a PFAS should be no more than _____.
   a. 3½ feet
   b. 6 feet
   c. 10 feet
   d. 12 feet

5. The organization that establishes many standards related to the performance of products and devices, including body harnesses, is the _____.
   a. ANSI
   b. AFPA
   c. NSTA
   d. NCCER

6. The three primary components of a personal fall arrest system are _____.
   a. anchor points, connecting devices, and a shock absorbing lanyard
   b. a body harness, connecting devices, and a shock absorbing lanyard
   c. anchor points, a body harness, and a vertical lifeline
   d. anchor points, a body harness, and connecting devices

7. When an anchor point rated at 5,000 pounds is not available, the climber must then ensure that the chosen anchor position can at least withstand the maximum load that could be placed on it during a fall arrest, multiplied by a factor of _____.
   a. 1.5
   b. 2
   c. 3
   d. 5

8. Anchor points that are used as fall restraint anchors, and *not* as fall arrest anchors, must be load rated to handle _____.
   a. 2,500 pounds
   b. 3,000 pounds
   c. 5,000 pounds
   d. 7,500 pounds

9. The climber's personal fall arrest harness must be connected to the anchor point using _____.
   a. at least two lanyards
   b. the chest D-ring only
   c. both the chest and back D-ring
   d. the back D-ring only

10. Groin straps are an integral part of the full body harness and *not* an optional accessory.
    a. True
    b. False

11. A properly connected positioning lanyard should *not* allow a climber to fall more than _____.
    a. 18 inches
    b. 2 feet
    c. 3½ feet
    d. 6 feet

12. Self-retracting lanyards _____.
    a. are available in lengths up to 200 feet
    b. are limited to a length of 10 feet
    c. should never be used without adding a shock absorbing lanyard
    d. should stop a climber before the fall distance reaches 10 feet

13. It is suggested that climbers wear _____.
   a. short sleeve shirts whenever weather allows
   b. short sleeve shirts without a collar
   c. long sleeve shirts without a collar
   d. long sleeve shirts with a collar

14. Personal safety equipment that has been involved in a fall arrest must be _____.
   a. taken out of service
   b. inspected by the manufacturer before reusing
   c. field-tested at the appropriate load rating before reusing
   d. tagged and used only for fall restraint

15. Tower climbing should never be attempted when _____.
   a. winds are above 5 knots
   b. lightning is in the area
   c. the tower is wet
   d. thunderstorms are in the forecast

16. The minimum approach distance is based upon the _____.
   a. height of the tower
   b. work to be done and number of crew members
   c. potential current flow in energized electrical components
   d. voltage of energized electrical components

17. It is recommended that climbers complete a course in tower rescue procedures _____.
   a. once a month when actively climbing
   b. every 6 months
   c. once per year
   d. only once if they are to be assigned rescue responsibilities

18. When you are an experienced climber, if you do not agree with a particular climbing policy of the employer, you should _____.
   a. do what you know to be safer
   b. discuss it with your supervisor
   c. write to OSHA for an opinion
   d. report your feelings to ANSI

19. While climbing, you should always maintain _____.
   a. three points of contact
   b. two or more lanyard connections to anchor points
   c. visual contact with the horizon
   d. visual contact with the ground below

20. The maximum number of climbers that can be connected to a single anchor point is based on the _____.
   a. type of harnesses used by each climber
   b. load rating of the anchor point
   c. load rating of the carabiner
   d. load rating of the chosen shock-absorbing lanyards

Trade Terms Quiz

Fill in the blank with the correct term that you learned from your study of this module.

1. A device shaped like the link of a chain that can be securely closed once a line is inserted into it is called a(n) ____________.
2. The ____________ is a measure force.
3. Devices used to attach the PFAS to anchor points are referred to as ____________.
4. The point where a tower component meets the soil is called the ____________.
5. The attaching point for climbing or rigging systems is called a(n) ____________.
6. A system used to control a fall in progress is called a(n) ____________.
7. A component that locks onto a cable when a too-rapid descent occurs is a(n) ____________.
8. A system of straps and rings that distributes the body weight evenly over the body is a(n) ____________.
9. The device used to prevent a fall from occurring is the ____________.
10. The ____________ is the area in space within which the body would swing in a fall.
11. A rapid descent involving a series of short leaps is referred to as a(n) ____________.

Trade Terms

Anchor point
Body harness
Cable grabs
Carabiner
Connecting devices
Fall arrest
Fall restraint
Newtons
Point of daylight
Rappel
Swing zone

Cornerstone of Craftsmanship

Robert (Bob) Groner

Safety Manager Southern Division
MasTec Energy North America,
Asheboro, NC

Years ago, Bob Groner had the good fortune to be laid off from the aircraft industry. Instead of going back when the work picked up, he joined the local utility company and found a rewarding job that has lasted 34 years. After an apprenticeship and continuing training over the years, he became a training supervisor, where he's been able to pass along the job knowledge, training, and skills gained over his long career.

How did you get started in the construction industry?
I was laid off from the aircraft industry and found a job with the local utility company (Florida Power & Light) that lasted 34 years. During that time I was trained in many aspects of distribution utility construction and maintenance, including a 4-year apprentice lineman training program.

Who inspired you to enter the industry?
I found the vocation to be challenging and rewarding along with providing a good income for my family.

What do you enjoy most about your job?
All of the above, but the most rewarding has been passing along the job knowledge, training, and skills I gained over many years in the industry.

Do you think training and education are important in construction?
I think they are the most important part, other than experience. A worker who has been properly trained has many opportunities for advancement.

How important are NCCER credentials to your career?
Until recently they were not part of any training I received. However, since being part of the training development team as a subject matter expert, I have come to realize the importance of the role NCCER is playing in the distribution-transmission industry. Utility workers who complete this training are better prepared to work safely and efficiently when they go on the job.

How has training/construction impacted your life and your career?
Without training over the length of my career with a major utility, I would not have been able to rise to the position of Training Supervisor and go on after retirement to work for a major contractor as a trainer and safety manager.

Would you suggest construction as a career to others?
Only to those who truly want the challenge of hard work and continuing education to keep step with the industry.

How do you define craftsmanship?
A craftsman is a person who takes pride in his or her work and never settles for less than the best, in both safety and product.

Trade Terms Introduced in This Module

Anchor point: The location of attachments on a structure for all types of climbing and/or rigging systems.

Body harness: A system of straps and rings worn on the body with the intent of distributing weight and force applied evenly across the shoulders, chest, waist, thighs, and pelvic area. The body harness must not be confused with a body belt, which is simply worn around the waist and is not approved as a fall arrest device.

Cable grabs: Components which ride along the length of a cable smoothly when moved casually, but lock onto the cable sharply when movement becomes too rapid or violent.

Carabiner: A chain-link shaped device which can be opened on one side for the insertion of a line, then closed securely. They are usually rated in Newtons, since they are tested by impact to determine their strength.

Connecting devices: Devices used to connect the PFAS and positioning belts to anchor points and positioning points.

Fall arrest: A means of stopping or controlling a fall in progress without injury to the climber.

Fall restraint: A means to prevent a fall from occurring.

Newtons: A measure of force applied, equal to the amount of force required to accelerate a mass of one kilogram at a rate of one meter per second, per second. One pound of force = 4.45 Newtons.

Point of daylight: The point where anchor assemblies or tower components meet the soil and they are no longer exposed to daylight.

Rappel: Descent of a vertical surface by sliding down a rope, typically while facing the surface and performing a series of short backward leaps to control the descent.

Swing zone: The area in space where the momentum and inertia of a fall would cause the body or protected object to swing until a center of gravity is stabilized (hanging straight down).

Additional Resources

This module presents thorough resources for task training. The following resource material is suggested for further study.

OSHA Regulation 1926, Subpart M, Fall Protection, Latest edition. Occupational Health and Safety Administration. **www.osha.gov**.

Figure Credits

Tony Vazquez, Module opener

Topaz Publications, Inc., Figure 1

TIC, The Industrial Company, Figure 2

Fall protection materials provided courtesy of Miller Fall Protection, Franklin, PA, Figures 3–23

NASA, Figure 24

© iStockphoto.com/Curt_Pickens, Figure 25

© 2010 Photos.com, a division of Getty Images. All rights reserved, Figure 26

Centers for Disease Control and Prevention, National Institute for Occupational Safety and Health, SA01

CONTREN® LEARNING SERIES — USER UPDATE

NCCER makes every effort to keep its textbooks up-to-date and free of technical errors. We appreciate your help in this process. If you find an error, a typographical mistake, or an inaccuracy in NCCER's Contren® materials, please fill out this form (or a photocopy), or complete the online form at www.nccer.org/olf. Be sure to include the exact module number, page number, a detailed description, and your recommended correction. Your input will be brought to the attention of the Authoring Team. Thank you for your assistance.

Instructors – If you have an idea for improving this textbook, or have found that additional materials were necessary to teach this module effectively, please let us know so that we may present your suggestions to the Authoring Team.

NCCER Product Development and Revision

3600 NW 43rd Street, Building G, Gainesville, FL 32606

Fax: 352-334-0932
Email: curriculum@nccer.org
Online: www.nccer.org/olf

☐ Trainee Guide ☐ AIG ☐ Exam ☐ PowerPoints Other ____________

Craft / Level: Copyright Date:

Module Number / Title:

Section Number(s):

Description:

Recommended Correction:

Your Name:

Address:

Email: Phone:

Tools of the Trade

49107-11

Trainees with successful module completions may be eligible for credentialing through NCCER's National Registry. To learn more, go to **www.nccer.org** or contact us at **1.888.622.3720.** Our website has information on the latest product releases and training, as well as online versions of our *Cornerstone* newsletter and Pearson's Contren® product catalog.

Your feedback is welcome. You may email your comments to **curriculum@nccer.org**, send general comments and inquiries to **info@nccer.org**, or use the User Update form at the back of this module.

V.1 7/11

49107-11

Tools of the Trade

Objectives

When you have completed this module, you will be able to do the following:

1. Identify and explain the use of common insulated hand tools.
2. Identify and explain the use of line workers' ladders.
3. Identify and explain the use of line workers' specialty tools.
4. Demonstrate the ability to use line workers' tools specified by the instructor.

Performance Tasks

Under the supervision of the instructor, you should be able to do the following:

1. Demonstrate the ability to use five line worker tools specified by the instructor. These tools may include the following:
   - Clamp stick
   - Loadbuster®
   - Web hoist (Jack strap)
   - Hand line
   - Crimping tool

Trade Terms

Chuck
Circular mil
Conductor
Cutout
Insulated

Contents

Topics to be presented in this module include:

Figures and Tables

1.0.0 Introduction

The power line worker uses a variety of tools on the job every day. It is very important to understand how to properly inspect, maintain, and use these tools safely. Power line workers use the common hand and power tools that were previously covered in the *Core Curriculum* portion of this training program. This module describes the additional tools that are more specific to power line workers. These are tools line workers use on a daily basis. The type of truck you work from and the nature of the job you are doing will determine which tools you will need. This module identifies the more common tools found on the power line service vehicles.

2.0.0 Insulated Tools

Historically, electrocutions have been among the leading causes of occupational fatalities. In addition to the thousands of fatalities annually, there are also a large number of electrical-related disabling injuries. In an effort to reduce the number of fatalities and injuries, OSHA published the initial regulations for electrical safety in the workplace in 1990. One of these regulations was a policy that requires the use of insulated tools when working near energized circuits. *OSHA Standard 29 CFR 1910.335(a)(2)(i)* reads as follows: "When working near exposed energized conductors or circuit parts, each employee shall use insulated tools or handling equipment if the tools or handling equipment might make contact with such conductors or parts . . ."

Insulated hand tools are individually tested and certified by the manufacturer to be suitable for specific working conditions. They must also be periodically inspected to ensure that they remain safe to use. Generally, the maximum rated voltage for insulated hand tools, such as pliers and screwdrivers, is 1,000 volts AC and 1,500 volts DC. However, power line workers must obey the minimum approach distance guidelines, which means there will be many times they cannot get close enough to their specific work area to use traditional hand tools. To enable power line workers to maintain the components on energized lines, a variety of long-reach insulated tools are available. Requirements for live-line tools are found in *OSHA Standard 29 CFR 1910.269(j)*. Live-line tools made of fiberglass-reinforced plastic (FRP) must be designed and built to withstand test voltage of 100,000 volts per foot of length for five minutes per *ASTM F711, Standard Specification for Fiberglass-Reinforced Plastic (FRP) Rod and Tube Used in Live Line Tools*. OSHA requires that live-line tools be cleaned and inspected daily for defects.

Follow these guidelines when using any type of insulated tool:

- Keep tools clean and dry.
- Always handle insulated tools with care.
- Inspect insulation prior to each use.
- If you doubt the integrity of the insulation, destroy the tool or have it retested.
- Follow the manufacturer's temperature recommendations for use.
- Have a qualified person inspect and recertify tools annually for safe use.
- Store the tools in a protective sleeve or tube to prevent damage and to protect against condensation and dust.
- Use required personal protective equipment with insulated tools.

The following sections describe the insulated tools that power line workers use on a regular basis.

2.1.0 Hot Sticks

A hot stick is an insulated pole, usually made of fiberglass. It is used by power line workers when working on energized high-voltage electric power lines and components. Hot sticks have adapters on the working end that allow a variety of tools to be attached. Depending on the tool attached, it is possible to test for voltage, apply tie wires, replace fuses, tighten nuts and bolts, open and close switches, place insulating sleeves on wires, and perform other tasks without exposing the worker to the risk of electric shock.

Hot sticks are made in different lengths, from a few feet long up to telescoping types (extendo poles) that can be extended out to 40' long. Because the fiberglass provides electrical insulation, the hot stick allows utility workers to safely perform operations on power lines without de-energizing them. This is important because operations such as opening or closing combination fuse/switches must often be performed on an energized line. Additionally, after a fault occurs, the exact state of a line may not be certain. In this case, the power line workers must treat the line as though it was energized until it can be proven that it is not, and safely attach grounding cables to the line. If power tools are fitted to the end of the hot stick, they are usually powered hydraulically rather than electrically because hydraulic fluid is also a good insulator. The hydraulic power is commonly supplied from the bucket truck supporting the workers or a utility truck if the workers are on the ground.

The hot stick not only insulates the worker from the energized conductor, it also provides physical separation from the device being operated. This helps reduce the chance of burns that might result from electrical arcing if there is a malfunction of the device being operated.

In the United States, *ASTM F711* specifies stringent requirements for the use of hot sticks. OSHA standards require that they be inspected and electrically tested every two years. *Figure 1* shows a standard hot stick.

Many hot sticks are equipped with a universal tool head (*Figure 2*) to allow you to connect various tools to the end of the stick to match the job you are performing. These universal head sticks provide a swivel head design that allows various tools to be easily connected at the proper angle to efficiently perform a particular job.

2.2.0 Clamp Stick (Shotgun Stick)

A clamp stick (*Figure 3*), commonly referred to as a shotgun stick, contains an easy-to-control hook on the end of an insulated hot stick that can be extended and retracted to grab objects. These tools are used to apply or remove devices such as clamps on electrical lines or conductors. The stick includes a support rod with an insulated head at one end and a movable actuating handle at the other end. The head houses a hook-type claw that is extended and retracted using the sliding hand grip at the bottom end of the stick. The sliding hand grip is raised to extend the hook, and pulled down after securing the device, to retract the hook into the end of the stick, tightly securing the device or tool head. A thumb latch must then be pressed to release the locked hand grip so it can open the hook. The shotgun stick is used most often for installing and removing grounds and mechanical jumpers, but it can be used with various end fittings to open and close overhead and underground circuits.

Shotgun sticks can also be equipped with a universal tool head (*Figure 4*) at the bottom end of the stick. This head allows various tools to be connected.

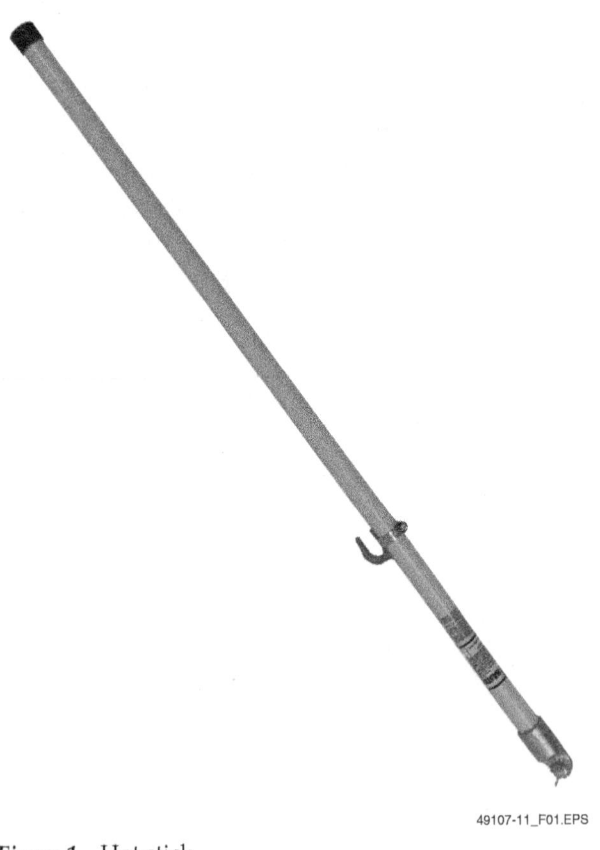

49107-11_F01.EPS

Figure 1 Hot stick.

49107-11_F02.EPS

Figure 2 Hot stick with universal head.

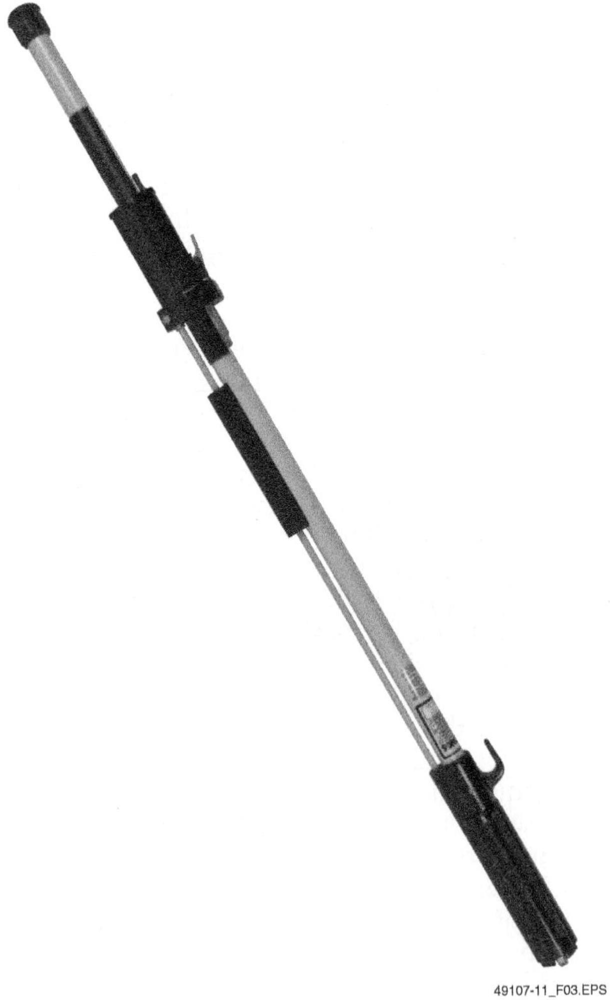

49107-11_F03.EPS

Figure 3 Shotgun stick.

49107-11_F04.EPS

Figure 4 Shotgun stick with universal head.

2.3.0 Extendo Stick

Extendo sticks (*Figure 5*) are telescoping sticks that offer maximum convenience for line workers because they can be used to perform many overhead tasks from ground level. These poles have universal end fittings that accept a variety of attachments so that the power line worker can disconnect switches, replace cutout tubes, remove pole covers, prune trees, and perform many other overhead tasks. These sticks retract to approximately 4' or 5' for easy storage and portability.

Heavy-duty, spring-loaded plastic buttons are used to keep the extended portions of the stick securely locked into position. As each section is extended and slightly rotated, the buttons pop into place, providing a secure lock of each extended section of the pole.

2.4.0 Telescoping Measuring Sticks

Telescoping measuring sticks (*Figure 6*) are handy tools used to quickly determine ground clearances and measure pole heights. These sticks operate the same way as the extendo sticks, and are available with English and metric measuring scales. The insulated head of the measuring stick contains a hook that can be placed over lines or into conductor rings to measure clearance heights from the ground.

2.5.0 Insulated Rescue Hook

The insulated rescue hook (*Figure 7*) is used to withdraw an injured worker from a hazardous area such as inside a confined space, near an electrical cabinet, or within a vault. It features a foam-filled, fiberglass-reinforced handle and a coated, heat treated body hook with an 18" opening. Standard lengths are 6' and 8'; other lengths are available as needed.

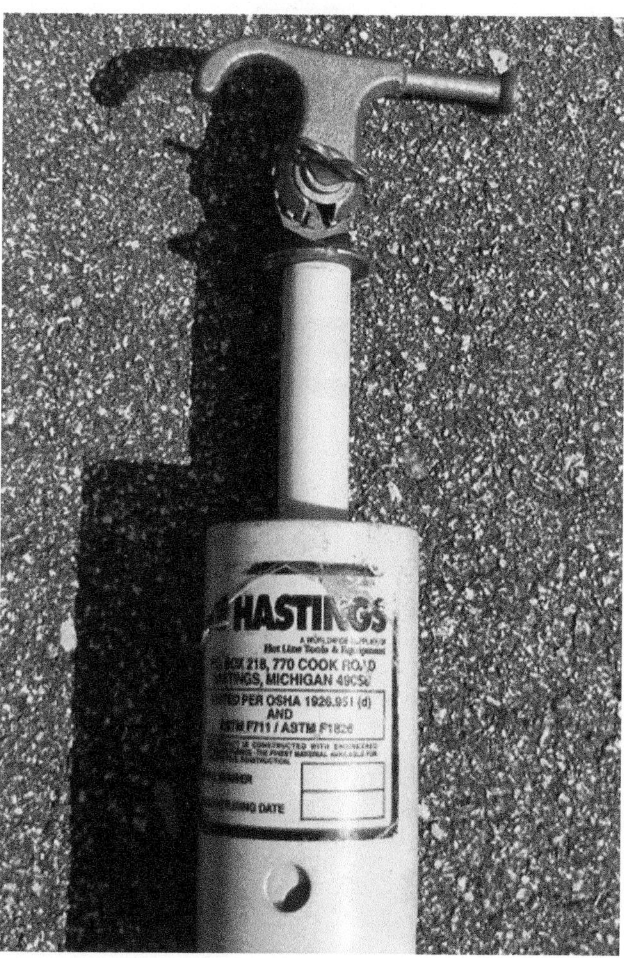

49107-11_F05.EPS

Figure 5 Telescoping extendo stick.

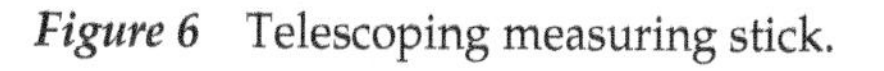

49107-11_F06.EPS

Figure 6 Telescoping measuring stick.

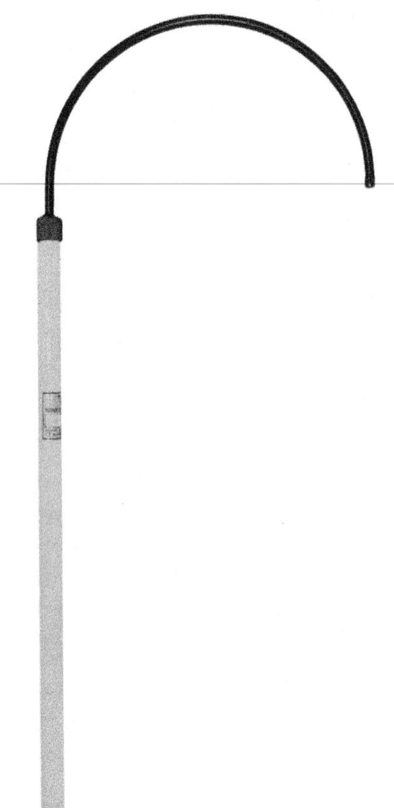

Figure 7 Insulated rescue hook.

2.6.0 Universal Tool Accessories

There are a wide variety of tool heads that attach to universal tool head adapters on hot sticks, clamp sticks, and extendo sticks. In addition, there are three types of tool head adapters used on these sticks.

The quick-change adapter has a ⅝" threaded stud that threads into a hexagon ferrule. A hexagon collar slides down over this ferrule to make a strong, positive connection.

The universal adapter allows a quick-change tool head to be attached to the fiberglass stick at several different angles. The tool head is secured to the universal adapter by tightening a wing nut.

The grip-all or eye screw adapter is used with shotgun sticks. The hook from the shotgun stick is looped through the eye hole of the adapter. The hook is then retracted into the clamp stick and secured.

The following sections describe some of the more common tool heads that can be attached to hot sticks and telescoping sticks. Although most of the tools have universal fittings, all three types of fittings are available for each.

2.6.1 Disconnect Tool

The disconnect tool (*Figure 8*) is used to open switches and enclosed cutouts. They are also used for installing and removing open-link fuse links.

2.6.2 Chuck Blank

The chuck blank (*Figure 9*) can hold screwdrivers, hack saws, wrenches, and other tools that are inserted into the device and soldered into place. Variations of the chuck blank include a wing nut that can be tightened to secure the tool in place.

2.6.3 Fixed-Prong Tie Stick Head

The fixed-prong tie stick head (*Figure 10*) is used to manipulate wires that have looped ends. This tool is helpful when working in close quarters where the loose ends of the wire must be rolled up to prevent contact with cross arms or other hardware.

Figure 8 Disconnects.

Figure 9 Chuck blank.

49107-11_F10.EPS

Figure 10 Fixed-prong tie stick head.

2.6.4 Ratchet Wrench

The ratchet wrench (*Figure 11*) is typically equipped with a ½" drive for accepting sockets; other sizes are available. This tool is handy when tightening bolts in substation equipment and hardware bolts on distribution lines.

2.6.5 Spiral Disconnect

The spiral disconnect (*Figure 12*), commonly referred to as a pigtail disconnect, is an extremely useful tool for opening switches and removing and installing cutout doors on porcelain type cutouts. A pointed disconnect tool is also available.

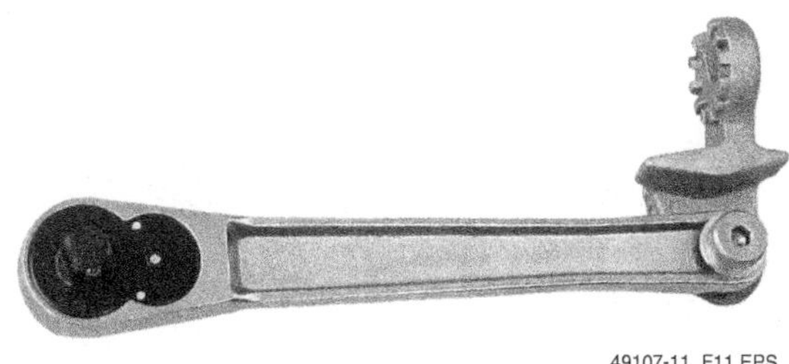
49107-11_F11.EPS

Figure 11 Ratchet wrench.

49107-11_F12.EPS

Figure 12 Spiral disconnect.

2.6.6 Fixed-Blade Tie Stick Head

The fixed-blade tie stick head (*Figure 13*) can be used to manipulate tie wires with or without looped ends. The V-notched blade is designed to fit into tight places in order to loosen wires. The head is also set at a 60-degree angle from the pole, to make it easier to operate.

2.6.7 Flexible Wrench Head

The flexible wrench head with ½" square drive (*Figure 13*) is designed to fit standard wrench sockets. The heavy-duty spring body allows flexibility to permit use at various angles.

2.6.8 Rotary Prong Tie Stick Head

The rotary-prong tie stick head (*Figure 14*) is used for placing insulator ties with looped ends. The prong swivels freely, allowing a full turn on the tie wire without releasing contact. This minimizes the possibility of kinking or burning wires.

2.6.9 Rotary- and Stationary-Blade Tie Stick Heads

The rotary-blade tie stick head (*Figure 14*) is used to manipulate tie wires with or without looped ends. A V-notched blade securely grasps the wire. The body design permits a swivel action so that

FIXED-BLADE TIE STICK HEAD

FLEXIBLE WRENCH HEAD

49107-11_F13.EPS

Figure 13 Fixed blade tie stick head and flexible wrench head.

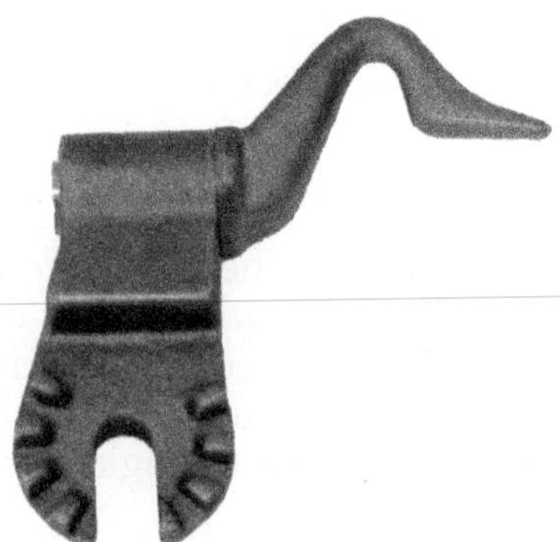

ROTARY-PRONG TIE STICK HEAD

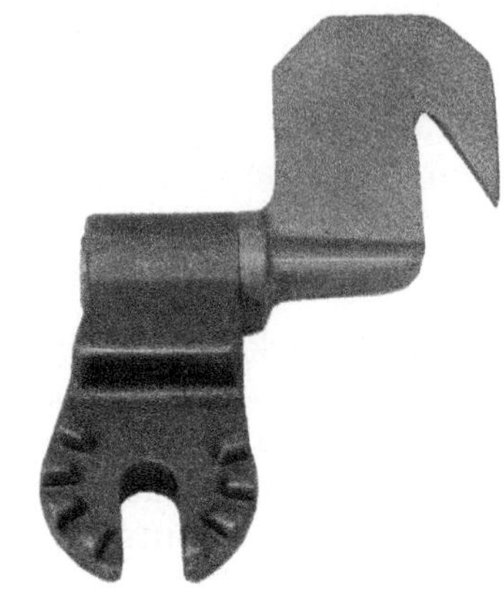

ROTARY-BLADE TIE STICK HEAD

49107-11_F14.EPS

Figure 14 Rotary-prong and rotary-blade tie stick heads.

the wire can be wrapped or unwrapped without turning the universal pole.

A stationary-blade tie stick head is also used for manipulating tie wires, but has a fixed head that does not rotate.

2.6.10 Skinning Knife

The skinning knife (*Figure 15*) is used for cutting or scraping insulation and cleaning conductors before making splices.

2.6.11 Conductor Cleaning Brush

The conductor cleaning brush (*Figure 15*) consists of two brushes in a V-shape. This shape that permits cleaning of conductors by entrapping the conductor between the brushes and working the tool back and forth. When the brushes wear down, they can be rotated and repositioned so that unused bristles come in contact with the conductor. The individual brushes can also be replaced without having to replace the entire tool.

2.6.12 Tubular Line Cleaner

The tubular line cleaner (*Figure 16*) allows the entire circumference of a conductor to be cleaned.

SKINNING KNIFE

CONDUCTOR CLEANING BRUSH

49107-11_F15.EPS

Figure 15 Skinning knife and conductor cleaning brush.

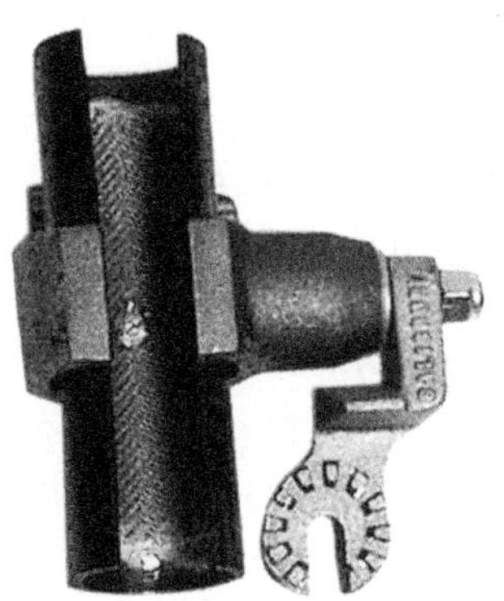

TUBULAR LINE CLEANER

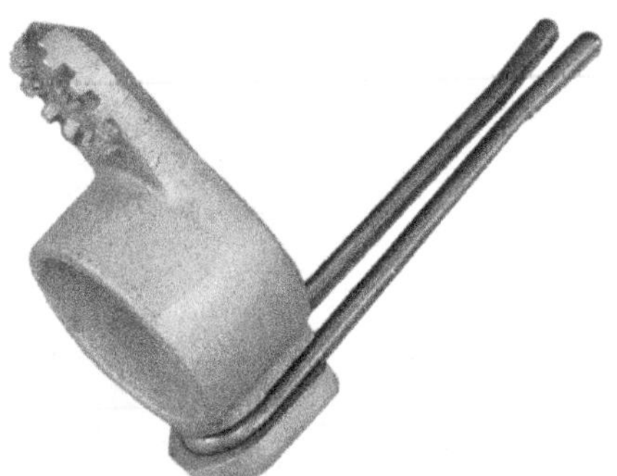

BLANKET PIN TOOL

49107-11_F16.EPS

Figure 16 Tubular line cleaner and blanket pin tool.

2.6.13 Blanket Pin Tool

The blanket pin tool (*Figure 16*) attaches to the end of the hot stick and allows securing pins to be placed on insulating blankets used on live conductors.

2.6.14 Pruning Saw

Pruning saws (*Figure 17*) are available for attachment on universal-head hot sticks and are used for trimming small tree limbs that may come into contact with power lines.

2.6.15 Tree Pruner

The tree pruner (*Figure 17*) contains an integrated non-stick blade and a hook that is used for trimming a portion of a limb. The typical tree pruner attachment can cut up to a 1½" diameter limb. The pruning component operates using a lever, pulley, and cord. The blade is housed inside of hook. This blade is attached to a cord that is pulled to move the blade from its housing inside the hook to make a cut. To cut a limb, just place the hook over the area to be cut and pull the cord. Releasing the cord returns the blade into its housing.

2.7.0 Loadbuster® Tool

A Loadbuster® tool (*Figure 18*) is a load break tool that is used to quickly and safely open energized disconnects, cutouts, power fuses, capacitor banks, and fuse limiters. The Loadbuster® accomplishes this by containing the external electrical charge that occurs when a fuse under a load is opened.

The tool is positioned across the front of the device to be opened, with the tool's anchor placed on the attachment hook on the far side of the device. The device's pull-ring is engaged, with the pull-ring hook on the Loadbuster®. A firm, steady, downward pull is used to extend the Loadbuster® to its maximum length. This will open the device and divert the current through the Loadbuster® arc-extinguishing chamber, or silencer. At the same time, the Loadbuster®'s internal operating spring is charged. Then, at a predetermined time during the operating stroke, its internal trigger trips, the charged operating spring is released, the internal circuits are separated, and the circuit is positively interrupted. Before using it again, the tool must be reset, and then it is ready to break the next circuit. *Figure 19* shows how the circuit is energized through the Loadbuster®.

Figure 17 Pruning saw and tree pruner.

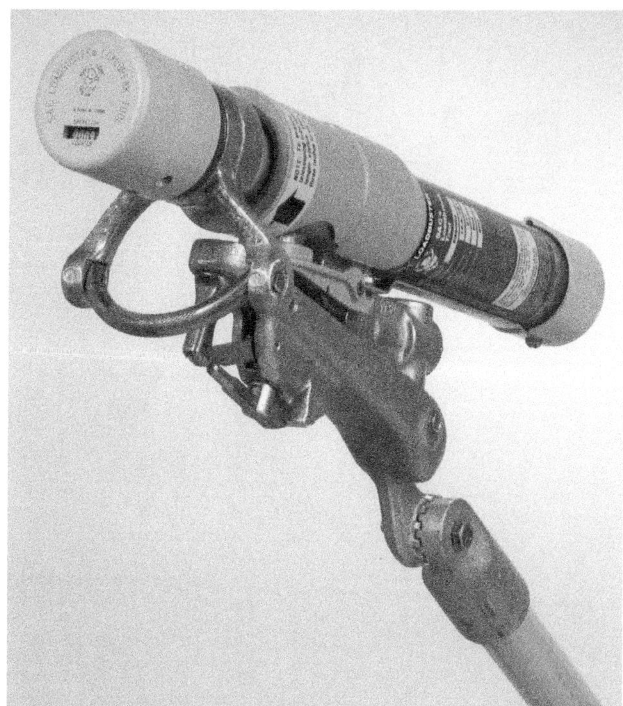

Figure 18 Loadbuster® tool.

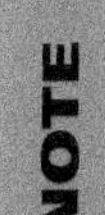

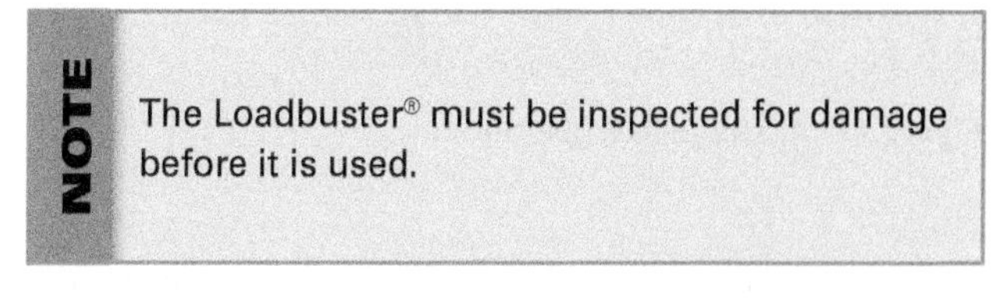

The Loadbuster® must be inspected for damage before it is used.

Loadbuster® operation can be achieved using the following steps:

Step 1 Make sure that the load break mechanism is locked in place prior to use.

Step 2 Attach the Loadbuster® to the cutout. Reach across the front of the cutout and attach the Loadbuster®'s anchor to the attachment hook on the far side of the cutout.

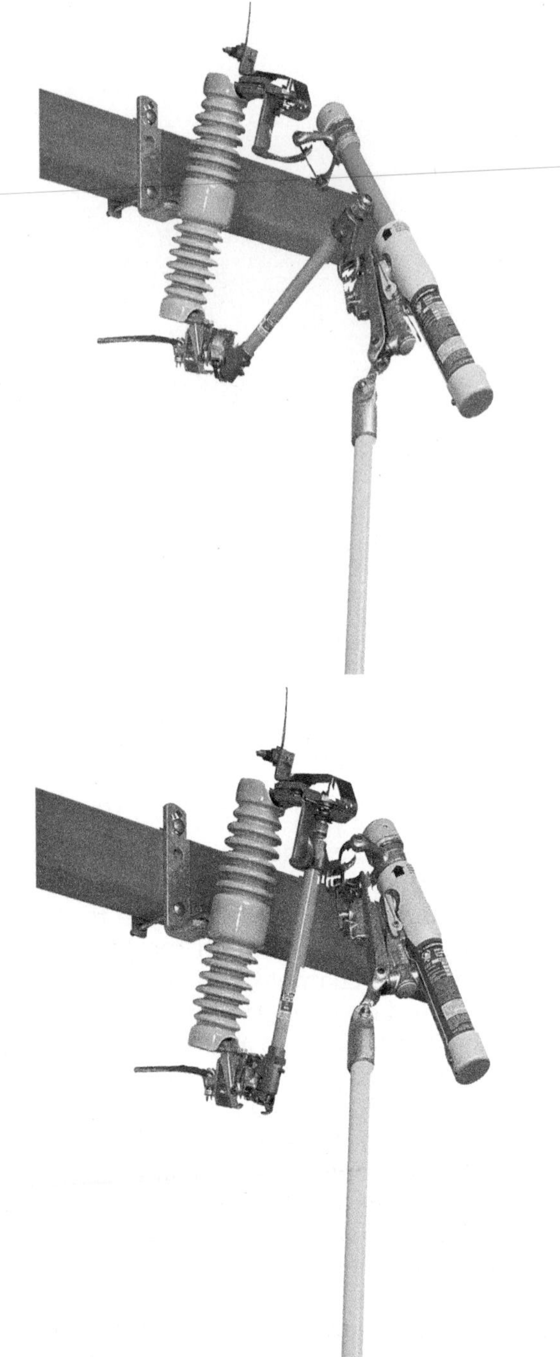

49107-11_F19.EPS

Figure 19 Loadbuster® operation.

Step 3 Engage the cutout's pull ring with the Loadbuster®'s pull-ring hook. The Loadbuster®'s pull-ring latch prevents inadvertent disengagement of the cutout pull ring and the Loadbuster®'s pulling hook.

Step 4 Pull down on the Loadbuster®'s handle using a firm, steady stroke. Pulling the Loadbuster® to its maximum extended length opens the cutout in the normal manner as the current is diverted through the Loadbuster®. At a predetermined point in the opening stroke, the Loadbuster® trips, breaking the circuit positively.

Step 5 With the blade in the open position, remove the Loadbuster® from the pull ring with a simple roll-off motion. Remove the Loadbuster® from the cutout. Disengage the Loadbuster® by removing its anchor from the cutout attachment hook. If more than one device is being opened, recock and check the mechanism.

3.0.0 Ladders and Work Platforms

Power line workers generally use bucket trucks to gain access to the devices on a utility pole. When this is not possible, work platforms and ladders are used to provide a stable work surface for the power line worker. Normally, the power lines are de-energized and grounded before a line worker sets up a ladder at the top of a pole to perform maintenance activities. This is not always the case, however. Power line workers use ladders vertically and horizontally as aerial platforms. There are specially designed ladders that the power line workers use to safely and efficiently gain access to power line components.

Anytime a ladder is used near energized power lines, the following general safety requirements must be observed:

- Keep all ladders and other tools within the minimum approach distance, at least 10' from any power lines. For higher voltage lines, the required distance is even greater (see *Table 1*). When uncertain of a power line's voltage, stay 18' away.
- Never count on a power line to be insulated, no matter what it looks like. Most utility power lines are not insulated.
- Use only wood or fiberglass ladders, but don't count on a wooden ladder to protect you from electric shock. Wood conducts electricity. Fiberglass ladders are best, but even they are no guarantee of safety, especially if they are wet. Wet, dirty, and defective ladders of any kind conduct electricity.

Table 1 Minimum Approach Distance

Power Line Voltages	OSHA Minimum Approach Distance
0 to 69,000 volts	10'
115,000 to 138,000 volts	11'
230,000 volts	13'
500,000 volts	18'

- Before using a ladder, check the area carefully for power lines or other electrical equipment. Pick a safe route to carry the ladder to the work area, and then carry it horizontally–never upright. Set it up only at the work area. Always make sure that if the ladder falls, it will not contact any power lines or other electrical equipment.
- When working from a ladder, balance and control can sometimes be difficult. Be careful with pipes, conduits, gutters, antennas, and other long objects. Never hold them in a position where they could fall onto a power line. Remember that distances are deceiving from the top of a ladder.

3.1.0 Insulated Work Platforms

Electrically insulated aerial work platforms (*Figure 20*) are used by line workers to gain access to utility poles devices when the use of a bucket truck is not an option. The platforms must be constructed and maintained for an electrically insulated condition. They are typically made of laminated wood or fiberglass with a non-skid coating to prevent slippage. Line workers may stand, sit, or kneel on these platforms. Most platforms fold flat for easier storage and to make them easier and safer to transport to the work site on the pole. A typical platform is between 10" and 30" wide, can be 3' to 12' in length, and weighs 45 to 100 pounds. Smaller platforms are also available.

3.2.0 Hook Ladders

Hook ladders (*Figure 21*) have many uses in power line installation, repair, and maintenance. They are used to position line workers in the most advantageous working location, making line repairs possible at otherwise inaccessible places. Hook ladders can be used vertically, at an angle, and horizontally. Two basic styles of hook ladders include regular-duty ladders and heavy-duty ladders. Regular-duty ladders typically have 2" side rails and are normally used for vertical suspension applications. Heavy-duty ladders, with 2½" side rails are recommended where power line workers are required to work from the ladder in a tagged-out position.

49107-11_F20.EPS

Figure 20 Work platform.

49107-11_F21.EPS

Figure 21 Hook ladder.

The hooks at the top of the ladder are formed from 1" diameter tempered steel and can be swiveled to best fit various angles on the structure. Ladders are normally furnished with 8" diameter hooks but 14" or 18" diameters can be ordered for other structural applications. Steel chains clip to the hooks to assist in securing the ladder to the support. There are also eye rings at the bottom of the ladder so that the bottom can be tied off for stability.

3.3.0 Three-Rail Ladders

Three-rail ladders (*Figure 22*) offer greater mechanical strength and less deflection of the ladder when working in a tagged-out position. The third rail also provides a convenient, centered,

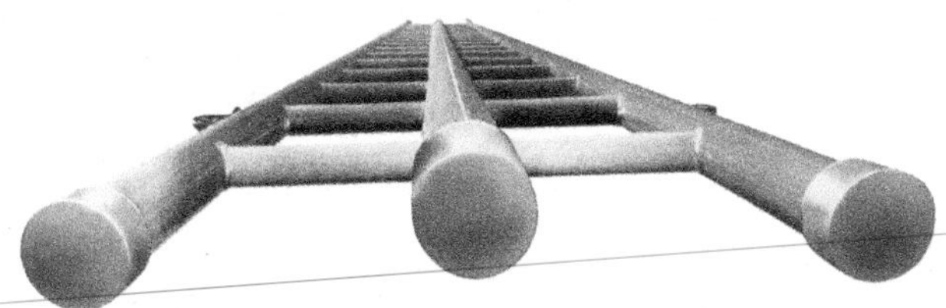

Figure 22 Three rail ladder.

safety belt tie, and divides the ladder rung into natural positions for placement of each foot.

Typically, the center rail is slightly larger than the side rails. A regular-duty three-rail ladder will have a 2½" diameter center rail, capped on both ends, and 2" diameter side rails. These are also available with 8" swivel hooks on the top.

3.4.0 Spliced Ladders

Spliced ladders (*Figure 23*) are available in several different lengths to provide ease in transporting and storage. They offer extended reach capabilities after being pieced together at the job site. These ladders typically use 1"-diameter steel rods to tie-off or tag-out the ladders on the utility pole. These rods have hooks and steel chains to securely fasten them to the supporting structure.

The splice area consists of a steel collar that fits over both ends of the ladder sections and is secured using click pins and eye pins.

Figure 23 Spliced ladder.

4.0.0 Specialty Tools

Power line workers use a variety of specialty tools specific to the power line industry. These tools can be powered using batteries, pneumatic pressure, hydraulic fluid, or powder-actuated controlled explosives.

4.1.0 Battery-Powered Tools

In addition to common battery-powered tools, such as screw drills, hammer drills, reciprocating saws, nail guns, and flashlights, power line workers also use battery-powered crimpers and cable cutters.

> **WARNING!**
>
> Always disconnect the batteries before changing accessories, servicing, or working on battery operated tools to prevent injury from accidental startup.

4.1.1 Crimpers

Battery-powered crimpers (*Figure 24*), or compression tools, are designed to squeeze commonly used service entrance connectors, including lugs

Figure 24 Battery-powered crimper.

and sleeves, and to crimp cables together for non-tension splices and terminations. Some must be used with die heads between the jaws and the item being crimped, while others are designed to be used without dies. High-end models include features such as pressure monitoring, which alerts the operator if the crimping force of the tool does not meet specifications. Most of these tools have an automatic retraction stop feature that automatically retracts the ram just enough after a crimp so that the tool can be moved into position for the next crimp cycle. They are rated according to the output pressure, typically between 4 and 12 tons (8,000 to 24,000 pounds) and the size of wire they can crimp, as measured in circular mils (MCM). A typical 6-ton crimper can crimp cables and lugs up to 750 MCM.

4.1.2 Cable Cutters

Battery-powered cable cutters (*Figure 25*) are used to cut copper cables up to 1,000 MCM in diameter and aluminum cables up to 1,500 MCM in diameter. These cutters have a powerful gear reduction motor that pulls a moveable blade through copper and aluminum cables. They cannot be used to cut steel strand cable. Most of these devices provide one-handed operation and up to 35 cuts on one battery charge, depending on the size of cable being cut.

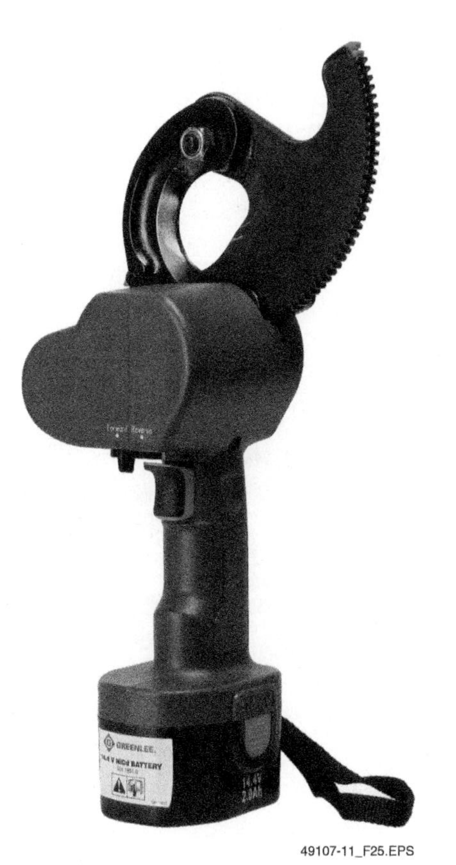

49107-11_F25.EPS

Figure 25 Battery-powered cable cutter.

4.2.0 Pneumatic-Powered Tools

Screw drills, hammer drills, impact wrenches, and nail guns are all commonly used pneumatic powered tools. In addition to these tools, power line workers also use pneumatic underground piercing tools, commonly referred to as a Hole-Hog®. Many times pneumatic tools are preferred over hydraulic tools because they are cleaner to operate, with no possibility of fluid leaks and contamination.

The pneumatic Hole-Hog® (*Figure 26*) is used to place underground cables under driveways, sidewalks, foundations, and other obstacles in short-run applications. As the Hole-Hog® travels underneath these obstructions, it makes a clean compacted tunnel to receive utility service lines, conduit, or pipeline. This allows utilities to be placed underground without disturbing existing surface structures and landscaping with open-cuts, trenching, or augering.

In the event of impassable soil conditions, the Hole-Hog® has a reverse mechanism. The positive locking three-quarter turn reverse permits the

49107-11_F26.EPS

Figure 26 Hole-Hog® underground piercing tool.

Hole-Hog® to be easily removed from impassable soil conditions. The reverse feature also permits easy direction changes and will not slip unexpectedly back into forward.

A puller adapter allows a sleeve to easily follow the Hole-Hog® wherever it goes. The sleeve encases and protects the air supply line making it easy to stop and change direction if necessary. When the Hole-Hog® reaches the target pit, the hex flush mount bolts can be removed, the air line uncoupled, and the Hole-Hog can be removed from the pit. The air line is pulled back through the sleeve, and utility lines can then be installed inside the sleeve.

4.3.0 Hydraulic-Powered Tools

Power line workers use a variety of hydraulic tools to maintain overhead and underground power line installations. Most bucket trucks and pole trucks (digger derricks) have hydraulic lines with quick-release connection points to make hydraulic power accessible when on site. The typical hydraulic tools include rotary drills, saws, pumps, impact wrenches, tamps, and jackhammers. Specialty tools include cable crimpers, cable cutters, and cable spikes.

4.3.1 Hydraulic Cable Crimpers

Manual and remote-powered hydraulic cable crimpers (*Figure 27*) are used to squeeze commonly used connectors, including lugs and sleeves, and to crimp cables together for splices and terminations. Some must be used with die heads between the jaws and the item being crimped, while others are designed to be used without dies. Hydraulic crimpers are available hand-held manual crimpers that offer 6 to 12 tons of crimping force, as well as remote-powered hydraulic crimpers that require an external source of hydraulic pressure. These crimpers offer up to 15 tons of crimping pressure and operate using 10,000 psi of hydraulic pressure.

4.3.2 Hydraulic Cable Cutters

Hydraulic cable cutters (*Figure 28*) are also available to cut through copper and aluminum cables. These tools can be manually powered, powered by remote cylinder attached to the cutter, or remotely powered from a pressurized hydraulic source, such as the connection on a bucket truck or digger derrick. They typically offer higher cutting pressures than battery-powered cutters. Always check the manufacturer's literature for cutting pressures and ratings.

4.3.3 Hydraulic Cable Spike

Hydraulic cable spikes (*Figure 29*) are used to verify that underground cables are de-energized before cutting, repairing, splicing, or replacing. The design includes a solid brass body with stainless steel piercing tip that is used to penetrate the cable to test for voltages. The unit is fitted to hydraulic compression tools so that the cable can be spiked from safe distances. Always follow the manufacturer's instructions for use and ensure that all safety guidelines are followed before using the hydraulic spike.

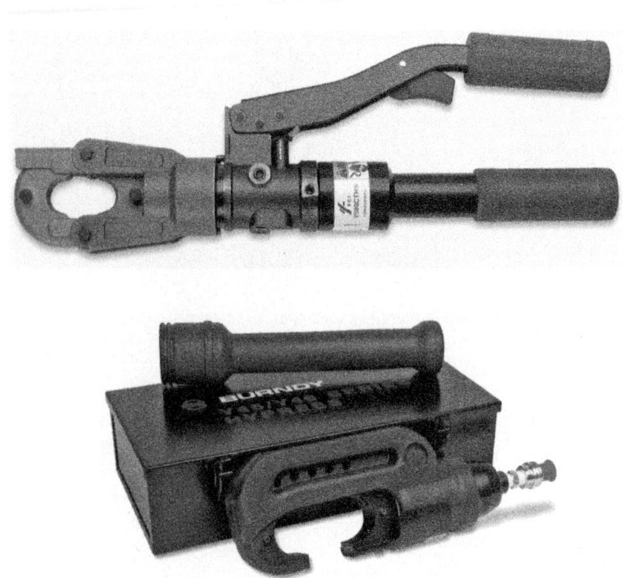

49107-11_F27.EPS

Figure 27 Hydraulic crimpers.

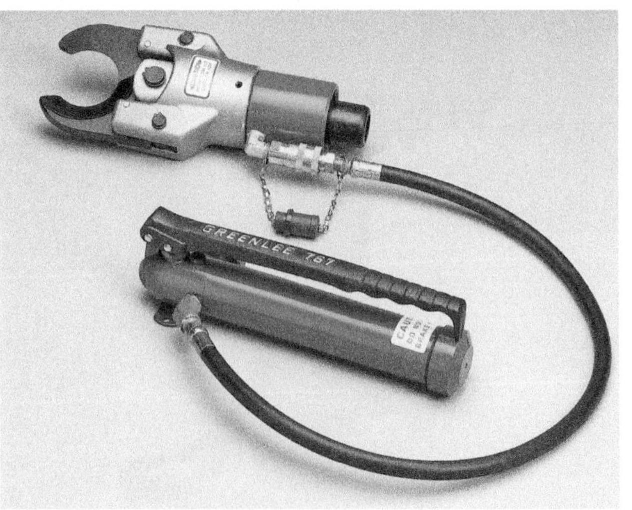

49107-11_F28.EPS

Figure 28 Hydraulic cutter.

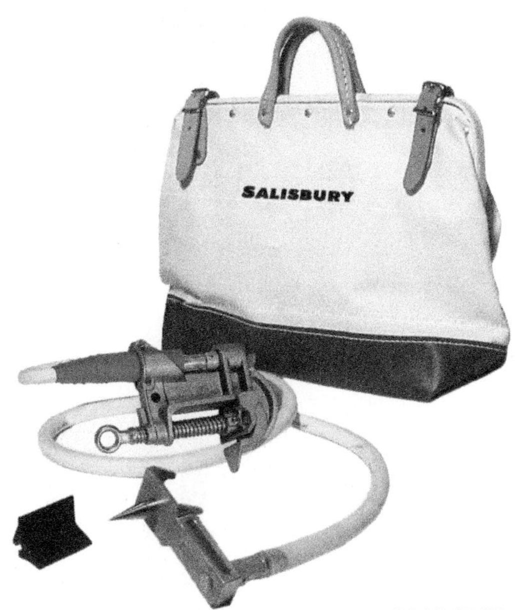

49107-11_F29.EPS

Figure 29 Hydraulic spike.

4.4.0 Powder-Actuated Tools

The most commonly used powder-actuated tool that a power line worker will use is the AMPACT™ cable splicer. The AMPACT™ tool sets a copper tap between two wire cables or between a wire cable and a ground rod when a permanent splice or connection is to be made. The copper tap consists of a pure copper wedge and an aluminum bronze alloy "C" member (*Figure 30*).

The "C" member is placed over the cables to be spliced and the AMPACT™ tool is used to force the wedge between the cables. This forms a secure connection that maintains constant mechanical pressure between the components being spliced. When properly matched, these taps meet or exceed the current-carrying capacity of the conductors being connected. These taps can be used on stranded wire or solid rod.

The AMPACT™ tool is used to drive the wedge in place between the wires. It is a powder-actuated tool that clamps around the "C" member with the wedge pressed into place. A firing cartridge is placed in the handle of the AMPACT™ tool (*Figure 31*) and is actuated by striking the end of the handle with a hammer.

The firing cartridges are color coded and should be matched with the color code on the wedge and cartridge clips (*Figure 32*). This ensures proper firing pressure for the size of wedge being used.

4.5.0 Capstan Winch

Capstan winches are used for industrial pulling and hoisting applications. While capstan winches can be used for hoisting, their main function is pulling. Capstan winches are commonly installed on the front or rear of the cab on a digger derrick. The winches are typically used for pulling poles during installations or removals. Most capstan winches are available with a hydraulic brake and can be configured to pull rope or cable (wire rope).

4.6.0 Strap Hoists

Strap hoists (*Figure 33*) are ratchet-style lifting tools equipped with a durable and rugged webbing strap. The hoist attaching hooks and web lifting hooks typically swivel and have safety latches. Strap hoists commonly are designed with a 4:1 mechanical factor. Interlocking pawls

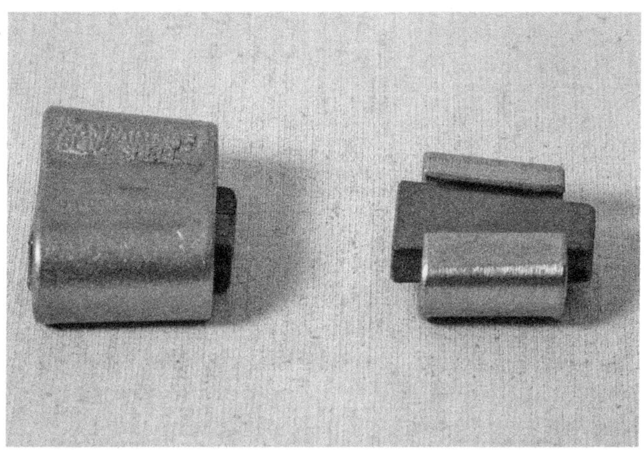

49107-11_F30.EPS

Figure 30 Copper wedges and "C" members.

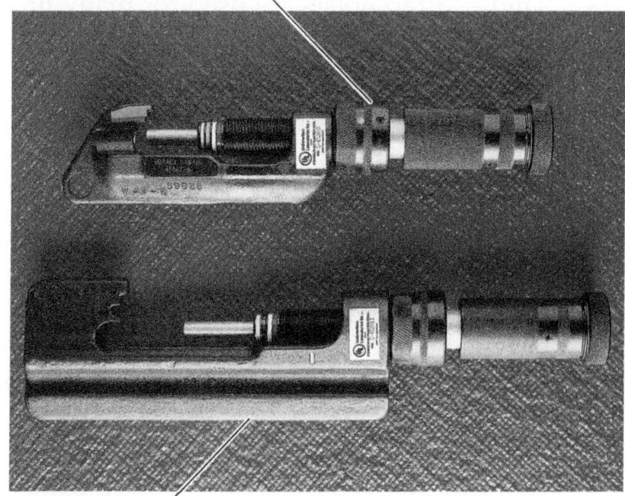

49107-11_F31.EPS

Figure 31 AMPACT™tool.

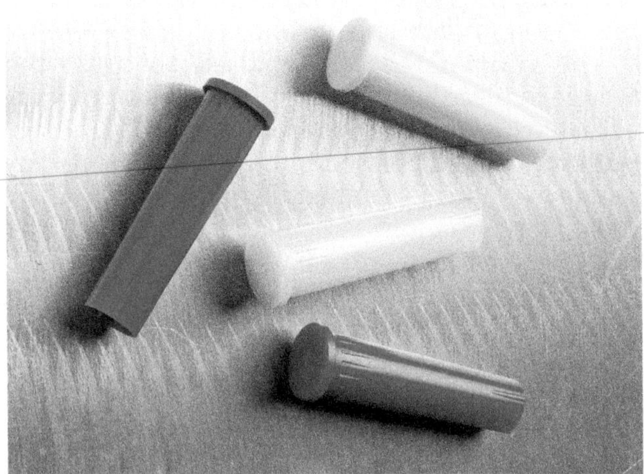

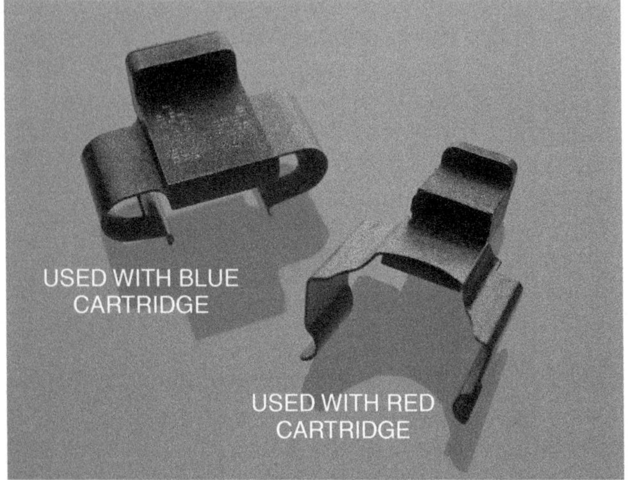

49107-11_F32.EPS

Figure 32 AMPACT™ firing cartridges and cartridge clips.

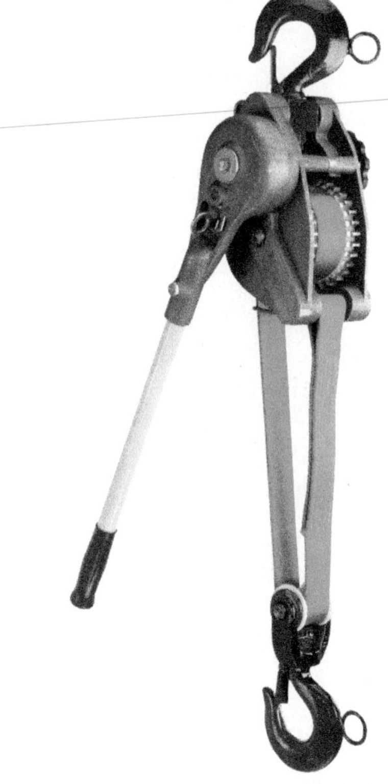

49107-11_F33.EPS

Figure 33 Strap hoist for lineman work.

provide positive load control and the 4 to 1 gear ratio provides extremely low handle effort. Most are equipped with a solid non-conductive fiberglass handle. Strap hoists are used for many lifting applications, most often, for hoisting and positioning components.

4.7.0 Chain Saws

Chain saws are commonly used by power line workers. Chain saws are used to trim trees and branches away from power lines and poles and for cutting poles, cross-arms, and other timbers as necessary during pole removals. Small capstan winch attachments (*Figure 34*) can even be installed on chain saws for special pulling applications.

4.8.0 Hand Lines

Hand lines are used for practically every installation. Hand lines are typically a ½" diameter rope that can be used to lift some components and tools. Hand lines are tied around hoisted components and are maneuvered by a worker on the ground to help control component swing during lifting. Hand lines also allow the worker on the ground to assist with positioning cross-arms during installations. Hand lines can be used in conjunction with a block and tackle, snatch block, or other lifting tools.

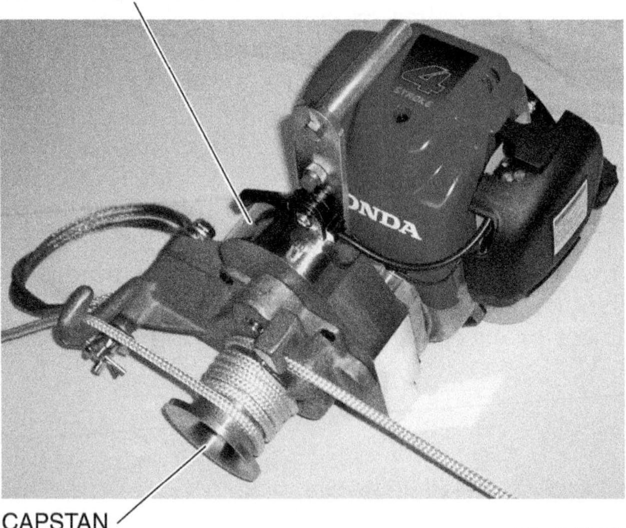

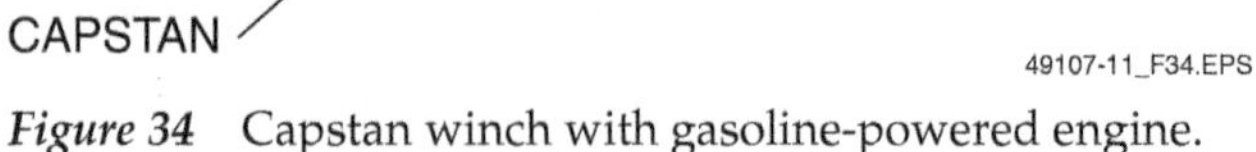

49107-11_F34.EPS

Figure 34 Capstan winch with gasoline-powered engine.

SUMMARY

This module introduced some of the tools that you will use on a daily basis during your training and after you become a power line worker. This is an introduction to some of the more common tools you are likely to use on a day-to-day basis. These tools fall into three basic categories: insulated hand tools for working on or around live circuits; ladders and platforms; and specialty tools, such as those powered pneumatically or hydraulically. These tools are designed and manufactured based on the needs of power line workers to ensure that the tasks required to install and maintain power line components can be carried out in a safe and effective manner. Always follow the manufacturer's recommendations when using any type of hand or power tool. Your safety and the safety of others working around you depend on proper handling and operation of these tools.

Review Questions

1. OSHA published the initial regulations for electrical safety in the workplace in _____.
   a. 1980
   b. 1985
   c. 1990
   d. 1995

2. According to *ASTM F711*, live-line tools made of fiberglass-reinforced plastic (FRP) must be designed and constructed to withstand test voltage of _____.
   a. 100,000 volts per foot of length for five minutes
   b. 100,000 volts per foot of length for ten minutes
   c. 200,000 volts per foot of length for five minutes
   d. 200,000 volts per foot of length for ten minutes

3. A clamp stick is commonly referred to as a(n) _____.
   a. hot stick
   b. shotgun stick
   c. extendo stick
   d. telescoping stick

4. Extendo sticks are _____.
   a. hot sticks
   b. shot guns
   c. rescue hooks
   d. telescoping sticks

5. If you need to withdraw an injured worker from a hazardous area such as inside a confined space, near an electrical cabinet, or within a vault, you would use a _____.
   a. hot stick
   b. shot gun
   c. rescue hook
   d. telescoping stick

6. The spiral disconnect is commonly referred to as a _____.
   a. tie stick head
   b. pointed disconnect
   c. pigtail disconnect
   d. fixed prong disconnect

7. Work platforms have a non-skid coating to prevent slippage and are typically made of laminated wood or _____.
   a. fiberglass
   b. aluminum
   c. steel
   d. copper

8. A battery-powered compression tool is also known as a _____.
   a. cable cutter
   b. crimper
   c. spiking tool
   d. Hole-Hog®

9. A tool used to place underground cables under driveways, sidewalks, foundations, and other obstacles in short-run applications is called a _____.
   a. cable cutter
   b. crimper
   c. spiking tool
   d. Hole-Hog®

10. A tool used to set a copper tap between two wire cables or between a wire cable and a ground rod when you need to make a permanent splice or connection is called a(n) _____.
    a. AMPACT™ tool
    b. crimper
    c. spiking tool
    d. Hole-Hog®

Trade Terms Quiz

Fill in the blank with the correct term that you learned from your study of this module.

1. A specialized type of clamp used to hold an object is a(n) ____________.
2. A term that refers to a combination fuse and knife switch used on power poles is a(n) ____________.
3. A material that resists the flow of electric current is said to be ____________.
4. A unit of area referring to the cross section of a wire or cable with a circular cross section is called a(n) ____________.
5. A material allowing the flow of electric current is known as a(n) ____________.

Trade Terms

Chuck
Circular mil
Conductor
Cutout
Insulated

Cornerstone of Craftsmanship

Joe Holley

Senior Instructor
Baltimore Gas and Electric

Joe Holley started out at 19 as a groundhand and worked his way up through the ranks to linesman A, linesman B, the control room, and now to senior instructor. He loves coaching new people and helping them to develop their skills.

How did you get started in the construction industry?
I worked as a laborer in general construction and then worked as a plumber's helper. A gentleman asked if I wanted to work for BG&E. I was not even looking for a job at the time, but I was tired of sweeping floors. My dad encouraged me, so I put in an application and I have been here for 38 years. I took a pay cut, but it turned out for the best. On the application, there were options for underground or overhead. I said I would do either one. They picked overhead for me because that's where the need was. I started out as a linesman trainee and worked up to the position I hold today.

Who inspired you to enter the industry?
No one really. As a kid I marveled at the guys climbing the pole. At 9 or 10, I followed the linesmen around the neighborhood watching them restring the lines. I thought that would be a neat thing to do. I forgot about that until I was in training. When talking with some of the other guys, I recalled that this is one of things I wanted to do as a kid.

What do you enjoy most about your job?
Currently, teaching new people. Watching a new person come here who knows nothing, they don't know how to climb or anything, and taking that person and watching them progress. Seeing how well they do after they finish the training is pretty gratifying. I like coaching and teaching.

Do you think training and education are important in construction?
Yes. The smarter you are the more you can branch out. It is always good to start at the bottom and work your way up. In this day and age, there are many people looking for a job, so if people can better themselves through education, they will be better off. After the training, they can stay and be a big asset here, or they can take what they have learned and be a better asset to the company in a different role. Education is something that can't be taken away from you.

How important do you think NCCER credentials will be in the future?
It will be an intricate part of the future. The NCCER credentials are portable. If someone does not like it here, they can use their training to go someplace else. In the powerline program, NCCER training provides a well-rounded start for people to learn the trade. Learning is a foundation, but working on the job is the real classroom. You can be here 40 years and you will always be learning new things. However, a basic foundation is critical.

How has training/construction impacted your life and your career?
I am training now. I have always been training, both at work and at home. I am always coaching, always teaching. Even when I was in the field working, I would have new people on my truck and teach them. Now, I am doing it as my sole job. I am more like a coach—a basketball or baseball coach. I've always liked and it is important.

Would you suggest construction as a career to others?
Yes, because it is always going to be needed. If people want to get their hands dirty, if they want to work hard and earn a good living, this is the place to be. There will always be positions. Plumbers, bricklayers, and linemen will retire and there will be openings.

It is not a good place for someone who does not want to work. If you do, it is a good, rewarding job. You need a good work ethic—you need to want to learn and want to do. We try to teach that, too.

How do you define craftsmanship?
You have to be dedicated in your field and have a strong desire to do well. When you are finished you can sit back and be proud of it. A craftsman is like a Michelangelo or Van Gogh, especially in line work. They can prepare and do it and then sit back and know that there are very few people who can do what they do.

It takes awhile to be a craftsman. You have to be dedicated and want to do well. You have to learn your art or trade and how to do it in a safe manner, the safest you can be. Learning and dedication—that is what makes a craftsman.

Trade Terms Introduced in This Module

Chuck: A specialized type of clamp used to hold an object.

Circular mil: A unit of area equal to the area of a circle with a diameter of one mil. A mil is one thousandth of an inch. A circular mil is a unit for referring to the area of the cross section of a wire or cable with a circular cross section.

Conductor: A material allowing the flow of electric current.

Cutout: A term that refers to a combination fuse and knife switch used on power poles.

Insulated: A material that resists the flow of electric current.

Additional Resources

This module presents thorough resources for task training. The following resource material is suggested for further study.

Construction Equipment Guide. New York, NY: John Wiley & Sons.

Allied Construction Products, LLC, Cleveland, OH. www.alliedcp.com.

Hubbel Power Systems, Centralia, MO. www.hubbellpowersystems.com.

Salisbury by Honeywell, Bolingbrook, IL. www.whsalisbury.com.

Figure Credits

Federal Emergency Management Agency, U.S. Department of Homeland Security, Module opener

Hastings, Figures 1, 3, 4, and 20

Salisbury Electrical Safety, Figures 2, 7–17, and 29

Topaz Publications, Inc., Figures 5 and 6

S&C Electric Company, Figures 18 and 19

A. B. Chance Co./Hubbell Power Systems, Figures 21 –23

Greenlee Textron, Inc., a subsidiary of Textron Inc., Figures 24, 25, and 28

Allied Construction Products, LLC, Figure 26

Specialized Products Company, http://www.specialized.net/, Figure 27

Tyco Electronic Energy Division, Figures 30–32

Columbus McKinnon Corporation, Figure 33

Simpson Winch Inc., Figure 34

CONTREN® LEARNING SERIES — USER UPDATE

NCCER makes every effort to keep its textbooks up-to-date and free of technical errors. We appreciate your help in this process. If you find an error, a typographical mistake, or an inaccuracy in NCCER's Contren® materials, please fill out this form (or a photocopy), or complete the online form at www.nccer.org/olf. Be sure to include the exact module number, page number, a detailed description, and your recommended correction. Your input will be brought to the attention of the Authoring Team. Thank you for your assistance.

Instructors – If you have an idea for improving this textbook, or have found that additional materials were necessary to teach this module effectively, please let us know so that we may present your suggestions to the Authoring Team.

NCCER Product Development and Revision

3600 NW 43rd Street, Building G, Gainesville, FL 32606

Fax: 352-334-0932
Email: curriculum@nccer.org
Online: www.nccer.org/olf

☐ Trainee Guide ☐ AIG ☐ Exam ☐ PowerPoints Other ____________

Craft / Level: Copyright Date:

Module Number / Title:

Section Number(s):

Description:

Recommended Correction:

Your Name:

Address:

Email: Phone:

Aerial Framing and Associated Hardware

49108-11

Trainees with successful module completions may be eligible for credentialing through NCCER's National Registry. To learn more, go to **www.nccer.org** or contact us at **1.888.622.3720.** Our website has information on the latest product releases and training, as well as online versions of our *Cornerstone* newsletter and Pearson's Contren® product catalog.

Your feedback is welcome. You may email your comments to **curriculum@nccer.org**, send general comments and inquiries to **info@nccer.org**, or use the User Update form at the back of this module.

V.1 7/11

49108-11

Aerial Framing and Associated Hardware

Objectives

When you have completed this module, you will be able to do the following:

1. Describe the difference between single-phase and three-phase construction.
2. Identify the hardware used in aerial framing.
3. Using the standards manual, identify materials, assorted pole hardware, and support arms needed to perform aerial framing on:
   - A single cross-arm
   - A double cross-arm
   - A dead triple cross-arm set
   - An outrig arm
   - An alley arm
4. Describe, assemble, and install guys.
5. Perform an aerial framing procedure as defined by the instructor.
6. Hand-pull single-phase and three-phase primary conductors, dead end, and sag.
7. Explain how to install a transformer and connect conductors.

Performance Tasks

Under the supervision of the instructor, you should be able to do the following:

1. Assemble and install guys.
2. Perform an aerial framing procedure as defined by the instructor.
   - Single cross-arm
   - Double cross-arm
   - Triple dead-end cross-arm
   - Double alley arm
3. Hand-pull single-phase and three-phase primary conductors, sag, and dead end.

Trade Terms

Bullwheel
Chicago-style grip
Chinese finger
Conductor
Cross-arm
Cutout
Dead-end
Distribution capacitor
Guy anchor
Guy rod
Guys
Insulator
Jumper
Sag
Splice
Stringing block

Contents

Topics to be presented in this module include:

Figures and Tables

Figures and Tables (*continued*)

1.0.0 Introduction

Installing a utility pole is only half the job. The pole must be firmly supported and the top of the pole must be fitted-out with the hardware and supports needed to carry conductors and other components. Fitting-out the top of the pole is referred to as aerial framing. Aerial framing involves four major elements:

- Installation of pole support members, known as guys
- Installation of the cross-arms, insulators, and hardware used to support conductors and components
- Installation of conductors including: pulling, tensioning, sagging, dead-ending, and securing the conductors to the insulators
- Installation of transformers and protective devices

The installation of guys is performed after the pole is erected. When possible, cross-arms and related hardware are installed while the pole is on the ground. It is sometimes necessary to do this work on an upright pole, however. Transformers are usually installed while the pole is upright.

1.1.0 Three-Phase and Single-Phase Power

The rotors of power plant generators are divided into three segments that are 120 degrees apart (*Figure 1*).

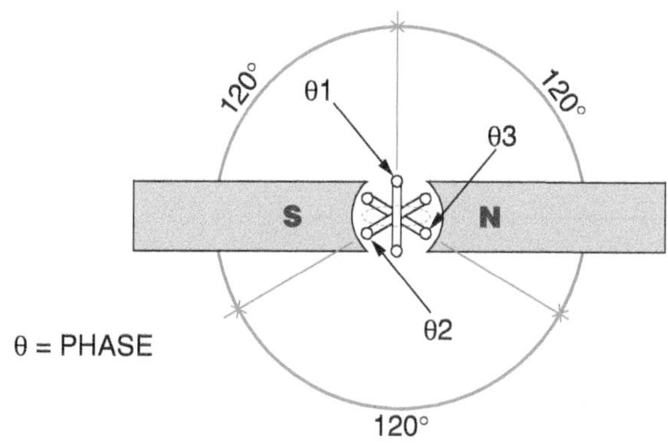

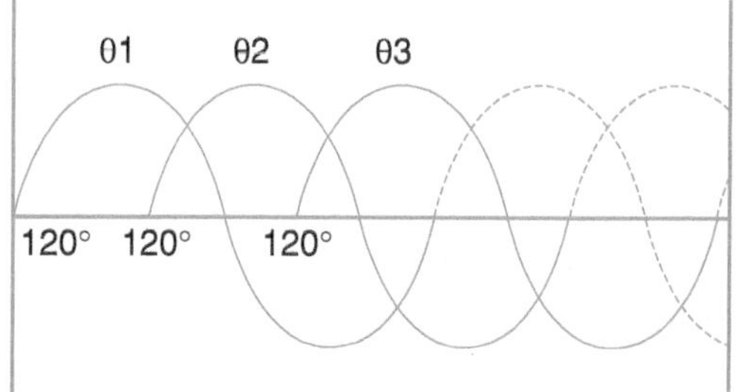

Figure 1 Three-phase generator.

Such generators produce three voltages referred to as a neutral or common point. The three voltages are electrically 120 degrees apart, or out of phase with each other. Substations that step up and step down the power use three-phase transformers to do so. It is three-phase power that traverses the transmission and large distribution lines (*Figure 2*). Primary conductors carry the high voltage (typically 7,200 volts) three-phase power and are normally installed at the top of the pole.

On Site

Learning About Aerial Framing

At this line worker school in Florida, students learn aerial framing skills on shortened utility poles set up in the course lab.

49108-11_SA01.EPS

49108-11_SA02.EPS

Figure 2 Three-phase distribution lines.

1.1.1 Industrial Users

All three legs of a three-phase power system are needed by industries to provide the power needed to operate heavy-duty motors and equipment. Power for these users can be tapped from the distribution system using a three-phase transformer setup (*Figure 3*). However, some larger industries have their own substations and transformers to step down the power, and so would not require a utility transformer. Typical power levels for these three-phase users are 240, 480, and 575 volts.

1.1.2 Residential Users

Homes and small businesses do not need three-phase power. They use a single transformer (*Figure 4*) that provides the 120 and 240 volts needed for lighting and appliances.

L TO L = 480V
L TO GROUND = 277V
G TO GROUND = 0V

FUSE
FUSE
FUSE
480V L
480V L
480V L
THREE PHASE
277V G
GROUNDED AT POLE
4800V THREE-PHASE PRIMARY
120/240V SINGLE PHASE
STEP-DOWN TRANSFORMER
THREE 277V SINGLE-PHASE FLUORESCENT LIGHTING CIRCUITS
480V THREE-PHASE POWER CIRCUITS
120/240V SINGLE PHASE
49108-11_F03.EPS

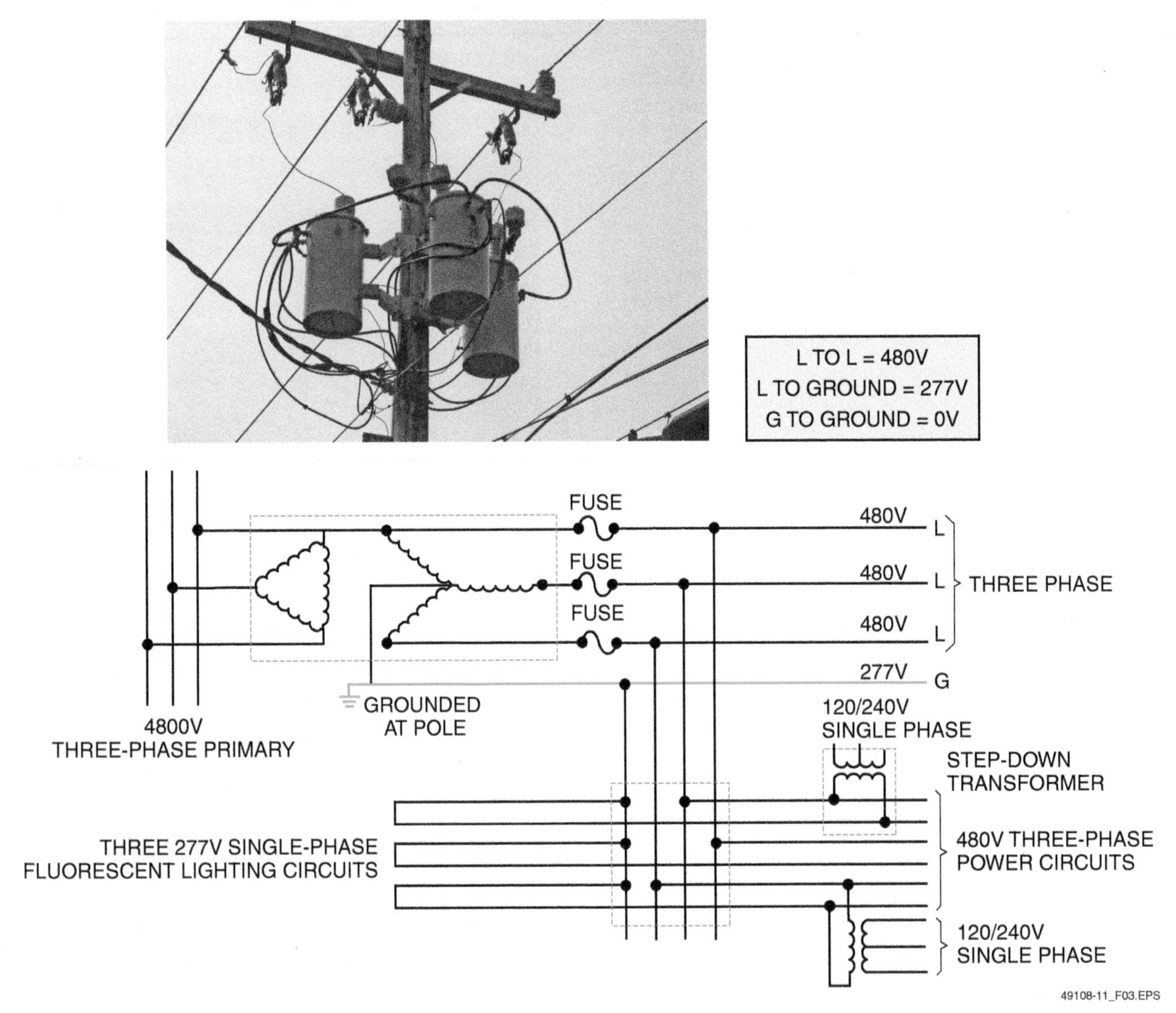

Figure 3 Three-phase power or industrial user setup.

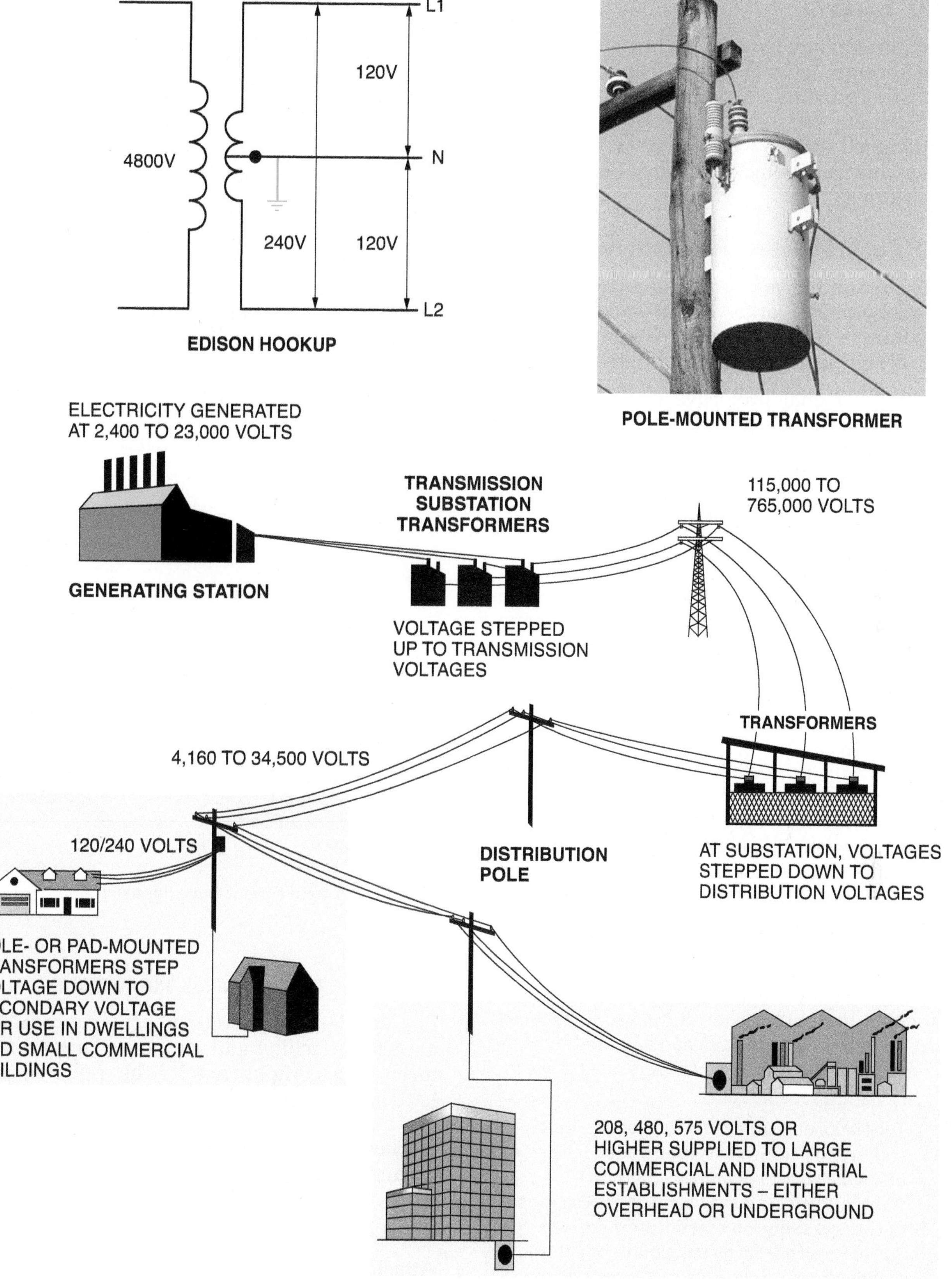

Figure 4 Residential user setup.

2.0.0 Safety

All required safety precautions must be followed when performing work on poles at heights with, or in close proximity to, energized components. Aerial framing and installations also should never be attempted during wet or bad weather conditions except in extreme emergency outages, such as ice storms.

2.1.0 General Safety Precautions

Safety precautions and procedures must be considered by each worker before attempting any aerial framing project. The following safety items must always be performed, at minimum:

- Proper personal protective equipment (PPE) must be used including gloves, hard hats, and eye protection. Insulated tools, such as hot sticks, must always be used when working from the pole. If working from an insulated bucket, hand tools may be used when wearing rubber gloves and sleeves.
- When poles are being installed or removed, care must be taken to prevent them from contacting any bare energized lines or components. Workers handling the grounded butt end of the pole must wear rubber gloves and/or use insulating tools. Any open pole hole must be covered or barricaded to prevent falling accidents.
- Proper fall protection equipment must be used when working above heights of 6 feet as required by OSHA regulations.
- At times when it is necessary to work with, or around, energized lines, all required PPE, such as rubber gloves and sleeves, must be worn. In addition, all required insulating equipment, such as conductor covers and rubber blankets, must be clamped in place to isolate the lineman from the voltage (*Figure 5*). Insulated tools, such as hot sticks, should be used whenever possible to keep workers at a safe distance and from contact with the energized lines.

WARNING!

The National Institute for Occupational Safety and Health (NIOSH) details and posts incidents of lineman electrocution accidents on its website at www.cdc.gov/niosh/face/. The cause of virtually every incident documented was due to improper safety procedures or not using mandated safety equipment. Always take time to properly install and use all necessary safety equipment and follow proper safety procedures. Remember, with high voltage accidents, you rarely get a second chance.

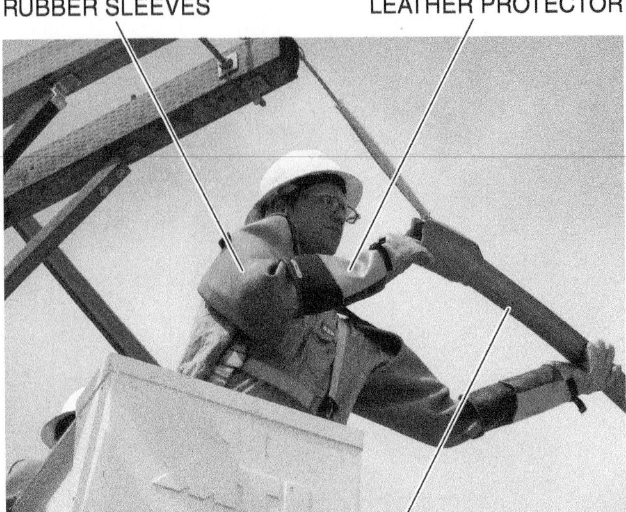

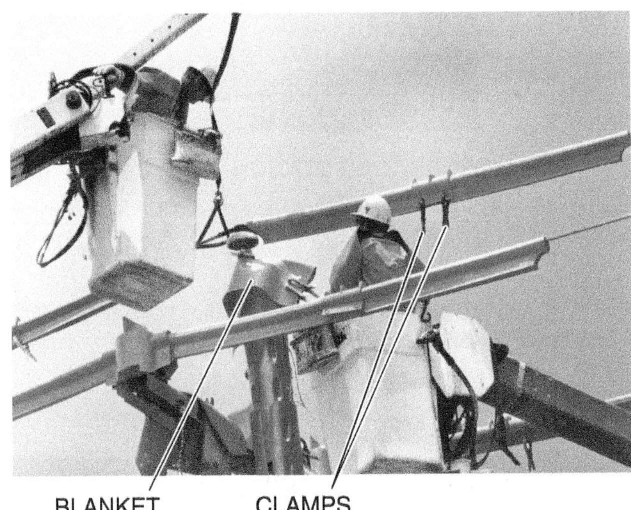

Figure 5 Protective insulating equipment.

3.0.0 Aerial Framing Hardware

Framing hardware consists of items commonly used for attaching and mounting support structures and components to the pole. Framing hardware includes attaching hardware, mounting hardware, and supporting hardware. Line workers may also need to use other hardware to repair supported components.

3.1.0 Attaching Hardware

Attaching hardware includes items used to secure mounting hardware components and support structures to the pole (*Figure 6*). These include such things as lag bolts, square washers, lock washers, square nuts, carriage bolts, and threaded rods.

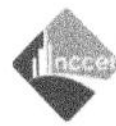

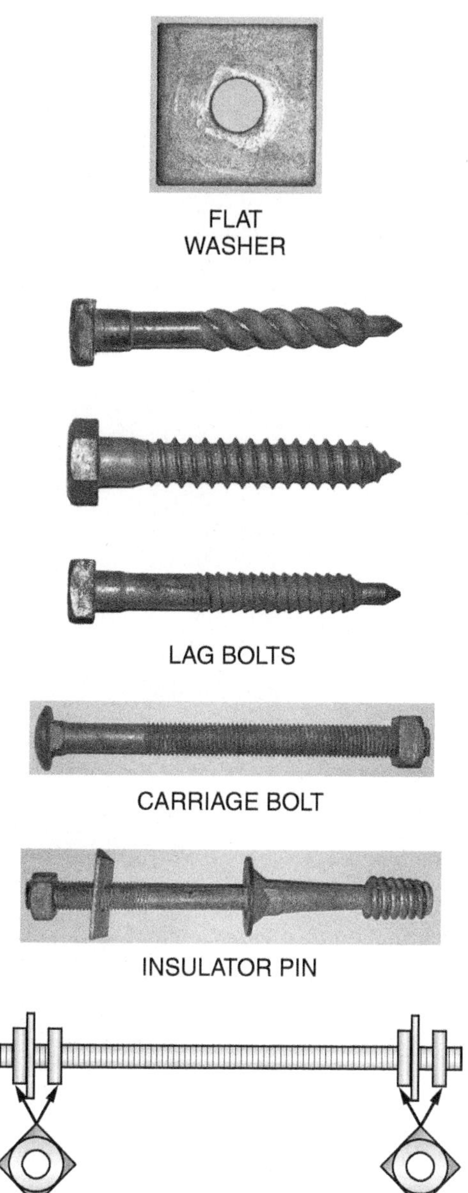

Figure 6 Attaching hardware.

- *Easy-out lag bolt* – The spiral grooves on this bolt make it easier to screw out of a pole. It has the same strength as other lag bolts.
- *Gimlet lag bolt* – This is a standard bolt used for installing braces and other hardware.
- *Pilot lag bolt* – This bolt is used for installing braces and other hardware. The point on the end aids in setting the bolt.
- *Carriage bolt* – This bolt is used to attach cross-arms to the pole and braces to the cross-arm.
- *Double-arm (DA) bolt* – This bolt is a fully-threaded (all-thread) rod. Two square nuts and a washer are installed at each end. Such bolts are used to attach the cross-arm to the pole.
- *Flat washer* – This washer is used for multiple mounting applications on the pole. Two washers are normally used per bolt unless a clevis is attached on one end.
- *Insulator pin* – This pin is used to attach a conductor insulator. The pin positions the insulator and conductor away from the pole.

3.2.0 Mounting Hardware

Mounting hardware is used to provide a secure connection point or support for components or supporting structures (*Figure 7*). These include such things as brackets, hooks, and clamps.

- *Hot clamp* – This clamp is used for connecting a wire to a primary conductor and is available in various sizes. The hole in the end is to accept an insulated hot stick.
- *Copper KVS clamp* – This is a grounding clamp used to join a ground wire to a grounding rod or to join two ground wires together.
- *Eye nut* – Available in ¾" and ⅝", the eye nut is used as an attaching point for various items and lines such as guys set low on the pole.
- *Four-way loop connector* – This connector is also known as a bird box. It is used as a bolt-on connector for secondary lines.
- *Wedge shoe* – This is an aluminum shoe used to secure aluminum wire. Commonly used as a dead-end clamp, the tension of the conductor secures the wedge onto the conductor.
- *Bolted shoe* – This is a bolted shoe used to secure aluminum wire. Similar to the wedge shoe, this shoe must be bolted onto the conductor to secure it.
- *Thimble eye* – This is a ¾" double groove eye nut used for 5⁄16" guy wire screws on the end of a ⅝" bolt.
- *Ground wire clamp* – This clamp is used to secure a ground wire to the pole grounding rod.
- *Guy clamp* – This is a two-bolt clamp used in place of a make-up for attaching a guy wire.
- *Fused cutout* – The gated cutout serves as a circuit overload protector for secondary conductors and components.
- *Guy hook* – The guy hook serves as an attaching point for an insulated guy rod and guy wire make-up.
- *Cross-arm bracket* – This bracket serves as a cross-arm mount on a pole. Cross-arm braces are not necessary with the use of this bracket.
- *Rams head P-eye* – This eye is a type of attachment for a guy wire make-up for ¼" to 5⁄16" guy wires.

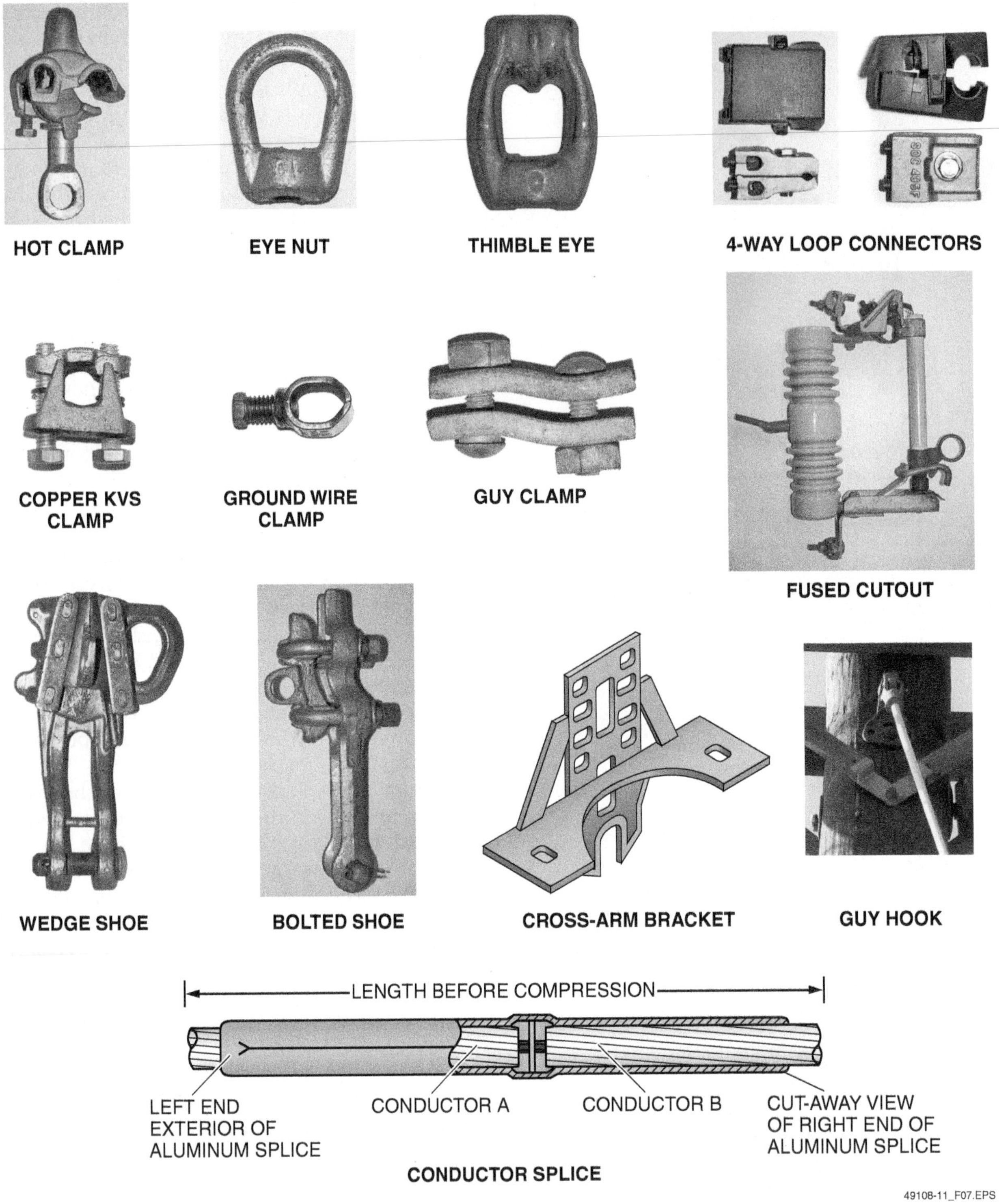

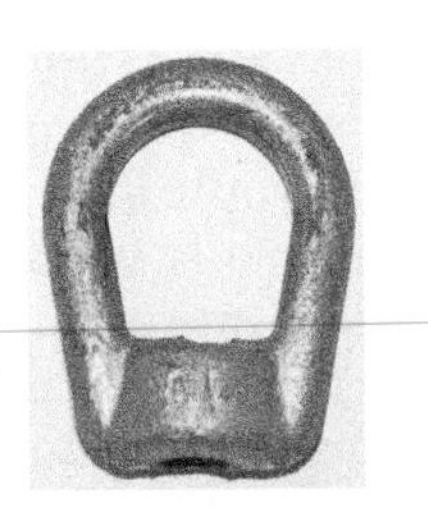

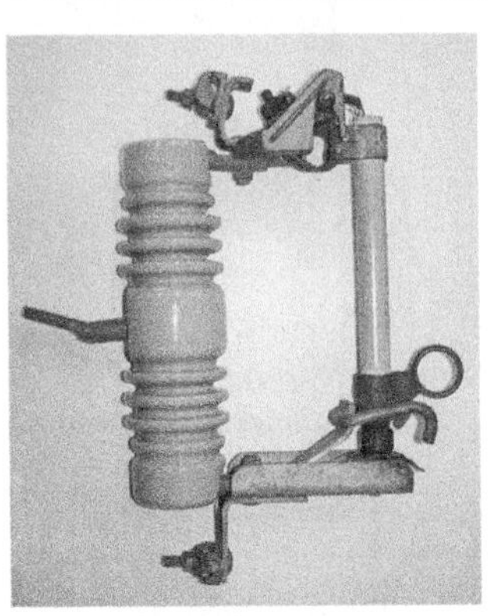

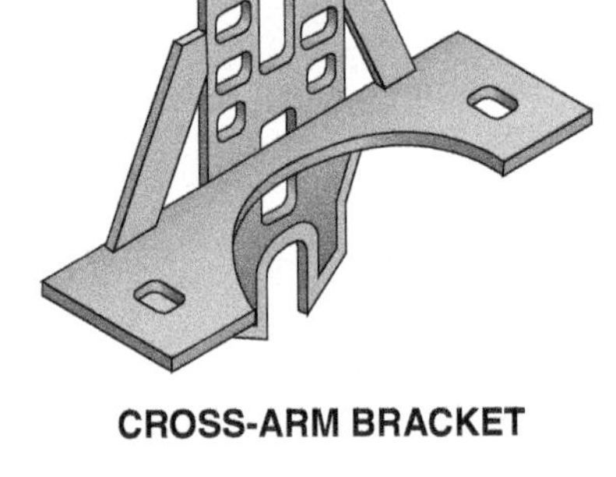

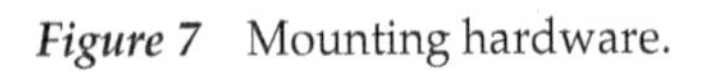

Figure 7 Mounting hardware.

3.3.0 Supporting Hardware

Supporting hardware is used to secure or support necessary components such as conductors. The supporting hardware most commonly used during aerial framing are various types of insulators (*Figure 8*). Insulators keep the conductors from contacting the pole or its cross-arms.

3.4.0 Repair Hardware

Conductor splices are used at times to repair damaged conductors. The most common splices are compression splices and helical rod type or automatic splices. With compression splices, the damaged section of conductor is cut out, the conductor ends are cleaned of any corrosion, and the

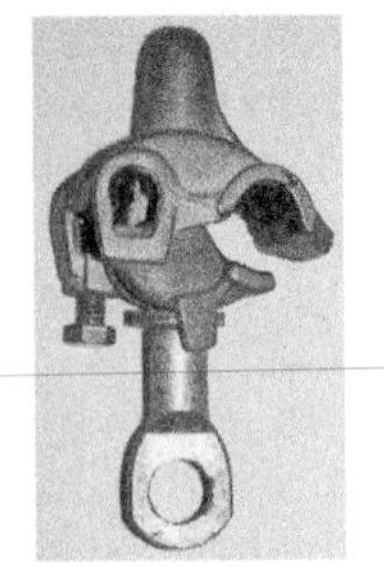

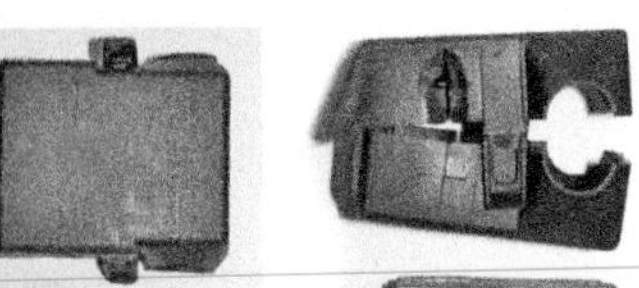

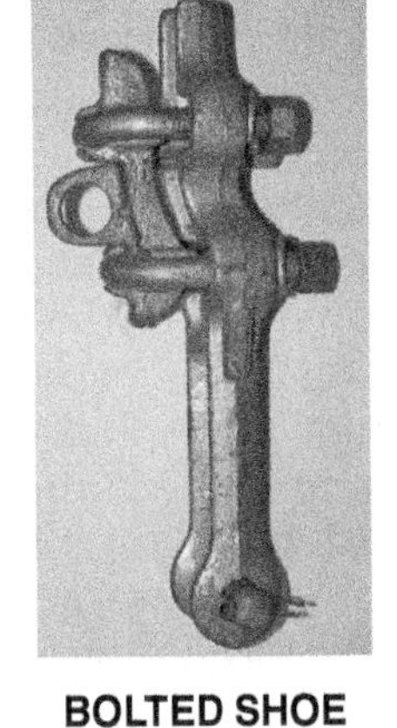

49108-11_F08.EPS

Figure 8 Conductor insulating hardware.

splice is crimped to the conductor core and the outer helical strands with a manual crimping tool or a mechanical tool. Although more expensive, automatic splices are much quicker and easier to use. Automatic splices are wire mesh or Chinese finger style splices that require no special tools for installation.

4.0.0 Aerial Framing Components

Framing components consist of items commonly installed during the installation of new poles (*Figure 9*). These include such items as cross-arms, insulators, primary and secondary conductors, fused cutouts, lightning arrestors, transformers, regulators, and capacitor banks.

Insulators are designed as pin type or suspension type. Pin insulators are mounted on the top of the pole and pole cross-arms and support the primary conductors. The conductors are either tied into the insulator with a wire tie or clamped in depending on the insulator type. Suspension insulators are commonly used for large transmission lines suspended from steel towers. Smaller suspension type insulators are also most often used at conductor dead-end terminations.

Distribution capacitors (*Figure 10*) are used to improve poor power quality or low voltage in specific areas. Fluctuating voltages can overheat conductors and other components. The installation of capacitor banks can maintain consistent voltage levels in areas of heavy loads long distances from substations or distribution transformers, both upstream and downstream.

Voltage regulators (*Figure 11*) are used to adjust and maintain a consistent voltage downstream through the primary conductors. Regulators are commonly used in areas a long distance from a

TRANSFORMERS

INSULATOR

LIGHTNING ARRESTORS

49108-11_F09.EPS

Figure 9 Pole-mounted components.

49108-11_F10.EPS

Figure 10 Capacitors banks.

substation where voltages can fluctuate. Similar in appearance to transformers, regulators are equipped with a dial adjustment to regulate and set the line voltage to the required level.

Primary conductors carry the voltage throughout the distribution network to each customer. Primary conductors are available in various sizes and construction:

- *ACSR* – Aluminum conductor steel reinforced is most often used because of its high strength and relatively low cost. It is constructed of aluminum strands helically wound around a steel core. The aluminum strands are used for their excellent conductivity, low weight, and low cost. The center steel strands provide the strength required to support the weight and prevent stretching of the aluminum. ACSR consists of one or more layers of aluminum strands surrounding a core of 1, 7, 19, or 37 galvanized steel strands. An ACSR conductor with 72 aluminum strands surrounding a core of 7 steel strands is identified by the numbers 72/7. As shown in *Table 1*, conductors are also identified by code word names to meet the requirements of the *Aluminum Association Code Words for Overhead Aluminum Electrical Conductors,* Seventh Edition (January 1999). ACSR is made in a wide range of sizes and strandings ranging from #6 AWG 6/1 (OD = 0.198 inches) up to 2,156 kcmil, 84/19 (OD = 1.762 inches). The most common sizes are 6/1, 26/7, and 54/7. Certain strandings are stronger than others; 36/1 ACSR is the weakest stranding, while 30/7 is the strongest.

49108-11_F11.EPS

Figure 11 Voltage regulators.

Table 1 Typical ASCE Conductors

Size (kcmil)	Stranding (Aluminum/Steel)	Code Word	Nominal Diameter (in)	Minimum Rated Strength (lbs)
397.5	18/1	Chickadee	0.743	09,900
397.5	26/7	Ibis	0.783	16,300
795	26/7	Drake	1.108	31,500
795	30/19	Mallard	1.140	38,400
954	45/7	Rail	1.165	25,900

- *ACSR/AW* – ACSR Alumoweld conductor is constructed with an aluminum-clad steel reinforced core and is most often used in corrosive environments.
- *ACSR/SD* – Self-damping conductor is constructed of two trapezoidal layers of aluminum strands around a steel core and is more expensive than regular ACSR. The strands are made of #6201 aluminum, are self-damping against aeolian vibration, and can be strung at very high tensions.
- *ACAR* – Aluminum alloy reinforced conductor is constructed of strands of #1350 aluminum around a core of #6201 aluminum alloy. Although more expensive, it is lighter than ACSR, but just as strong. It is most often selected for use in corrosive environments.
- *AAC-1350* – Aluminum conductor is constructed only of #1350 aluminum strands. It is most often used in applications that require good conductivity and over short spans.
- *AAAC-6201* – Aluminum alloy conductor is constructed only of #6201 aluminum alloy strands. Although more expensive, it is stronger and lighter than ACSR and is most often used for long spans in corrosive environments.

Some factors to be considered when selecting the line conductors include the following:

- Required sag and span between conductors
- Tension on the conductors
- Whether or not the atmosphere is corrosive
- Whether or not the line is prone to vibration
- Power loss allowed on the line
- Voltage loss allowed on the line

Conductor connectors are used to secure conductors at dead-ends and to attach supply lines to the conductors. These include such items as bolted shoes, wedge shoes, dead-end clamps, parallel clamps, and various hot clamps.

- Bolted shoes and bolted dead-end clamps are usually used to secure the conductor ends at a dead-end termination when it is desirable for the conductor not to be crimped.
- Wedge shoes are normally used at a permanent dead-end. The tension of the conductor causes the wedge shoe to clamp tightly onto the conductor end creating some crimping of the conductor.
- Parallel clamps are used to attach a second line parallel to the conductor. Parallel clamps are often used to attach such items as jumper lines.
- Hot clamps are used to attach offset lines to the conductor. These clamps can be installed with an insulated hot stick and the attached lines often supply components such as transformers.

5.0.0 Cross-Arms

Cross-arms are installed on poles to support the conductors and other necessary components. A traditional cross-arm is an 8' long wooden timber. Cross-arms are available in dimensions of 3½" × 5½" or 5½" × 5½". Most cross-arms are predrilled with mounting holes and additional component mounting holes. When installed, cross-arms are supported and stabilized with braces or brackets. Wooden braces are most often used, although metal and fiberglass braces are also available. Single cross-arms can be mounted or two cross-arms can be mounted as a double cross-arm arrangement for increased stability and area for component attachment. Various cross-arms are available for specific mounting applications and include the following types:

- Single cross-arms (*Figure 12*) are generally used to support two or more primary conductors. Single cross-arms can also be used for dead-ending conductors. Additionally, cutouts and lightning arrestors may also be attached to the cross-arm.
- Double cross-arms (*Figure 13*) are used when increased support is required for the tension or weight of conductors and components. Double cross-arms are typically joined together at the ends with threaded rods and nuts to stabilize the ends of the cross-arms.

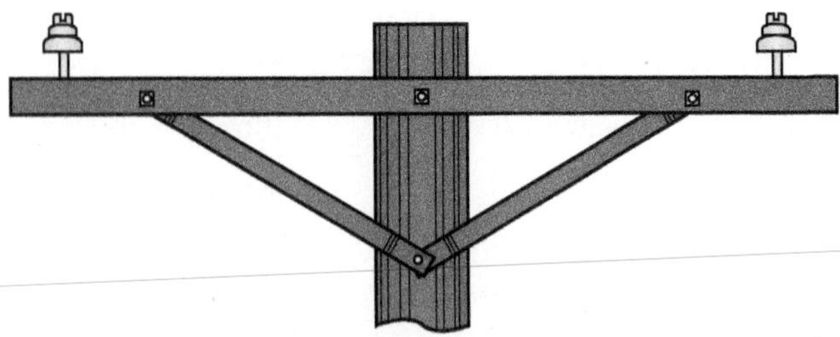
49108-11_F12.EPS

Figure 12 Single cross-arm.

49108-11_F13.EPS

Figure 13 Double cross-arm.

- Triple cross-arms (*Figure 14*) are sometimes used for dead-ending a long run of large primary conductors. With this arrangement two single cross-arms are joined together on the back of the pole and a single cross-arm is mounted on the front of the pole. The three arms are joined together at the ends with threaded rods and nuts. Triple cross-arms are used in place of extra-strength 5½" × 5½" cross-arms.
- Alley arms (*Figure 15*) are shortened cross-arms that only extend from one side of the pole. These cross-arms are used in limited clearance areas, such as alleys, where a full length cross-arm will not fit. They are supported with one brace, usually wooden, although metal braces can also be used. Some metal alley arm braces are equipped with a mid-point step to provide a foot-hold for the worker.
- Outrig arms (*Figure 16*) are long cross-arms, up to 12' long, used for temporary construction. This type of arm is used to temporarily mount conductors away from the pole during construction. The long outrig arm is supported with a lower brace and also a side brace.

5.1.0 Installing Cross-Arms

Whenever possible, cross-arms are attached to the pole while it is on the ground. The exception is when a cross-arm bracket is being used. When installing cross-arms, the distance between the brace mounting holes is known as the span. The distance between the center mounting hole and

49108-11_F14.EPS

Figure 14 Triple cross-arm.

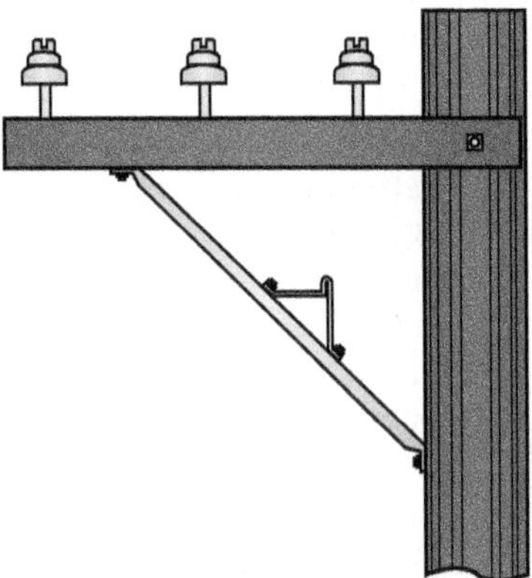

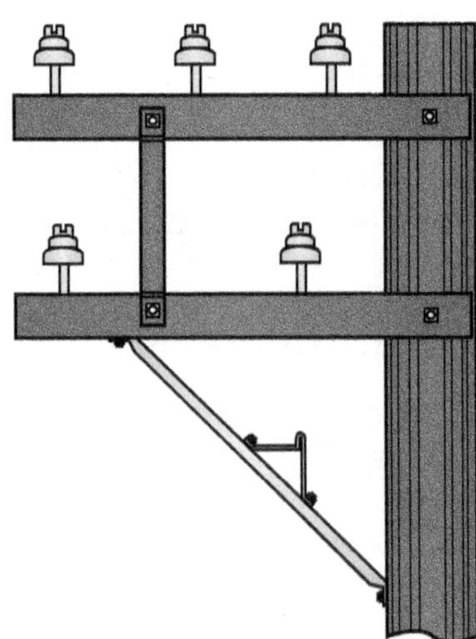

49108-11_F15.EPS

Figure 15 Alley arm.

the lower mounting hole in the pole for the braces is known as the drop (*Figure 17*). Some poles come pre-drilled or these holes can be drilled in the pole while it is on the ground.

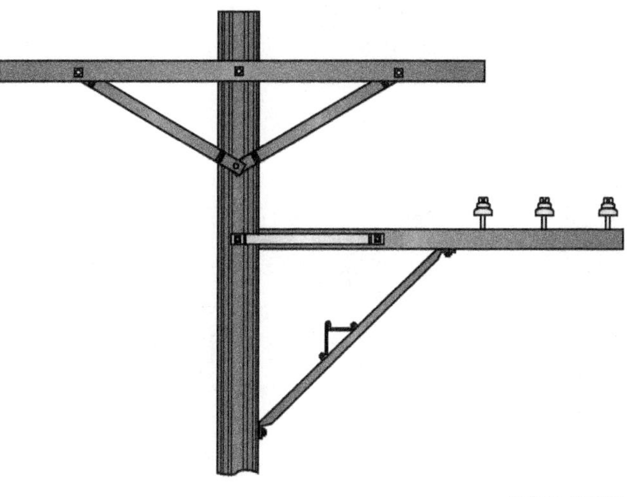

Figure 16 Outrig arm.

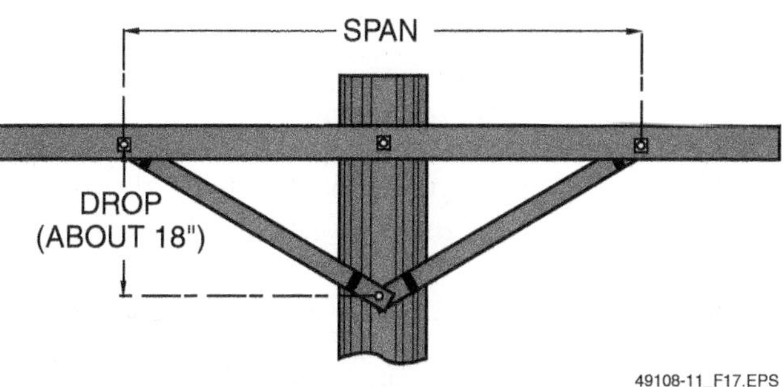

Figure 17 Cross-arm positioning.

The cross-arm is attached to the pole with a double-arm (DA) bolt but not completely tightened. Insulator pins can also be installed in the cross-arm before the pole is raised. Braces are attached to the arm but not tightened. The cross-arm is turned vertically parallel to the pole with the braces also hanging vertically (*Figure 18*).

A hand line is tied to both ends of the cross-arm and sent up vertically. When the pole is raised, set, and guyed, a worker on the ground can pull the hand line to aid in turning the 8' long arm to a level position. Once the lower ends of the braces are secured to the pole, all the bolts and hardware can be completely tightened and the hand line can be released. Insulators can now be screwed onto the insulator pins. A pole top bracket and insulator will also usually be installed.

If a cross-arm bracket (braceless gain) is used, the cross-arm must be hoisted up the pole, set in place on the bracket, and secured with a through-the-pole carriage bolt and cross-arm bolts (*Figure 19*). The use of braces is not required with cross-arm brackets.

Double cross-arms and alley arms can be installed in the same manner. Outrig arms are usually installed after the pole has been raised.

When work orders are given to install cross-arms and the components that will be attached to them, there will be drawings included that provide the details of the job. Those drawings may include the size and type of connecting hardware, along with the size of cross-arms and any braces that must be installed with them. The *Appendix* contains examples of actual cross arm drawings.

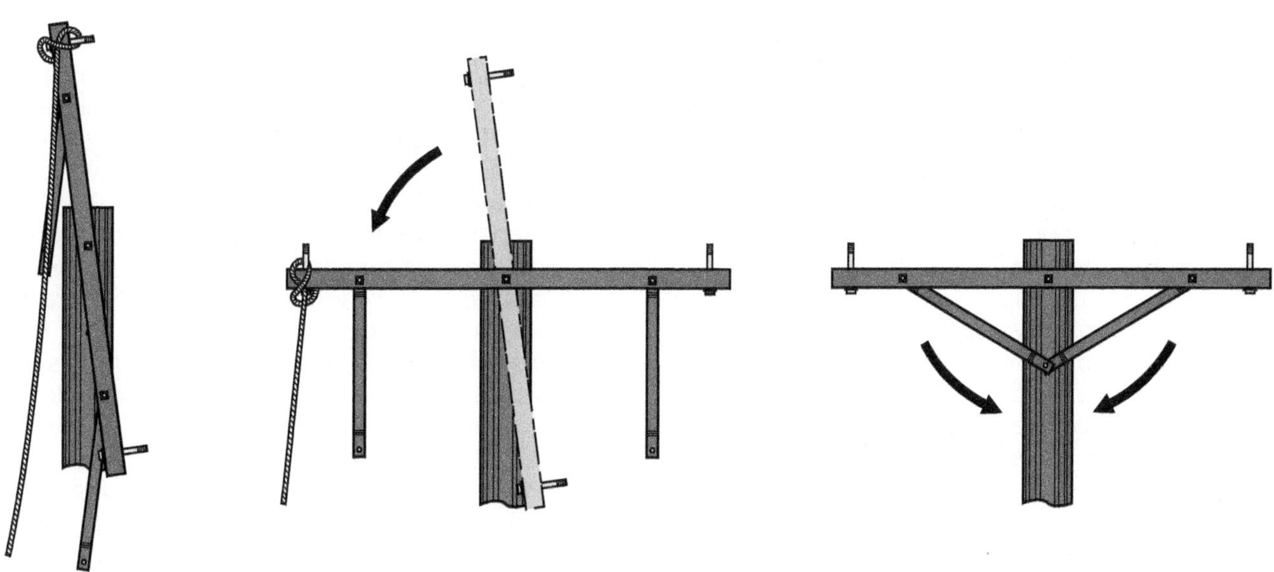

Figure 18 Mounting a cross-arm.

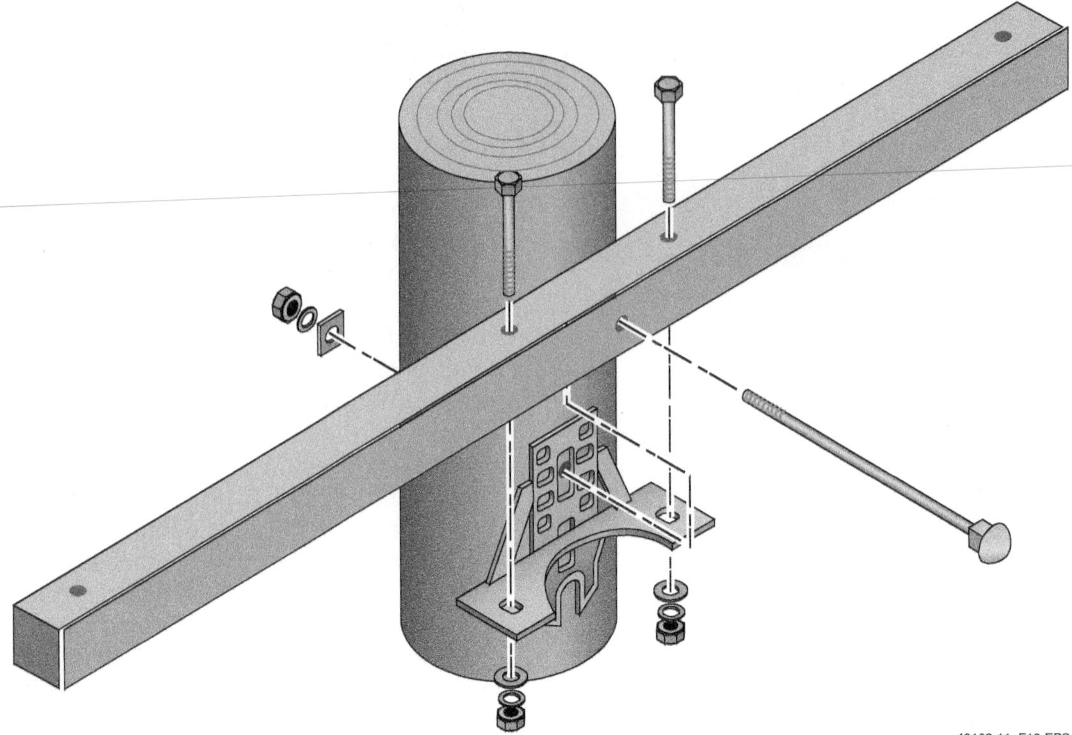

Figure 19 Mounting to a cross-arm bracket.

6.0.0 Guys

Guys, sometimes called guy wires or guy braces, are necessary to maintain the vertical stability of certain poles. Guys are used, for example, on dead-end poles, on poles with angled conductors, or on a pole set in hilly or uneven terrain. In such cases, the weight and tension of the conductors create a directional pull on the pole. Guys must be installed on the opposite side of the pole to counteract the pull of the conductors and keep the pole upright. They must be secured by approved guy anchors that are embedded in the ground. Various types of guys are used for specific applications and include the following:

- *Down guy* – The down guy is the most common type used (*Figure 20*). This guy is anchored in the ground near the pole and is generally attached to the upper third of the pole. Down guys are typically required for each level of conductors or other cabling attached to the pole. This means that a guy is necessary at the top primary conductor level, a guy is necessary at the lower secondary conductor level, and a guy is necessary for any lower telephone or other utility cables. The telephone or cable companies are generally responsible for installing the lower guys.
- *Span guy* – This is a guy wire connected from the top of one pole to the top of another (*Figure 21*). It is used to reduce the strain on conductors.
- *Stub guy* – The stub guy connects the pole to a shorter pole (*Figure 22*). It typically extends over a curve in a roadway. It is used to support the angled conductors when there is no clearance for a down guy to be used on the pole. A down guy will be attached to support the stub pole.
- *Head guy* – A head guy (*Figure 23*) runs from the top of one pole to a lower point on an adjoining pole. Head guys are often used on hilly, uneven terrain, and in locations where a down guy is not practical.
- *Arm guy* – An arm guy (*Figure 24*) connects the cross-arms of two adjoining poles. It is used as a counterbalance when one arm has more conductors than the other.
- *Sidewalk (vertical) guy* – This type of guy is used when the guy wire must be anchored close to the pole, typically over a sidewalk (*Figure 25*), due to limited ground clearance.
- *Push guy* – A push guy (*Figure 26*) is sometimes used as a diagonal brace to support a pole. Push guys are often cut from a good section of a removed pole.

49108-11_F20.EPS

Figure 20 Down guys.

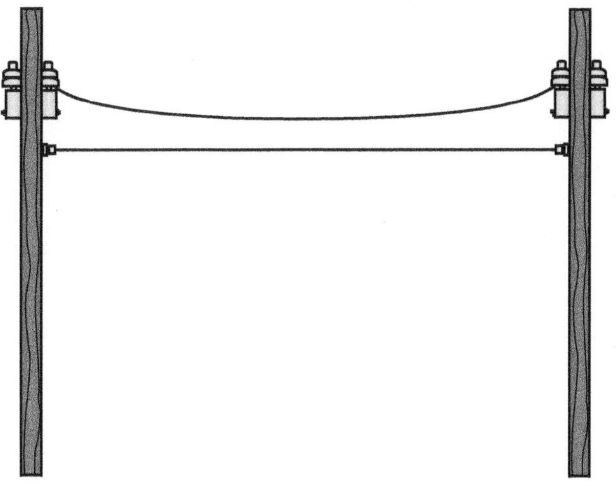

49108-11_F21.EPS

Figure 21 Span guy.

49108-11_F22.EPS

Figure 22 Stub guy.

6.1.0 Guy Anchors

The guys must be securely anchored in the ground to withstand the stresses created by the weight of the pole, the conductors, and all equipment mounted to the pole. Several types of anchors are available. The most common are expansion anchors and screw-type anchors.

Expansion anchors (*Figure 27*) are inserted into the hole in a collapsed state. When struck with a ram, the anchor expands open.

Screw-type anchors (*Figure 28*) are literally screwed into the ground. They can be installed using the auger arm of a digger derrick or other equipment with a power takeoff (PTO) or with a hand tool designed for the purpose (*Figure 29*). Different types of screw anchors are designed for different soil conditions, including very rocky

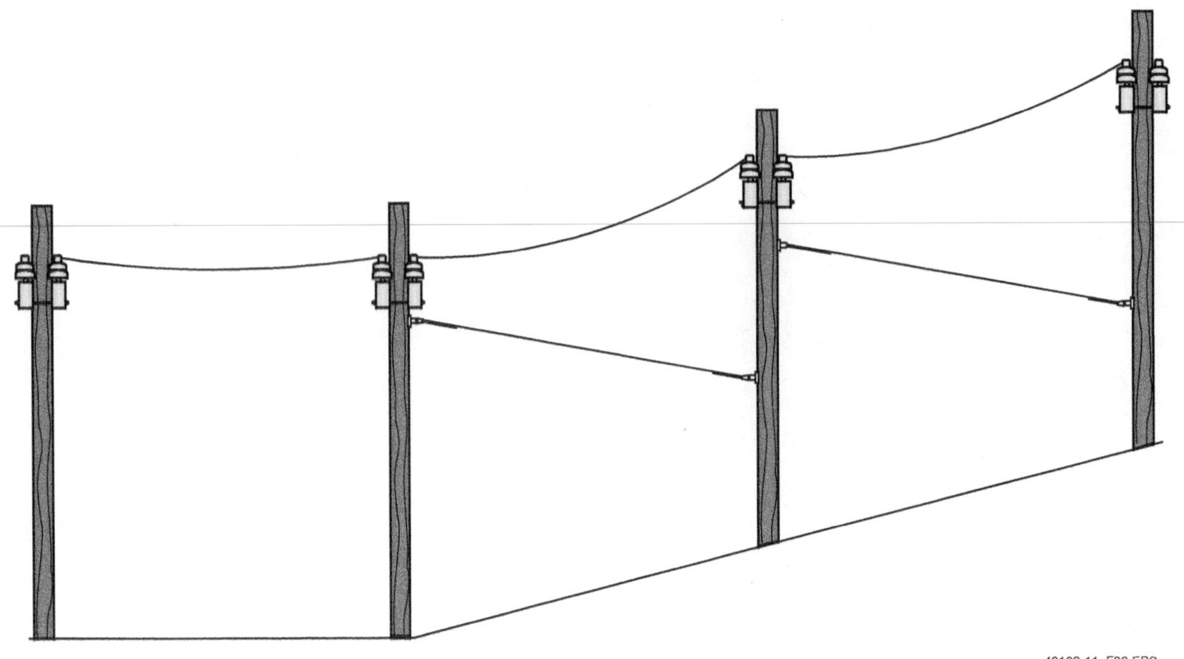

49108-11_F23.EPS

Figure 23 Head guy.

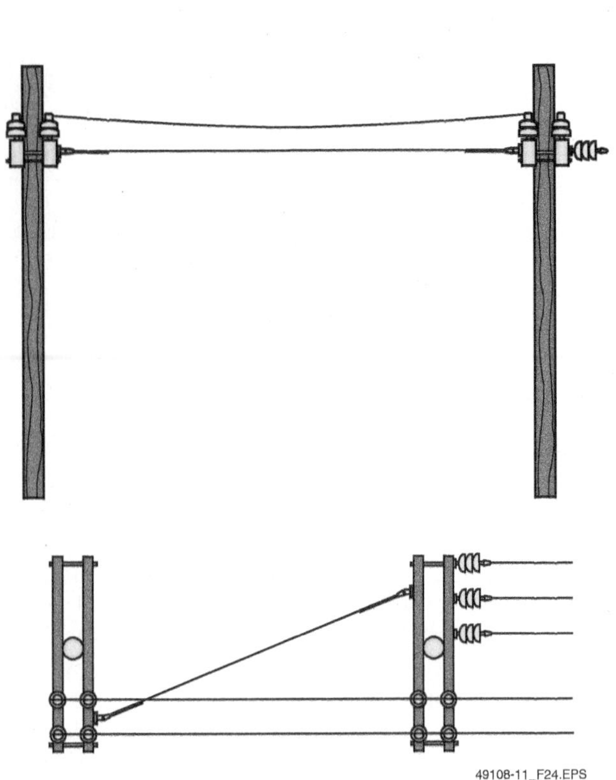

49108-11_F24.EPS

Figure 24 Arm guy.

49108-11_F25.EPS

Figure 25 Sidewalk guys.

49108-11_F26.EPS

Figure 26 Push guy.

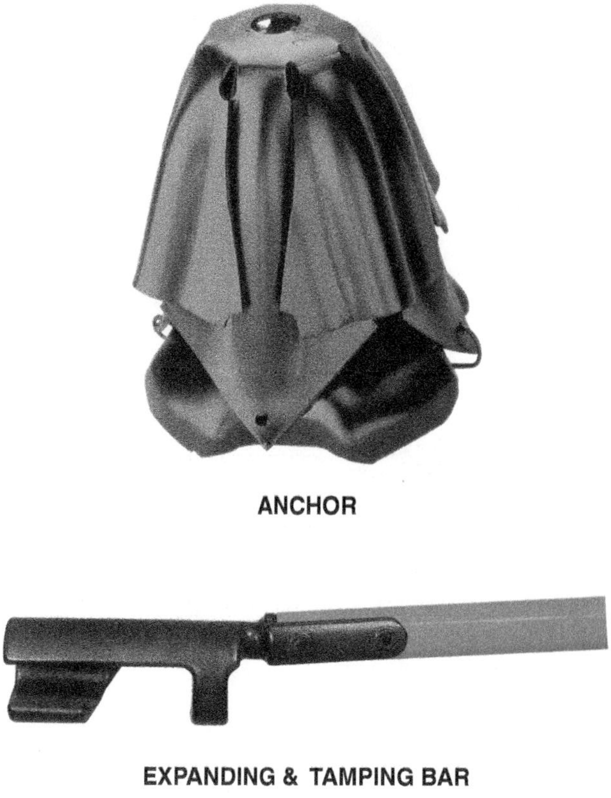

ANCHOR

49108-11_F27.EPS

Figure 27 Expansion anchor.

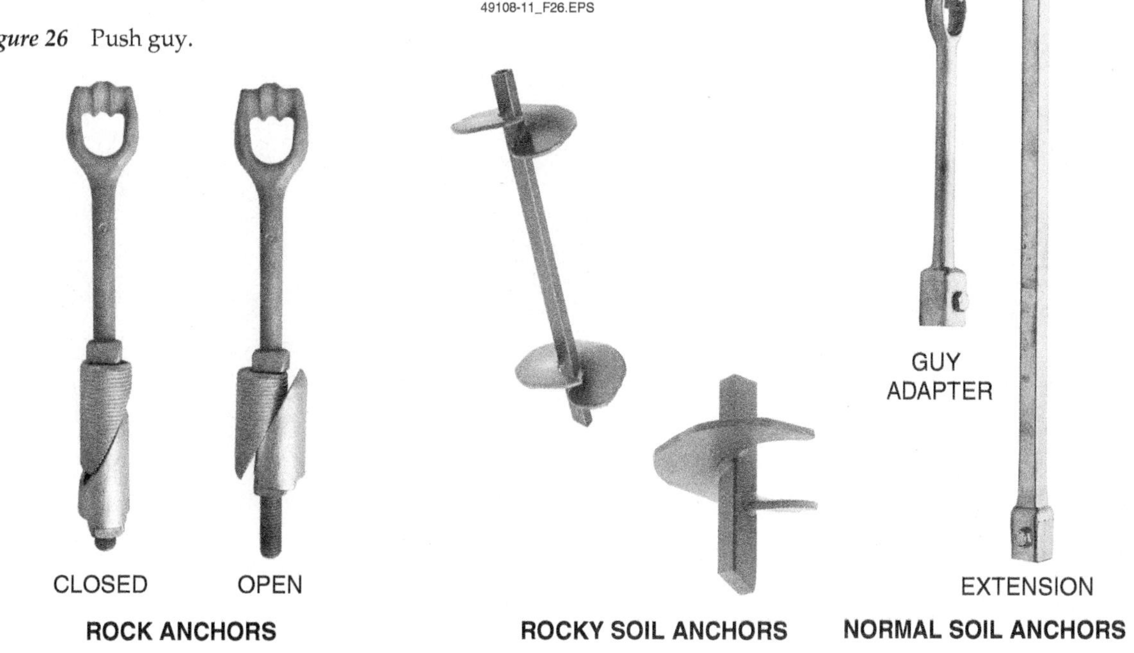

Figure 28 Screw-type anchors.

ANCHORING BY HAND

INSTALLING AN EXPANSION ANCHOR

POWERED INSTALLATION

49108-11_F29.EPS

Figure 29 Installing guy anchors.

On Site

The Deadman

Before manufactured guy anchors were readily available, it was fairly common for pole installers to bury logs in the ground for use as guy anchors. A typical log anchor was 8' long, had a 9" diameter, and was buried about 5' deep. The same thing can be done with a concrete deadman, which is simply a block of concrete with an upside-down U-bolt embedded in the top so it could be lifted and placed in the hole with a crane or other machine.

ground. When space allows, a down guy anchor for primary conductors should typically be set at 15 to 20 feet from the pole to provide an optimum angle to the pole.

6.2.0 Installing Guys

A guy bracket, or guy hook, must be installed with a lag bolt at each level of the pole requiring a guy wire connection. The top guy hook at the primary conductor level must also be equipped with an insulating guy rod. For a down guy, the guy rod is a 7' long fiberglass rod with attaching rings at both ends (*Figure 30*). This provides insulating protection in the event a high voltage primary conductor comes in contact with the guy. A guy rod is not required on single-phase construction or at the neutral line level.

49108-11_F30.EPS

Figure 30 Installing a down guy.

If the tensioned guy wire will be secured with guy clamps, the guy wire is passed through the guy anchor at the ground. The top end of the guy wire is passed through the end of the guy rod and both ends of the wire are clamped with bulldog-style wire clamps. A come-along is connected between the wire clamps, the wire is hand tensioned, and the ends are secured with guy wire clamps. The rule of thumb when tensioning the guy wire (*Figure 31*) is to tighten it as much as possible, wait a few minutes, and then tighten it one more click with the ratchet.

In most cases, the tensioned guy wire will be secured with helically wound wire wraps (*Figure 32*). In this case, a side clamp will be attached to the guy anchor. The guy wire will be passed through the side clamp and will be tensioned with the come-along. After tensioning, the guy wire will be secured to the guy anchor and to the guy rod with the wire wraps.

After the primary and secondary level guy wires are installed, a small copper grounding wire, sometimes called a stringer wire (*Figure 33*), is connected between the guy wires and then clamped to the pole ground.

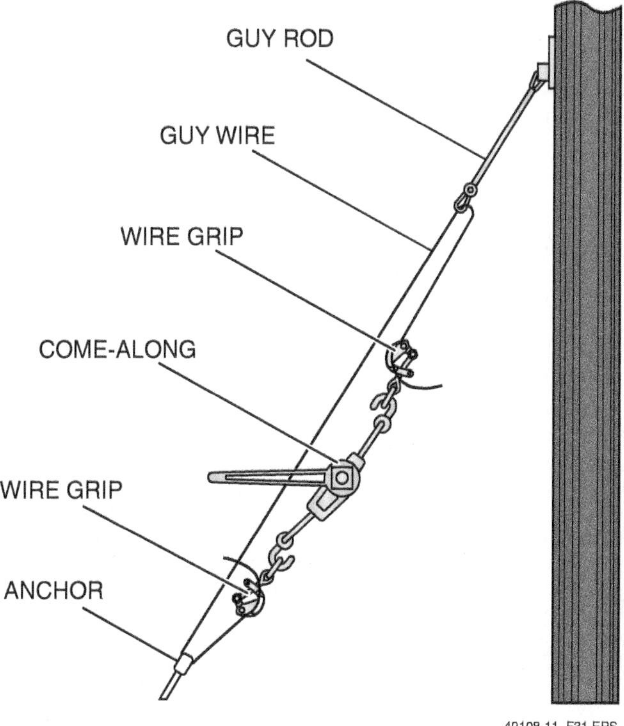

Figure 31 Tensioning the guy wire.

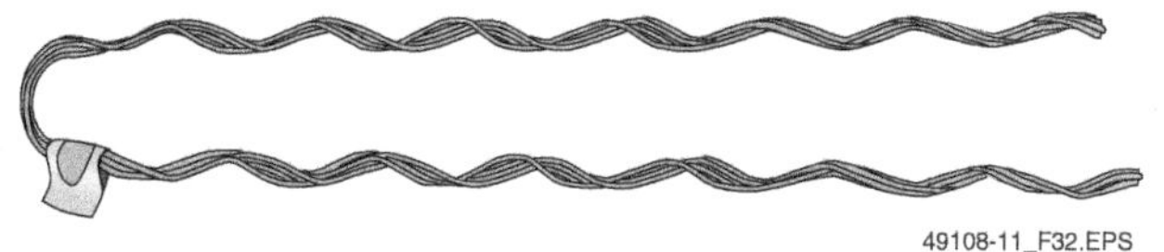

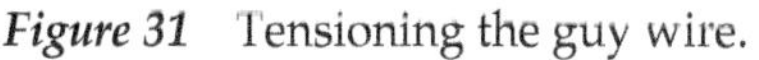

Figure 32 Helical wire wrap.

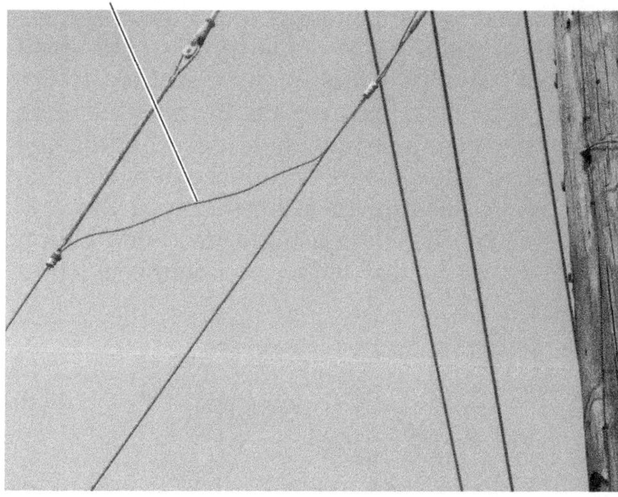

Figure 33 Guy grounding wire.

In residential or high traffic areas, guy wire covers are typically installed on the lower portion of the guy wires. The covers are often high-visibility yellow plastic, but can also be made of galvanized metal. The covers are used mainly for pedestrian protection.

7.0.0 Installing Primary Conductors

Primary conductors carry the high voltage (typically 7,200 volts) three-phase power and are normally installed at the top of the pole. Primary conductors are usually bare wire and are constructed typically of a steel core shrouded in layers of helically wound aluminum wire. The conductors are generally large gauge, heavy, with limited flexibility. For these reasons, primary conductors must be handled carefully during installation to prevent damage. The main causes of conductor damage are contact with or dragging along the ground, twisting, kinking, or bending. Severe twisting can cause the outer layers of wire in a section of conductor to unwind or delaminate from the core. This raised section of wire is known as birdcaging.

The steps involved in stringing or installing conductors include pulling or paying off the necessary length of conductor from the supply reel, raising the conductor to installation height, dead-ending the conductor to the first pole, tensioning, sagging, and dead-ending the conductor to the last pole. Finally, the new conductors must be tied in or clipped to the insulators.

When installing new conductors, the span length of the conductors must be secured to the poles at each end. Securing the terminating ends of the conductors is known as dead-ending the conductors. The starting end of the new conductor must be secured to a cross-arm at the first pole. This can be done using a dead-end clamp (*Figure 34*) or helical wire wrap connected to a dead-end insulator that has been secured to the cross-arm. Once the new conductor span has been tensioned and sagged, the opposite end will be secured at the last pole in the same manner (*Figure 35*).

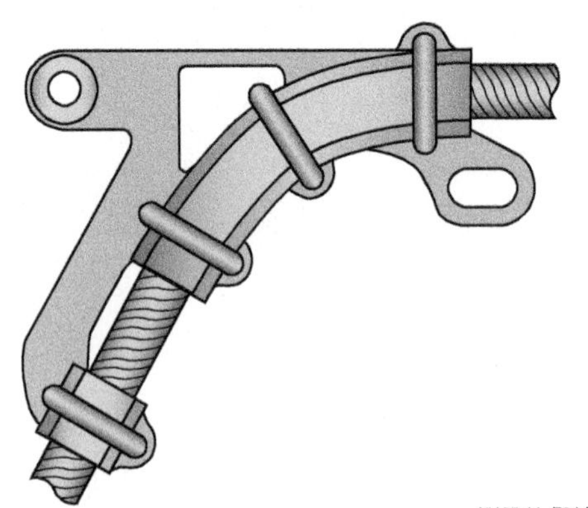

Figure 34 Dead-end clamp.

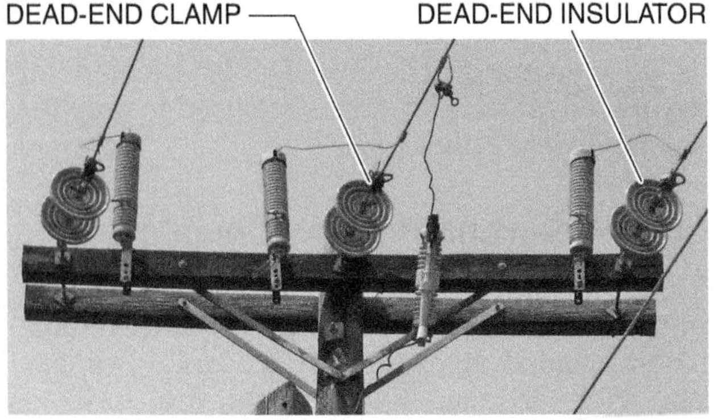

Figure 35 Dead-end connections.

The conductor tension is determined by how tightly the conductor is pulled between poles. The tension controls the conductor sag or the level it drops between the poles (*Figure 36*). Where minimum ground clearance is critical, the conductor sag must be set at the required height from the ground (*Table 2*).

The two most common stringing methods used are the slack/layout method and the tension method.

7.1.0 Slack/Layout Method

In this method, the required length of conductor is either pulled off the reel along the ground by a vehicle, or the reel is carried by a vehicle and the length of conductor is deposited on the ground as the vehicle travels down the line. Once the desired length is obtained, the conductor is raised into stringing blocks on the poles, dead-ended, tensioned, and sagged. This method presents a high chance for conductor damage due to the conductor being in contact with, or dragged along the ground. This method should only be used if the tension method is not possible.

Table 2 Typical Overhead Conductor Sag Clearances

Voltage Level	Clearance to Ground
Less than 66kV	20 feet (6.1m)
66 kV to 110kV	21 feet (6.4m)
110 kV to 165kV	22 feet (6.7m)
Greater than 165kV	23 feet (7.0m)

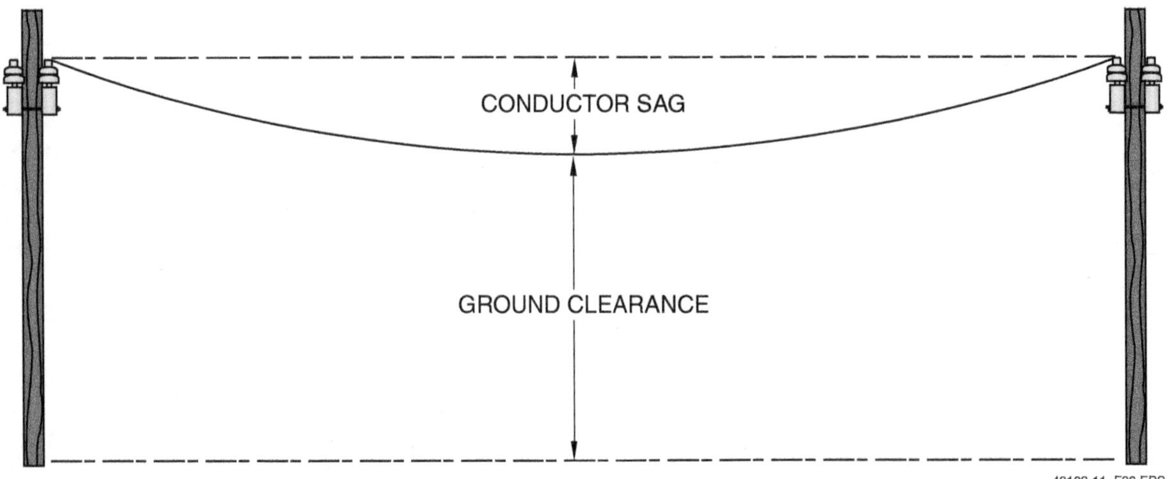

Figure 36 Conductor sag.

7.2.0 Tension Method

The tension method is preferred for stringing conductors. With this method, consistent tension is applied and the conductor never contacts the ground. Tensioning equipment also prevents conductors from accidentally sagging into any energized lines during removal or installation.

The tension is provided by a bullwheel tensioner at the conductor supply reel and by a drum-style pulling machine at the end of the line. The conductor, under steady tension from the bullwheel, is slowly pulled by the pulling machine through stringing blocks (wheels) mounted on the pole cross-arms.

> **WARNING!**
>
> Workers must never be present under the lines during conductor pulling and installation. Injuries could occur if the tensioning or pulling equipment fails, if the conductor grip slips, or if the conductor fails.

7.2.1 Stringing Blocks

When installing conductors, stringing blocks are used to smoothly guide the conductors between poles. The blocks consist of mountable pulley-style wheels compatibly sized to the conductor. Stringing blocks are available in various types with one wheel or multiple wheels. The blocks can be suspended under the cross-arms or mounted on top of the cross-arms. The top mounted blocks are more stable than the suspended blocks since they cannot swing back and forth as the conductor is pulled.

While conductors are usually pulled one at a time, for some large jobs, three conductors may be ganged to an equalizer bar and pulled together. In this case, a stringing block with three wheels may be used.

7.2.2 Conductor Grips

The new conductor must be safely secured by a pulling line. A wire mesh grip, sometimes called a Chinese finger or Kellems grip, must always be used to secure the conductor (*Figure 37*). Bulldog grips or Chicago-style grips cannot be used since they will damage the conductor. The wire mesh grip must be compatible in size to the conductor.

The wire mesh grip is slid over the conductor end and the grip end is wrapped with friction or plastic tape. When tension is applied, the grip cinches tightly around the conductor without damaging it. A pulling rope is secured with a smaller diameter wire mesh grip. The loop ends of the two grips are joined with a cylindrical swivel to prevent twists or kinks in the conductor. A smaller diameter rope, sometimes called a finger line, is joined to the end of the pulling rope. The finger line will guide the pulling rope and conductor through the stringing blocks to the puller.

7.2.3 Running Grounds

When new conductors are being pulled, temporary grounding grids are laid out and connected to the equipment being used. These grounds protect both personnel and the equipment from any kind of static or induced voltages that may be generated during the work. Running grounds are devices that must be installed on the new conductors that are being pulled. There are multiple reasons for such grounding. Running grounds and lines help dissipate static electricity that can be generated during pulling. Grounding lines help protect workers if the new conductor accidentally contacts an energized conductor. Grounds help protect from possible induced voltage from parallel energized conductors. Grounds must be installed on all terminating ends of the conductors and at each pole where

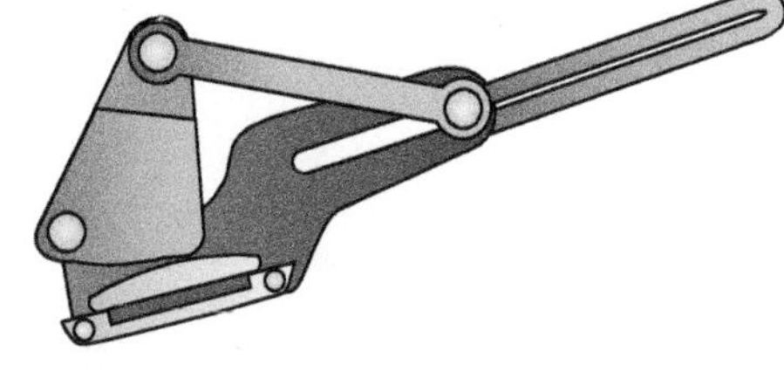

Figure 37 Wire grips.

conductors are being tied in. All ground lines must be connected to the pole ground at these locations. The running grounds must always be adjusted appropriately to the size of the conductor. The grounds must remain connected until all work is completed.

7.2.4 Bullwheel Tensioner and Drum Puller

The bullwheel provides positive tension and braking as the conductor is pulled through the stringing blocks. The bullwheel is equipped with grooved wheels that apply tension to the conductor as it is pulled from the supply reel. The supply reel should be positioned to rotate in the same direction as the bullwheel (*Figure 38*). The capacity of the bullwheel should be compatible to the conductor size, span length, and terrain and should be equipped with tension indicators. The tensioner must also be able to maintain the tension when pulling is stopped. The conductor puller at the end of the line must also be equipped with load indicators and load limiters to monitor and control the pulling tension.

CAUTION

During conductor installation, clear lines of communication must be maintained between workers at both ends of the line. In the event of a problem, the pulling operation must be stopped immediately.

The main considerations to prevent conductor damage during pulling are the tensile stress or pulling tension, and the bending stress exerted at the break-over point of the first stringing block. If the combination of these two stresses exceeds the conductor rated breaking strength, conductor damage is likely. Monitoring the bullwheel and puller tension will control the tensile stress. Use of a correctly sized wheel diameter at the first stringing block and the recommended bullwheel distance from the first stringing block will control bending stress. Typically, the bullwheel should be set in line at a distance of at least three times the height of the first stringing block.

For example, if the stringing block is mounted at 30', the bullwheel should be located at least 90' from the pole to minimize the bending angle of the conductor through the block. The conductor manufacturer's specifications must always be followed regarding these requirements.

7.3.0 Tensioning and Sagging Conductors

Once the required length of conductor has been strung between the poles, it must be secured and correctly tensioned and sagged. Use an extension of new lines being added to the dead-end of an existing line as an example. The new line will have a number of evenly spaced new poles on reasonably level ground with minimal curves. At the dead-end of the existing line, a second cross-arm must be installed if a double cross-arm is not present on the pole. A new dead-end will be installed on the cross-arm for the new conductors. The required insulator strings and dead-end clamps will be attached to the cross-arm and pole. The new conductor will be secured in the dead-end clamp and the conductor supply will be cut as close as possible to the end of the dead-end clamp. The new conductor is now dead-ended to the dead-end pole of the existing line.

The new length of conductor will now be tensioned at the last pole of the new line to achieve the desired sag. At this point, the new conductor should be allowed to set for at least two hours to

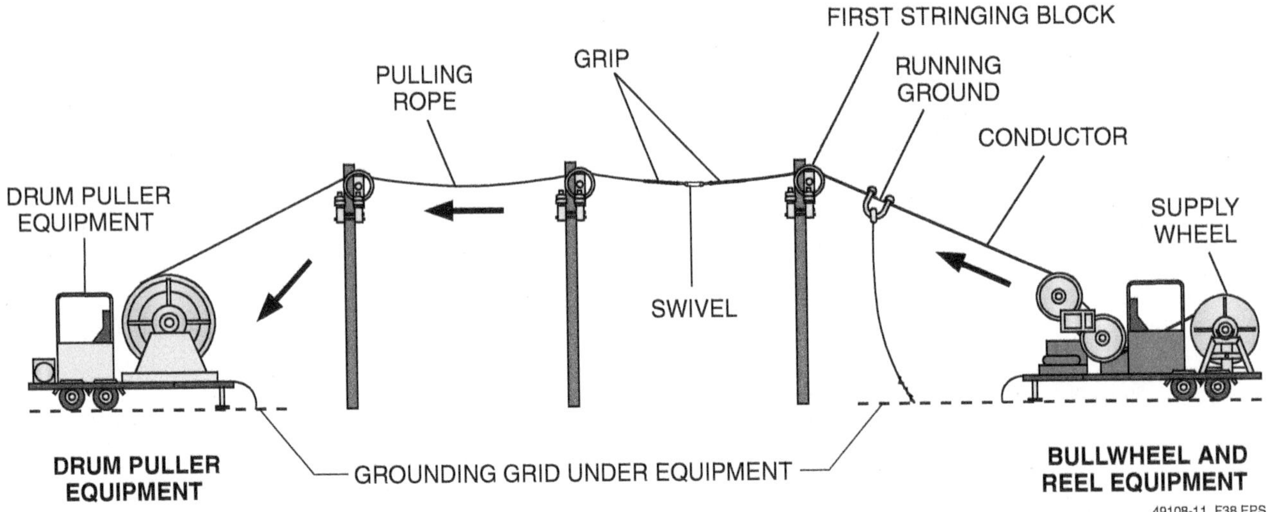

Figure 38 Pulling conductors.

creep. During this time, the conductor tension will slowly equalize and become evenly distributed between poles. The conductor will also relax and stretch a bit depending on the prevailing temperature. The conductor can then be tensioned back to the original sag height as necessary.

With evenly spaced poles on reasonably level ground, the conductor sag should automatically be evenly distributed and consistent between poles after creeping. If not, slight hand pulling adjustments may be made.

Special considerations must be noted when tensioning and sagging conductors on hilly terrain or an incline. In this situation, the tension and weight of the conductor will be increased on the slope, causing too much sag. At the crest and top of the hill the tension will be decreased, causing too little sag. In this case, pulling and tensioning adjustments must be made at the top of the hill to evenly distribute the sag to a uniform height between all poles.

The prevailing temperature affects the conductor sag (*Figure 39*). On hot days, the conductor will expand and the sag will increase, while on cold days, the conductor will retract and the sag will decrease. When a minimal ground clearance is

Gauge No. awg	Temperature °F	Approximate Sag in Inches by Span Lengths				
		100 ft	125 ft	150 ft	175 ft	200 ft
6	30	6	6	13	18	-
	60	8	12	18	24	-
	90	12	17	23	30	-
4	30	6	8	13	18	35
	60	8	12	18	24	42
	90	12	17	23	30	50
2	30	6	8	13	20	29
	60	8	12	18	26	36
	90	12	17	23	33	44
1	30	6	8	13	24	24
	60	8	12	18	28	31
	90	12	17	23	31	39
0	30	6	8	13	15	18
	60	8	12	18	20.5	23
	90	12	17	23	27	29
00	30	6	8	13	17	21
	60	8	12	18	22	27
	90	12	17	23	28	34
0000	30	6	8	13	14	16
	60	8	12	18	19	21
	90	12	17	23	25	27

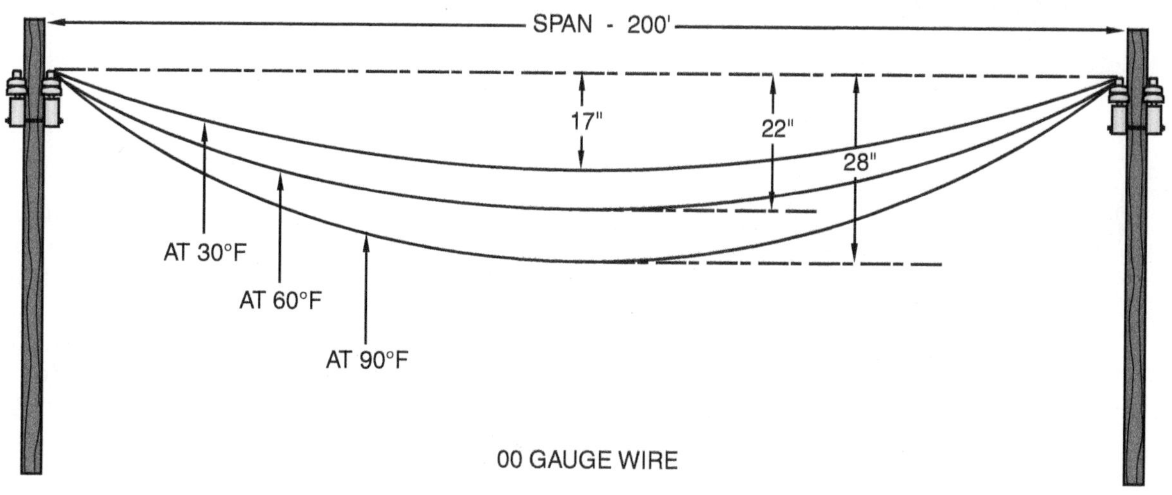

Figure 39 Effect of temperature on conductor sag.

critical, the ambient temperature and its effect on sag must always be considered.

For uniform sag consistency, a sag scope (a telescopic sight similar to a gun sight) can be mounted to an end pole (*Figure 40*) and the line-of-sight sag in the length of conductor line can be observed and adjusted as necessary.

A dynamometer, which measures the pulling force, may also be used to measure sag (*Figure 41*). The dynamometer is installed in the pulling line and the pull tension is referenced to a chart that gives the pull tension and sag for the conductor size, span length, and temperature.

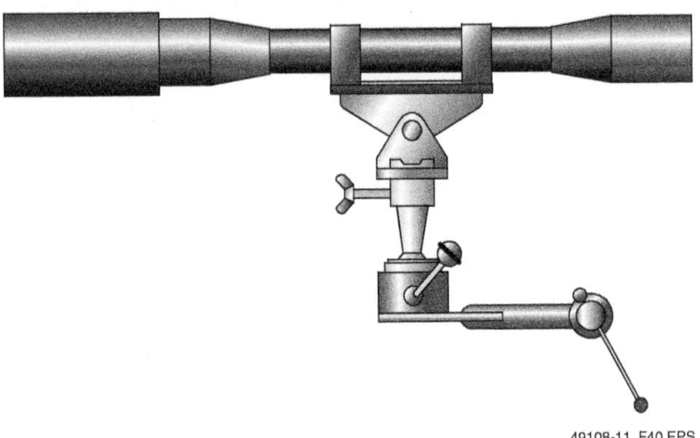

49108-11_F40.EPS

Figure 40 Sag scope.

49108-11_F41.EPS

Figure 41 Dynamometer.

In the past, small boards such as a lath were tacked at the desired sag height on each pole. A lineman on the end pole sighted down the line of boards and gave pulling instructions to linemen on the other poles.

Once the desired tension and sag is achieved on the new conductor, it will be secured to the dead-end clamp and insulator string at the last new pole. The excess conductor will be cut off at the clamp.

For three-phase primary conductor stringing, the remaining two conductors will be installed in the same manner.

7.3.1 Over-Tensioning

Over-tensioning is sometimes used to compensate for creep in new conductors. With this method, a dynamometer is attached in the pulling line and the conductor is tensioned 5 to 10 percent over the required tension for the prevailing temperature. The reasoning is that the conductor will creep back down to the correct tension and sag height after a period of time.

Over-tensioning can also be done visually by pulling the conductor slightly higher than the desired sag height. Over time the conductor creep will drop the conductor back to the desired level.

7.3.2 Pre-Stressing

Some companies may advocate pre-stressing as a way to deal with creeping of new conductors. During this process, the new conductor is pulled well above the desired tension but not to exceed 50 percent of the breaking strength. The tension is measured by a dynamometer and this increased tension is applied for 20 to 30 minutes. At the end of this time period, the tension is released allowing the conductor to drop. The relaxed conductor is then pulled to the correct sag height for the temperature.

7.4.0 Tying in the Conductors

Once the new conductors have been tensioned, sagged, and dead-ended, they must be tied in or clipped to the insulators. If a clamp-style insulator is used, the conductor is simply clamped to the top of the insulator. The most common way to tie the conductor to the insulator is with specially designed wire ties. Depending on the type of insulator used and the angle of the conductor to the insulator, the conductors can be tied in a top groove or tied in a side groove (*Figure 42*). The wire is wrapped around

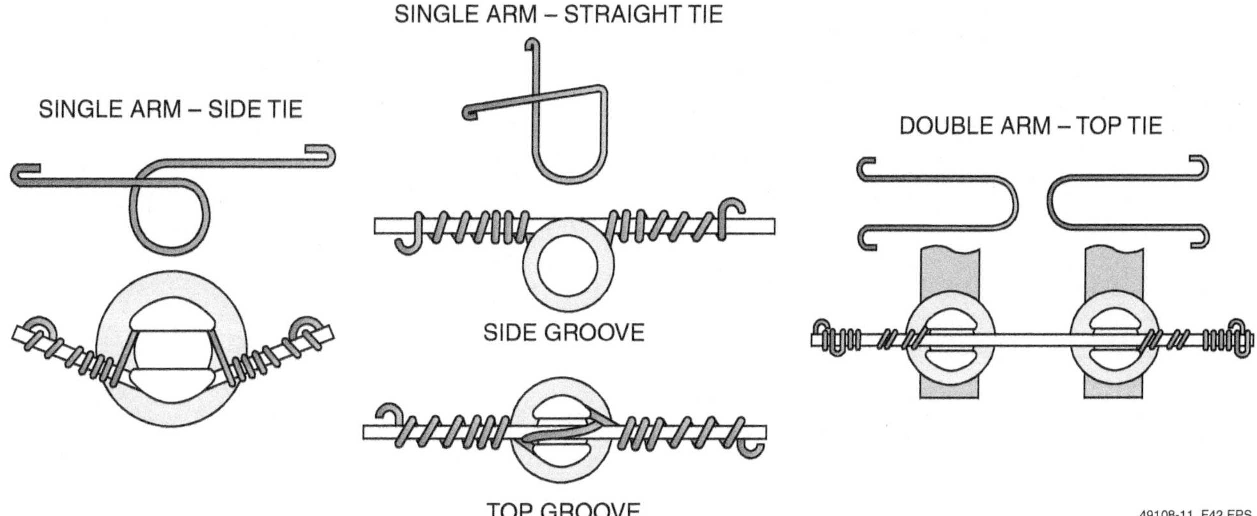

Figure 42 Insulator wire ties.

the conductor in a specific pattern to tightly secure it to the insulator.

7.5.0 Connecting the Conductors

After the new conductors have been strung, tensioned and sagged, dead-ended, and tied in to the insulators, they must be energized by connecting them to the existing energized conductors. This is done by clamping a jumper between the new conductor and the energized conductor (*Figure 43*).

The jumper is first clamped to the new conductor and then tied in to the insulator on the new cross-arm. While using the proper insulating PPE, the opposite end of the jumper is clamped to the energized conductor on the opposite side of the pole. The voltage then flows through the jumper and energizes the new conductors.

7.6.0 Neutral Line

The neutral line is attached to each pole and provides a universal ground that is connected throughout the system. The neutral line is tied to insulators below the primary conductors and cross-arms. All neutral conductors supplying the customers will be tied to this neutral line. When installing new poles and conductors, the neutral line must also be joined and continued along with the new span of conductors and poles. Most often, the neutral line is tied to spool-type insulators (*Figure 44*) attached to the side of the poles.

At the end of a span of conductors and poles, the neutral line must also be dead-ended (*Figure 45*). Helical wire wraps are commonly used to dead-end the neutral line.

Figure 43 Connecting the conductors with a jumper.

49108-11_F44.EPS

Figure 44 Neutral line insulators.

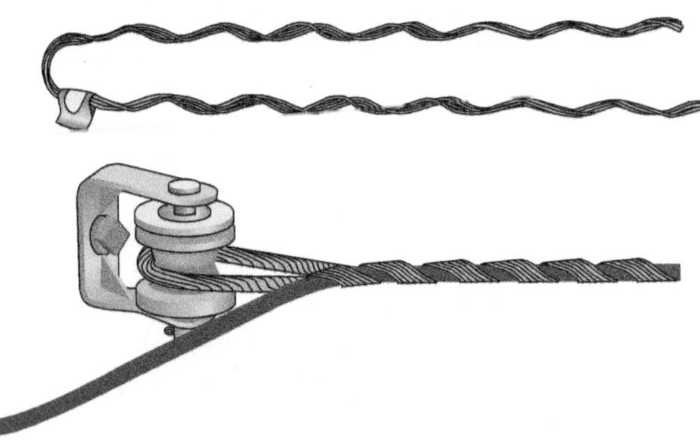

49108-11_F45.EPS

Figure 45 Neutral line dead-end.

8.0.0 Transformers

As described earlier, transformers are necessary for stepping down a high voltage supply into usable levels for the end users. Transformers are designed as single-phase and three-phase. Sizes and weights can vary from 275 lbs

49108-11_F46.EPS

Figure 46 Transformer mounting arrangements.

to well over 700 lbs (*Figure 46*). Properly rated lifting equipment must always be used as determined by the transformer size. Transformers can be raised into position by truck booms when they are available. Large transformers must be lifted with trucks or heavy-duty winches. Smaller single-phase transformers typically used for residential applications can be lifted by hand using a gin pole and block and tackle arrangement.

CAUTION

Transformers must always remain in an upright position during transporting and moving. Transformers should not be tipped over, rolled, or dragged into position. They should always be raised and moved into position using the top lifting lugs.

Depending upon the clearance and access to the mounting hardware, some transformer brackets can be installed to the transformer on the ground. All single-phase transformers are designed with the brackets as part of the housing. For the installation of multiple transformers on one pole, the brackets may need to be attached to the pole before lifting the transformers into mounting position.

8.1.0 Installing a Single-Phase Transformer with a Gin Pole

A gin pole provides an anchor point for attaching lifting equipment when raising components into place. The gin pole is securely mounted around the pole above the component installation height. The gin pole is designed with a large eye for attaching a block and tackle or snatch block. Gin poles are available in different types (*Figure 47*) and must be properly rated to the weight of the component. The block and tackle must also be correctly sized to the weight of the component. The lifting capacity and pulling ratio of the block and tackle is determined by the number of lifting lines or ropes, not including the hauling line.

Once the gin pole is secured, the block and tackle is suspended from the eye. A properly rated strap or rope must be secured around the transformer lifting lugs. The block and tackle is lowered and the lifting strap is placed in the lower hook. Using the block and tackle hauling rope, the transformer is slowly lifted into mounting position by workers on the ground or by a winch, if available.

Once at installation height, the worker on the pole secures the transformer bracket to the pole with through-the-pole machine bolts and attaching hardware. Some multiple transformer brackets can additionally be joined together with threaded rods and nuts (*Figure 48*). If the bracket was first attached to the pole, the worker secures the transformer to the bracket with attaching hardware.

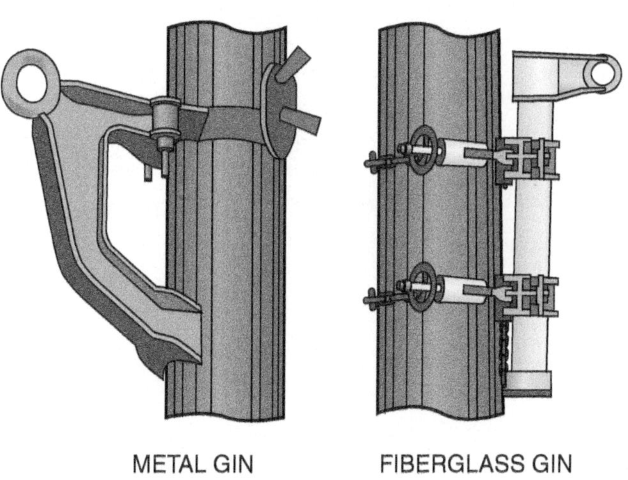

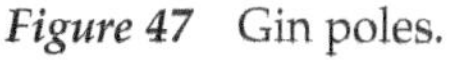

Figure 47 Gin poles.

8.2.0 Connecting the Transformer

After the transformer is mounted, it must be connected to the energized conductor. An infeed line is connected to the top of a cutout located on the cross-arm above the transformer or on the pole. The opposite end is connected to a hot clamp installed on the conductor. The cutout gate and fuse must be open during installation.

Figure 48 Transformer mountings.

The hot clamp end of the line (*Figure 49*) must be disconnected with a hot stick to completely de-energize the cutout during installation.

> **WARNING!**
>
> Workers must wear the required PPE, including rubber gloves and sleeves, when working around energized conductors and components.

The transformer voltage infeed line is also clamped to the bottom of the fused cutout. The opposite end of the infeed line must be attached to the bushing on top of the transformer. A lightning arrestor is normally included on the transformer. This component must also be jumpered to the transformer bushing. After these connections are complete, protective covers, sometimes called critter covers, are placed over them to prevent animals such as squirrels from shorting the connections and damaging the transformer.

8.3.0 Connecting Secondary Connectors

The 120V, 240V, and neutral secondary conductors (*Figure 50*) must be connected to the terminals on the side of the transformer. These secondary conductors must be bundled and then supplied to the customer location.

Once all lines and ground connections have been attached, the hot clamp end of the infeed line must be connected back into the conductor hot clamp using a hot stick. The fused cutout must now be closed to energize the transformer. This is also done with the hot stick. The hook on the end of the hot stick is placed in the eye of the cutout gate. After turning his head and looking away, the worker must thrust the gate closed with the hot stick. The fuse is now connected and the transformer and secondary conductors should be in an energized state.

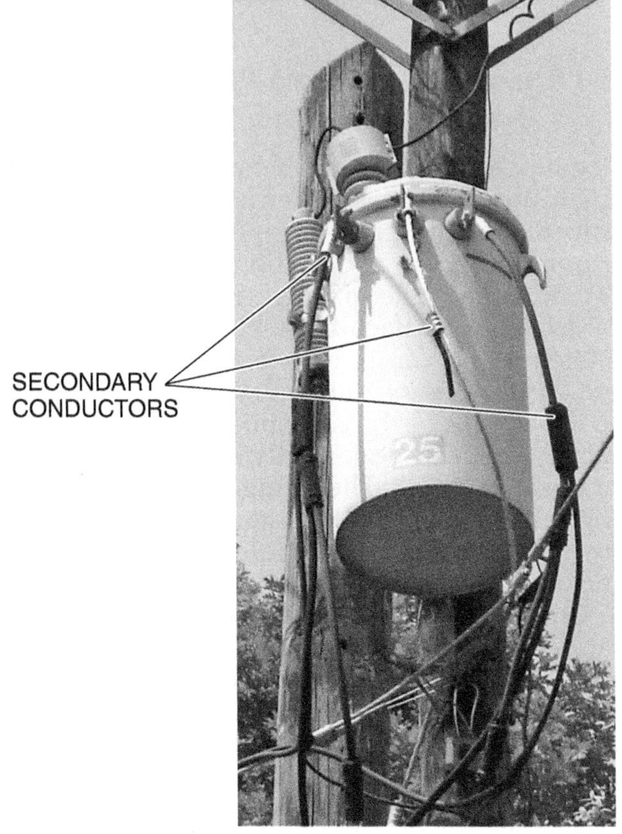

Figure 50 Secondary conductors.

Figure 49 Conductor hot clamp and transformer connections.

SUMMARY

Once a utility pole is erected, it must be firmly supported using an appropriate guying system that is firmly anchored in the ground. The top of the pole is fitted-out with the hardware, insulators, and supports needed to carry conductors and other components. Guys are installed after the pole is erected. When possible, cross-arms and related hardware are installed while the pole is on the ground. In some instances, however, this work is done on an upright pole. Conductor installation may also be started while the pole is being fitted-out on the ground. Transformers are usually installed while the pole is upright. Safety is extremely important when performing aerial framing, as this work can place installers in proximity to live high-voltage lines.

There are several types of cross-arms used to support conductors. The most common is an 8' long wooden timber. Most cross-arms are predrilled with mounting holes. Alley arms extend from one side of a pole. They are used when there is not enough space for cross-arms. Conductors are heavy-gauge uninsulated cables consisting of a steel core wrapped in layers of aluminum. These cables are strung using either the slack/layout method or the tension method. The latter is preferred. Once the conductors are strung, they must be properly tensioned and sagged.

Pole transformers are installed after the pole is erected. A bucket truck is commonly used to raise the transformer(s) into position. A block and tackle can be used if a bucket truck is not available.

Review Questions

1. Power supplied by transmission and large distribution primary conductors is _____.
   a. single-phase
   b. raw voltage
   c. three-phase
   d. four-phase

2. Typical power levels supplied by a three-phase transformer are _____.
   a. 208V, 480V, and 575V
   b. 200V, 400V, and 500V
   c. 240V, 480V, and 575V
   d. 240V, 480V, and 600V

3. Single-phase transformers supply the customer with _____.
   a. 120V and 220V
   b. 110V and 220V
   c. 110V and 208V
   d. 120V and 240V

4. The attaching hardware used to mount cross-arms and transformer brackets to the pole are _____.
   a. fully threaded rods with nuts
   b. heavy-duty toggle bolts
   c. pilot lag bolts
   d. carriage bolts with nuts

5. The special type of washer used when installing components to the pole is a _____.
   a. titanium washer
   b. Teflon®-coated washer
   c. flat washer
   d. brass washer

6. The mounting hardware used to isolate an energized conductor from the pole structure is a(n) _____.
   a. stand-off pin
   b. insulator pin
   c. plastic bushing
   d. insulated bolt

7. The length of a typical wooden cross-arm is _____.
   a. 6'
   b. 7'
   c. 8'
   d. 9'

8. The thickness dimensions available for wooden cross-arms are _____.
   a. 3½" × 5½" and 5½" × 5½"
   b. 6" × 6" and 5½" × 6½"
   c. 4" × 4" and 4" × 6"
   d. 3" × 4" and 5" × 6"

9. The distance between the brace holes on a cross-arm is called the _____.
   a. spread
   b. span
   c. drop
   d. miter

10. A guy anchored close to the pole due to limited ground clearance is called a _____.
    a. stub guy
    b. head guy
    c. span guy
    d. sidewalk guy

11. The number of down guys required on a pole is determined by the _____.
    a. levels of conductors, lines, and cables attached to the pole
    b. angle of the pole and conductors
    c. size of the primary conductors
    d. height of the pole

12. The guy commonly used on hilly terrain is the _____.
    a. span guy
    b. arm guy
    c. head guy
    d. down guy

13. Bare wire primary conductors are typically constructed of _____.
    a. helically wound steel, aluminum, and copper wires
    b. a steel core shrouded in layers of helically wound aluminum wire
    c. an aluminum core shrouded in layers of helically wound steel wire
    d. a steel core shrouded in layers of helically wound copper and steel wire

14. When installing conductors, the devices used to smoothly guide the conductor between poles are _____.
    a. sky hooks
    b. equalizer bars
    c. stringing blocks
    d. conductor grips

15. When installing conductors, a bullwheel is used to provide _____.
    a. conductor lubrication
    b. anti-static conditioning
    c. tension and braking
    d. load limiting

Trade Terms Quiz

Fill in the blank with the correct term that you learned from your study of this module.

1. The terminating arrangement at the end of a run of primary conductors is the ____________.
2. A device used to securely join two conductor ends together is a(n) ____________.
3. A length of fiberglass rod connected at the top of guy wires that extend into the primary conductor level on a pole is the ____________.
4. A motorized piece of equipment used to provide consistent tension and braking to a conductor supply during pulling and installation is the ____________.
5. The amount of drop in a span of mounted conductors is the ____________.
6. A term to describe wire mesh grips and splices that cinch around a conductor when tension is applied is ____________.
7. A wire grip used for securing guy wires during tensioning is a(n) ____________.
8. A pulley-style device used to smoothly guide conductors between poles during installation is a(n) ____________.
9. An automatic fused safety device located in the line between a primary conductor and a secondary conductor or other components is a(n) ____________.
10. A wire constructed line used to convey electric power from the point of generation to an end user is the ____________.
11. A length of wire used to connect two electrical distribution sources together is a(n) ____________.
12. A pole mounted structure used to support conductors and components is a(n) ____________.
13. A ceramic or polymer component used to support and isolate energized conductors from the pole structure is a(n) ____________.
14. A device well secured in the ground to provide an anchor point for a guy wire is a(n) ____________.
15. Various types of supports used to maintain the vertical stability of poles are known as ____________.
16. A bank of components connected to distribution lines to stabilize power where fluctuations could occur are known as ____________.

Trade Terms

Bullwheel
Chicago-style grip
Chinese finger
Conductor
Cross-arm
Cutout
Distribution capacitors
Guy anchor
Guy rod
Guys
Insulator
Jumper
Sag
Splice
Stringing block

Cornerstone of Craftsmanship

Terry G. Williams

Apprenticeship Training Manager
Quanta Services, Houston, TX

Terry Williams defines craftsmanship as pride—personal pride in your work, whatever it may be, knowing that you have done the best possible job to make everything perfect. He believes a major part of craftsmanship is the training, education, and experience gained from NCCER courses.

How did you get started in the construction industry?
After serving in the Marine Corps, I started in the construction industry in 1970. My first job was ironworker on a high-rise building. It was in the dead of winter, and I still remember the wind and the steel being very, very cold.

Who inspired you to enter the industry?
A friend of mine who had worked on the first Alaskan pipeline introduced me to the benefits of the construction industry. The high rate of pay and travel to new locations inspired me to join this industry.

What do you enjoy most about your job?
I enjoy the interaction with the diverse group of employees across the country. I get a great deal of personal satisfaction helping others know more about their jobs. The knowledge and training that I provide helps them work smarter, be more productive, rapidly advance in their careers, and most importantly, keeps them safe while at work.

Do you think training and education are important in construction?
Training and education are critical to the success of any construction business. If the accumulated knowledge of past master-level journeymen is not shared with the new workers, we will lose the craftsmanship that made America the number one nation in the world.

How important are NCCER credentials to your career?
NCCER credentials are very important for anyone's career. These credentials certify that a person has completed competency-based training that meets standards set by experts in the field.

How has training/construction impacted your life and your career?
Training and education have had a major impact on my life and my family's standard of living. I have always been a person with a thirst for knowledge. Early in my career I sought out training and learning opportunities. As I learned more I became a more valued employee, so raises and advancement in my profession came easily. Construction has provided my family and me with a great way of life.

Would you suggest construction as a career to others?
Construction is a wonderful career choice for several reasons. The pay is good, you get to travel to new places and meet a lot of great people, and you get personal satisfaction in seeing the finished results of your labor. As you become established in your career, you make lifelong business relationships.

How do you define craftsmanship?
I define craftsmanship as pride—personal pride in your work, whatever it may be, knowing that you have done the best possible job to make everything perfect. A major part of craftsmanship is the training, education, and experience gained from NCCER courses.

Trade Terms Introduced in This Module

Bullwheel: A motorized piece of equipment used to provide consistent tension and braking to a conductor supply during pulling and installation.

Chicago-style grip: A wire grip used for securing guy wires during tensioning. This type of grip cannot be used for gripping conductors.

Chinese finger: A term to describe wire mesh grips and splices that cinch around a conductor when tension is applied. This type of grip will not damage conductors.

Conductor: A wire constructed line used to convey electric power from the point of generation to an end user.

Cross-arm: A pole-mounted structure used to support conductors and components.

Cutout: An automatic fused safety device located in the line between a primary conductor and a secondary conductor or other components.

Dead-end: A term used to identify the terminating arrangement at the end of a run of primary conductors.

Distribution capacitors: A bank of components connected to distribution lines to stabilize power where fluctuations could occur.

Guy anchor: A device well-secured in the ground to provide an anchor point for a guy wire.

Guy rod: A length of fiberglass rod connected at the top of guy wires that extend into the primary conductor level on a pole.

Guys: Various types of supports used to maintain the vertical stability of poles.

Insulator: A ceramic or polymer component used to support and isolate energized conductors from the pole structure.

Jumper: A length of wire used to connect two electrical distribution sources together.

Sag: The amount of drop in a span of mounted conductors.

Splice: A device used to securely join two conductor ends together.

Stringing block: A pulley-style device used to smoothly guide conductors between poles during installation.

EXAMPLES OF AERIAL FRAMING DRAWINGS

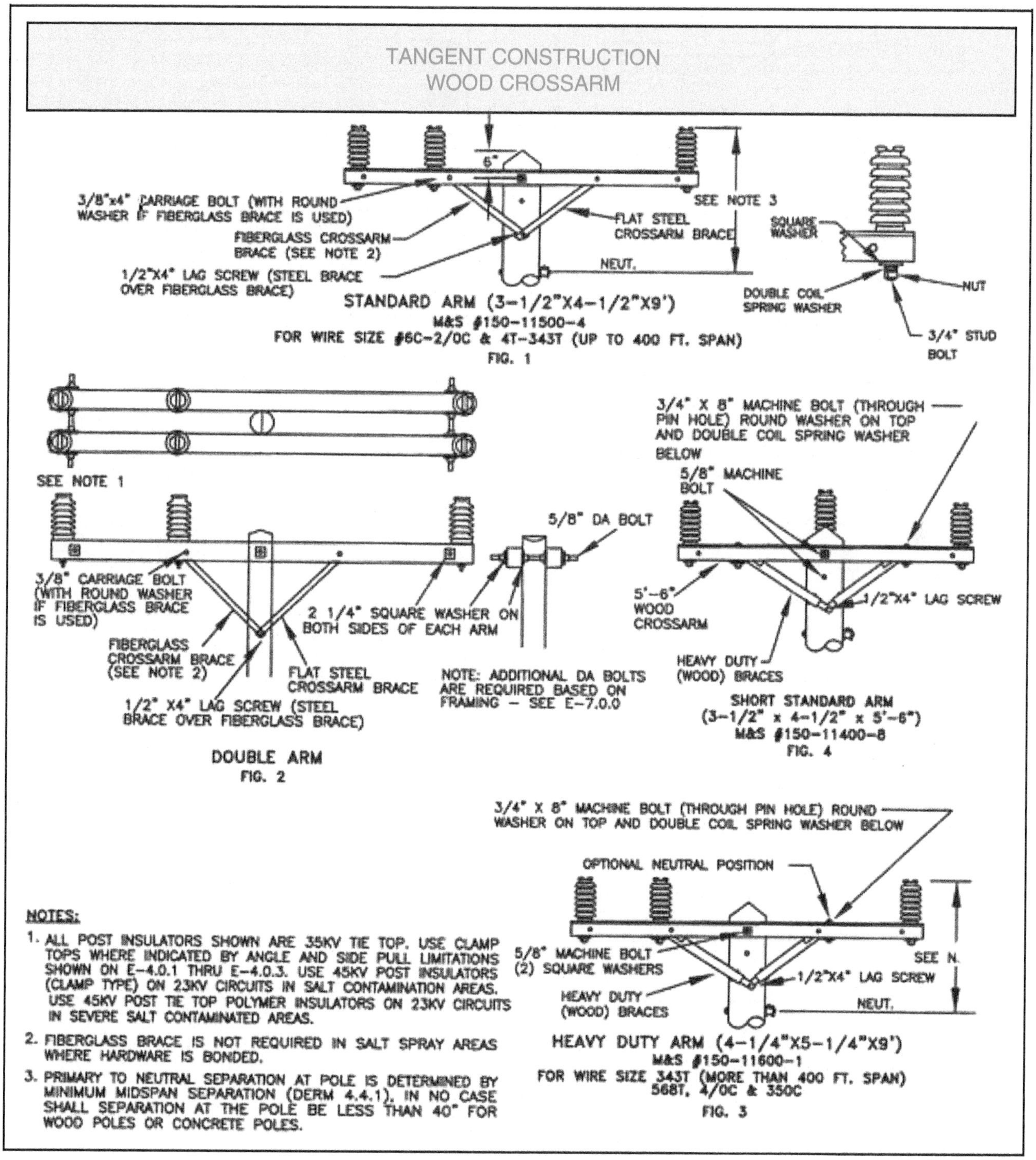

TANGENT CONSTRUCTION
WOOD CROSSARM

6"

3/8"x4" CARRIAGE BOLT (WITH ROUND WASHER IF FIBERGLASS BRACE IS USED)

FIBERGLASS CROSSARM BRACE (SEE NOTE 2)

1/2"X4" LAG SCREW (STEEL BRACE OVER FIBERGLASS BRACE)

FLAT STEEL CROSSARM BRACE

SEE NOTE 3

NEUT.

SQUARE WASHER

DOUBLE COIL SPRING WASHER

NUT

3/4" STUD BOLT

STANDARD ARM (3-1/2"X4-1/2"X9')
M&S #150-11500-4
FOR WIRE SIZE #6C-2/0C & 4T-343T (UP TO 400 FT. SPAN)
FIG. 1

SEE NOTE 1

3/8" CARRIAGE BOLT (WITH ROUND WASHER IF FIBERGLASS BRACE IS USED)

FIBERGLASS CROSSARM BRACE (SEE NOTE 2)

1/2" X4" LAG SCREW (STEEL BRACE OVER FIBERGLASS BRACE)

FLAT STEEL CROSSARM BRACE

2 1/4" SQUARE WASHER ON BOTH SIDES OF EACH ARM

5/8" DA BOLT

NOTE: ADDITIONAL DA BOLTS ARE REQUIRED BASED ON FRAMING – SEE E-7.0.0

DOUBLE ARM
FIG. 2

3/4" X 8" MACHINE BOLT (THROUGH PIN HOLE) ROUND WASHER ON TOP AND DOUBLE COIL SPRING WASHER BELOW

5/8" MACHINE BOLT

5'-6" WOOD CROSSARM

1/2"X4" LAG SCREW

HEAVY DUTY (WOOD) BRACES

SHORT STANDARD ARM
(3-1/2" x 4-1/2" x 5'-6")
M&S #150-11400-8
FIG. 4

3/4" X 8" MACHINE BOLT (THROUGH PIN HOLE) ROUND WASHER ON TOP AND DOUBLE COIL SPRING WASHER BELOW

OPTIONAL NEUTRAL POSITION

5/8" MACHINE BOLT
(2) SQUARE WASHERS

HEAVY DUTY (WOOD) BRACES

1/2"X4" LAG SCREW

SEE N.

NEUT.

HEAVY DUTY ARM (4-1/4"X5-1/4"X9')
M&S #150-11600-1
FOR WIRE SIZE 343T (MORE THAN 400 FT. SPAN)
568T, 4/0C & 350C
FIG. 3

NOTES:

1. ALL POST INSULATORS SHOWN ARE 35KV TIE TOP. USE CLAMP TOPS WHERE INDICATED BY ANGLE AND SIDE PULL LIMITATIONS SHOWN ON E-4.0.1 THRU E-4.0.3. USE 45KV POST INSULATORS (CLAMP TYPE) ON 23KV CIRCUITS IN SALT CONTAMINATION AREAS. USE 45KV POST TIE TOP POLYMER INSULATORS ON 23KV CIRCUITS IN SEVERE SALT CONTAMINATED AREAS.
2. FIBERGLASS BRACE IS NOT REQUIRED IN SALT SPRAY AREAS WHERE HARDWARE IS BONDED.
3. PRIMARY TO NEUTRAL SEPARATION AT POLE IS DETERMINED BY MINIMUM MIDSPAN SEPARATION (DERM 4.4.1). IN NO CASE SHALL SEPARATION AT THE POLE BE LESS THAN 40" FOR WOOD POLES OR CONCRETE POLES.

49108-11_A01.EPS

WOOD ALLEY ARM CONSTRUCTION DETAILS ALTERNATE METHODS

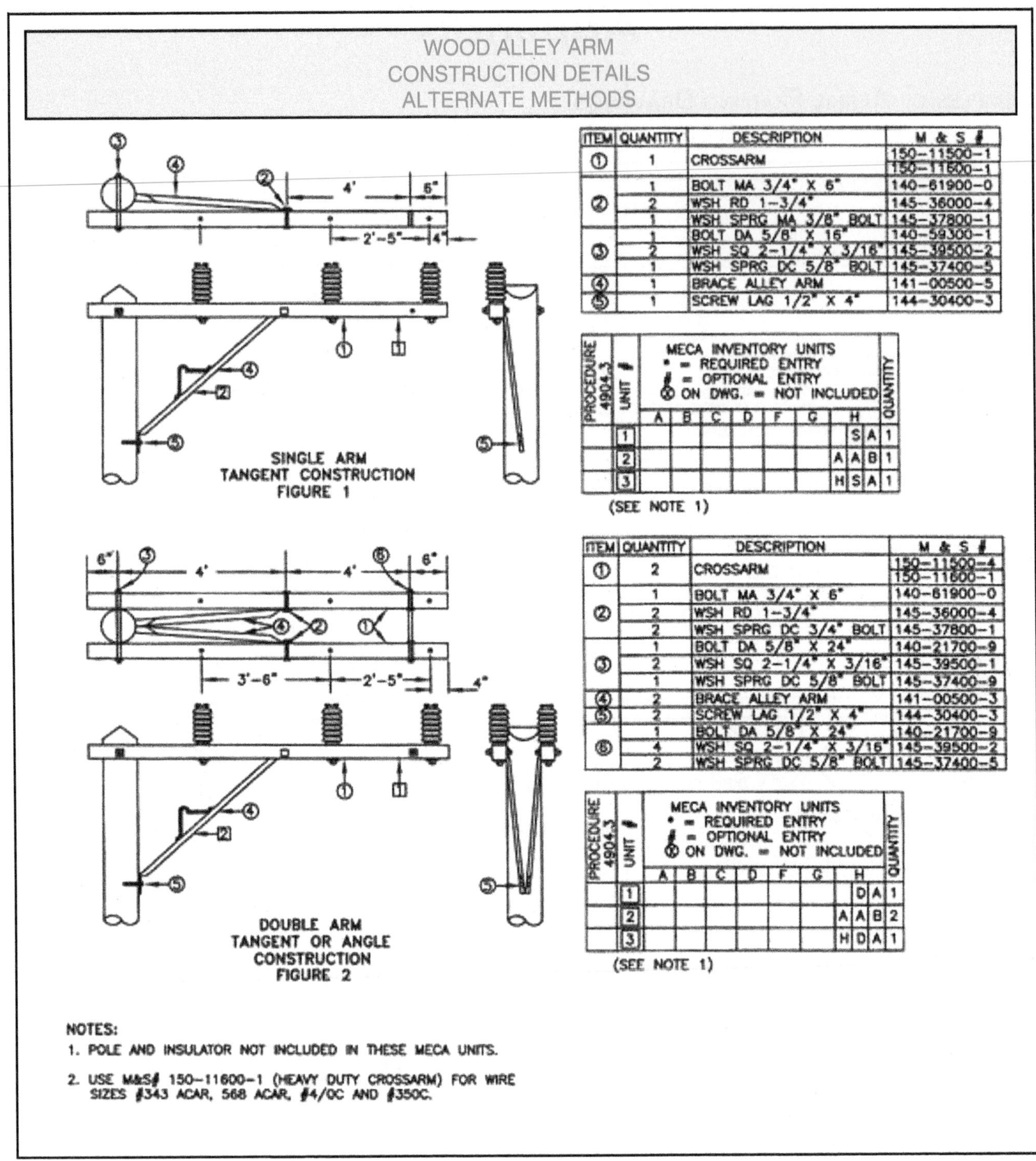

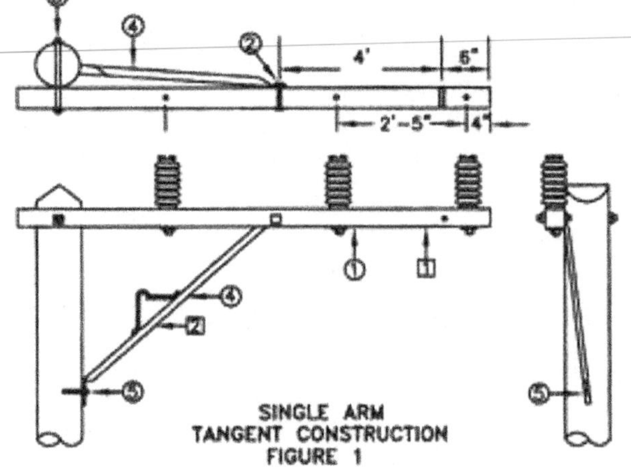

SINGLE ARM TANGENT CONSTRUCTION FIGURE 1

ITEM	QUANTITY	DESCRIPTION	M & S #
①	1	CROSSARM	150-11500-1 150-11600-1
②	1	BOLT MA 3/4" X 6"	140-61900-0
	2	WSH RD 1-3/4"	145-36000-4
	1	WSH SPRG MA 3/8" BOLT	145-37800-1
③	1	BOLT DA 5/8" X 16"	140-59300-1
	2	WSH SQ 2-1/4" X 3/16"	145-39500-2
	1	WSH SPRG DC 5/8" BOLT	145-37400-5
④	1	BRACE ALLEY ARM	141-00500-5
⑤	1	SCREW LAG 1/2" X 4"	144-30400-3

PROCEDURE 4904.3	UNIT #	MECA INVENTORY UNITS * = REQUIRED ENTRY # = OPTIONAL ENTRY ⊗ ON DWG. = NOT INCLUDED								QUANTITY
		A	B	C	D	F	G	H		
	1								S	A	1
	2							A	A	B	1
	3							H	S	A	1

(SEE NOTE 1)

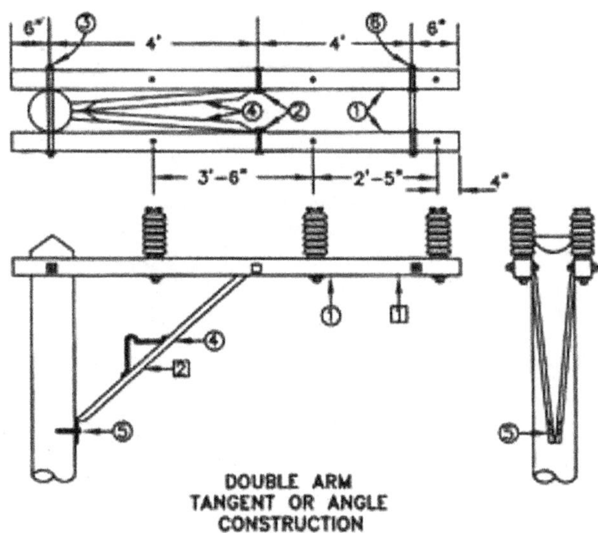

DOUBLE ARM TANGENT OR ANGLE CONSTRUCTION FIGURE 2

ITEM	QUANTITY	DESCRIPTION	M & S #
①	2	CROSSARM	150-11500-4 150-11600-1
②	1	BOLT MA 3/4" X 6"	140-61900-0
	2	WSH RD 1-3/4"	145-36000-4
	2	WSH SPRG DC 3/4" BOLT	145-37800-1
③	1	BOLT DA 5/8" X 24"	140-21700-9
	2	WSH SQ 2-1/4" X 3/16"	145-39500-1
	1	WSH SPRG DC 5/8" BOLT	145-37400-9
④	2	BRACE ALLEY ARM	141-00500-3
⑤	2	SCREW LAG 1/2" X 4"	144-30400-3
⑥	1	BOLT DA 5/8" X 24"	140-21700-9
	4	WSH SQ 2-1/4" X 3/16"	145-39500-2
	2	WSH SPRG DC 5/8" BOLT	145-37400-5

PROCEDURE 4904.3	UNIT #	MECA INVENTORY UNITS * = REQUIRED ENTRY # = OPTIONAL ENTRY ⊗ ON DWG. = NOT INCLUDED								QUANTITY
		A	B	C	D	F	G	H		
	1								D	A	1
	2							A	A	B	2
	3							H	D	A	1

(SEE NOTE 1)

NOTES:

1. POLE AND INSULATOR NOT INCLUDED IN THESE MECA UNITS.
2. USE M&S# 150-11600-1 (HEAVY DUTY CROSSARM) FOR WIRE SIZES #343 ACAR, 568 ACAR, #4/0C AND #350C.

49108-11_A02.EPS

TRANSFORMER INSTALLATION
SINGLE PHASE PRIMARY CONSTRUCTION
& MODIFIED VERTICAL CONSTRUCTION

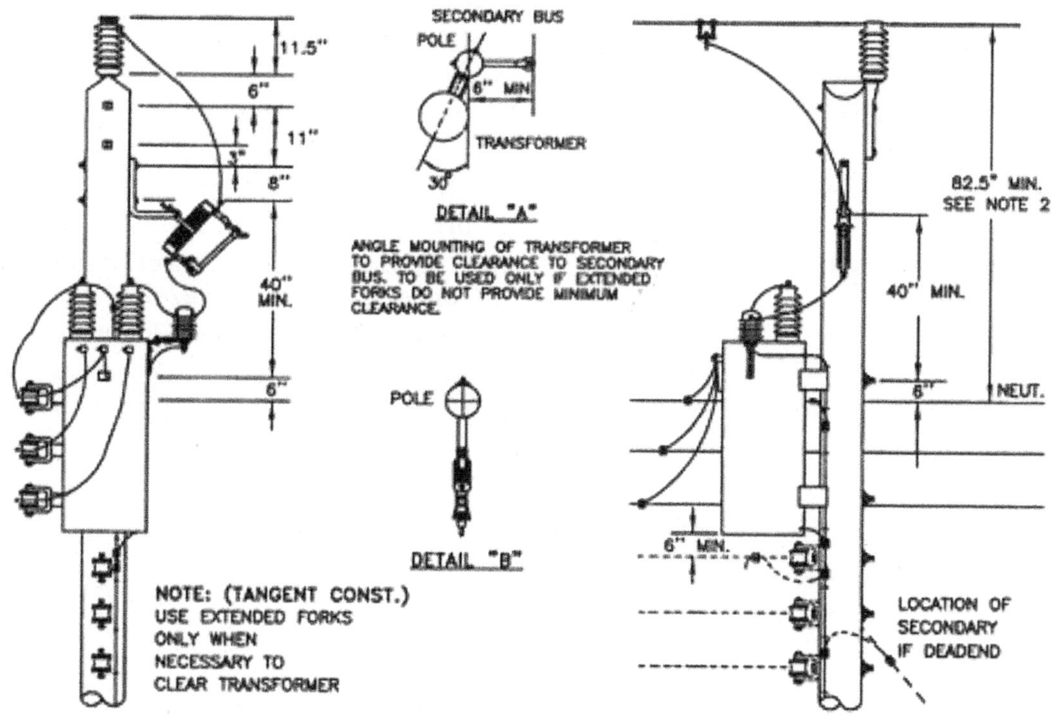

TANGENT TRANSFORMER POLE

NOTES:

1. IN SALT SPRAY AREAS USE SALT SPRAY TRANSFORMER, SALT SPRAY ARRESTERS, AND SALT SPRAY CUTOUTS.
2. PREFERRED LOCATION FOR NEUTRAL IS 6" MINIMUM BELOW TOP TRANSFORMER MOUNTING BRACKET BOLT, BUT IN NO CASE HIGHER THAN THE TOP OF TRANSFORMER TANK.
3. SEE SHEET I–53.1.1 FOR SCHEMATIC DIAGRAM OF TRANSFORMER CONNECTIONS.
4. DOUBLE PRIMARY BUSHING TRANSFORMERS ILLUSTRATED, FOR DETAILS OF SINGLE OR DOUBLE BUSHING GROUNDING, SEE SHEET I–5.0.0.
5. FOR ARRESTER INSTALLATION SEE I–6.0.0.

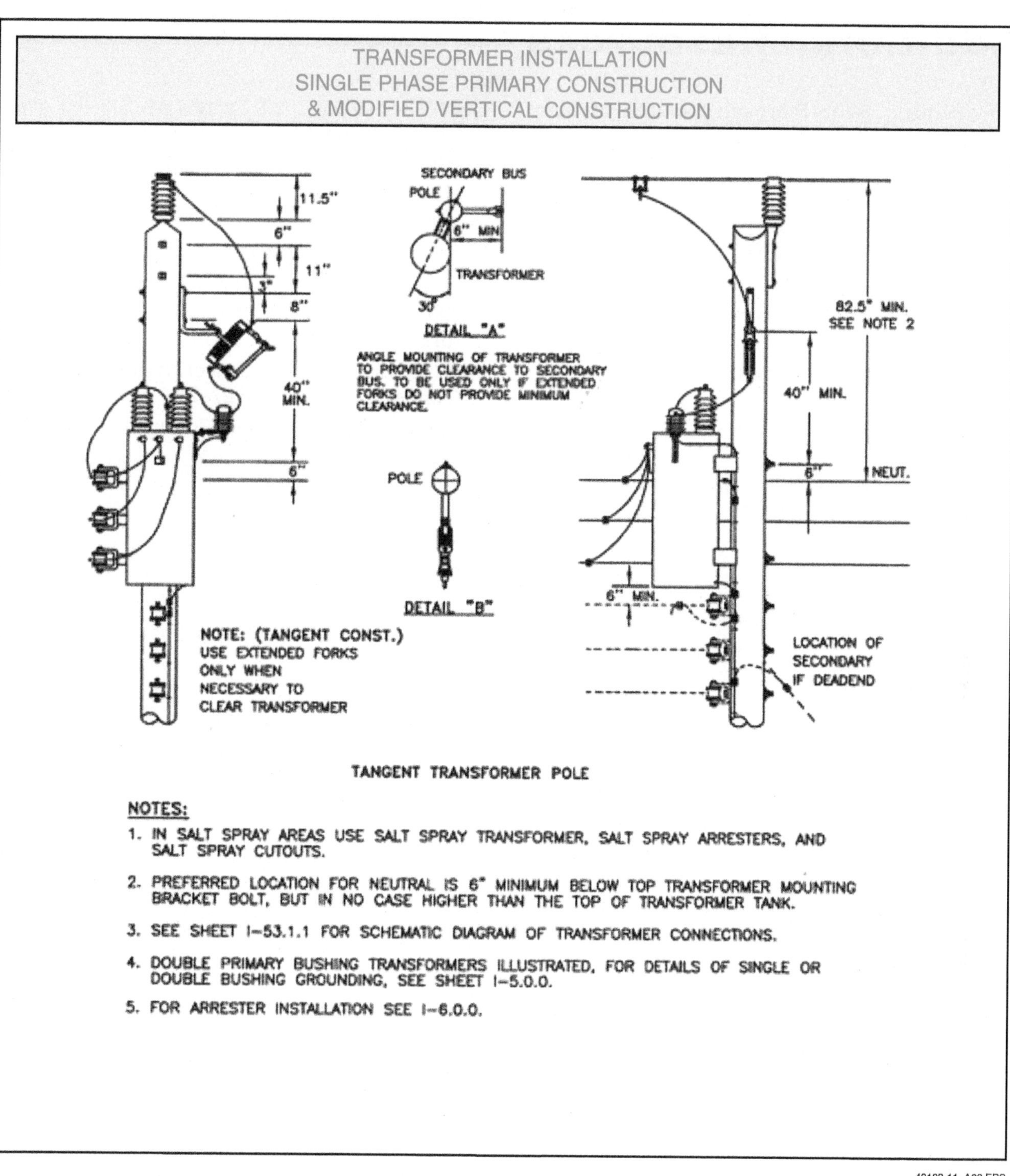

49108-11_A03.EPS

Additional Resources

This module presents thorough resources for task training. The following resource material is suggested for further study.

OSHA Regulation 1926, Subpart M, Fall Protection, Latest edition. Occupational Health and Safety Administration. **www.osha.gov**.

Figure Credits

Tony Vazquez, Module opener

Topaz Publications, Inc., SA01, SA02, Figures 2, 4 (photo), 13, 26, 30, and 48 (threaded rods and nuts)

Bob Groner, Figures 6 (photos), 7 (photos), and Appendix

Jim Mitchem, Figures 3 (photo), 7 (cross-arm bracket), 8–12, 14–25, 31–39, 42–47, 49, and 50

Salisbury Electrical Safety, Figures 5 and 48 (brackets part of transformer housing)

A. B. Chance/Hubbell Power Systems, Figures 7 (connector splice) and 27–29

Dillon Force Measurement, Figure 41

CONTREN® LEARNING SERIES — USER UPDATE

NCCER makes every effort to keep its textbooks up-to-date and free of technical errors. We appreciate your help in this process. If you find an error, a typographical mistake, or an inaccuracy in NCCER's Contren® materials, please fill out this form (or a photocopy), or complete the online form at www.nccer.org/olf. Be sure to include the exact module number, page number, a detailed description, and your recommended correction. Your input will be brought to the attention of the Authoring Team. Thank you for your assistance.

Instructors – If you have an idea for improving this textbook, or have found that additional materials were necessary to teach this module effectively, please let us know so that we may present your suggestions to the Authoring Team.

NCCER Product Development and Revision

3600 NW 43rd Street, Building G, Gainesville, FL 32606

Fax: 352-334-0932
Email: curriculum@nccer.org
Online: www.nccer.org/olf

☐ Trainee Guide ☐ AIG ☐ Exam ☐ PowerPoints Other ______________

Craft / Level: Copyright Date:

Module Number / Title:

Section Number(s):

Description:

Recommended Correction:

Your Name:

Address:

Email: Phone:

Utility Service Equipment

49109-11

Trainees with successful module completions may be eligible for credentialing through NCCER's National Registry. To learn more, go to **www.nccer.org** or contact us at **1.888.622.3720.** Our website has information on the latest product releases and training, as well as online versions of our *Cornerstone* newsletter and Pearson's Contren® product catalog.

Your feedback is welcome. You may email your comments to **curriculum@nccer.org**, send general comments and inquiries to **info@nccer.org**, or use the User Update form at the back of this module.

V.1 7/11

Utility Service Equipment

Objectives

When you have completed this module, you will be able to do the following:

1. Identify the types of bucket trucks and digger derricks used by power line workers.
2. Identify the operator safety requirements that must be followed when operating a bucket truck or digger derrick.
3. Explain and demonstrate how to perform a pre-start inspection on a service vehicle.
4. Describe the safety considerations associated with setting up a service vehicle at a job site.
5. Describe and demonstrate the safety considerations and basic operations procedures associated with using a bucket truck at a job site.
6. Describe and demonstrate the safety considerations and basic operations procedures associated with using a digger derrick at a job site.
7. Describe ways that a crew can prepare for and react to a bucket truck or digger derrick related emergency.
8. Inspect, set up, and operate utility service equipment:
   - Bucket truck
   - Digger derrick
   - Crane truck
   - Aerial lift

Performance Tasks

Under the supervision of the instructor, you should be able to do the following:

1. Inspect, set up, and operate a bucket truck.
2. Inspect, set up, and operate a digger derrick.
3. Inspect, set up, and operate a crane truck.
4. Inspect, set up, and operate an aerial lift.

Trade Terms

Aerial device
Anchorage point
Articulating boom
Boom
Equipotential region
Insulated aerial device
Jib
Lanyard
Outrigger
Pathogen
Platform
Telescopic boom

Contents

Topics to be presented in this module include:

Figures and Tables

1.0.0 Introduction

Power line workers use bucket trucks (also referred to as boom trucks and cherry pickers) to gain access to overhead transmission and distribution lines. They also use digger derrick trucks to erect the poles that support the lines. Crane trucks can also be used to lift materials and larger components. Aerial lifts are also available for lifting workers to the required working height. Bucket trucks are mainly used to airlift a line worker to safely install, maintain, and repair power transmission and distribution lines and components at the top of utility poles. Digger derrick trucks are used in digging as well as placement of utility poles that support these components and transmission lines. There are different kinds of specialized vehicles that companies use depending on the nature and services performed by the business. Power line workers use these vehicles on a regular basis.

2.0.0 Bucket Truck Overview

Bucket trucks are large utility trucks with a pedestal-mounted boom that lifts an aerial platform (referred to as a bucket) into the air so that a line worker can gain access to aerial power lines and power line components. Bucket trucks come in several configurations and sizes. The main differences are in the type and height of the boom, which relates directly to the size of the truck. They are also available in single-bucket and double-bucket configurations equipped with or without lifting jibs to lift materials. All bucket trucks contain controls within the bucket so that the technician working from the bucket can control its movement. When double-bucket trucks are used, the controls must be easily accessible from both buckets. They also contain controls on the truck, or remote controls, so that an operator on the ground can control the movement of the bucket, if necessary.

Bucket trucks are normally classified as insulated or non-insulated. This classification refers to the electrical conductivity of the boom and personnel bucket. An insulated bucket truck contains a fiberglass bucket and a fiberglass upper boom section. The lower boom section has a fiberglass insert to provide an insulation gap between the steel sections. To be considered insulated, the fiberglass boom and personnel bucket must be certified in accordance with the *American National Standards Institute (ANSI) A92.2* dielectric rating requirements.

> **WARNING!**
> No bucket truck, aerial lift, or other type of aerial platform, whether insulated or not, provides any electrical protection to the person working in the platform or bucket if there is direct contact with an energized power source. Such contact will cause serious injury or death.

Bucket trucks are also classified by the type of boom they use. Types of booms include the telescopic boom and the articulating boom. Some trucks use a combination of both.

2.1.0 Telescopic Boom Bucket Trucks

The telescopic boom consists of straight shafts that fit inside each other and extend to raise an aerial platform. They are designed to reach vertically or horizontally. Telescopic boom trucks (*Figure 1*) are used primarily by power line technicians, cable installers, telephone line workers, arborists, and technicians who maintain signs and lighting. These booms rarely exceed 45 feet in working height and can usually be mounted on a lighter truck. Depending on the size of the truck, they can offer working heights between 20 and 55 feet.

2.2.0 Articulating Boom Bucket Trucks

Power line workers also use articulating boom bucket trucks. An articulating boom bucket truck (*Figure 2*) is an aerial device with two or

On Site

What About Rentals?

Many times, due to unforeseen circumstances, you may have to rent a bucket truck or digger derrick for a day or longer while yours is being repaired or maintained. Therefore, you might not be familiar with the controls and other key features on the rental model. You will also be unfamiliar with the maintenance history of the vehicle. How can you ensure safety while operating an unfamiliar vehicle? Here are some guidelines:

- Before leaving the rental site with the vehicle, ensure that the company renting the vehicle has properly serviced the vehicle.
- Inspect the vehicle with the rental company representative before leaving their site.
- Ensure that all operating and maintenance manuals are available and match the equipment being rented.
- Make sure that the operator controls are easy to reach and properly marked.

Figure 1 Telescoping bucket truck.

Figure 2 Articulating bucket truck.

more hinged boom sections. They are designed to reach up and over obstacles. Articulated aerial lift devices can be equipped with either a single bucket or double bucket mounted at the end of the upper boom section. Insulated versions have a lower articulating arm that is made of steel with an electrically insulated fiberglass insert. The upper arm is made of fiberglass or similar non-conductive product. This reduces the potential for electrical shock in case the boom comes into contact with a live wire. Variations of the articulating boom bucket truck range from smaller gas engine units that provide a working height of 30 feet, to large diesel-engine units that have a working height of up to 185 feet. The larger units have outriggers that extend out from the sides of the truck to provide stability to the truck when in service.

WARNING!

Bucket trucks with insulated booms can only isolate the operator from grounding through the boom and vehicle. They cannot provide phase-to-phase or phase-to-ground protection from contacts occurring at the boom tip. The boom tip has metal components that will conduct electricity if the boom comes into contact with an energized conductor. Electricity can even arc through air if the metal object gets close enough. If the boom tip contacts an energized circuit, the entire boom tip, including the controls, will become energized. The only way to ensure your safety is to maintain required clearances and wear required PPE.

2.3.0 Combination Boom Bucket Trucks

The combination boom bucket truck (*Figure 3*) is similar to the telescoping boom truck except that it has an articulating lower boom.

Figure 3 Combination boom bucket truck.

GOING GREEN

New Hybrid Bucket Trucks Gaining Popularity

Utility companies across the nation are starting to use hybrid bucket trucks in an effort to reduce vehicle emissions, conserve fossil fuels, and provide quieter working environments. Variations of the new hybrid vehicles include biofuel trucks, diesel-electric trucks, natural gas trucks, and all-electric trucks.

The biofuel trucks run on a blend of 80-percent diesel fuel and 20-percent biofuel made from soy. The diesel-electric trucks alternate between diesel fuel and battery power to offer up to 60-percent in fuel savings. Natural gas trucks and all-electric trucks use alternative power sources to operate the truck engine and power the hydraulic pump for the bucket.

As these new hybrid trucks enter the work force, many states are also providing means to support the trucks, such as complimentary electric vehicle charging stations at rest stops.

The lower boom gives the operator extra height while also offering additional horizontal side reach, to enable working at heights over obstructions. Depending on the size of the truck, they can offer working heights between 30 and 85 feet.

Bucket trucks are available in overcenter and non-overcenter designs. The articulating boom of a non-overcenter truck can extend and rotate from horizontal to 90 degrees vertical. The boom of an overcenter truck can rotate from horizontal, to 90 degrees vertical, plus horizontal in the opposite direction for a full 180 degrees range of motion. An overcenter truck provides a much larger operating range.

Jibs are also available and are installed at the end of the boom. A jib is a short lifting arm equipped with cabling and a lifting hook and/or eye. Jib cables are winch operated and the winches are usually either hydraulic or powered. Depending on the size and rating of the boom, winch, and cabling, jibs can safely lift from 1,000 to 2,000 pounds of material or components.

3.0.0 Digger Derrick Truck Overview

The digger derrick truck (*Figure 4*) is a multipurpose utility vehicle that power line workers use to dig holes for new utility poles, place utility poles into the holes, and hold the poles while they are being secured in the ground. The main components of the digger derrick are the telescoping boom and pedestal, the hole auger, a powered winch and cable, and the pole claw.

Like the bucket truck, the digger derrick uses a pedestal-mounted extension boom to raise the hydraulically powered auger to dig the hole for the utility pole. Augers used to dig holes for utility poles range in size from 16 to 48 inches in diameter. There are also different designs for digging in normal soil or rock (*Figure 5*).

The telescoping boom can extend to dig the hole several feet from the base of the truck (*Figure 6*). When setting up the digger derrick truck at the worksite, always ensure that you have enough room to lift the pole and set it in the hole to be dug. Controls allow the auger to

Figure 4 Digger derrick.

49109-11_F05.EPS

Figure 5 Auger.

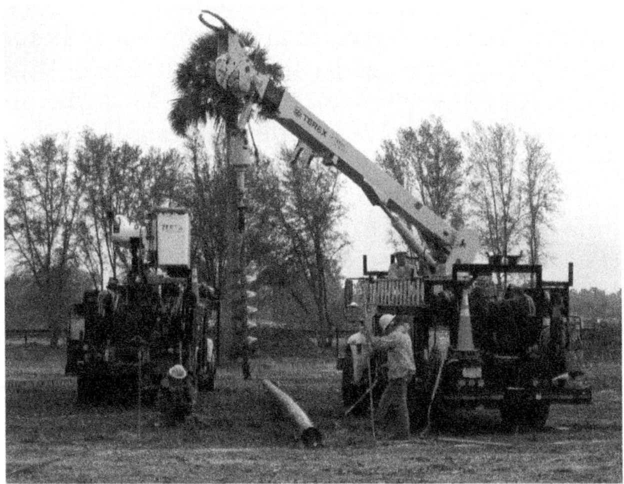

49109-11_F06.EPS

Figure 6 Digger extended.

be operated clockwise and counterclockwise to speed up the digging process. To begin the hole, the auger should be placed touching the ground where the pole is to be set. Start the auger slowly turning to prevent tilt of the auger as it enters the ground. When the auger becomes full of dirt, or gets stuck, the rotation should be stopped, and the auger should be lifted from the ground to clear the dirt. Clear the dirt from the auger using a shovel. If the auger will not lift from the ground, reverse the rotation to assist in lifting the auger out of the ground. Continue this process until the hole is dug to the proper depth for the pole to be set.

After digging the hole to the proper depth, the pole is ready to be set. Lifting charts (covered later in this module) indicating weight versus boom angle must be referenced to ensure that the lift will not tip over the digger derrick. To initially lift the pole, the winch line with hook is lowered from the top of the boom. The pole should be secured slightly above the balance point of the pole using the hook and line attached to an approved sling so that the butt (bottom) of the pole outweighs the top end. The powered winch is then used to reel in the winch line, lifting the pole top to the top of the boom. Once the pole is at the top of the boom, the pole claw (*Figure 7*) is opened to accept the pole, and then closed around the pole. The boom is typically equipped with a pole claw, pole guides, or boom flares that are used primarily for aligning and positioning poles.

WARNING!

Severe injury or death can result if the pole claw or pole guides are used for pole lifting or pole pulling operations. Pole weight must be supported at all times by the winch load line.

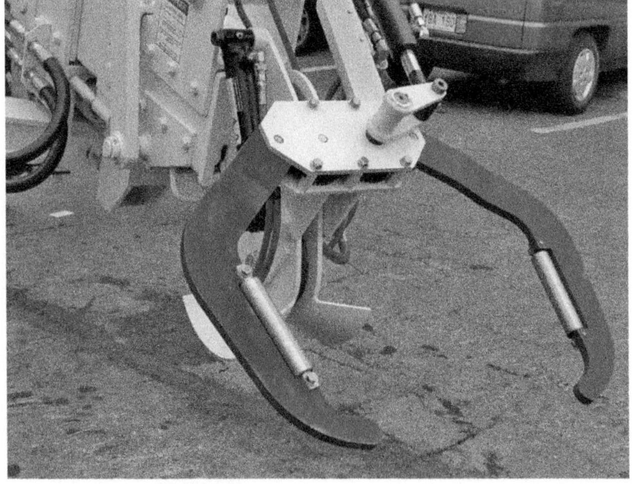

49109-11_F07.EPS

Figure 7 Pole claw.

4.0.0 Operator Safety Requirements

Employers must ensure that operators are trained by a qualified person on the specific bucket truck, digger derrick, crane truck, or aerial lift they are using and must understand and adhere to all requirements of the manufacturer's operations manual. Before operating one of these specialized vehicles, the operator must learn and be aware of the safety requirements necessary to use the vehicle and be able to demonstrate the proper handling guidelines.

4.1.0 Federally Mandated Safety Requirements

According to the Occupational Safety and Health Administration (OSHA), the leading cause of accidental power line contacts involve heavy equipment (cranes, bucket trucks, digger derricks); long-handled tools; and ladders and other items carried by workers coming in contact with energized power lines. OSHA issues regulations addressing safety in the workplace, but they also rely on national standards for certain requirements. For electrical safety, OSHA recognizes several standards from the National Fire Protection Agency (NFPA). The American National Standards Institute (ANSI) has also developed standards that govern the safe work practices for close proximity to power lines and other sources of electricity.

An elevated work platform can be an efficient and safe substitute for other methods of working at heights, such as ladders and scaffolds. However, the lift must be carefully selected to match the work being performed and must be maintained and operated in strict conformity with the manufacturer's manual and with OSHA, NFPA and ANSI standards. The OSHA, NFPA, and ANSI standards include:

- *OSHA 1910.269, Electric Power Generation, Transmission, and Distribution*
- *OSHA 1917.45, Cranes and Derricks*
- *OSHA 1910.67, Vehicle Mounted Elevating and Rotating Work Platforms*
- *OSHA 29 CFR 1926, Subpart N, Cranes, Derricks, Hoists, Elevators, and Conveyors*
- *NFPA 70E, Standard for Electrical Safety in the Workplace*
- *ANSI A10.31, Safety Requirements, Definitions and Specifications for Digger Derricks*
- *ANSI A92.2, Vehicle-Mounted Elevating and Rotating Aerial Devices*

NOTE

Adherence to ANSI is voluntary. Compliance with OSHA is mandatory.

Some of the requirements for using a bucket truck, digger derrick, crane truck, or aerial lift under these rules include the following:

- Only trained, qualified employees can operate the bucket truck, digger derrick, crane truck, or aerial lift.
- Each employee shall be protected from hazards that might arise from equipment contact with the energized lines. This will be accomplished by using the best available ground to minimize the time the lines remain energized, providing ground-gradient mats to extend areas of the equipotential region, and employing insulating protective equipment or barricades to guard against any remaining hazardous potential differences.
- The safe work load capacity may not be exceeded. Always consider the weight of the worker, tools, and any materials to be transported by the bucket when calculating the load. Load rating charts must always be followed to make sure the specific lift is safe according to the manufacturer's recommendations.
- Bucket trucks, digger derricks, or crane trucks may be modified for other than the intended use, but only if approval to the modification is received in writing by the manufacturer or an equal authority.
- For utility trucks to be used near electrical power lines, strict requirements must be followed. These include maintaining a specific distance from the power source, performing the necessary de-energizing procedures, and grounding.
- Employees not engaged in high-voltage work must remain a minimum of 10 feet from overhead power lines. High-voltage workers must de-energize or insulate power lines while using specialized personal protective equipment and tools.
- Controls must be clearly and visibly marked and defined by function.
- Controls must be tested every day before the utility truck is used to ensure that the controls are in safe operating condition.

4.2.0 Bucket Truck Operator Safety Requirements

An insulated bucket truck has three components that will provide some protection if they are properly maintained by being kept clean, dry, and periodically tested per *ANSI A92.2*:

1. An insulated basket liner will protect only that portion completely inside the liner. Anything conductive that extends outside the liner will conduct electricity into the liner and make it ineffective.
2. The insulating section of the upper boom will prevent current flow from the boom tip through the boom to the elbow only.
3. The lower boom insert will provide an insulating section between the elbow and the truck chassis.

The boom tip does not provide insulation because it contains metal components to provide structural support. Manufacturers place a band of arrows (*Figure 8*) on the upper boom to show the end of the insulated section. Past the band of arrows on the boom, any part of the boom tip that contacts an energized line will become energized at that potential. Also, any part of the boom tip that contacts a grounded component will ground the entire boom tip. This includes the controls and all other components that are past the band of arrows. Covers and guards may provide limited protection but you cannot depend only on the arrows on the boom. Contact of fiberglass covers and guards with energized parts may arc along the surface or through the fasteners to metal under the cover and energize the entire boom tip.

WARNING!

Even when working from an insulated bucket attached to an insulated boom, you must maintain proper equipotential and ground clearance from all sources of electricity.

Just because you are in an insulated aerial device, you are not protected from all contact with an energized object. If you touch or become part of a path between two objects at different potential, you can be electrocuted. If the uninsulated portion of the boom tip comes into contact with an energized line, the controls in the bucket can become energized, sending deadly current through the operator touching the controls. The aerial device only prevents one energized source from having a path to the ground through the boom. The operator safety requirements include the following:

- Any employee operating the bucket truck must wear required PPE at all times.
- Bucket truck workers must wear a full-body harness fall arrest system (*Figure 9*) with a shock-absorbing lanyard. The safety harness and lanyards must be connected to the manufacturer provided anchorage point, and the lanyard cannot exceed 6 feet in length.
- The truck must be properly grounded.

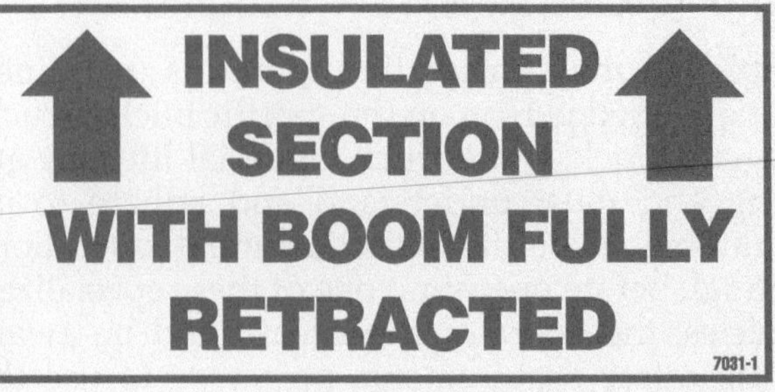

49109-11_F08.EPS

Figure 8 Arrows indicating end of insulated section of boom.

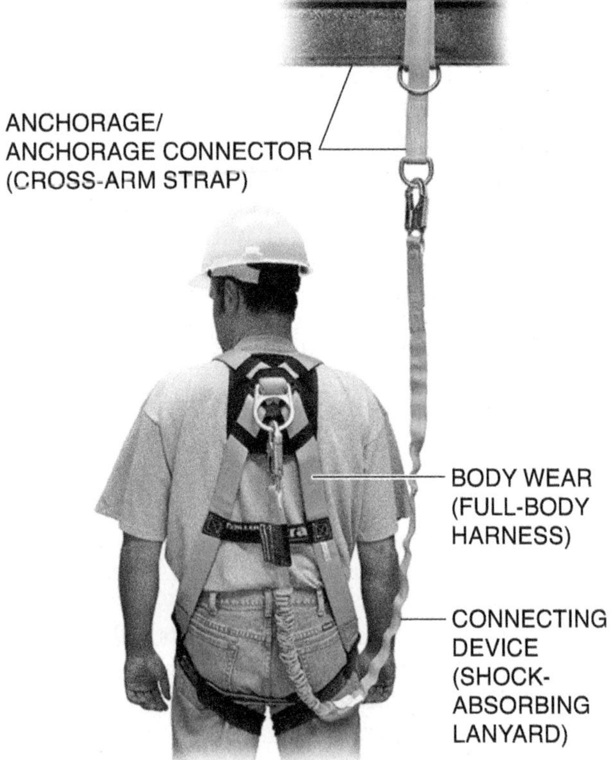

49109-11_F09.EPS

Figure 9 Fall restraint safety belt.

- The sides of the bucket must be at least 39 inches high so that the top edge will be higher than the waists of most workers and reduce the risk of falling out.
- Never allow the boom or pedestal to enter traffic lanes without proper barricading.
- Park the vehicle to block traffic, use appropriate barricades and safety signs, and use a trained traffic control person when necessary.
- Refer to and follow the safety regulations as set forth by the Department of Transportation *Work Zone Safety Manual* as required by local, state, and federal regulations when setting up a bucket truck on a publicly travelled road.
- When moving the bucket, always look in the direction of the boom travel. Remember that when rotating an articulating boom, the lower section and hinge is moving in the opposite direction that the bucket is moving.
- Do not move the bucket truck while the bucket or platform is extended. Bucket trucks must never be driven while employees are still in the bucket.
- Maintain a safe distance from overhead obstacles.
- Do not operate the bucket unless the vehicle's brakes are set, wheels are chocked, the outriggers are in place with the outrigger pads firmly resting on the ground, and the vehicle is level.
- Do not operate the bucket in strong or gusty winds.
- Do not place ladders or scaffolds inside the bucket or against any part of the machine.
- Do not exit the bucket or lift platform once it is in the elevated position.
- Do not climb down from the bucket when it is raised and do not get on the boom for any reason.
- Ensure that the area below the bucket and boom is clear of any personnel and obstructions before lowering the boom.
- Do not operate the lower controls without the consent of the employee in the bucket except in the case of an emergency.

4.3.0 Digger Derrick Operator Safety Requirements

Only personnel who have received general instructions and operations training covering the inspection, application, and operation of the specific digger derrick truck to be operated may operate a digger derrick truck. This training must include the recognition and avoidance of hazards associated with their operation. Such items covered must include the following guidelines.

- Locate and properly use the equipment operating manuals.
- Perform prestart inspections according to the manufacturer's recommendations.
- Make sure that the area is clear of all people before operations begin.
- Make sure the truck is properly grounded.
- Never allow the boom to enter traffic lanes without proper barricading.
- Park the vehicle to block traffic, use appropriate barricades (*Figure 10*) and safety signs, and use a trained traffic control person when necessary.
- Any vehicle over 10,000 lbs must be chocked.
- Refer to and follow the safety regulations as set forth by the *Manual on Uniform Traffic Control Devices* (*MUTCD*) as required by local, state, and federal regulations when setting up a bucket truck on a publicly travelled road.
- Any employee operating the digger derrick must wear required PPE at all times.
- Read and interpret vehicle load charts and determine load lifting capacity for a given load. Never exceed the load limits as stated on the load rating charts.
- Understand the purpose and function of all operator controls.
- Ensure that the area below the auger (*Figure 11*) is clear of any personnel and obstructions before lowering the auger from its stowed position.
- Always look in the direction of the boom travel when moving the auger or utility pole.
- Maintain a safe distance from overhead obstacles.
- Maintain the specified distance from all electrical lines and energized utility pole components. Never approach energized power lines with any part of your machine.

49109-11_F10.EPS

Figure 10 Traffic control.

On Site

Safety Gear for Flaggers

For daytime work, the flagger's vest, shirt, or jacket must be either orange, yellow, yellow-green, or a fluorescent version of these colors. For nighttime work, the vest, shirt, or jacket must be reflective. The reflective material must be either orange, yellow, white, silver, yellow-green, or a fluorescent version of these colors.

Source: MUTCD.

49109-11_F11.EPS

Figure 11 Auger in stowed position.

- Do not operate the digger derrick unless the vehicle's brakes are set, the outriggers are in place with the outrigger pads firmly resting on the ground, and the vehicle is level.
- Do not operate the digger derrick in strong or gusty winds.
- Ensure that the area below the boom is clear of any personnel and obstructions before lowering the boom.
- Do not allow anyone near the digger derrick during operation except the machine operator.
- The machine operator must remain on the truck at all times during operation.
- Use safety barricades and flaggers as required by your company.

5.0.0 Inspection and Maintenance

The bucket truck, digger derrick, crane truck, or aerial lift must be thoroughly inspected each day before it is put into use. The daily prestart vehicle inspection contains the more common items that should be inspected and documented by the operator. The periodic inspection and the hydraulic systems inspections contain preventive maintenance (PM) activities that should be performed and documented on a regular basis as determined by company policy and manufacturer's recommendations.

5.1.0 Daily Prestart Vehicle Inspections

The utility truck must be inspected for defects prior to each operation. In addition to these items, the operator should also follow the manufacturer's recommendations for each specific vehicle. These inspection items include the following:

- Ensure that all operating controls and emergency controls are functioning properly.
- Ensure that all safety devices are functioning properly (foot pedal, spring lock, boom angle indicator).
- Check the seatbelts for proper operation and condition.
- Ensure that all personnel protective safety devices are in place and functioning properly.
- Ensure that all fiberglass and other insulating materials are securely in place and show no signs of visible damage or contamination.
- Check the hydraulic, air, fuel, and electrical systems for wear, leakage, excessive dirt or contamination build up, moisture, or any other condition that may impair the use of these systems.
- Check to make sure that all warning placards and decals, control indicator markings, operational notices, and informational placards are in place and legible.
- Ensure that all lighting is operational including headlights, brake lights, signals, flashers, and other emergency lighting.
- Perform a visual inspection of all mechanical fasteners.
- Check all cables and wiring harnesses to ensure they are connected, in good condition, and free of nicks, cuts, and other defects.
- Check the wheels and tires for sufficient air pressure, sufficient thread, tight lug nuts, and good overall condition with no tire bulges or cuts.
- Check the oil level in the vehicle engine.
- Ensure that the vehicle operating manuals are in place, current, and protected inside waterproof containers on the lift or in the cab of the truck.
- Ensure that the outriggers or stabilizers are in good working condition.
- Ensure that the aerial lift has a current electrical insulation resistance test (verify by checking inspection sticker) and a current DOT inspection sticker.

- Ensure that the fuel tank is full of fuel before heading out on a job. If there is an auxiliary motor to power the lifting boom, also ensure that motor has plenty of fuel.

5.2.0 Periodic Vehicle Inspections

The periodic vehicle inspection activities include detailed inspection and PM activities. When these procedures are completed on a regular basis, it will ensure the maximum life of the vehicle components. These items should be performed every one to twelve months depending on the frequency of operation, severity of environmental and operating conditions, and manufacturer's PM charts. In addition to these items, the operator should also follow the manufacturer's recommendations for each specific vehicle.

- Inspect and test the liner and insulated boom section(s) of your aerial device regularly per *ANSI A92.2* (latest revision).
- Inspect all structural members to ensure there are no signs of deterioration, cracks, or corrosion.
- Inspect critical pins, bushings, bearings, shafts, gears, rollers, locking devices, chains, sprockets, wire rope, and sheaves for wear, cracks, or deformations.
- Inspect the brake pads.
- Inspect the weld integrity ensuring there are no cracks, rust, or signs of damage.
- Inspect the electrical systems and components for deterioration or wear, including those that are not typically looked at during the prestart inspections.
- Check the condition and tightness of all critical bolts and fasteners.
- Perform a performance test of all boom/auger movements.

5.3.0 Hydraulic Systems Inspections

Hydraulic system PM inspections should be performed regularly to ensure that the system components are in proper working order and to maintain a long working life for components. Company policies and procedures, along with the manufacturer's recommendations, usually specify how often PM inspections should be performed. The hydraulic system PM inspection includes inspecting the following components:

- Hydraulic fluid reservoirs
- Filters and strainers
- Hydraulic pumps
- Hydraulic motors
- Control valves
- Check valves
- Cylinders
- Hoses, fluid lines, and gauges
- Auxiliary line quick-disconnects
- Controls for automatic operations

Never inspect the hydraulic system for leaks with your hands or fingers with the system under pressure (*Figure 12*). Use a piece of cardboard or wood to find leaks but do not use your bare hand.

> **WARNING!** Hydraulic fluid leaking from a pinhole in a hydraulic line under pressure can pierce the skin, causing a serious injury or infection that will require medical attention. If any fluid is injected into the skin, it must be surgically removed within a few hours by a doctor familiar with this type of injury.

5.3.1 Inspecting Hydraulic Fluid Reservoirs

A reservoir in a hydraulic system stores the fluid used in the system. In the reservoir, air can separate from the fluid, and particles of contamination

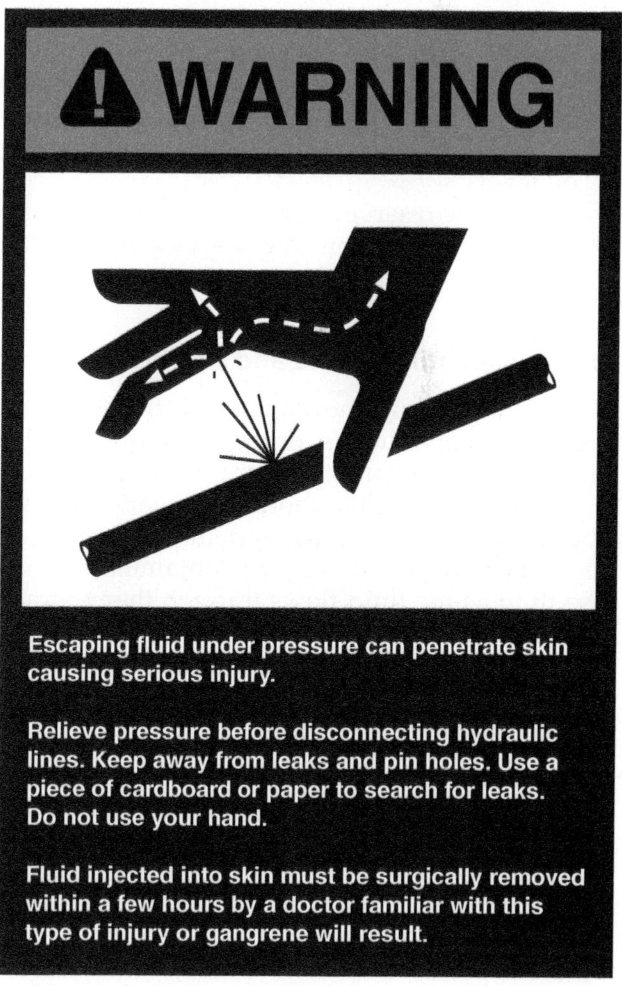

Figure 12 Hydraulic warning.

can settle. The working fluid is cooled by passing through the reservoir, where the heat dissipates from the fluid into the atmosphere, keeping the operating temperature of the fluid within allowable limits. Reservoirs are usually made of welded steel plate. A baffle plate in the center of the reservoir helps returning fluid circulate and cool before being pumped back into the system. Drain plugs and removable covers help drain and clean the reservoir. Follow these guidelines when inspecting a hydraulic fluid reservoir:

- Inspect the reservoir tank to ensure that there are no leaks.
- Inspect the reservoir tank to ensure that it is clean.
- Inspect the reservoir air breather to ensure that it is clean.
- Inspect all line connections to and from the reservoir to ensure that they are tight and that there is no fluid leaking from the system or air leaking into the system.
- Examine the maintenance records to determine if the reservoir is due for a scheduled fluid change and cleaning.
- Look at the reservoir sight glass to ensure that the fluid is at the proper operating level, and add fluid as necessary. When checking the sight glass, be aware that a false line may appear due to oil stains or lines from previous oil levels.
- Inspect the reservoir for water contamination.
- Inspect the reservoir tank for evidence of corrosion.

5.3.2 Inspecting Filters and Strainers

As hydraulic fluid moves through the system, it picks up contaminants, or particles of unwanted materials. If contaminants are allowed to build up in the system, they shorten the working life of the fluid and damage the system parts. Strainers and filters trap and remove contaminants from the fluid as the fluid flows through them. Strainers remove the larger particles from the fluid. Filters remove smaller particles that slip through the strainers. Follow these guidelines when inspecting filters and strainers:

- Inspect the filter or strainer housing for leaks.
- Inspect the filter or strainer housing for cleanliness and clean it as necessary.
- Examine the filter or strainer maintenance records to determine if the filter or strainer is due for scheduled cleaning or replacement.
- Clean and/or replace the filter or strainer as necessary.

5.3.3 Inspecting Hydraulic Pumps

Pumps in hydraulic systems change the mechanical energy produced by the pump drive motor into hydraulic energy. The pumps do this by exerting force on the fluid and creating flow in the hydraulic system. Follow these guidelines when inspecting a hydraulic pump:

- Inspect the pump casing for leaks.
- Inspect the shaft seal for leakage.
- Inspect all fluid line connections to ensure that they are tight and that there is no fluid leakage from the system or air leakage into the system.
- Listen for unusual noises from the pump.
- Inspect the pump spring if the pump is pressure-compensated.

A noisy pump may indicate pump wear or cavitation. Cavitation occurs when the pump does not receive adequate fluid because of the low fluid level in the reservoir or a plugged strainer or filter. Examine the pump and drive motor for an unusually high temperature using a pyrometer or other temperature-measuring device.

An excessively high temperature may indicate damaged motor bearings, binding due to worn parts, or misalignment. It may also indicate that the fluid in the system is not being properly cooled or that the flow is restricted.

Excessive vibration in a pump or drive motor may indicate misalignment of the pump-to-motor coupling, worn bearings in the pump or motor, or restricted suction.

5.3.4 Inspecting Cylinders

Cylinders in a hydraulic system are actuators/operators. They change hydraulic energy into mechanical energy. The pressure and movement of the hydraulic fluid move the cylinder rod to perform work. Cylinders are called linear actuators/operators because they produce motion and force in a straight line. Follow these guidelines to inspect the hydraulic cylinders:

- Inspect the cylinder to ensure that there are no leaks.
- Inspect the cylinder to ensure that it is clean, and clean it as necessary.
- Inspect the cylinder shaft to ensure that it has no scratches or burrs, if applicable.
- Examine the cylinder mounting bracket to ensure that the bolts are in good condition and properly adjusted.
- Listen for unusual noises from the cylinder.
- Monitor the cylinder for jerky, rough, sluggish, or otherwise irregular operation.

A noisy cylinder or the irregular operation of a cylinder may indicate bad seals, a bent cylinder shaft, clogged ports, or scratched cylinder walls. Test the cylinder for excessively high temperature using a pyrometer or other temperature measuring device.

An excessively high cylinder temperature may indicate worn internal parts, an overloaded cylinder, or cylinder binding.

5.3.5 Inspecting Hoses, Fluid Lines, and Gauges

Hoses, fluid lines, and gauges make up the part of a hydraulic system that carries the fluid to the different system components. The type and size of the hoses, fluid lines, and gauges used in a system depend on the requirements of the system. Follow these guidelines when inspecting hoses, fluid lines, and gauges:

- Inspect all hose, fluid line, and gauge connections to ensure that they are tight and that there is no leakage.
- Inspect all gauges to ensure that they are in good operating condition.
- Examine all flexible hoses closely to ensure that there are no areas of dry rot, worn spots, kinks, or thin spots.
- Examine all quick-disconnects on auxiliary lines to ensure that they are in good condition. These can include lines in buckets for operating hand tools and lines at the back of digger derricks for operating tamping devices and pole pullers.
- Examine all fluid lines to ensure that they are properly supported by clamps.

6.0.0 Driving and Setting Up at Worksites

When in transit to the jobsite, the vehicle operator must have a valid driver's license for the particular class of vehicle being operated and should always practice safe, courteous driving manners. Before traveling, ensure that the boom, auger, and/or bucket are secured in the transport position. Also make sure that the outriggers are up and locked in the transport position and that all auxiliary equipment is properly stored and secured. Always follow the posted speed limits and warning signs en route to the jobsite. While most power line installation and repair work is performed on the side of a road or parking lot, it is possible to have to set up on a wide variety of ground terrains. Always consider any additional hazards that may be caused by the terrain that you are working on. Also consider the current weather conditions.

You must also know your vehicle weight and height when planning your route to the jobsite. The gross vehicle weight (GVW) should be posted on the truck along with the travel height. Always ensure that any bridges that you must cross can withstand the weight of your truck and any underpasses or overhead obstacles are tall enough for your truck to travel under.

6.1.0 Work-Site Preparation

Follow these guidelines when preparing the work site to operate a bucket truck or digger derrick:

- All personnel must wear hard hats and orange safety vests during setup operations performed in or next to a road or parking lot.
- Ensure that proper traffic control is used including the proper placement of warning cones, flashing arrows, signage, and traffic control personnel as required by local, state, and federal guidelines.
- Try not to park on uneven ground and be aware of holes, bumps, and debris.
- Pay special attention to any trenches, standing or flowing water, embankments, or other obstacles that may affect the compaction of the soil and the ground conditions where the vehicle will be set up.
- Position the vehicle as close to the work area as possible to minimize reach and maximize lifting ability. *Table 1* shows limited, restricted, and prohibited approach boundaries for energized electrical conductors and circuit parts.
- Make sure there are no obstructions between you and the work area that might prevent safe and proper lift and rotation.
- When setting up the digger derrick truck at the worksite, always ensure that you have enough room to lift the pole and set it in the hole to be dug.
- Shift the transmission to neutral and apply the vehicle emergency brakes (if applicable). Hydraulic-braked trucks with boom equipment have a mechanical brake and a hydraulic locking brake. Ensure that both are applied.
- Once the vehicle is in place, activate the emergency lights and engage the power takeoff (if applicable) before exiting the cab of the truck.
- Chock the vehicle's wheels.
- Properly ground the truck.

Table 1 Approach Boundaries to Energized Electrical Conductors or Circuit Parts for Shock Protection

Nominal System Voltage Range, Phase to Phase[2]	Limited Approach Boundary[1]		Restricted Approach Boundary[1] Includes Inadvertent Movement Adder	Prohibited Approach Boundary[1]
	Exposed Movable Conductor[3]	Exposed Fixed Circuit Part		
Less than 50V	Not specified	Not specified	Not specified	Not specified
50V to 300V	3.05m (10 ft 0 in)	1.07m (3 ft 6 in)	Avoid contact	Avoid contact
301V to 750V	3.05m (10 ft 0 in)	1.07m (3 ft 6 in)	304.8mm (1 ft 0 in)	25.4mm (0 ft 1in)
751V to 15kV	3.05m (10 ft 0 in)	1.53m (5 ft 0 in)	660.4mm (2 ft 2in)	177.8mm (0 ft 7 in)
15.1kV to 36kV	3.05m (10 ft 0 in)	1.83m (6 ft 0 in)	787.4mm (2 ft 7in)	254mm (0 ft 10 in)
36.1kV to 46kV	3.05m (10 ft 0 in)	2.44m (8 ft 0 in)	838.2mm (2 ft 9in)	431.8mm (1 ft 5 in)
46.1kV to 72.5kV	3.05m (10 ft 0 in)	2.44m (8 ft 0 in)	1.0m (3 ft 3 in)	660mm (2 ft 2 in)
72.6kV to 121kV	3.25m (10 ft 8 in)	2.44m (8 ft 0 in)	1.29m (3 ft 4 in)	838mm (2 ft 9 in)
138kV to 145kV	3.36m (11 ft 0 in)	3.05m (10 ft 0 in)	1.15m (3 ft 10 in)	1.02m (3 ft 4 in)
161kV to 169kV	3.56m (11 ft 8 in)	3.56m (11 ft 8 in)	1.29m (4 ft 3 in)	1.14m (3 ft 9 in)
230kV to 242kV	3.97m (13 ft 0 in)	3.97m (13 ft 0 in)	1.71m (5 ft 8 in)	1.57m (5 ft 2 in)
345kV to 362kV	4.68m (15 ft 4 in)	4.68m (15 ft 4 in)	2.77m (9 ft 2 in)	2.79m (8 ft 8 in)
500kV to 550kV	5.8m (19 ft 0 in)	5.8m (19 ft 0 in)	3.61m (11 ft 10 in)	3.54m (11 ft 4 in)
765kV to 800kV	7.24m (23 ft 9 in)	7.24m (23 ft 9 in)	4.84m (15 ft 11 in)	4.7m (15 ft 5 in)

Note: For arc flash protection boundary, see ***70E Section 130.3(A)***.

1. See definition in ***70E Article 100*** and text in ***70E Section 130.2(D)(2)*** and ***Annex C*** for elaboration.
2. For single-phase systems, select the range that is equal to the system's maximum phase-to-ground voltage multiplied by 1.732.
3. A condition in which the distance between the conductor and a person is not under the control of the person. The term is normally applied to overhead line conductors supported by poles.

49109-11_T01.EPS

CAUTION

Hydraulic brakes can bleed off pressure over time. Ensure that the wheels are chocked and the mechanical brake is applied, because after the job is complete and the outriggers are raised, the truck may roll freely. Visually inspect the boom, pedestal, chassis, and entire truck for damage or hydraulic leaks that may have occurred in transit.

6.2.0 Stabilize the Vehicle for Operation

After the truck is in place, the outriggers must be extended to provide stability to the truck and prevent tip-over. *OSHA Standard 1910.269(p)(2)(i)* governing the use of outriggers states the following:

> Vehicular equipment, if provided with outriggers, shall be operated with the outriggers extended and firmly set as necessary for the stability of the specific configuration of the equipment. Outriggers may not be extended or retracted outside of clear view of the operator unless all employees are outside the range of possible equipment motion.

The bucket truck or digger derrick will have two or four outriggers, depending on the size of the truck. Follow these guidelines when stabilizing the vehicle for operation:

- When extending the outriggers, make sure that the outriggers extend smoothly.
- Check for any dents or deformity in the tubes, as this could indicate internal damage to the outrigger beams.
- Use outrigger pads, blocking, or cribbing under the outriggers (*Figure 13*) to provide a solid surface for the outriggers.
- Extend the outriggers to lift all of the weight that is supported by the springs. Some outriggers have a mark that shows the minimum extension to provide stability.
- When operating a truck with the turret located directly behind the cab and only two outriggers, many times the front wheels of the vehicle will be lifted off the ground when the outriggers are extended. Place planking or some other type of support under the front wheels if they are lifted off the ground.
- If you are setting up on a sloping ground, extend the low-side outrigger first and then extend the other outrigger(s) until the vehicle is level.

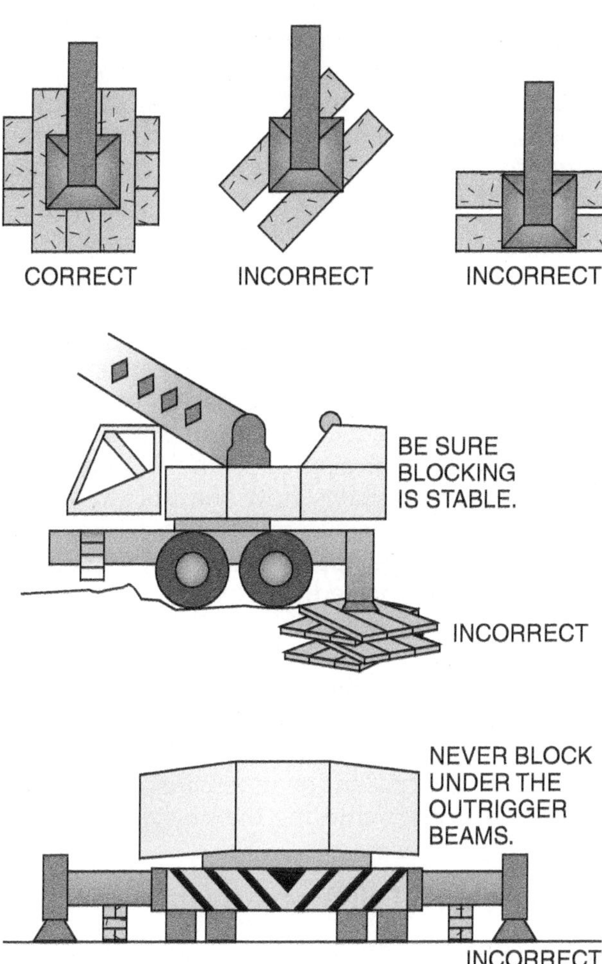

Figure 13 Correct and incorrect use of outrigger blocking.

- If the slope is greater than 5 degrees, you may be required to build up the low side with extra timbers or dig out on the high side to keep the vehicle level.

6.3.0 Installing the Vehicle Ground

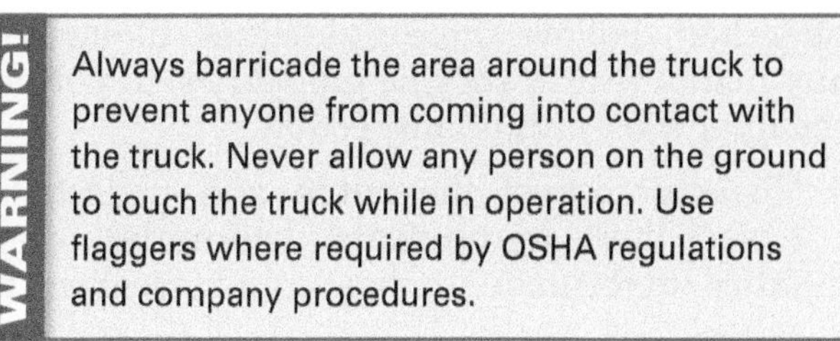

Before the boom is raised, the truck must be properly grounded in order to provide a quick trip-out of the circuit if the vehicle were to become energized from an uninsulated part of the boom coming in contact with a live wire. It is important to know that the ground does not completely protect a worker touching the truck; it simply provides a quicker route for the electrical current to go through the vehicle, and will take most of the fault current into the ground. The trouble is that the electricity takes all available paths to the ground. Depending on the voltage, it will try to go through the tires (typically melting them), the outriggers, and also the person who may be touching the truck. The installed ground will simply limit the time and amount of current that would flow through the person touching the truck because it provides an easier path for the electricity to follow.

> **WARNING!** Always barricade the area around the truck to prevent anyone from coming into contact with the truck. Never allow any person on the ground to touch the truck while in operation. Use flaggers where required by OSHA regulations and company procedures.

There are three methods used to ground the vehicle. Listed in order of most preferable to use, they are as follows:

1. A permanent ground network such as a station ground, a system neutral, or a pole ground
2. An existing approved ground rod already located in the earth
3. A temporary ground rod driven into the earth, or portable screw anchor that is screwed into the earth

The system neutral is the best ground point, as it offers the lowest resistance path for fault current to return. Regardless of the type of ground used, if you have a retractable reel for your ground, always make sure that the ground cables are pulled completely off the reels. Under a fault condition the reel can self-destruct with the ground still partially coiled, resulting in total failure of the ground (*Figure 14*).

6.4.0 Final Walk-Around Inspection

Before operating the bucket truck or digger derrick, you must perform a final walk-around inspection to ensure that the truck is ready for

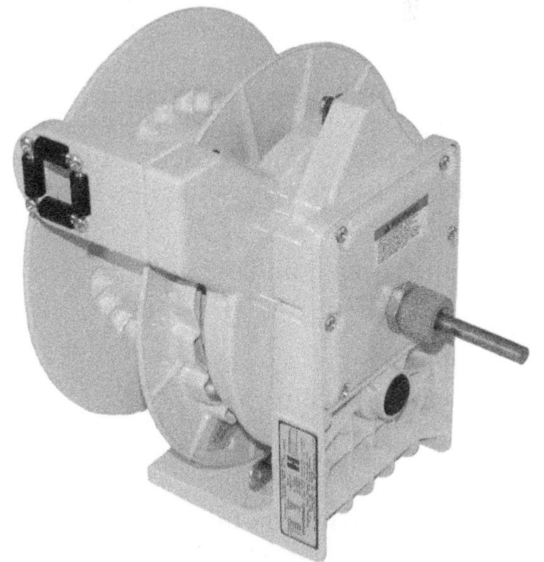

Figure 14 Truck grounding equipment.

operation. Follow these guidelines, along with any manufacturer specific guidelines, to perform the final walk-around inspection:

- Visually inspect the boom and pedestal to ensure there are no dents, deformities, or any other deformations that could weaken the pieces.
- Ensure there are no leaks in the hydraulic cylinders or hoses.
- At the main controls on a digger derrick or the ground controls on a bucket truck, activate the Emergency Stop pushbutton. Try to activate the other controls with the Emergency Stop set to ensure they are working properly.
- Reset the Emergency Stop pushbutton.
- Activate each of the ground controls to ensure they are functioning properly.
- Test any auxiliary controls.
- In the bucket controls on a bucket truck, test the Emergency Stop pushbutton.
- Test the horn, foot controls, and any auxiliary controls located in the bucket.

7.0.0 Bucket Truck Operations

The operator must perform all prestart inspection activities as specified before operating the bucket truck. Setup and operation must be performed in accordance with the manufacturer's requirements.

7.1.0 Weight Limitations

If a bucket truck is equipped with a jib and winch for lifting loads, it is required to have a load chart showing how much load can be lifted at what angle or load radius. Boom and platform/bucket load limits specified by the manufacturer may not be exceeded. The load chart must be positioned so that the operator can see it while at the controls. If the winch is to be operated from the lower controls, a load chart must be available there also.

To determine the lifting capacity of the truck, the operator must identify some key lift indicators, including:

- Weight of the load and rigging (include personnel, tools, and materials)
- Load radius
- Boom angle and configuration
- Center point of the lift

CAUTION

Lifts performed over the truck's side or rear can change the rated load capacity.

Always reference the truck's load chart rating under the required lift indicators. The load chart matrix identifies the truck's lifting capacity up to 85 percent of the rated lift capacity. This is a safety standard in all load charts for all utility truck makes and models.

7.2.0 Boarding the Truck and Bucket Properly

Always board the truck and bucket using the steps or ladder built into the design of the truck (*Figure 15*). Different configurations will provide different means of access, so the type of truck you are using dictates how the bucket should be boarded. Never climb into the bucket from the boom or on the opposite side from where the access steps are located. Always enter the bucket by placing one leg into the bucket, securely planting your foot in the bottom of the bucket, and pulling your other leg in. Do not jump into the bucket, throwing both feet over the edge at once. Carelessness at the beginning of the job can cause an accident or injury just as easily as carelessness after entering the bucket.

7.3.0 Using a Fall Arrest System

After entering the bucket, the first thing you must do before raising the bucket is attach your safety harness lanyard to the anchorage point provided for the vehicle. A fall arrest system must be worn at all times by any person working from an aerial lift. Several circumstances will cause a worker to fall out of the basket when

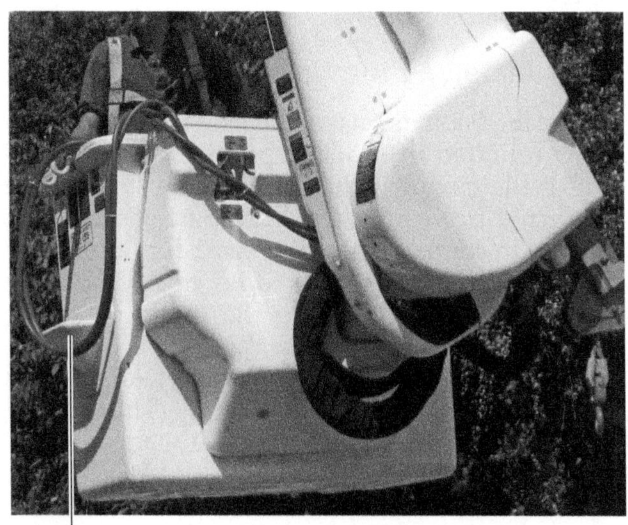

49109-11_F15.EPS

Figure 15 Bucket access.

working from a bucket truck. Some common examples are as follows:

- Over-reaching
- Swinging under overhanging limbs, structures, or lines
- Stumbling over tools inside the basket
- Suddenly reversing boom direction without stopping first

Follow these safety requirements to prevent being injured by a fall from a raised bucket:

- The operator's feet must be firmly planted on the floor of the bucket at all times.
- Bucket truck workers must wear a fall-restraint safety belt or full-body harness fall arrest system.
- The operator must have a safety harness and lanyards connected to the manufacturer provided attachment (*Figure 16*). Belting off to an adjacent pole, structure, or equipment while working from an aerial lift is not permitted.
- Use a lanyard that will limit your free fall to a maximum of 4 feet.
- Do not tie off to an existing structure that is not part of the aerial platform.
- Do not wear climbers while performing work from an aerial lift.
- Do not sit or climb on the edge of the basket.
- Do not use planks, ladders, or other devices to stand on while in the bucket.

7.4.0 Operating the Bucket

Before entering the bucket, release the boom strap (if so equipped). After entering the bucket and attaching the safety harness lanyard to the anchorage point, unlock the bucket from the stowed position. This allows the bucket to remain horizontal while it is raised using the provided controls. The bucket should only be operated from the bucket controls and never by a co-worker using the ground controls, except in the case of an emergency. The bucket controls will vary between different manufacturers, but typically consist of four basic controls that are normally joysticks on older model trucks and pistol-type grips on newer trucks. These four basic controls consist of the following:

1. *Boom (pedestal) rotation* – CW-CCW
2. *Boom (lower) elevation* – Up-Down
3. *Boom extension* – Extend-Return
4. *Arm (upper)* – Raise-Lower

If your truck has the pistol-type grips, squeeze and hold the trigger with your index finger before moving the control. When initially raising the bucket from the stowed position, start by raising the lower boom and then the upper boom. Move the controls smoothly; avoid jerking the controls as this would cause the bucket to bounce as it travels. Jerking the controls will cause sudden starts and stops of the bucket and could cause you to lose your balance. This may also cause damaging shock loads to the boom components. Always look in the direction of your movement and stay clear of all obstructions above or beside your bucket. As you are raising the bucket, attach the truck ground to the system neutral. As you are approaching your final working position, gently nudge the controls to carefully locate the bucket within the acceptable clearances of the live wires and in close enough proximity to your working area that you do not have to reach to perform the job.

When using an older truck with the joystick-type controls, be extremely cautious not to inadvertently bump the controls after locating the bucket in your final working position. Also secure your tools so they do not come into contact with the controls. Bumping the controls on the newer trucks with the pistol-type controls will not cause the bucket to move unless the trigger is squeezed.

49109-11_F16.EPS

Figure 16 Properly tied off.

8.0.0 Digger Derrick Operations

Following the pre-use inspection and before operating the digger derrick, ensure that all underground utilities have been properly located and marked. If an underground utility is ruptured, take the appropriate action as explained in the following sections. Digger derrick operations include using the operating controls, operating the auger, lifting and setting poles, and possibly removing existing poles (*Figure 17*).

49109-11_F17.EPS

Figure 17 Lifting a pole.

8.1.0 Locate Underground Utilities

Buried power lines, regardless of voltage, present a different kind of hazard to utility line workers. Every utility company is required to belong to a local One-Call system that is specific to the state where the utility is located. A utility company not belonging to the One-Call system, or failure for a contractor to call the system to mark the underground utilities, can cost thousands of dollars in fines. In most states, the contractor must notify the One-Call system 48 hours in advance of the actual dig, but the lead time for marking utility lines may vary from state to state. After calling the local One-Call system, the various utility companies that have buried lines in the area will come out to mark their lines. The lines should be marked using the color code system developed by the American Public Works Association (APWA), as follows:

- *Red* – Electrical power lines, cables, conduit, and lighting cables

On Site

Get a Dig Permit

One-Call is a service that tracks underground utilities. After calling One-Call, utility workers from each of the utility companies with underground lines in the area will come out and mark the locations of their underground pipes and cables using the APWA underground color codes. This action is sometimes referred to as getting a dig permit.

Dig Safely.
New York

- Call Before You Dig
- Wait The Required Time
- Confirm Utility Response
- Respect The Marks
- Dig With Care

800-962-7962
www.digsafelynewyork.com

i-notice

- Create your own stake out-request on the internet, 24 hours a day, 7 days a week, 365 days a year.
- The i-notice will satisfy all the requirements of an excavation notification as mandated by Code Rule 753.
- No waiting on the phone.
- Electronic maps are available to help pin point accurate dig site locations.
- The only requirement is access to the internet and Microsoft Internet Explorer 4.0 or newer.
- Verbal transcription errors are eliminated.
- Self-paced on line tutorials and site training is available.
- The i-notice service is FREE!!

Simply send an e-mail to: register@digsafelynewyork.com expressing your interest in participating in this innovative, time efficient and user friendly method to create your OWN stake-out notice.

FRONT

BACK

49109-11_SA01.EPS

- *Yellow* – Gas, oil, steam, petroleum, or gaseous materials
- *Orange* – Communication, alarm, or signal lines, cables or conduit
- *Blue* – Potable water
- *Green* – Sewers and drain lines
- *Purple* – Reclaimed water, irrigation, and slurry lines
- *White* – Proposed excavation
- *Pink* – Temporary survey markings

8.2.0 Rupturing Underground Utilities

If you come into contact with an underground utility, contact the utility company that owns the line that has been ruptured, and in some instances (as with gas lines) call 911. Do not resume the work until the utility company that owns the line has instructed you that it is okay to do so. The following guidelines describe the general actions to take in the event of rupturing an underground utility with the digger auger. Always follow your company policies and procedures and all local, state, and federal ordinances should a utility line be ruptured.

- Electrical utility line:
  - If you are on the digger derrick, do not leave the machine.
  - If you are on the ground remotely operating the digger, stay where you are and do not move. Do not touch the digger derrick or anything connected with the machine.
  - Warn others that an electrical strike has occurred and to stay away.
  - Raise the digger auger to break contact with the electrical line.
- Gas line:
  - Warn others that a gas strike has occurred and to stay away.
  - Immediately shut off all engine(s) and remove any ignition sources.
  - Do not smoke or do anything that could cause a spark.
  - Evacuate and secure the work site so that no one will enter into the hazardous area.
  - Call 911 and other local emergency personnel to report the leakage.
- Fiber-optic line:
  - Do not look into the cut ends of a fiber-optic line or any other unidentified cable. A cut fiber-optic cable can cause severe eye injury if you look into the damaged end of the cable.
  - Warn others that a fiber-optic line rupture has occurred.
- Water or sewage line
  - Warn others that a water or sewage strike has occurred.
  - Try to contain the seeping water if possible and slow down the leak.
  - Do not come in contact with leaking sewage as it may contain pathogens. It may be advisable to seek medical attention for personnel coming in contact with sewage.

8.3.0 Digger Derrick Controls

Digger derrick operations are typically controlled from a control station that is normally perched above the pedestal at the base of the boom. Some have remote controls that can be operated away from the machine. A typical setup for a derrick with three boom sections will have six basic controls and an engine stop and start button (*Figure 18*). These six basic controls consist of:

1. *Boom (pedestal) rotation* – CW-CCW
2. *Boom elevation* – Up-Down
3. *Second boom extension* – Extend-Return
4. *Third boom extension* – Extend-Return
5. *Winch* – Up-Down
6. *Auger* – CW-CCW
7. *Auger lock/unlock control*

8.4.0 Operating the Auger

After grounding the truck, you are ready to start the digging process. Raise the boom and release the auger from its stowed position.

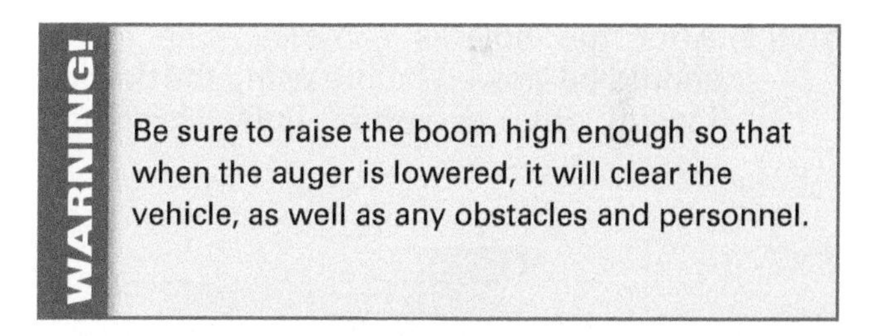

WARNING!

Be sure to raise the boom high enough so that when the auger is lowered, it will clear the vehicle, as well as any obstacles and personnel.

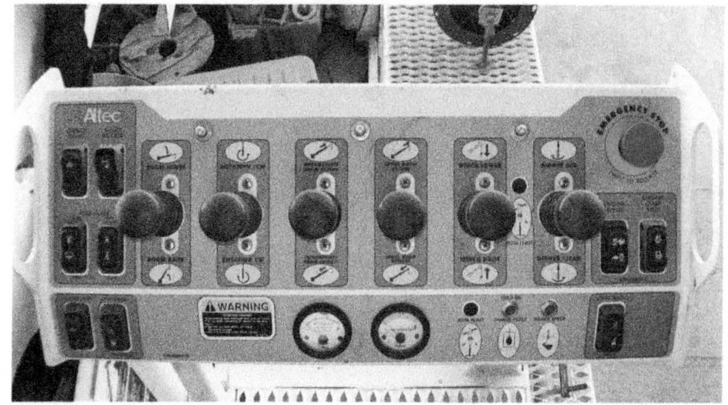

49109-11_F18.EPS

Figure 18 Digger derrick controls.

Always follow manufacturer procedures for unstowing and setting up the auger for use. The following is a general procedure for using the auger:

Step 1 With the digger set at its slowest speed, move the auger rotation control slightly to unwind the auger and release it from the stow latch.

Step 2 Retract the stow latch from the auger tube and slowly rotate the auger to lower it (*Figure 19*). When the auger tube has cleared the stow latch, return the control to the locked position. Continue to lower the auger until it is in the vertical position. Stop rotating the auger and remove the windup cable from the auger flight.

Step 3 Using the pedestal rotation and second boom extension controls, position the auger directly over the hole location. Start clockwise rotation of the auger and slowly lower the boom so that the auger enters the ground. As the auger is lowered, use the boom extension controls to keep it vertical at all times.

Step 4 As dirt is removed and clogs the auger, raise the auger up and shake it or use a shovel to remove excess dirt (*Figure 20*). If the auger will not lift from the ground, reverse the rotation to assist in lifting the auger out of the ground. Continue this process until the hole is dug to the proper depth for the pole being set.

Step 5 After the hole is complete, the auger should be stowed before using the derrick for any other operation. Fully retract the intermediate boom section and raise the boom high enough so that the auger will clear the vehicle and all obstacles when it is raised to the stowed position. Attach the auger wind-up cable to the auger according to the manufacturer's instructions.

Step 6 Making sure the digger is set in its slowest speed, slowly rotate the auger to lift it into its stowed position on the boom. Securely latch the auger in place according to the manufacturer's instructions.

49109-11_F19.EPS

Figure 19 Lowering the auger.

49109-11_F20.EPS

Figure 20 Removing dirt from the auger.

Case History

Underground Utilities

While installing a utility pole anchor in a city sidewalk in December 1998, a St. Cloud, Minnesota construction company struck and ruptured a plastic gas pipeline. While utility workers and emergency response personnel were evaluating the situation and taking the initial steps to stop the gas leak, an explosion occurred. Because of the explosion, four people died; one person was seriously injured; and ten people, including two firefighters and a police officer, received minor injuries. In addition, six buildings were destroyed.

The Bottom Line: Always contact the local One-Call office or local utility companies before starting any digging project.

Source: Occupational Safety and Health Administration (OSHA).

8.5.0 Lifting a Load Safely

You must ensure that the truck is set up in a location that provides enough room to raise the pole and set it in the hole. One of the most important considerations when lifting a load is to review and understand the load capacity chart (*Figure 21*) for the machine's boom. The maximum weight capacity will be shown on the manufacturer's load capacity chart. The load capacity of the boom depends on the distance from the center line of the boom turret and the boom angle. The boom angle is indicated by a pointer located on the side of the boom (*Figure 22*).

8.6.0 Installing Poles and Anchors

When poles are installed near exposed energized overhead conductors, the pole must not contact the conductors. Workers on the ground should guide the pole using tag lines as the pole is being lifted and moved toward the hole.

Figure 21 Load capacity chart.

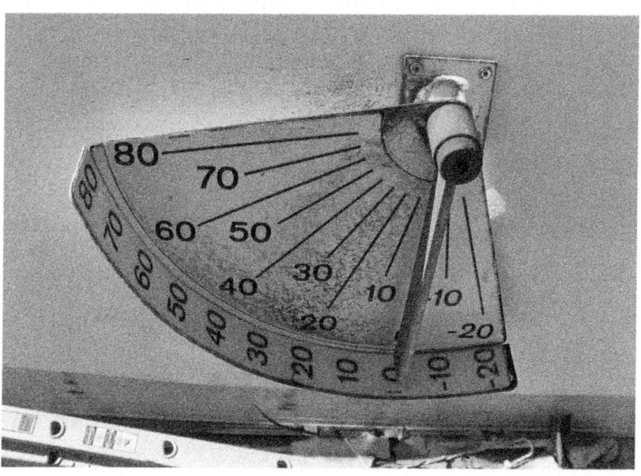

Figure 22 Boom angle indicator.

When a pole is installed near an exposed energized overhead conductor, the ground workers must wear electrical protective equipment and use insulated devices when handling the pole so that no worker contacts the pole with uninsulated parts of his or her body. To protect workers from falling into holes where poles are to be placed, the holes must be attended by personnel or barricaded whenever anyone is working nearby.

8.7.0 Pulling Existing Poles from Ground

The digger derrick is often used to replace existing utility poles that have become damaged or eroded. The winch of a digger derrick is designed to lift utility poles, but is not designed to remove a pole that is already installed in the ground. If the derrick is to be used to lift a pole from the ground, use a pole puller (*Figure 23*) to break the pole free before attempting to lift the pole with the winch line.

Before using the pole puller, attach the winch line to the top of the pole and grasp the top of the pole with the pole claw to support the pole when it is broken loose from the ground. Follow the manufacturer's operating instructions for setting up and operating the pole puller. Most are hydraulically driven and attach to the hydraulic lines from the digger derrick truck. When finished

Figure 23 Pole puller.

Case History

Be Alert

A 23-year-old apprentice lineman died from injuries he received after being run over by the tandem dual rear tires of a digger derrick truck. He was part of a five-person crew that had been working on the job for three days. At the time of the incident, the victim and two other crew members had just finished setting and back-filling around a utility pole. They proceeded to the next pole requiring framing and setting, walking 30 feet ahead of the digger derrick in one lane of the road. The digger derrick moved slowly in reverse to the same pole. Approximately midway between the two poles, the victim knelt with his back to the truck to apparently inscribe a word or initials into some seal coating on the roadway. He was hit and run over by the backing truck's passenger-side tandem dual rear tires. The digger derrick truck did not have a back-up alarm system.

The Bottom Line: Keep your mind on the job. Communication with other workers can save your life. Check equipment for warning signals before you use it. If it does not have warning signals, establish hand or radio signals.

Source: The National Institute for Occupational Safety and Health (NIOSH).

using the pole puller, bleed off all hydraulic pressure from the tool and detach the hydraulic lines. All hoses and tools should then be returned to their stowed position on the truck. After the pole is broken free from the ground, the winch line can be used to lift the pole from the hole and set the pole in a safe area on the ground.

9.0.0 Job Completion and Preparation for Transport

After completing work at a work site, follow the specific instructions in your operating manual for proper stowage of equipment and attachments. While each unit and work site is slightly different, the following guidelines should apply to most situations.

9.1.0 Bucket Truck Preparations

- Remove all tools, equipment, and other loose objects from the bucket.
- Gently lower the boom into its supports.
- Secure the boom and bucket in its travel position according to the manufacturer's recommendations.
- Cover the bucket with a bucket cover, if so equipped.
- Place the boom rotation brake (if so equipped) in the travel position.
- Clean and stow all tools and equipment so no loose objects remain in the truck bed.
- Retract the outriggers to the stowed position.
- Replace wheel chocks and barriers in their travel positions.
- Check the work area for tools and equipment.
- Make note of any operating occurrence that indicates a need for maintenance or repair.

9.2.0 Digger Derrick Preparations

- Stow the auger and secure it in its transport position.
- Return the load line, hook, and boom attachments to their stowed position.
- Return the boom to its travel position.
- Fasten all straps and locks.
- Place the boom rotation brake (if so equipped) in the travel position.
- Clean and stow all tools and equipment so no loose objects remain on the truck bed.
- Retract the outriggers to the transport position.
- Disengage the PTO and return the engine to idle.
- Replace wheel chocks and barriers in their travel positions.
- Check the work area for tools and equipment.
- Make note of any operating occurrence that indicates a need for maintenance or repair.

10.0.0 In Case of Emergency

Two types of emergencies can occur when operating a bucket truck. One is when the bucket truck becomes inoperable and the line worker must exit the bucket while it is still raised. The other type of emergency is when the line worker in the bucket becomes incapacitated due to illness or injury and the bucket must be lowered so the worker can be extracted from the bucket.

10.1.0 Exiting Raised Inoperable Bucket

There are many different types of bucket evacuation devices on the market. All buckets should be equipped with some type of evacuation device per company policy. OSHA does not mandate the availability or use of an escape system. However, it is good practice, especially when working alone, to have one of these systems within easy access of the bucket. Various types (*Figure 24*) include those with rope ladders, rappelling ropes and containment bags for storage. Always follow the manufacturer's instructions when using a bucket evacuation device.

10.2.0 Emergency Bucket Truck Rescue

An emergency rescue from an aerial lift on a bucket truck is necessary when the worker in the bucket becomes ill, is injured, or loses consciousness. Injuries and loss of consciousness can occur if the worker is electrocuted; falling or hanging from the bucket; hit by an object or structure outside the bucket; or sustains an injury using hand tools while performing work from the bucket. Any time a line worker is injured inside the bucket to the point where he or she cannot operate the controls, a bucket rescue must be initiated by the co-workers on the ground. Employees qualified and trained to operate bucket trucks should be trained annually in rescue techniques. Always follow company procedures.

Any time an accident occurs, reaction time is critical.

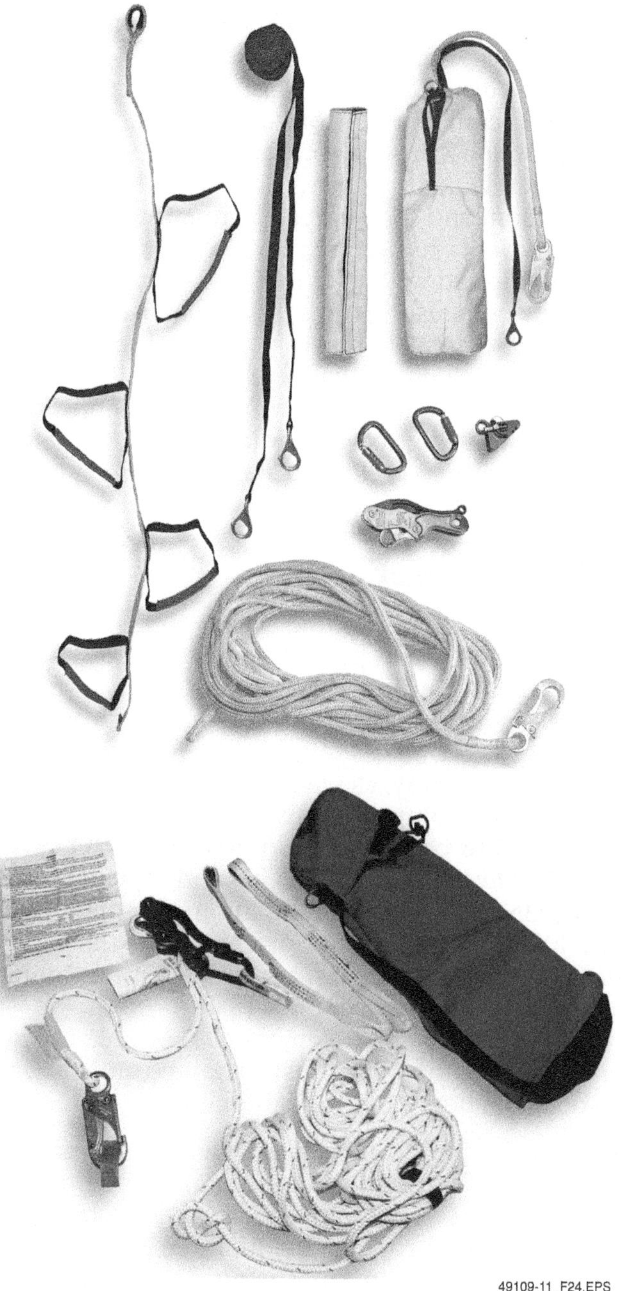

49109-11_F24.EPS

Figure 24 Bucket evacuation systems.

10.2.1 Lowering the Bucket

The first order of business is to call for assistance, then lower the bucket to the ground. The lower controls are used to lower the bucket to the ground in an emergency. The lower controls override the upper (in-bucket) controls, so if the worker in the bucket is unconscious and has fallen onto the upper control board activating a control, the lower bucket controls should override this function. Always ensure that there are no obstacles under the bucket that may interfere with lowering the bucket.

WARNING!

In case of an electrocution emergency involving any energized electrical source, do not approach or enter the vehicle unless you are certain the vehicle is not energized. If the vehicle is energized, or you are not sure, stand on an insulated pad or blanket and use a long insulated pole to operate the lower controls to bring the bucket down.

10.2.2 Removing the Worker from the Bucket

The bucket on the aerial truck may be designed to tilt in order to safely remove an incapacitated worker from the bucket. If this is the case, release the locking lever on the bucket, tilt the bucket over, unhook the safety lanyard from the worker's belt, and carefully remove the worker from the bucket.

If the bucket does not tilt, appropriate rigging may have to be set up to lift the worker up and over the edge of the bucket. Always secure the rigging device securely, unhook the lanyard from the worker, attach the rigging to the worker, and lift the worker out of the bucket.

10.3.0 Performing First Aid

Once removed, lay the injured worker on his or her back and observe if the victim is conscious. Great care should be taken to protect the victim from further injury. Give necessary first aid and call for help. If the victim is not conscious and not breathing, tilt his or her head back to provide an open airway. Perform mouth-to-mouth resuscitation and call immediately for help.

11.0.0 Other Utility Service Equipment

Additional commonly used utility service equipment includes crane trucks and aerial lifts. Crane trucks are designed with a telescoping crane boom used to lift large loads of materials or heavy components. Aerial lifts are used to lift workers to the required working height typically in indoor spaces. Aerial lifts, sometimes called man lifts, can also be used outdoors and are often used at job sites during installations. Aerial lifts may also be used in substations where access by bucket truck is not possible.

11.1.0 Crane Trucks

A crane truck is a type of boom truck designed with a crane boom installed on a truck chassis (*Figure 25*). This design allows crane trucks to be driven on public streets to work sites without the need for separate hauling trucks and trailers. Crane trucks are available in different sizes and load capabilities and are capable of lifting heavy components and materials and setting them into position.

As a rubber-tire vehicle, the tires determine the amount of weight the chassis can support. To eliminate this problem, crane trucks are often equipped with hydraulic jacks at the corners of the chassis. When the jacks are activated, the chassis is raised and the weight of the load is not applied to the tires. The crane truck is then capable of lifting and supporting much heavier loads than the tires would allow. Most crane trucks are equipped with jacks or outriggers to support the load and stabilize the truck.

49109-11_F25.EPS

Figure 25 Small crane truck.

Depending on the size and rating, crane trucks can lift various tonnage loads and are considered stationary when lifting. Most crane truck booms are hydraulically powered and are of a retractable or telescoping design. Typically, a gear pump pressurizes the hydraulic fluid to extend and retract the crane boom while a rotary gear allows the boom to be rotated.

11.1.1 Crane Truck Operator Controls

All parts of the crane are controlled by the operator and many devices may be installed in the cab, or remotely, to operate the crane. A crane truck, for example, may have two joystick controls. One controls the left-to-right movements, and the other controls the forward and back movements of the crane. Foot pedals in the cab are often used to control the extending and retracting of the boom and the degree of pressure generated from the boom hydraulic pump. Crane trucks can come equipped with load indicators that warn the operator with a series of lights if the load is being moved off course. In this case, the operator enters information, such as, the weight, height, and path of the load into a computer before initiating the lift. The signal lights will indicate to the operator if the crane truck is close to exceeding its capacity during the lift.

11.1.2 Crane Truck Safety

Crane truck operators must learn to inspect and operate crane trucks properly to avoid costly accidents and serious injuries to themselves or others. OSHA outlines crane truck safety measures and regulates acceptable crane design, boom design, crane/boom safety, as well as crane/boom inspection. Some companies require operators and inspectors to be trained and certified by OSHA. OSHA *Crane Standard 1926.550* and *Rigging Standard 1926.251*, detail the regulations for safe lifting operation. The National Commission for the Certification of Crane Operators (NCCCO) also trains and certifies crane operators.

All operators and personnel working on or around the crane truck must be trained and understand the following safety guidelines:

- Standard PPE, including hard hats, safety shoes, and barricaded work areas are among the important safety requirements at any job site.
- Ensure that properly maintained, good condition rigging equipment is available. Use only slings or other adequate lifting devices.

- Never work under a suspended load.
- Never leave a load suspended in the air when the hoisting equipment is unattended.
- Only trained and/or certified operators may use crane trucks.
- Ensure that the ground can support the crane truck and the load.
- Ensure that the crane truck's jacks or outriggers are properly deployed to a firm base.
- Always stay clear of all moving equipment.
- Ensure that the swing path or load path of the crane boom is clear of personnel and obstructions.
- Keep all workers from around the crane truck when in close proximity to power lines.
- Be aware of the location and voltage of any overhead power lines. If possible, de-energize the lines and erect insulated barriers as necessary. Determine a safe travel route and operate at a slower speed. If the operators view of the power lines or clearance is hindered while moving the crane truck, a trained signal person should be used for guidance.
- Be prepared for an emergency and know what to do in case of electrical, mechanical, or power failures.
- Know how to rescue an injured operator from the cab if necessary.
- Use an adequate number of workers with the crane truck for the job. This can include a crew member responsible for maintenance, a signal person, and the operator.
- Before using the crane, all moving parts of the crane should be inspected to note whether they are properly functioning. Test all control buttons and switches to ensure they work correctly and immediately release without sticking.
- Inspect the crane boom regularly to ensure that it is free of corrosion, cracks, and excessive wear. This includes regular inspection and maintenance of the hydraulics involved in telescoping booms.
- Avoid the possibility of the crane truck tipping forward by positioning counterweights as necessary.
- Clear the area surrounding the crane truck of all workers, debris, and any potential hazards that could interrupt the safe movement of the crane truck. Clear the path of the crane truck of any obstacles and people prior to moving a load.

11.1.3 Crane Truck Operation

The operator must follow the manufacturer's recommendations and any locally established restrictions during lifting operations. The operator must also understand and follow all required manufacturer-supplied operation instructions. Depending on the truck, crane trucks may be operated remotely or from the cab. Any employees who have not received proper safety and functional training must not be allowed to operate the crane truck until the certification requirements have been met. Operators are responsible for performing the following tasks during lifting operations:

- The operator must inspect the crane truck, rigging, and terrain before starting the equipment and beginning operation.
- All operators must be trained and proficient with the use of the pedals, levers, joysticks, and other controls necessary for operating the crane truck.
- All operators and signal persons must be trained and knowledgeable in the use of approved operator hand signals.
- Operators must calculate and position any necessary counterweight and use the mathematical formula based on the radius of the boom, the operating angle, and the load's weight.
- Operators must know the load capacity of the crane truck. If the load is over the limit, turn off the truck and lighten the load.
- Operators must not attempt to move loads exceeding the vehicle's limits.
- Before lifting, operators must ensure that all loads are securely slung and properly balanced to prevent shifting of any part.
- Operators must use one or more tag lines as necessary to keep the load under control.
- Operators must ensure that the load landing zone is clear of personnel before setting the load.

11.2.0 Aerial Lifts

Aerial lifts are used to raise and lower workers to and from elevated job sites. There are two main types of lifts: boom lifts and scissor lifts. Both types are made in various models. Some are transported on a vehicle to a job site where they are unloaded. Others are trailer-mounted and towed to the job site by a vehicle, and some are permanently mounted on a vehicle. Depending on their design, they can be used for indoor work, outdoor work, or both. *Figure 26* shows two commonly used types of aerial lifts.

Boom lifts, often called cherry pickers, are designed for both indoor and outdoor use. Boom lifts have a single arm that extends a work platform/enclosure capable of holding one or two workers. Some models have a jointed (articulated) arm that allows the work platform to be positioned both horizontally and vertically. Scissor

Figure 26 Aerial lifts.

lifts raise a work enclosure vertically by means of crisscrossed supports.

Most models of aerial lifts are self-propelled, allowing workers to move the platform as work is performed. The power to move these lifts is provided by several means, including electric motors, gasoline or diesel engines, and hydraulic motors.

11.2.1 Aerial Lift Assemblies

Aerial lifts normally consist of three major assemblies: the platform, a lifting mechanism, and the base.

On Site

Scissor Lift Tires

To prevent surfaces from being marked or scuffed, non-marking tires are available for scissor lifts used indoors. This feature is particularly desirable when using the lift on wooden or tile floor surfaces.

The platform of an aerial lift is constructed of a tubular steel frame with a skid-resistant deck surface, railings, toe board, and midrails. Entry to the platform is normally from the rear. The entry opening is closed either with a chain or a spring-returned gate with a latch. The work platform may also be equipped with a retractable extension platform. The lifting mechanism is raised and lowered either by electric motors and gears or by one or more single-acting hydraulic lift cylinder(s). A pump, driven by either an AC or DC motor, provides hydraulic power to the cylinder(s). The base provides a housing for the electrical and hydraulic components of the lift. These components are normally mounted in swingout access trays. This allows easy access when performing maintenance or repairs to the unit. The base also contains the axles and wheels for moving the assembly. In the case of a self-propelled platform, electrical or hydraulic motors drive two or more of the wheels to allow movement of the lift from one location to another. Brakes are incorporated on one or more of the wheels to prevent inadvertent movement of the lift.

On Site

Aerial Lift Operator Controls

Operator controls for aerial lifts differ depending on the manufacturer and model of the lift. Shown here are the ground control station and platform control box for an electric-drive scissor lift produced by one major lift manufacturer.

GROUND CONTROL STATION

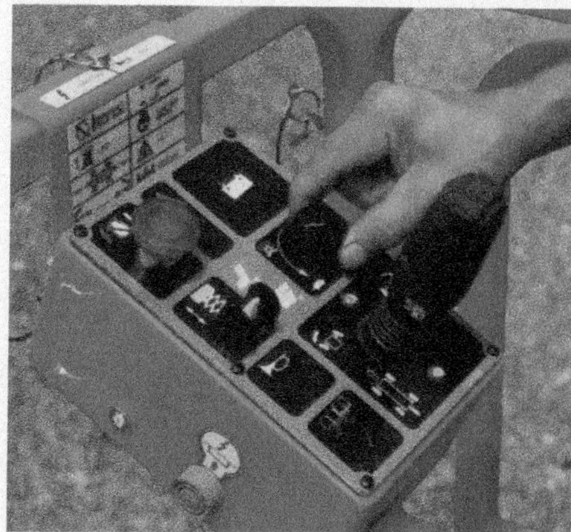

PLATFORM CONTROL BOX

49109-11_SA02.EPS

11.2.2 Aerial Lift Operator Qualifications

Only trained and authorized workers may use an aerial lift. Safe operation requires the operator to understand all limitations and warnings, operating procedures, and operator requirements for maintenance of the aerial lift. The following is a list of requirements the operator must meet:

- Understand and be familiar with the associated operator's manual for the lift being used.
- Understand all procedures and warnings within the operator's manual and those posted on decals on the aerial lift.
- Be familiar with the employer's work rules and all related government (OSHA) safety regulations.
- Demonstrate this understanding and operate the associated model of aerial lift during training in the presence of a qualified trainer.

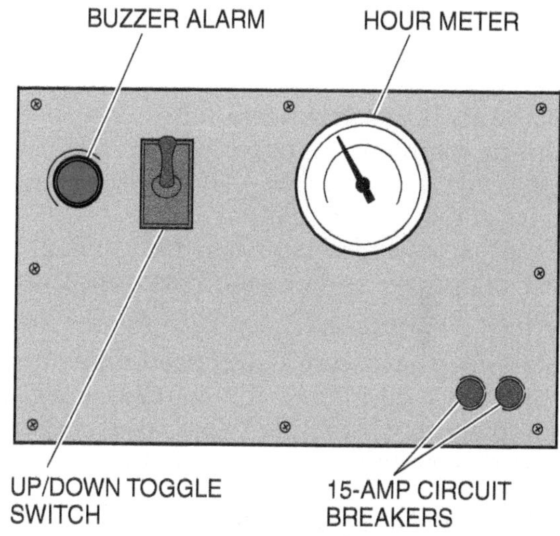

Figure 27 Aerial lift electrical panel.

11.2.3 Typical Aerial Lift Controls

Figure 27 shows an example of an aerial lift electrical panel and its associated controls. This is only an example, and other models of lifts may have different control configurations. This electrical panel contains the following switches and controls:

- *Up/Down toggle switch* – By holding this switch in the Up position, the platform can be raised to the desired level. The platform will stop moving when the switch is returned to the center position. By holding this switch in the Down position, the platform can be lowered.
- *Buzzer alarm* – An audible alarm sounds when the platform is being lowered. On some models, this alarm may sound when any control function is being performed.
- *Hour meter* – This meter records the number of hours the platform has been operating. The

meter will only register when the electric or hydraulic motor associated with operating the aerial lift is running.

One other control that may be available to the operator is an emergency battery disconnect switch. When this switch is placed in the Off position, it will disconnect power to all control circuits.

The basic controls for an aerial lift with a hydraulic system would generally include:

- *Emergency lowering valve* – This valve allows for platform lowering in the event of an electrical/hydraulic system failure.
- *Free wheeling valve* – Opening this valve allows hydraulic fluid to flow through the wheel motors. This allows the aerial lift to be pushed by hand, and prevents damage to the motors when the aerial lift is moved between job site locations. There are usually strict limits to how fast the aerial lift may be moved without causing damage to hydraulic system components.

Other types of controls that may be available on aerial lifts include:

- *Parking brake manual release* – This control allows manual release of the parking brake. This control should only be used when the aerial lift is located on a level surface.
- *Safety bar* – The safety bar is used to support the platform lifting hardware in a raised position during maintenance or repair.
- *Up/Down selector switch* – This switch is used to raise and lower the platform.
- *Emergency stop button* – This button is used to cut off power to both the platform and base control boxes.

Stabilizer hardware is required on some aerial lifts. At a minimum, this hardware includes a stabilizer leg, stabilizer lock pin and cotter key, and a stabilizer jack at each corner of the aerial lift.

Most aerial lifts are equipped with a base control box. The base control box includes a switch to select whether operation will be controlled from the platform or the base control box. The base control overrides the platform control.

The worker on the platform designated as the operator uses the platform controls. In some aerial lifts, operation from the platform is limited to raising and lowering the platform. On others, the aerial platform is designed to be driven by the operator using controls provided on the platform. In either case, the operator will have an operator control box located on the platform assembly. *Figure 28* shows a typical platform control box for a self-propelled aerial lift.

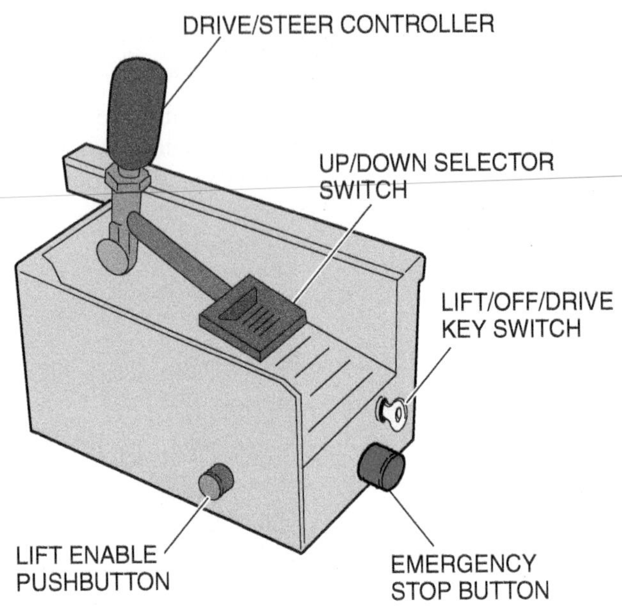

Figure 28 Aerial lift platform control box.

Controls available for the operator include (but are not limited to) the following:

- *Drive/steer controller* – This is a one-handed toggle-type (joystick) lever control for controlling speed and steering of the aerial platform. This is usually a deadman switch that returns to neutral and locks when released. The handle is moved forward to drive the aerial platform forward. The platform speed is determined by how far forward the handle is moved. The handle is moved backward to drive the aerial platform backward; speed is selected as before. Releasing the stick stops the motion of the aerial lift. Steering is performed by depressing a rocker switch on the top of the stick in the desired direction of travel, either right or left.
- *Up/Down selector switch* – Placing the switch in the Up position raises the platform; placing the switch in the Down position lowers the platform. When released, the switch returns to the middle position and stops the movement of the platform.
- *Lift/Off/Drive selector switch* – The Lift position energizes the lift circuit; the Off position removes power to the control box; the Drive position energizes the drive and steering controls.
- *Emergency stop pushbutton* – When pushed, this button disconnects power to the platform control circuit. In an emergency, push the button in. To restore power, pull the button out.
- *Lift enable button* – When pushed and held down, this button enables the lift circuit. The button must be held down when raising or lowering the platform. Releasing the button stops the motion of the platform.

Different aerial lifts have different controls. It is important that the operator fully understand all controls and their functions before operating the aerial lift. The operator must read and fully understand all control and safety information provided by the manufacturer before attempting to operate the equipment.

11.2.4 Aerial Lift Safety Precautions

Safety precautions unique to aerial lifts are listed here. Remember that the other safety precautions already discussed in this module also apply. Each manufacturer will provide specific safety precautions in the operator's manual provided with their equipment. Specifically, *OSHA Standard 1926.453* defines and governs the use of aerial lifts.

- Avoid using the lift outdoors in stormy weather or in strong winds.
- Prevent people from walking beneath the work area of the platform.
- Use personal fall arrest equipment (body harness and lanyard) as required for the type of lift being used. Use approved anchorage points.
- Do not use an aerial lift on uneven ground.
- Lower the lift and lock it into place before moving the equipment. Also, lower the lift, shut off the engine, set the parking brake, and remove the key before leaving it unattended.
- Stand firmly on the floor of the basket or platform. Do not lean over the guardrails of the platform, and never stand on the guardrails. Do not sit or climb on the edge of the basket or use planks, ladders, or other devices to attain additional height.

On Site

Aerial Lift Safety

A study by the Center to Protect Worker's Rights, a research arm of the Building and Construction Trades Department, AFL-CIO, covering a period from 1992 to 1999, found that aerial lifts used by the construction trades in the U.S. have been involved in a total of 207 deaths. Five percent of these deaths (about 10) involved carpenters. Of the total deaths, 33 percent were caused by electrocutions, 31 percent from falls, and 22 percent from lift collapses or tipovers. The remaining 14 percent resulted from other causes.

It should be pointed out that the deaths from electrocutions involved mostly electricians or electrical repair personnel. About 70 percent of the deaths involved the use of boom lifts. Another 25 percent involved the use of scissor lifts. Falls from boom lifts occurred mainly because the lift was struck by vehicles, cranes or crane loads, by falling objects, or when the lift was suddenly jerked. Falls from scissor lifts were mainly caused because the lift was struck by an object or because the safety chains or guardrails were removed. Some were caused because the occupant was standing on or leaning over the lift railings. The operation of any aerial lift is subject to hazards that can be protected against by following all safety rules, being careful, and using common sense.

11.2.5 Aerial Lift Operating Procedure

This section describes operations for a scissor-type, self-propelled aerial lift. Remember that operating procedures vary from model to model and the operator must become familiar with all operating procedures, controls, safety features, and associated safety precautions before operating any type of aerial lift.

A typical proportional control procedure involves the operation of a lever or foot pedal to cause the aerial lift to move. The further the control is moved, the more power is applied to the motor and the faster the aerial lift will move.

The following tasks must be performed before operating the aerial lift:

- Carefully read and fully understand the operating procedures in the operator's manual and all warnings and instruction decals on the work platform.
- Check for obstacles around the work platform and in the path of travel such as holes, dropoffs, debris, ditches, and soft fill. Operate the aerial lift only on firm surfaces.
- Check overhead clearances. Make sure to stay at least 10' away from overhead power lines.
- Make sure batteries are fully charged (if applicable).
- Make sure all guardrails are in place and locked in position.
- Perform an operator's checklist.
- Never make unauthorized modifications to the components of an aerial lift.

Figure 29 shows an example of an operator's checklist.

The following is an example of the operator controls used to drive the aerial lift forward. The operator selects the drive position with the lift drive select switch. The operator then lifts the handle lock ring and moves the drive/steer controller forward. The speed can be adjusted by continuing to move the controller forward until the desired speed is reached. By releasing

OPERATOR'S CHECKLIST

INSPECT AND/OR TEST THE FOLLOWING DAILY OR AT BEGINNING OF EACH SHIFT

___ 1. OPERATING AND EMERGENCY CONTROLS

___ 2. SAFETY DEVICES

___ 3. PERSONNEL PROTECTIVE DEVICES

___ 4. TIRES AND WHEELS

___ 5. OUTRIGGERS (IF EQUIPPED) AND OTHER STRUCTURES

___ 6. AIR, HYDRAULIC, AND FUEL SYSTEM(S) FOR LEAKS

___ 7. LOOSE OR MISSING PARTS

___ 8. CABLES AND WIRING HARNESSES

___ 9. DECALS, WARNINGS, CONTROL MARKINGS, AND OPERATING MANUALS

___ 10. GUARDRAIL SYSTEM

___ 11. ENGINE OIL LEVEL (IF SO EQUIPPED)

___ 12. BATTERY FLUID LEVEL

___ 13. HYDRAULIC RESERVOIR LEVEL

___ 14. COOLANT LEVEL (IF SO EQUIPPED)

49109-11_F29.EPS

Figure 29 Example of an aerial lift operator's checklist.

the controller, the forward motion of the lift is stopped. To drive in reverse, the lever is moved in the opposite direction.

11.2.6 Aerial Lift Operator's Maintenance Responsibility

The operator must be assured that the work platform has been properly maintained before using it.

WARNING!

Death or injury to workers and damage to equipment can result if the aerial platform is not maintained in good working condition. Inspection and maintenance should be performed by personnel who are authorized for such procedures.

As a minimum, if the operator is not responsible for the maintenance of the aerial lift, the operator should perform the daily checks. *Figure 30* shows an example of a maintenance and inspection schedule for a typical aerial lift. The daily checks should be performed at the beginning of each shift or at the beginning of the day if only one shift is worked. The aerial lift must never be used until these checks are completed satisfactorily. Any deficiencies found during the daily check must be corrected before using the aerial lift.

Figure 30 is broken into four major categories. Each category is further broken down into the components that should be checked daily, weekly, monthly, every three months, every six months, and yearly. Footnotes (the numbers in parentheses) following each component tell the operator what type of inspection is required. For example, under the Electrical category, battery fluid level should be checked daily or at the beginning of each shift. The (1) refers the operator to the Notes portion of the schedule. Note 1 tells the operator to perform a visual inspection of the battery fluid level.

Different models of aerial lifts have different maintenance requirements. Always refer to the operator's manual to determine exactly what checks are required before operation.

General maintenance rules applicable to any aerial lift include the following:

- Disconnect the battery ground negative (–) lead before performing any maintenance.
- Properly position safety devices before performing maintenance with the work platform in the raised position.

Preventive maintenance is easier and less expensive than corrective maintenance.

	Daily	Weekly	Monthly	3 Months	6 Months	*Annually
Mechanical						
Structural damage/welds (1)	✓					✓
Parking brakes (2)	✓					✓
Tires and wheels (1)(2)(3)	✓					✓
Guides/rollers/slides (1)	✓					✓
Railings/entry chain/gate (2)(3)	✓					✓
Bolts and fasteners (3)	✓					✓
Rust (1)			✓			✓
Wheel bearings (2) King pins (1)(8)	✓					✓
Steer cylinder ends (8)				✓		✓
Electrical						
Battery fluid level (1)	✓					✓
Control switches (1)(2)	✓					✓
Cords and wiring (1)	✓					✓
Battery terminals (1)(3)	✓					✓
Terminals and plugs (3)	✓					✓
Generator and receptacle (2)	✓					✓
Limit switches (2)	✓					✓
Hydraulic						
Hydraulic oil level (1)	✓					✓
Hydraulic leaks (1)	✓					✓
Lift/lowering time (10)				✓		✓
Hydraulic cylinders (1)(2)		✓				✓
Emergency lowering (2)	✓					
Lift capacity (7)			✓			✓
Hydraulic oil/filter (9)					✓	✓
Miscellaneous						
Labels (1)(11) Manual (12)	✓					✓

Notes:

(1) Visually inspect.
(2) Check operation.
(3) Check tightness.
(4) Check oil level.
(5)(6) N/A.
(7) Check relief valve setting. Refer to serial number nameplate.
(8) Lubricate.
(9) Replace.

(10) General specifications.
(11) Replace if missing or illegible.
(12) Proper Operating Manual *must* be in the manual tube.
*Record Inspection Date

49109-11_F30.EPS

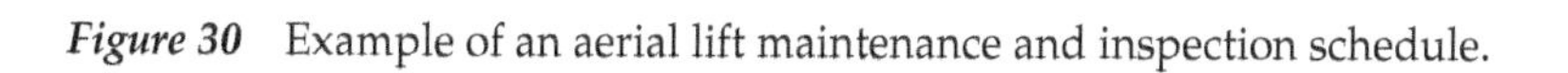

Figure 30 Example of an aerial lift maintenance and inspection schedule.

SUMMARY

Power line technicians use bucket trucks and digger derricks on a daily basis. These vehicles can be extremely dangerous if they are not operated properly. All operators must complete thorough training on the specific vehicle that they will be operating. Only trained, qualified operators are allowed to operate these vehicles. In addition to following all of the safety rules when operating these vehicles, operators must also know how to perform daily, weekly, and monthly inspections to ensure that the equipment is properly maintained and safe to use. Since the operator is in direct control of the bucket truck or digger derrick being operated, it is his or her responsibility to always use good judgment, exercise extreme caution, and conform to all safety rules to ensure his or her own safety and the safety of others on the job site.

Review Questions

1. An aerial device with two or more hinged boom sections is known as a(n) _____.
   a. articulating boom bucket truck
   b. telescopic boom bucket truck
   c. combination boom bucket truck
   d. derrick boom bucket truck

2. To dig holes for placement of utility poles, a digger derrick uses a(n) _____.
   a. boom
   b. bucket
   c. winch
   d. auger

3. The boom on a digger derrick is typically equipped with a pole claw, pole guides, or boom flares that are used for primarily _____.
   a. supporting the weight of the pole
   b. aligning and positioning poles
   c. lifting the pole from the ground
   d. pressing the pole into the hole

4. Adherence to ANSI is mandatory, while compliance with OSHA is voluntary.
   a. True
   b. False

5. To make sure that a specific lift is safe according to the manufacturer's recommendations, always refer to _____.
   a. *NFPA 70E®*
   b. the preventive maintenance chart
   c. the load rating charts
   d. *National Electric Code®*

6. While working in the bucket, workers must wear a safety harness with an attached lanyard, the length of which cannot exceed _____.
   a. 1 foot
   b. 2 feet
   c. 3 feet
   d. 6 feet

7. Manufacturers place a band of arrows on the upper boom to show the _____.
   a. location of the bucket
   b. direction of boom travel
   c. end of the insulated section
   d. safety harness anchorage point

8. At all times during operation, the digger derrick operator must remain _____.
   a. beside the truck
   b. on the truck
   c. inside the cab of the truck
   d. on the ground behind the truck

9. To ensure that the aerial lift has a current electrical insulation resistance test, check the _____.
   a. smoothness of the fiberglass insert
   b. maintenance log book in the office
   c. date on the inspection sticker
   d. expiration date on the license plate

10. Vehicle operating manuals should be stored in waterproof containers and left _____.
    a. in the job box
    b. by the lower controls
    c. by the upper controls
    d. on the lift or in the cab of the truck

11. When setting up a vehicle on a slope, the low side may need to be built up with timbers if the slope is more than _____.
    a. 5 degrees
    b. 6 degrees
    c. 7 degrees
    d. 10 degrees

12. Although the lead time for marking utility lines may vary from state to state, in most states, the contractor must notify the One-Call system _____.
    a. 12 hours in advance of the actual dig
    b. 24 hours in advance of the actual dig
    c. 36 hours in advance of the actual dig
    d. 48 hours in advance of the actual dig

13. After utility lines have been marked, blue markings indicate _____.
    a. potable water lines
    b. fiber-optic lines
    c. gas lines
    d. sewer lines

14. If you are operating a digger auger and rupture an underground electrical power line, _____.
    a. do not leave the machine
    b. slowly leave the operator control seat
    c. jump off and run
    d. call your supervisor

15. To break an existing pole free from the ground, always use _____.
    a. an auger
    b. a pole puller
    c. the winch line
    d. a pole claw

Trade Terms Quiz

Fill in the blank with the correct term that you learned from your study of this module.

1. Any personnel-carrying device, basket, or bucket, which is a component of an aerial device is known as a(n) ____________.
2. Any of various small cords or ropes used for securing or suspending is called a(n) ____________.
3. An aerial device with two or more hinged boom sections that are designed to reach up and over obstacles is called a(n) ____________.
4. An aerial device designed for work in close vicinity of energized lines and apparatus is called a(n) ____________.
5. An extension added to the boom and mounted to the boom tip, which may be in line with the boom or offset from it is referred to as the ____________.
6. A bacteria, chemical, or virus that is known to cause disease is called a(n) ____________.
7. A place for properly securing a lanyard is known as a(n) ____________.
8. A boom that is designed to reach vertically or horizontally and consists of straight shafts that fit together and extend to raise an aerial platform is called a(n) ____________.
9. A spar or beam projecting from the mast of a derrick for supporting and guiding an aerial platform to be lifted is called the ____________.
10. Any vehicle mounted device, telescoping or articulating, or both, which is used to raise and position personnel to work at heights is called a(n) ____________.
11. An extension that projects from the main chassis of the utility truck to add stability and support is the ____________.
12. A region of constant voltage that contains no electric field is called the ____________.

Trade Terms

Aerial device
Anchorage point
Articulating boom
Boom
Equipotential region
Insulated aerial device
Jib
Lanyard
Outrigger
Pathogen
Platform
Telescopic boom

Cornerstone of Craftsmanship

James Anthony

Substation Maintenance Supervisor
Baltimore Gas and Electric

James Anthony has worked for the same company his entire career. He keeps developing his skills to take on new positions and challenges. Starting as an entry-level laborer, he moved into Underground Line, where he learned about underground cables and the network system. He then transferred to the Station Service Group, where he learned about DC battery systems and backup power. When his group merged into Substation Maintenance, he learned about substation equipment. After that, he became an instructor for DC power systems and generator maintenance. Later, he returned to Substation Maintenance, where he has excelled into a leadership position.

How did you get started in the construction industry?
I always was one for taking apart radios, record players, and other electronic equipment. I was very interested in electronics and how things operated. The more I learned, the more interested I became.

My high school was a vocational/technical high school. I studied electrical maintenance and industrial electronics in high school, so I really started off with a good foundation. I was hired right out of high school into the industry. My high school studies in electronics prepared me for work. Right after I graduated, my first job was as a laborer at a power plant. It was an entry-level position to get started in the industry.

Who inspired you to enter the industry?
My personal interest in electronics, maintenance, and equipment operations inspired my interest in this field.

What do you enjoy most about your job?
I enjoy the satisfaction of making repairs, restoring service, and reinforcing the integrity of our system. You go to a job site where the integrity of the equipment or service is compromised and you make the necessary repairs to restore service. It's very satisfying to walk away from a job knowing that you have addressed the problem and restored service.

Do you think training and education are important in construction?
Education and training are the keys to being successful in this industry. It's a priceless foundation that will really help you get started. Knowledge is power and the best way to get that is through education and training.

How important do you think NCCER credentials will be in the future?
I think that NCCER credentials will be very important. They clearly show potential employers, and the industry as a whole, that you have the solid foundation they're looking for. There are not a lot of places you can go and get that training except in the industry. So it is great for employers to see that you have that necessary training even if you don't have the work experience yet.

How has training/construction impacted your life and your career?
It gave me the jump-start that I needed to move along in the maintenance/construction industry. I started off with minimal work experience, but I had an educational background that gave me a good foundation.

Would you suggest construction as a career to others?
I think is an extremely good career for the right person. It's not for everybody. If you have the knack for electricity, maintenance, and construction, it can be a great career. The sky is the limit, if you apply yourself.

How do you define craftsmanship?
Craftsmanship means practicing a craft with great skill. It's being a professional who considers the quality, effectiveness, and efficiency of the work being performed. Craftsmen have a level of commitment when performing their duties and don't see their work as just another task. Craftsmen maintain a positive attitude and take pride in their job assignments.

Trade Terms Introduced in This Module

Aerial device: Any vehicle mounted device, telescoping or articulating, or both, which is used to raise and position personnel to work at heights.

Anchorage point: A place for properly securing a lanyard.

Articulating boom: An aerial device with two or more hinged boom sections. They are designed to reach up and over obstacles.

Boom: A spar or beam projecting from the mast of a derrick for supporting and guiding an aerial platform to be lifted.

Equipotential region: A region of constant voltage that contains no electric field. A charged particle in an equipotential region experiences no electric force, so there will be no electric current between two points of equal voltage because there is no force to drive the electrons.

Insulated aerial device: An aerial device designed for work in close vicinity of energized lines and apparatus.

Jib: An extension added to the boom and mounted to the boom tip, which may be in line with the boom or offset from it.

Lanyard: Any of various small cords or ropes used for securing or suspending.

Outrigger: An extension that projects from the main chassis of the utility truck to add stability and support.

Pathogen: A bacteria, chemical, or virus that is known to cause disease.

Platform: Any personnel-carrying device, basket or bucket, which is a component of an aerial device.

Telescopic boom: A boom that consists of straight shafts that fit together and extend to raise an aerial platform. They are designed to reach vertically or horizontally.

Additional Resources

This module presents thorough resources for task training. The following resource material is suggested for further study.

OSHA 1910.269, Electric Power Generation, Transmission, and Distribution

OSHA 1917.45, Cranes and Derricks

OSHA 1910.67, Vehicle Mounted Elevating and Rotating Work Platforms

OSHA 29 CFR 1926, Subpart N, Cranes, Derricks, Hoists, Elevators, and Conveyors

ANSI A10.31, Safety Requirements, Definitions and Specifications for Digger Derricks

ANSI A92.2, Vehicle Mounted Elevating and Rotating Aerial Devices

www.buckettrucks.org for information and safety updates on bucket trucks.

www.diggerderricks.org for information and safety updates on digger derrick trucks.

www.photolibrary.fema.gov for examples of where line workers may have to work.

www.genielift.com for miscellaneous types of lifts.

www.jlg.com for miscellaneous types of lifts.

Figure Credits

Federal Emergency Management Agency, U.S. Department of Homeland Security, Module opener

Photo courtesy of Altec Industries, Figures 1 and 4

New York Power Authority, Figure 2

Topaz Publications, Inc., Figures 3, 10, 11, SA01, and 18

Bob Groner, Figures 5, 7, 17, and 19–22

Tony Vazquez, Figure 6, 15, and 16

Fall protection materials provided courtesy of Miller Fall Protection, Franklin, PA, Figure 9

DUECO, Figure 12

Table 1 reproduced with permission from *NFPA 70E®, Electrical Safety in the Workplace*, Copyright © 2008, National Fire Protection Association. This reprinted material is not the complete and official position of the National Fire Protection Association on the referenced subject, which is represented only by the standard in its entirety.

Hastings, Figure 14

Tiiger, Inc., Figure 23

Photos provided courtesy of Buckingham Manufacturing Company, Inc., Figure 24

ETI, Figure 25

JLG Industries, Inc., Figure 26 and SA02

CONTREN® LEARNING SERIES — USER UPDATE

NCCER makes every effort to keep its textbooks up-to-date and free of technical errors. We appreciate your help in this process. If you find an error, a typographical mistake, or an inaccuracy in NCCER's Contren® materials, please fill out this form (or a photocopy), or complete the online form at www.nccer.org/olf. Be sure to include the exact module number, page number, a detailed description, and your recommended correction. Your input will be brought to the attention of the Authoring Team. Thank you for your assistance.

Instructors – If you have an idea for improving this textbook, or have found that additional materials were necessary to teach this module effectively, please let us know so that we may present your suggestions to the Authoring Team.

NCCER Product Development and Revision

3600 NW 43rd Street, Building G, Gainesville, FL 32606

Fax: 352-334-0932
Email: curriculum@nccer.org
Online: www.nccer.org/olf

☐ Trainee Guide ☐ AIG ☐ Exam ☐ PowerPoints Other ______________

Craft / Level: Copyright Date:

Module Number / Title:

Section Number(s):

Description:

Recommended Correction:

Your Name:

Address:

Email: Phone:

Rigging

49110-11

Trainees with successful module completions may be eligible for credentialing through NCCER's National Registry. To learn more, go to **www.nccer.org** or contact us at **1.888.622.3720.** Our website has information on the latest product releases and training, as well as online versions of our *Cornerstone* newsletter and Pearson's Contren® product catalog.

Your feedback is welcome. You may email your comments to **curriculum@nccer.org**, send general comments and inquiries to **info@nccer.org**, or use the User Update form at the back of this module.

V.1 7/11

Rigging

Objectives

When you have completed this module, you will be able to do the following:

1. Describe and demonstrate hand signals and other communication methods used in rigging work.
2. Describe safety hazards and safety practices associated with rigging work.
3. Identify the safety procedures associated with the use of cranes in rigging work.
4. Describe how cranes are used to lift and move loads.
5. Tie knots used in rigging.
   - Square
   - Figure 8
   - Clove hitch
   - Double half hitch
   - Bowline
   - Bowline on a bight
   - Timber hitch
   - Sheet bend
   - Running bowline
   - Back splice
   - Sheep shank
6. Reeve a set of blocks.

Performance Tasks

Under the supervision of the instructor, you should be able to do the following:

1. Demonstrate hand signals and other communication methods used in rigging work.
2. Tie a minimum of six of the following knots:
   - Square
   - Figure 8
   - Clove hitch
   - Double half hitch
   - Bowline
   - Bowline on a bight
   - Timber hitch
   - Sheet bend
   - Running bowline
   - Back splice
   - Sheep shank
3. Reeve a set of blocks.

Trade Terms

Anti-two-blocking devices
Blocking
Center of gravity
Cribbing

Contents

Topics to be presented in this module include:

Figures and Tables

1.0.0 Introduction

Rigging involves the moving of equipment or material that is too large or heavy to move by hand (see *Figure 1*). Turbine rotors, for example, can weigh tens of thousands of pounds. Lifting such devices requires that workers involved in the lift understand what equipment to use and how to properly use it. Most importantly, they must ensure that any equipment used in a rigging operation is in good condition and is rated for the load it will have to handle.

One important job that a rigger performs is guiding the crane operator. This is often done using a standard set of hand signals, which everyone involved in rigging must memorize.

49110-11_F01.EPS

Figure 1 Moving a turbine rotor.

On Site

Truck-Mounted Cranes

One advantage of truck-mounted cranes is that, unlike crawler cranes, they can be driven to a job site. Truck-mounted cranes do not have the lifting capacity of large crawler cranes, however, so they may not be suitable for all jobs. One important thing to remember about truck-mounted cranes is that the outriggers must be deployed any time a load is being lifted. Otherwise, the crane could tip.

49110-11_SA01.EPS

Anyone doing rigging work must know how to select the right equipment, with the right capacity, for every rigging job. Failure to use the right equipment and correct rigging method can lead to serious consequences, including death or injury to workers and damage to equipment or materials.

Every crane and lifting device has capacity limits and conditions in which they are designed to operate safely. If you are not aware of these limits, you could make a fatal error. If you make the effort to learn the right and safe way to do the job, you will be successful. Always err on the side of caution.

This module is intended to instruct you on the basic safety precautions and rigging practices that will enable you to assist in rigging equipment and materials related to your craft.

2.0.0 Methods and Modes of Communication

The methods and modes of communication vary widely in lifting operations. The method refers to whether the communication is verbal (spoken) or nonverbal. The mode is what is used to facilitate the communication. This can include, for example, a bullhorn, a radio, hand signals, or flags.

2.1.0 Verbal Modes of Communication

Verbal modes of communication vary depending on the situation. One of the most common modes is a portable radio (walkie-talkie). Compact, low-power, inexpensive units enable the crane operator and signal person to communicate verbally. These units are rugged and dependable and are widely used on construction sites and in industrial plants.

There are some disadvantages, however, to using low-power and inexpensive equipment in an industrial setting. One disadvantage is interference. With low-power, budget units, the frequency used to carry the signal may have many other users. This crowding of the frequency could disrupt the signal from a more powerful unit. Another disadvantage is high background noise. In attempting to send a signal in a high-noise area, the person sending the signal may transmit unintended noise, resulting in a garbled signal for the receiver. On the receiving end, the individual may not be able to hear the transmission due to a high level of background noise in the cab of the crane.

There are several solutions to the problems associated with radio use. To overcome the shortcomings of low-power units, more expensive units with the ability to program specific frequencies and transmit at a higher power level may be needed. Some of these units may require licensing. To overcome the background noise problem, the use of an ear-mounted, noise-canceling microphone/headphone combination may be required (*Figure 2*).

Another solution is the use of an optional throat microphone. This device feeds the transmitted sound directly to the ear. This prevents noise from entering the microphone and blocks out

VOICE-OPERATED (VOX) OR PUSH-TO-TALK (PTT) RADIO SYSTEM WITH THROAT MICROPHONE

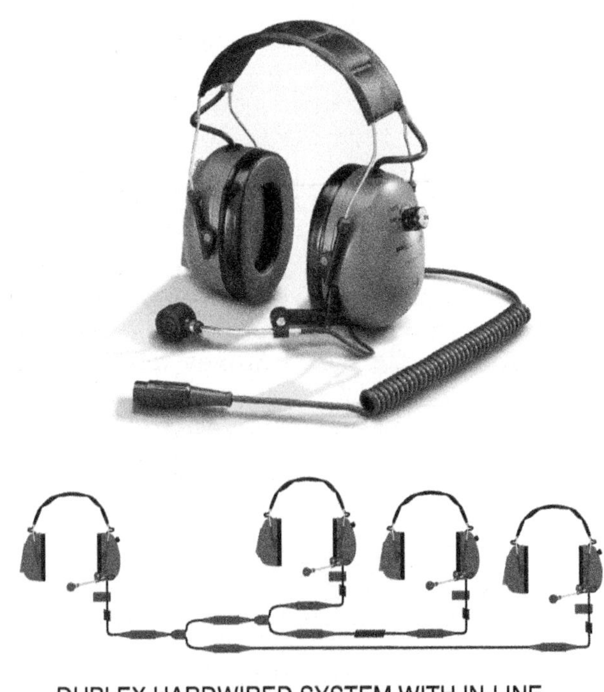

DUPLEX HARDWIRED SYSTEM WITH IN-LINE AMPLIFIER AND EXTENSION CORDS FOR SIMULTANEOUS TWO-WAY COMMUNICATION

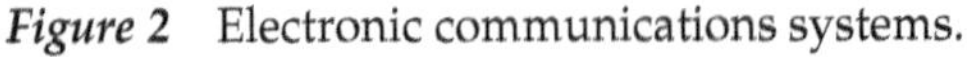

49110-11_F02.EPS

Figure 2 Electronic communications systems.

any background noise when listening. To avoid missed communication, the signal person's radio is usually locked in transmit so that the crane operator can tell if the unit is not transmitting. In any event, a feedback method should be established between the signal person and the crane operator so that the signal person knows the crane operator has received the signal.

Here is an example of a verbal communication between the signal person and the crane operator:

Rigger: "Is equipment ready to lift?"
Crane operator responds: "Equipment is ready to lift."
Rigger clears the area and communicates: "Proceed with the lift."

Another mode of verbal communication is a hardwired system (*Figure 2*). These units overcome some of the problems with radio use. When using this type of system, interference from another unit is unlikely because this system does not use a radio frequency to transmit information. Like a telephone system, occasional interference may be encountered if the wiring is not properly shielded from very strong radio transmissions. A hardwired unit is not very portable or practical when the crane is moved often. These units can also use an ear-mounted noise-canceling microphone/headphone combination to reduce the background noise.

2.2.0 Nonverbal Modes of Communication

Nonverbal communication is the most common type used when performing crane operations. Several modes are available for use under this method. One mode is the use of signal flags, which may mean different colored flags or a specific positioning of the flags to communicate the desired message. Another mode is the use of sirens, buzzers, and whistles in which the number of repetitions and duration of the sound convey the message. The disadvantage of these two modes is that there is no established meaning to any of the distinct signals unless they are prearranged between the sender and receiver. Also note that when sirens, buzzers, and whistles are used, background noise levels can be a problem.

The most common mode of communication reference used during crane operations is the *ASME B30.5 Consensus Standard* of hand signals (see *Figures 3* through *19*). In accordance with *ASME B30.5*, the operator must use standard hand signals unless voice communication equipment is used. It also requires that the hand signal chart be posted conspicuously at the job site.

RAISE LOAD OR HOIST UP
Fist up with pointer finger pointing straight up. Move hand in small horizontal circles.

EXPECTED MACHINE MOVEMENT
The load attached to the block or ball rises vertically, accelerating and decelerating smoothly.

49110-11_F03.EPS

Figure 3 Hoist.

LOWER LOAD OR HOIST DOWN
Fist down with pointer finger pointing straight down. Move hand in small horizontal circles.

EXPECTED MACHINE MOVEMENT
The load block or ball smoothly lowers vertically.

49110-11_F04.EPS

Figure 4 Lower.

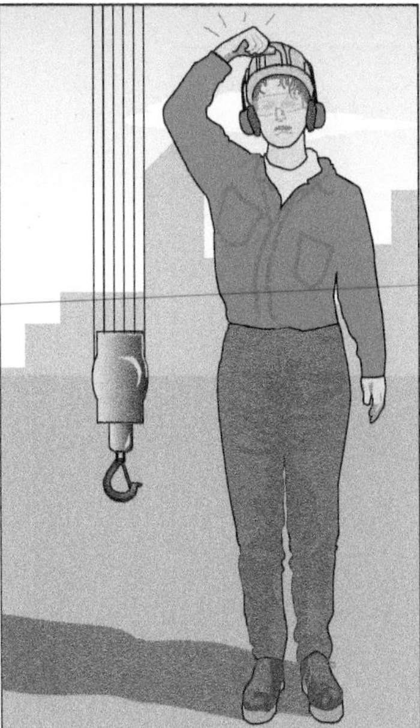

USE MAIN HOIST
Rap on hard hat with closed fist.

EXPECTED MACHINE MOVEMENT
None. This signal is used only to inform the operator that the signal person has chosen the main hoist for the action to be performed as opposed to the auxiliary hoist.

49110-11_F05.EPS

Figure 5 Use main hoist.

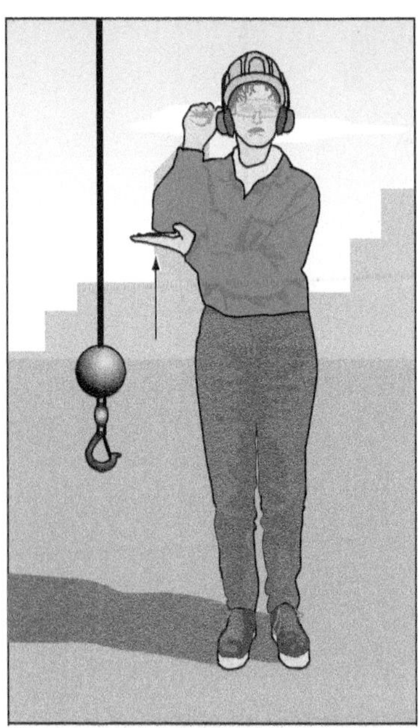

USE WHIP LINE (AUXILIARY HOIST)
Tap elbow with open palm of one hand, then use regular hand signal.

EXPECTED MACHINE MOVEMENT
None. This signal is used only to inform the operator that the signal person has chosen the auxiliary hoist for the action to be performed as opposed to the main hoist.

49110-11_F06.EPS

Figure 6 Use whip line.

RAISE BOOM
Arm extended, fingers closed, thumb pointing upward.

EXPECTED MACHINE MOVEMENT
The boom rises, increasing the hook height and reducing the overall machine height clearance. The operating radius is slowly decreased, thus possibly increasing machine capacity and stability.

49110-11_F07.EPS

Figure 7 Raise the boom.

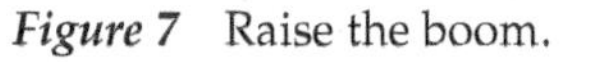

LOWER THE BOOM
Arm extended, fingers closed, thumb pointing downward.

EXPECTED MACHINE MOVEMENT
The boom will lower, decreasing the hook height and reducing the overall machine horizontal clearance. The operating radius is slowly increased, thus possibly decreasing machine capacity and stability.

49110-11_F08.EPS

Figure 8 Lower the boom.

MOVE SLOWLY
Use one hand to give any motion signal and place the other hand motionless over the hand giving the motion signal.

EXPECTED MACHINE MOVEMENT
Machine movement will vary depending on the signal being given.

49110-11_F09.EPS

Figure 9 Move slowly.

RAISE BOOM AND LOWER LOAD
Arm extended, fingers closed, thumb pointing up, flex fingers in and out as long as load movement is desired.

EXPECTED MACHINE MOVEMENT
The boom rises, reducing overall machine height clearance, as the load moves horizontally toward the crane. The operating radius is slowly decreased, thus possibly increasing machine capacity and stability.

49110-11_F10.EPS

Figure 10 Raise the boom and lower the load.

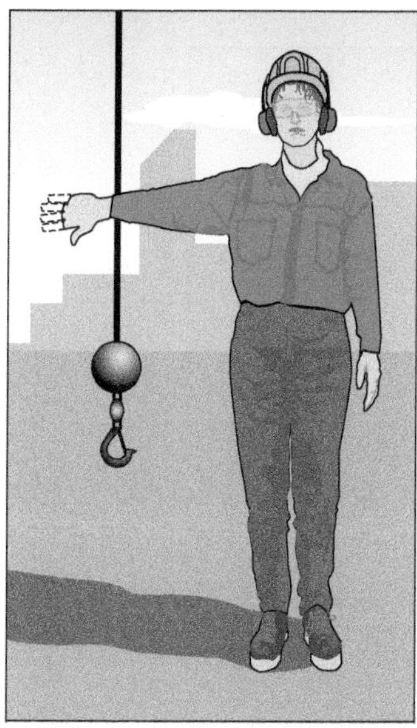

LOWER BOOM AND RAISE LOAD
Arm extended, fingers closed, thumb pointing down, flex fingers in and out as long as load movement is desired.

EXPECTED MACHINE MOVEMENT
The boom lowers, increasing overall machine height clearance, as the load moves horizontally away from the crane. The operating radius is slowly increased, thus possibly reducing both machine capacity and stability.

49110-11_F11.EPS

Figure 11 Lower the boom and raise the load.

SWING
Arm extended, point with fingers in direction of boom swing. (Swing left is shown as viewed by the operator.) Use appropriate arm for desired direction.

EXPECTED MACHINE MOVEMENT
The boom moves about the center of rotation with the load (block or ball) swinging in an arc, either toward the right or left, while remaining approximately equidistant to a level plane.

49110-11_F12.EPS

Figure 12 Swing.

STOP
Arm extended, palm down, move arm back and forth horizontally.

EXPECTED MACHINE MOVEMENT
None. All movement of the machine ceases.

49110-11_F13.EPS

Figure 13 Stop.

EMERGENCY STOP
Both arms extended, palms down, move arms back and forth horizontally.

EXPECTED MACHINE MOVEMENT
None. All movement of the machine ceases.

49110-11_F14.EPS

Figure 14 Emergency stop.

DOG EVERYTHING
Clasp hands in front of body.

EXPECTED MACHINE MOVEMENT
None.

49110-11_F15.EPS

Figure 15 Dog everything.

TRAVEL (BOTH TRACKS)
Position both fists in front of body and move them in a circular motion, indicating the direction of travel (forward or backward).

EXPECTED MACHINE MOVEMENT
Machine travels in the direction chosen.

49110-11_F16.EPS

Figure 16 Travel (both tracks).

TRAVEL (ONE TRACK)
Lock track on side of raised fist. Travel opposite track in direction indicated by circular motion of other fist, rotated vertically in front of body.

EXPECTED MACHINE MOVEMENT
Machine turns in the direction chosen.

49110-11_F17.EPS

Figure 17 Travel (one track).

The advantage to using these hand signals is that they are well established and published in an industry-wide standard. This means that these hand signals are recognized by the industry as the standard hand signals to be used on all job sites. This helps ensure that there is a common core of knowledge and a universal meaning to the signals when lifting operations are being conducted. Agreed-upon signals eliminate significant barriers to effective communication.

> **NOTE**
> The hand signals shown in *Figures 3 through 19* are the same as those previously covered in the Core module *Basic Rigging,* with the exception that *Figures 18* and *19* include signals for a telescoping boom. Also, the machine movement expected in response to the signal is shown in each figure.

Additions or modifications may be made for operations not covered by the illustrated hand signals, such as deployment of outriggers. The operator and signal person must agree upon these special signals before the crane is operated, and

A. TWO-HANDED

EXTENDING BOOM (TELESCOPING BOOM)
Both fists in front of body with thumbs pointing outward.

EXPECTED MACHINE MOVEMENT
Boom sections telescope out. The load radius is increased, possibly decreasing machine capacity and stability. The load (block or ball) rises.

NOTE: For digger derricks, the number of fingers is used to indicate which stage is to be extended.

B. ONE-HANDED

EXTENDING BOOM (TELESCOPING BOOM)
One fist in front of body with thumb pointing toward body.

EXPECTED MACHINE MOVEMENT
Boom sections telescope out. The load radius is increased, possibly decreasing machine capacity and stability. The load (block or ball) rises.

49110-11_F18.EPS

Figure 18 Extend boom.

A. TWO-HANDED

RETRACT BOOM (TELESCOPING BOOM)
Both fists in front of body with thumbs pointing toward each other.

EXPECTED MACHINE MOVEMENT
Boom sections retract. The load radius is decreased, possibly increasing machine capacity and stability. The load (block or ball) lowers.

B. ONE-HANDED

RETRACT BOOM (TELESCOPING BOOM)
One fist in front of body with thumb pointing outward.

EXPECTED MACHINE MOVEMENT
Boom sections retract. The load radius is decreased, possibly increasing machine capacity and stability. The load (block or ball) lowers vertically.

49110-11_F19.EPS

Figure 19 Retract boom.

these signals should not be in conflict with any standard signal.

If you need to give instructions verbally to the operator instead of by hand signals, all crane motions must be stopped before doing so.

NOTE

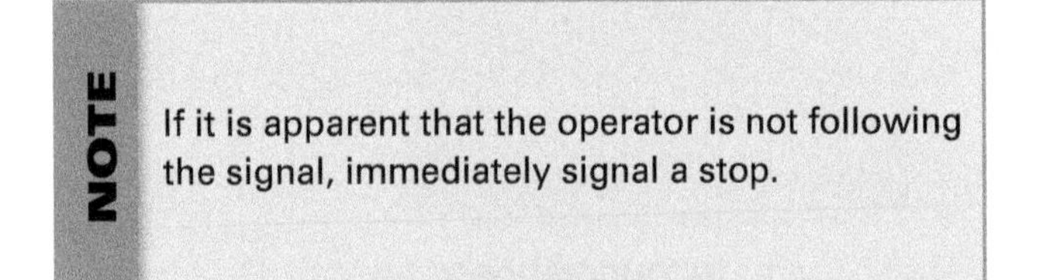

If it is apparent that the operator is not following the signal, immediately signal a stop.

When a mobile crane is being moved without direction from the rigger, audible travel signals must be given using the crane's horn:

- *Stop* – One audible signal
- *Forward* – Two audible signals
- *Reverse* – Three audible signals

There are certain requirements that mandate the presence of a signal person. When the operator of the crane can't see the load, the landing area, or the path of motion, or can't judge distance, a signal person is required. A signal person is also required if the crane is operating near power lines or another crane is working in close proximity.

Not just anyone can be a signal person. Signal persons must be qualified by experience, be knowledgeable in all established communication methods, be stationed in full view of the operator, have a full view of the load path, and understand the load's intended path of travel in order to position themselves accordingly. In addition, they must wear high-visibility gloves and/or clothing, be responsible for keeping everyone out of the operating radius of the crane, and never direct the load over anyone.

NOTE

Do not give signals to the crane operator unless you have been designated as the signal person.

Although personnel involved in lifting operations are expected to understand these signals when they are given, it is acceptable for a signal person to give a verbal or nonverbal signal to an operator that is not part of the *ASME B30.5* standard. In cases where such non-standardized signals are given, it is important that both the operator and the signal person have a complete understanding of the message that is being sent.

3.0.0 General Rigging Safety

As a rigging worker, you need to be aware of the hazards. You will be directing the movement of loads above and around other workers, where falling equipment and material can present a grave safety hazard (*Figure 20*). You may also work during extreme weather conditions where winds, slippery surfaces, and unguarded work areas exist. When working near cranes, always look up and be mindful of the hazards above and around you.

Safety consciousness is the key to reducing accidents, injuries, and deaths on the job site. Accidents can be avoided because most result from human error. Mistakes can be reduced by developing safe work habits derived from the principles of on-the-job safety.

49110-11_F20.EPS

Figure 20 Overhead hazards.

3.1.0 Personal Protection

Workers on the job are responsible for their own safety and the safety of their fellow workers. Management is responsible to ensure that the workers who prepare and use the equipment, and who work with or around it, are well-trained in operating procedures and safety practices.

Always be aware of your surroundings when working with cranes. Stay alert and know the location of equipment at all times when moving about and working within the job area.

Standard personal protective equipment, including hard hats, safety shoes, and barricaded work areas are among the important safety requirements at any job site.

3.2.0 Equipment and Supervision

Your employer is responsible for ensuring that all hoisting equipment is operated by experienced, trained operators. Riggers must also be capable of selecting suitable rigging and lifting equipment, and directing the movement of the crane and the load to ensure the safety of all personnel. Rigging operations must be planned and supervised by competent personnel who ensure the following:

- Proper rigging equipment is available.
- Correct load ratings are available for the material and rigging equipment.
- Rigging material and equipment are well maintained and in good working condition.

A rigging supervisor is responsible for the following functions:

- Proper load rigging
- Crew supervision
- Ensuring that the rigged material and equipment meet the required capacity and are in safe condition
- Ensuring that the lifting bolts and other rigging materials and equipment are installed correctly
- Guaranteeing the safety of the rigging crew and other personnel

3.3.0 Rigging Precautions

The overriding safety concern in rigging is to first determine the weight of the load and the load radius, and then to make sure that all equipment being used in the lift is in good condition and is rated for that load. When the assessment of the weight load is difficult, safe-load indicators or weighing devices should be attached to the rigging equipment. It is equally important to rig the load so that it is stable and the center of gravity is below the hook.

The personal safety of riggers and hoisting operators depends on common sense. Always observe the following safety practices:

- Always read the manufacturer's literature for all equipment with which you work. This literature provides information on required start-up checks and periodic inspections, as well as inspection guidelines. Also, this literature provides information on configurations and capacities in addition to many safety precautions and restrictions of use.
- Determine the weight of loads, including rigging and hardware, before rigging. Site management must provide this information if it is not known.
- Know the safe working load of the equipment and rigging, and never exceed the limit.
- Examine all equipment and rigging before use. Discard all defective components.
- Immediately report defective equipment or hazardous conditions to the supervisor. Someone in authority must issue orders to proceed after safe conditions have been ensured.
- Stop hoisting or rigging operations when weather conditions present hazards to property, workers, or bystanders, such as when winds exceed manufacturer's specifications; when the visibility of the rigger or hoist crew is impaired by darkness, dust, fog, rain, or snow; or when the temperature is cold enough that hoist or crane steel structures could fracture upon shock or impact. Never attempt to lift a load that is frozen to the ground.
- Recognize factors that can reduce equipment capacity. Safe working loads of all hoisting and rigging equipment are based on ideal conditions. These conditions are seldom achieved under typical working conditions.
- Remember that safe working loads of hoisting equipment are applicable only to freely suspended loads and plumb hoist lines. Side loads and hoist lines that are not plumb can stress equipment beyond design limits, and structural failure can occur without warning.
- Ensure that the safe working load of equipment is not exceeded if it is exposed to wind. Avoid sudden snatching, swinging, and stopping of suspended loads. Rapid acceleration and deceleration greatly increase the stress on equipment and rigging.
- Follow all manufacturer's guidelines, and consult applicable standards to stabilize mobile cranes properly (*Figure 21*).

3.4.0 Load Control

Tag lines typically are synthetic rope lines used to control the swinging of the load in hoisting activities (*Figure 22*). Improper use of tag lines can turn the simplest hoisting operation into a dangerous situation. Tag lines should be used to control swinging of the load when a crane is traveling. Tag lines should also be used when the

49110-11_F21.EPS

Figure 21 Mobile crane with outriggers.

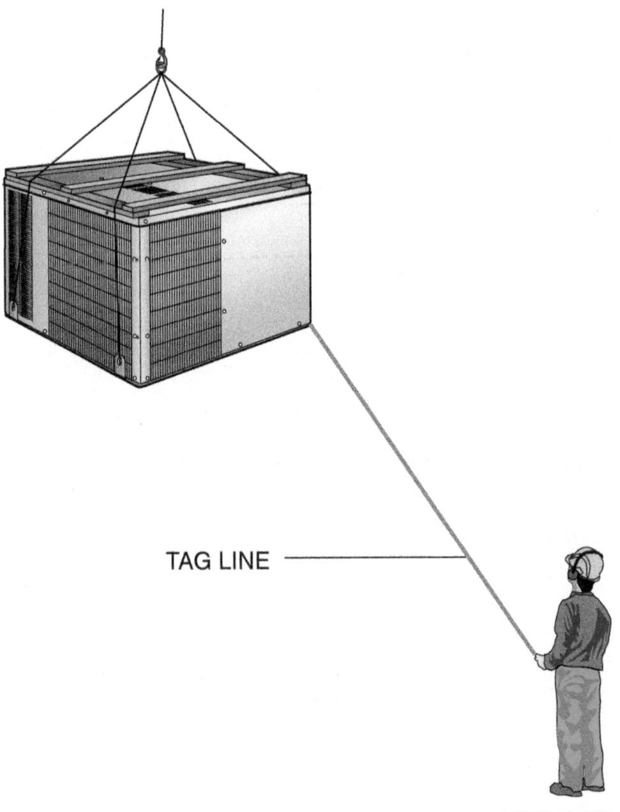

49110-11_F22.EPS

Figure 22 Use of a tag line to control swinging of a suspended load.

crane is rotated if rotation of the load is hazardous. A non-conductive tag line is required during operation of a mobile crane within the vicinity of power lines.

Tag lines made of synthetic ropes are light and strong for their size, and most are resistant to chemicals. When dry, they are poor conductors of electricity, although some synthetic ropes will readily absorb water and conduct electricity. Dry polypropylene line is preferred for use as a tag line.

The diameter of a tag line should be large enough so that it can be gripped well even when wearing gloves. Rope with a diameter of ½ inch is common, but ¾-inch and 1-inch diameter rope is sometimes used on heavy loads or where the tag line must be extremely long. Tag lines should be of sufficient length to allow control of the load from its original lift location until it is safely placed or control is taken over by co-workers. Special consideration should be given to situations where a long tag line would interfere with safe handling of loads, such as steel erection or catalyst pours.

Tag lines should be attached to loads at a location that gives the best mechanical advantage in controlling the load. Long loads should have tag lines attached as close to the ends as possible. The tag line should be located in a place that allows personnel to remove it easily after the load is placed. Knots used to attach tag lines should be tied properly to prevent slipping or accidental loosening, but they need to be easily untied after the load is placed. Some recommended knots are the clove hitch with overhand safety and bowline. Riggers use different knots for different purposes. *Figure 23* shows several knots commonly used in rigging. As much as possible, tag lines should be of one continuous length, free of knots and splices, and seized on both ends. If joining two tag lines together is necessary, it should be done by splicing the lines. Knots tied in the middle or with free ends of tag lines can create difficulties.

To properly handle a tag line, determine the mechanical advantage intended by the tag line and stand away from the load in order to have a clear area. When possible, stand where you can see the crane being used for lifting and never get under the load. Keep yourself and the tag line in view of the crane operator. Stay alert. Do not become complacent during the lift. Be aware of the location of any excess rope, and do not allow it to become fouled or entangled on anything.

Large loads often require the use of more than one tag line. In such cases, tag line personnel must work as a team and coordinate their actions.

WARNING!

Make sure that the tag line is not wrapped around any part of your body.

3.5.0 Barricades

Barricades should always be used to isolate the area of an overhead lift. This will reduce the possibility of injuring personnel who may walk into the area. Always follow the individual site requirements for barricade erection. If in doubt as to the proper procedure, ask your supervisor for guidance before proceeding with any overhead lifting. It is important to remember that accessible areas within the swing radius of the rear of the crane's rotating structure must be barricaded in such a way as to prevent an employee or others from being struck by the crane.

3.6.0 Load-Handling Safety

The safe and effective control of the load involves the rigger's strict observance of load-handling safety requirements. This includes making sure that the swing path or load path is clear of personnel and obstructions (*Figure 24*). Keep the front and rear swing paths of the crane clear for the duration of the lift. Most people watch the load when it is in motion, which prevents them from seeing the back end of the crane coming around.

Make sure the landing zone is clear of personnel, with the exception of the tag line tenders. Also make sure that the necessary blocking and cribbing for the load are in place before you position the load for landing. The practice of lowering the load just above the landing zone and then placing the cribbing and blocking can be dangerous. No one should work under the load. The layout of the cribbing can be completed in the landing zone before you set the load. Blocking of the load may have to be done after the load is set. In this case, do not take the load stress off the sling until the blocking is set and secured. Do not attempt to position the load onto the cribbing by manhandling it.

After the rigging has been set and whenever loads are to be handled, follow these guidelines:

- Before lifting, make sure that all loads are securely slung and properly balanced to prevent shifting of any part.
- Use one or more tag lines to keep the load under control.

BOWLINE

CLOVE HITCH

RUNNING BOWLINE

SQUARE KNOT

SHEET BEND
(BRACKET HITCH)

TIMBER HITCH

FIGURE EIGHT

SHEEP SHANK

49110-11_F23.EPS

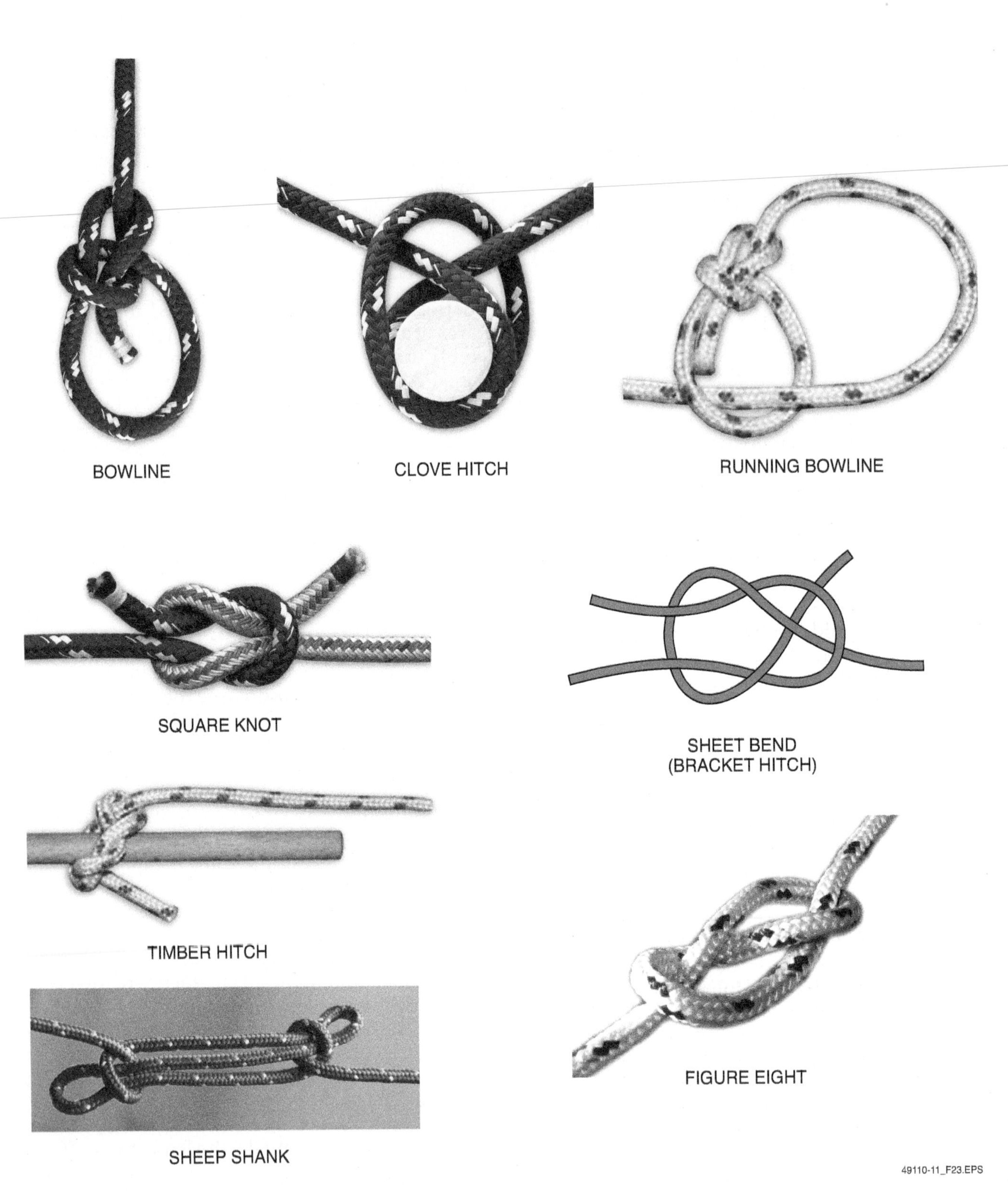

Figure 23 Common rigging knots.

- Safely land and properly block all loads before removing the slings.
- Only use lifting beams for the purpose for which they were designed. Their weight and working load abilities must be visible on the beams.
- Never wrap hoist ropes around the load. Use only slings or other adequate lifting devices.
- Do not twist multiple-part lines around each other.
- Bring the load line over the center of gravity of the load before starting the lift.

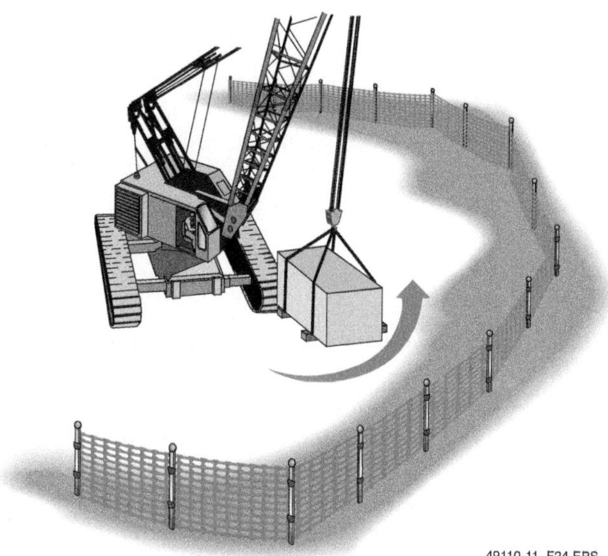

Figure 24 Personnel must be protected.

- Make sure the rope is properly seated on the drum and in the sheaves if there has been a slack rope condition.
- Load and secure any materials and equipment being hoisted to prevent movement which, in turn, could create a hazard.
- Keep hands and feet away from pinch points as the slack is taken up.
- Wear gloves when handling wire rope.
- See that all personnel are standing clear while loads are being lifted and lowered or when slings are being drawn from beneath the load. The hooks may catch under the load and suddenly fly free. It is prohibited to pull a choker out from under a load that has been set on the choker.
- Never ride on a load that is being lifted.
- Never allow the load to be lifted above other personnel.
- Never work under a suspended load.
- Never leave a load suspended in the air when the hoisting equipment is unattended.
- Never make temporary repairs to a sling.
- Never lift loads with one or two legs of a multi-leg sling until the unused slings are secured.
- Ensure that all slings are made from the same material when using two or more slings on a load.
- Remove or secure all loose pieces from a load before it is moved.
- Lower loads onto adequate blocking to prevent damage to the slings.

3.6.1 Blocking

Blocks should be made of hardwood that is dry and free of defects, oil, and grease. When used with jacks, they should provide a level foundation for the base of the jack. When a jack is being used to lift a metal object, a block should be placed between the object and the head of the jack to prevent slippage. When chocking is used to support a circular object, the general rule for safety is that 1 inch of chocking is required for each foot of diameter of the object.

3.6.2 Cribbing

Cribbing is formed by stacking timbers in alternate tiers (*Figure 25*). It is used to support a heavy weight or to raise an object to a height at which simple blocking would be unstable.

The most important factor to consider when cribbing is the foundation. The boom pieces must rest firmly and evenly on the ground. Dig away any high spots rather than filling low areas with loose material.

Cribbing can be used to jack a load in successive stages. The jacks are placed on blocking on the ground. The load is then lifted to the maximum height of the jacks, and one or more layers of cribbing are placed directly under the load. The cribbing is built up until there is enough space between the upper layer of cribbing and the load for an additional layer. At this point, the jacks are lowered until the load rests firmly on the cribbing. The blocking under the jacks is then built up so that the jacks touch the load in their lowered position. The process is then continued as before until the load is jacked to the desired height.

The size of the wood used in both blocking and cribbing usually starts at 4" × 4' but may be as large as 14" × 14', depending on the application. Hardwood is generally used. The following rules

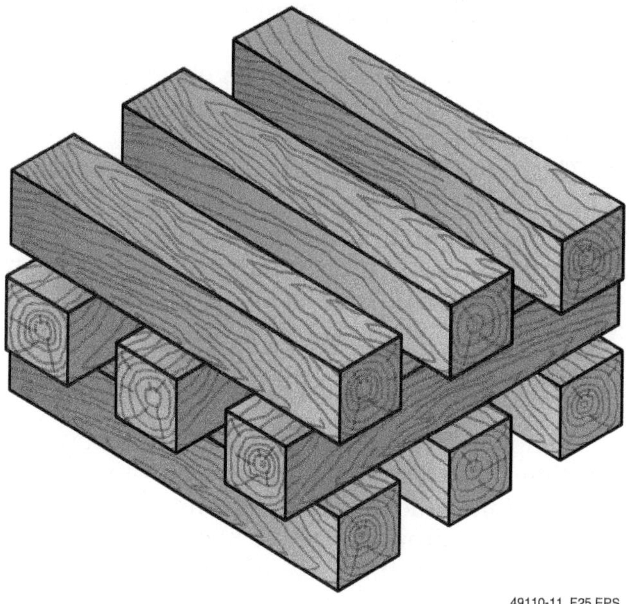

Figure 25 Cribbing.

On Site

Spreader Beams

Spreader beams are used to distribute the weight, reduce sling tension, and balance the load when lifting large, unwieldy objects. The beams help prevent the load from tipping or sliding. Adjustable spreader beams like the one shown are popular because they allow one beam to be used for a variety of lifting situations. Before adjustable beams, riggers needed to keep a variety of spreader beams on hand.

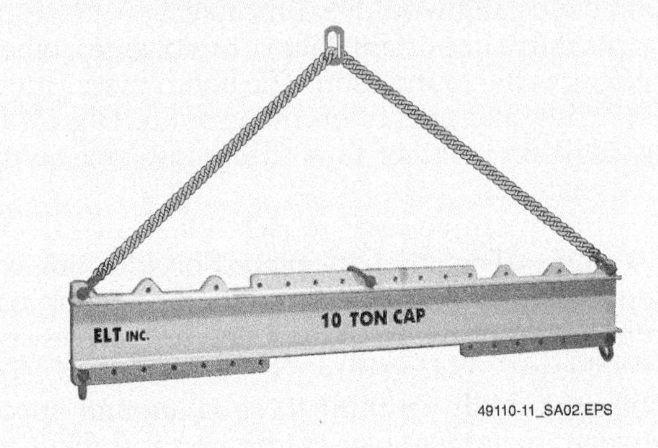

49110-11_SA02.EPS

concerning the safe working load (SWL) of wooden beams are necessary information every rigger should know:

- The greater the depth of a timber, the greater the SWL of the beam will be. Since SWL varies according to the square of a beam's depth, doubling the depth of a beam increases SWL by four.
- Two similar beams placed together are twice as strong as a single beam. To ensure the strength, secure the ends with nails or dowels.
- The more uniformly distributed a load, the greater the load that can be carried by a beam of a given size, material, and length. A concentrated load applied at the center of a beam reduces the SWL of that beam by one-half.

4.0.0 Working Around Power Lines

A competent signal person must be stationed at all times to warn the operator when any part of the machine or load is approaching the minimum safe distance from the power line. The signal person must be in full view of the operator at all times. *Table 1* and *Figure 26* show the minimum safe distance from power lines.

NOTE

Qualified employees are authorized to operate closer than the distances shown in *Table 1*, which are for non-qualified personnel.

WARNING!

The most frequent cause of death of riggers and material handlers is electrocution caused by contact of the crane's boom, load lines, or load with electric power lines. To prevent personal injury or death, stay clear of electric power lines. Even though the boom guards, insulating links, or proximity warning devices may be used or required, these devices do not alter the precautions given in this section.

The preferred working condition is to have the owner of the power lines de-energize and provide grounding of the lines that are visible to the crane operator. When that is not possible, observe the following procedures and precautions if any part of your boom can reach the power line:

Table 1 High-Voltage Power Line Clearances for Non-Qualified Electrical Employees

CRANE IN OPERATION[1]	
POWER LINE (kV)	BOOM OR MAST MINIMUM CLEARANCES (feet)
0 to 50	10
50 to 200	15
200 to 350	20
350 to 500	25
500 to 750	35
750 to 1,000	45
CRANE IN TRANSIT (with no load and the boom or mast lowered)[2]	
POWER LINE (kV)	BOOM OR MAST MINIMUM CLEARANCES (feet)
0 to 0.75	4
0.75 to 50	6
50 to 345	10
345 to 7500	16
750 to 1000	20

Note 1: For voltages over 50kV, clearance increases 5 feet for every 150 kV.
Note 2: Environmental conditions such as fog, smoke, or precipitation may require increased clearances.

49110-11_T01.EPS

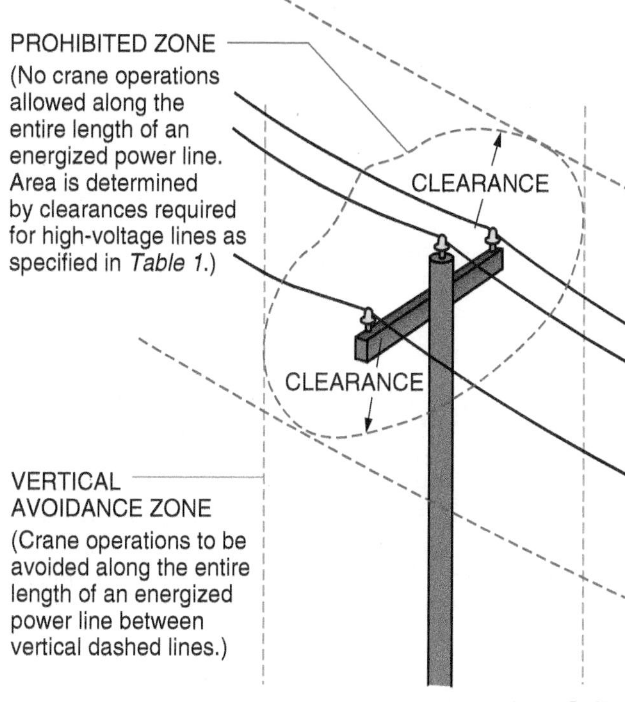

Figure 26 Prohibited and avoidance zones.

- Make sure a power line awareness permit (see *Appendix*) or equivalent has been prepared.
- Erect non-conductive barricades to restrict access to the work area.
- Use tag lines of a non-conductive type if necessary for load control.
- Make sure a qualified signal person(s), whose sole responsibility is to verify that the proper clearances are established and maintained, is in constant contact with the crane operator.
- The person(s) responsible for the operation must alert and warn the crane operator and all persons working around or near the crane about the hazards of electrocution or serious injury, and instruct them on how to avoid these hazards.
- All non-essential personnel must be removed from the crane work area.
- No one is permitted to touch the crane or load unless the signal person indicates it is safe to do so.

If a crane or load comes in contact with or becomes entangled in power lines, assume that the power lines are energized unless the lines are visibly grounded. Any other assumption could be fatal.

The following guidelines should be followed if the crane comes in contact with an electrical power source:

- The operator should stay in the cab of the crane unless a fire occurs.
- Do not allow anyone to touch the crane or the load.
- If possible, the operator should reverse the movement of the crane to break contact with the energized power line.
- If the operator cannot stay in the cab due to fire or arcing, the operator should jump clear of the crane, landing with both feet together on the ground. Once out of the crane, the operator must take very short steps or hops with feet together until well clear of the crane.
- Call the local power authority or owner of the power line.
- Have the lines verified as secure and properly grounded within the operator's view before allowing anyone to approach the crane or the load.

5.0.0 Site Safety

It takes a combined effort by everyone involved in crane operations to make sure no one is injured or killed. Site hazards and restrictions as well as crane manufacturer's requirements must be observed for safe crane operation.

WARNING!

Transmitters such as radio towers can also represent a hazard. These transmitters generate high-power radio frequency (RF) energy that can induce a hazardous voltage into the boom of a nearby crane. OSHA requires that the transmitter be de-energized in such situations, or that appropriate grounding methods be used to protect workers. Always consult your supervisor or site safety officer in such situations.

Induced RF voltages from transmitters are not an electrocution hazard like power line voltages. However, they can result in sparks. The sparks can cause fires or explosions of combustible or flammable materials. They can also result in painful burns and shocks to persons that come in close contact with the crane or the load being lifted by the crane.

5.1.0 Site Hazards and Restrictions

There are many site hazards and restrictions related to crane operations. These hazards include the following:

- Underground utilities such as gas, oil, electrical, and telephone lines; sewage and drainage piping; and underground tanks

- Electrical lines or high-frequency transmitters
- Structures such as buildings, excavations, bridges, and abutments

WARNING!

Power lines and environmental issues such as weather are common causes of injury and death during crane operations.

The operator and riggers must inspect the work area and identify hazards or restrictions that may affect the safe operation of the crane. This includes the following actions:

- Ensuring the ground can support the crane and the load
- Checking that there is a safe path to move the crane around on site
- Making sure that the crane can rotate in the required quadrants for the planned lift

The operator must follow the manufacturer's recommendations and any locally established restrictions placed on crane operations, such as traffic considerations or time restrictions for noise abatement.

6.0.0 Emergency Response

Operators and riggers must react quickly and correctly to any crane malfunction or emergency situation. They must learn the proper responses to emergency situations. The first priority is to prevent injury and loss of life. The second priority is to prevent damage to equipment and surrounding structures.

6.1.0 Fire

Judgment is crucial in determining the correct response to fire. The first response is to cease crane operation and, if time permits, lower the load and secure the crane. In all cases of fire, evacuate the area even if the load cannot be lowered or the crane secured. After emergency services have been notified, a qualified individual may judge if the fire can be fought with a fire extinguisher. A fire extinguisher can be used to fight a small fire in its beginning stage, but a fire can get out of control very quickly. Do not be overconfident, and keep in mind that priority number one is preventing loss of life or injury to anyone. Even trained firefighters using the best equipment can be overwhelmed and injured by fires.

6.2.0 Malfunctions During Lifting Operations

Mechanical failures during a lift can be very serious. If a failure causes the radius to increase unexpectedly, the crane can tip or the structure could collapse. Loads can also be dropped. A sudden loss of load on the crane can cause a whiplash effect that can tip the crane or cause the boom to fail.

The chance of these types of failures occurring in modern cranes is greatly reduced because of safety backups. However, failures do happen, so stay alert at all times.

If a mechanical problem occurs, the operator should lower the load immediately. Next, the operator should secure the crane, tag the controls out of service, and report the problem to a supervisor. The crane should not be operated until it is repaired by a qualified technician.

6.3.0 Hazardous Weather

Mobile crane operations generally take place outdoors. Under certain conditions, such as extreme hot or cold weather or in high winds, work can become uncomfortable and possibly dangerous. For example, snow and rain can have a dramatic effect on the weight of the load and on ground compaction. During the winter, the tires, outriggers, and crawlers can freeze to the ground. This may lead the operator or rigger to the false conclusion that the crane is on stable ground. In fact, as weight is added during the lifting operation, it may cause an outrigger float, tire, or crawler to sink into the ground below the frozen surface. Heavy rain can cause the ground under the crane to become unstable, resulting in instability of the crane.

High winds and lightning may cause severe problems on the job site (*Figure 27*). They are major weather hazards and must be taken seriously. Crane operators and riggers must be prepared to handle extreme weather in order to avoid accidents, injuries, and damages. It is very rare for high winds or lightning to arrive without some warning. This gives operators and riggers time to react appropriately.

6.3.1 High Winds

High winds typically start out as less dramatic gusts. Operators and riggers must be aware of changing weather conditions, such as worsening winds, to determine when the weather becomes hazardous. With high winds, the operator must secure crane operations as soon as it is practical. This involves placing the boom in the lowest possible position and securing the crane. Once this

On Site

Wind Effect on a Load

High winds can cause the load to swing, which can cause the crane to tip or the boom to collapse.

Forward stability is affected by wind on the rear of the boom. Wind applies a force to the boom and load that adds to the crane's overturning moment. This action has the same effect as adding load to the hook.

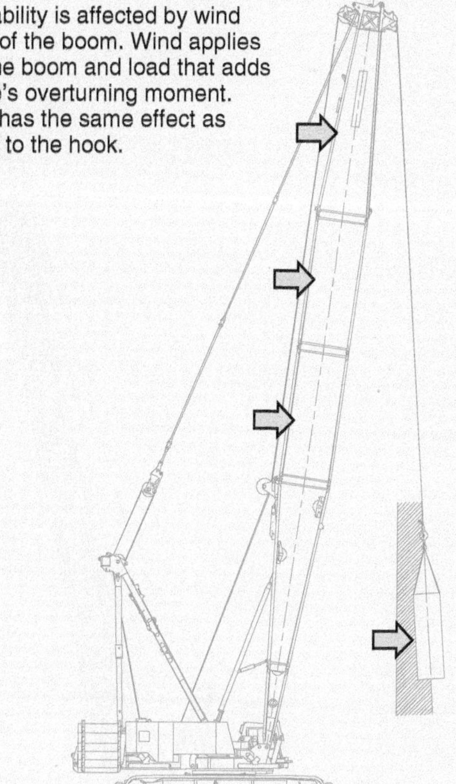

The wind's effect on the rear of the load increases the load radius. This condition can result in an overload hazard, possibly causing the crane to tip or the bottom to collapse.

To avoid this hazard, reduce operating speeds and load.

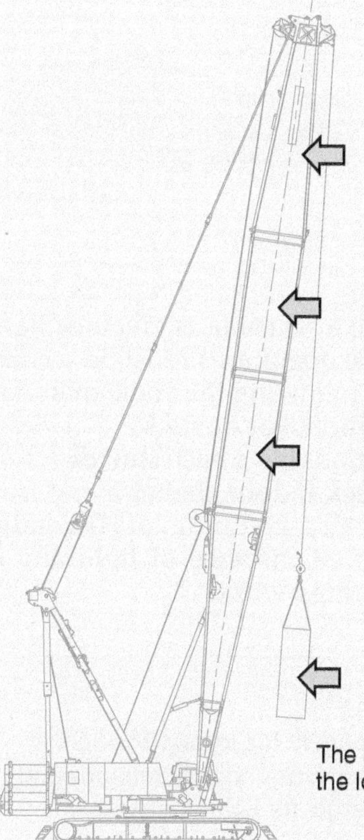

Backward stability is affected by wind on the front of the boom. This condition is especially dangerous when the boom is at or near the maximum angle when operating without load.

Wind forces on the front of the boom reduce the normal forward tipping effect of the boom. The crane can tip or the boom can collapse if this condition is not avoided.

The boom can buckle and collapse if the load contacts the boom.

Boom strength is affected the most when the wind acts on the side of the boom.

The wind's effect on the side of the load can cause the load to swing out past the boom hinge pin. This condition can result in excessive side load forces on the boom, possibly causing the crane to tip or the boom to collapse.

To avoid the hazard, reduce operating speeds and load.

49110-11_SA03.EPS

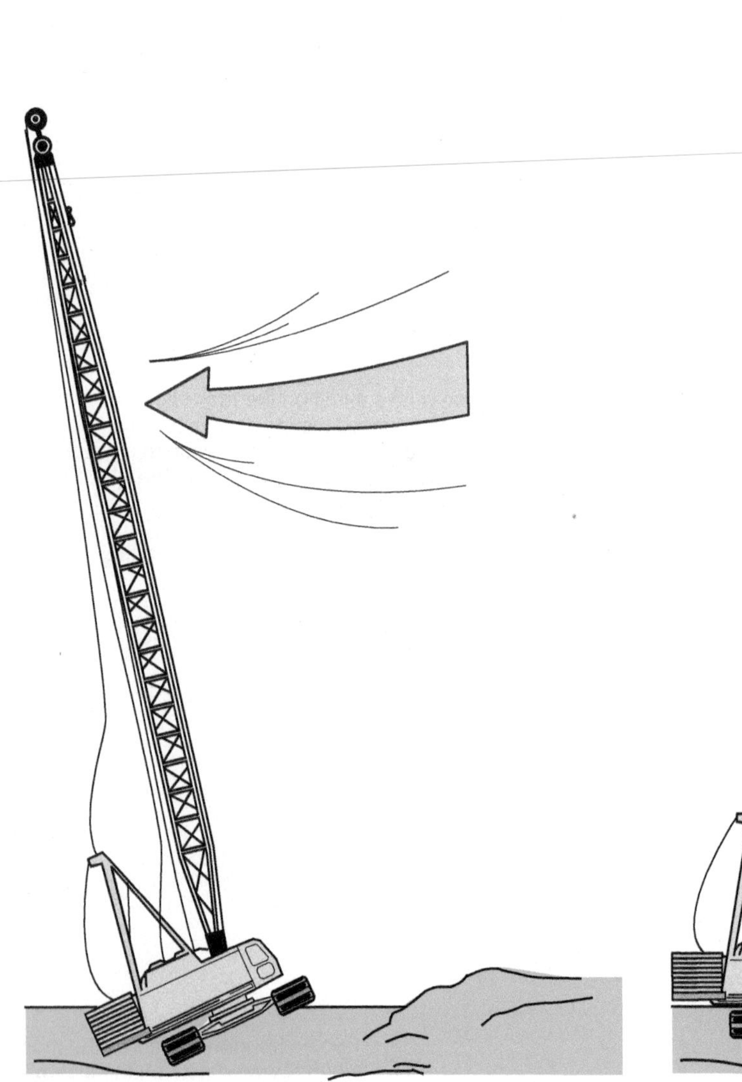

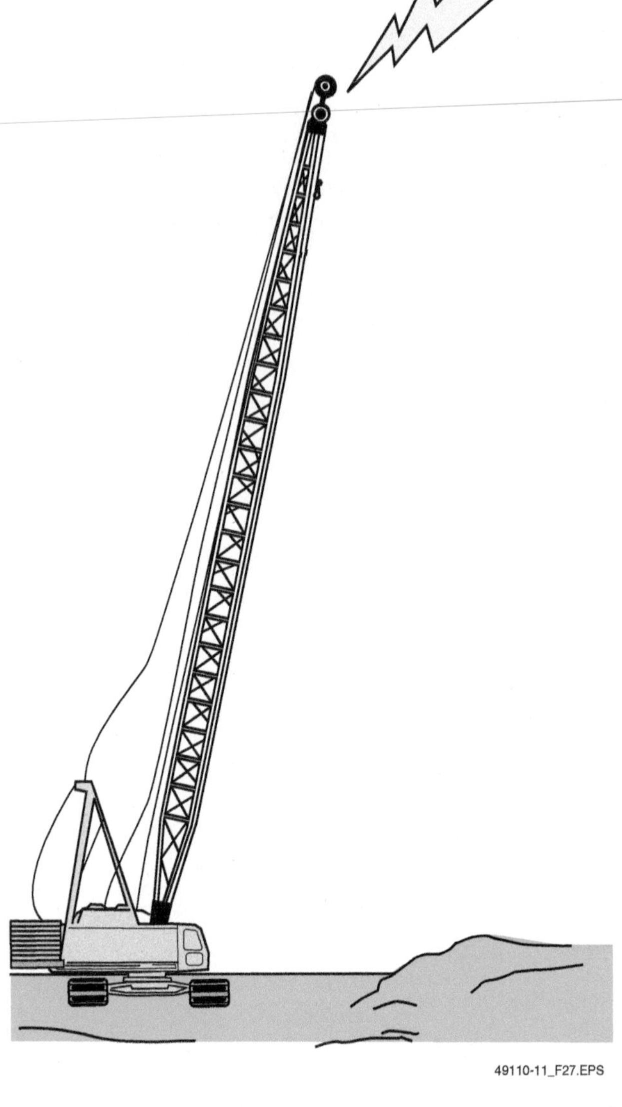

Figure 27 Wind and lightning hazards.

is done, all personnel should seek indoor shelter away from the crane.

6.3.2 Lightning

Because crane booms extend so high and are made of metal, they are easy targets for lightning. Operators and riggers must be constantly aware of this threat. Lightning can usually be detected when it is several miles away. As a general rule of thumb, when you hear thunder, the lightning associated with it is six to eight miles away. Be aware, however, that successive lightning strikes can touch down up to eight miles apart. That means once you hear thunder or see lightning, it is close enough to present a hazard.

In some high-risk areas, proximity sensors provide warnings when lightning strikes within a 20-mile radius. Once a warning is given or lightning is spotted, crane operations must be secured as soon as practical, following the crane manufacturer's recommendations for doing so.

Once crane operations have been shut down, all personnel should seek indoor shelter away from the crane. Always wait a minimum of 30 minutes from the last observed instance of lightning or thunder before resuming work.

WARNING!

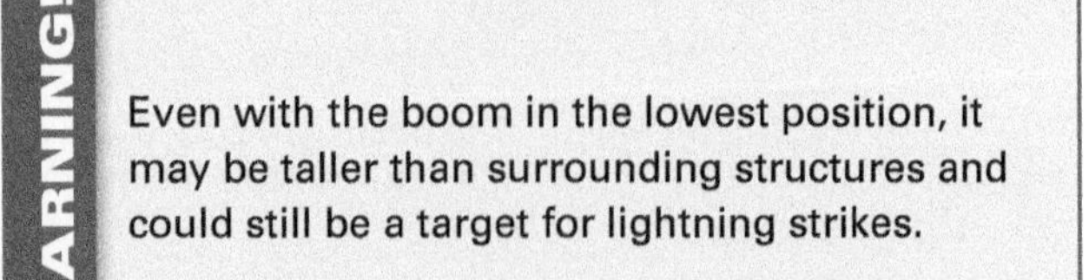

Even with the boom in the lowest position, it may be taller than surrounding structures and could still be a target for lightning strikes.

7.0.0 Using Cranes to Lift Personnel

Although using cranes to lift people was common in the past, OSHA regulations, as spelled out in *29 CFR 1926.550*, now discourage the practice. Using a crane to lift personnel is not specifically prohibited by OSHA, but the restrictions are such that it is only permitted in special situations where no other method is suitable. When it is allowed, certain controls must be in place, including the following:

- The rope design factor is doubled.
- No more than 50 percent of the crane's capacity, including rigging, may be used.
- Free-falling is prohibited.
- Anti-two-blocking devices are required on the crane boom.
- The platform must be specifically designed for lifting personnel.
- Before the personnel platform is used, it must be tested with appropriate weight, and then inspected.
- Every intended use must undergo a trial run with weights rather than people.

Figures 28 and *29* show equipment used for personnel lifts.

49110-11_F29.EPS

Figure 29 Suspended personnel platform.

7.1.0 Personnel Platform Loading

The personnel platform must not be loaded in excess of its rated load capacity. When a personnel platform does not have a rated load capacity, it must not be loaded in excess of its maximum intended load.

The number of employees, along with material, occupying the personnel platform must not exceed the limit established for the platform and the rated load capacity or the maximum intended load.

Personnel platforms must be used only for employees, their tools, and the materials necessary to do their work. They must not be used to hoist only materials or tools when not hoisting personnel.

Materials and tools for use during a personnel lift must be secured. These items must be evenly distributed within the confines of the platform while the platform is suspended.

49110-11_F28.EPS

Figure 28 Examples of a personnel platform.

7.2.0 Personnel Platform Rigging

When a wire-rope bridle sling is used to connect the personnel platform to the load line, each bridle leg must be connected to a master link or shackle in such a manner as to ensure that the load is evenly divided among the bridle legs (*Figures 30* and *Figures 31*).

Hooks on headache ball assemblies, lower load blocks, or other attachment assemblies shall be

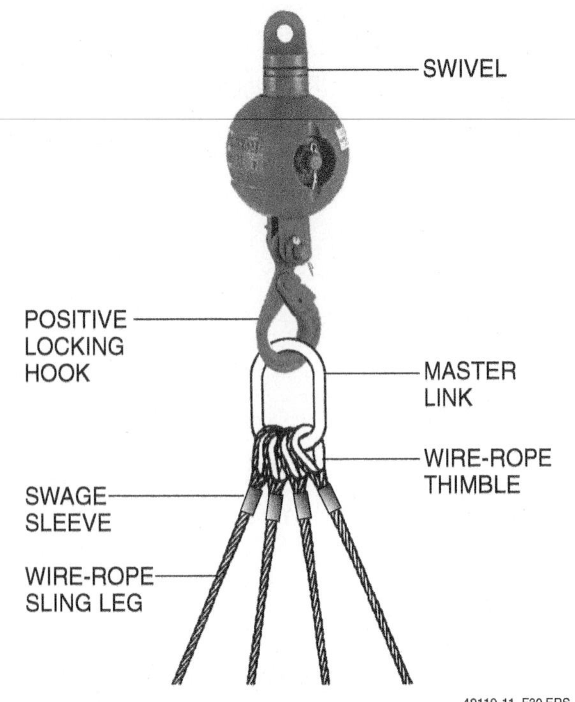

Figure 30 Bridle sling using a master link.

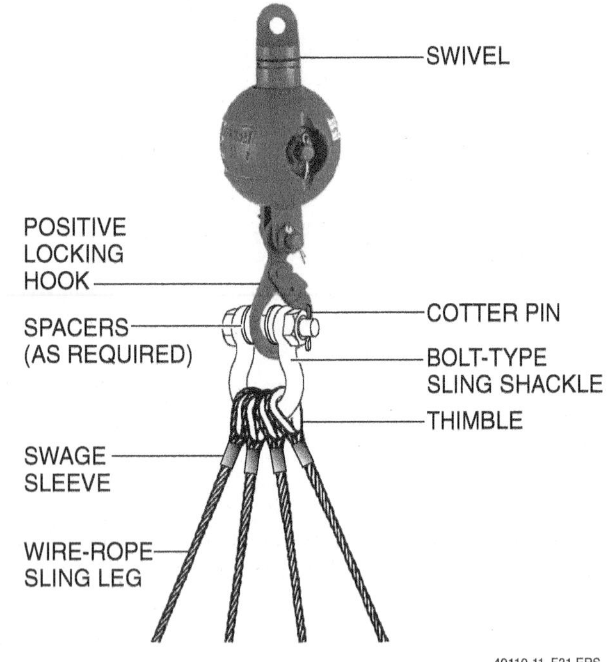

Figure 31 Bridle sling using a shackle.

of a type that can be closed and locked. Alternatively, an alloy anchor shackle with a bolt, nut, and retaining pin may be used.

Wire rope, shackles, rings, master links, and other rigging hardware must be capable of supporting, without failure, at least five times the maximum intended load applied or transmitted to that component. Where rotation-resistant rope is used, the slings must be capable of supporting without failure at least ten times the maximum intended load. All eyes in wire rope slings must be fabricated with thimbles.

Bridles and associated rigging for attaching the personnel platform to the hoist line must be used only for the platform and the employees, their tools, and the materials necessary to do their work. They must not be used for any other purpose when not hoisting personnel.

8.0.0 Block and Tackle

The block and tackle is the most basic lifting device. It is used to lift or pull light loads. A block consists of one or more sheaves or pulleys fitted into a wood or metal frame with a hook attached to the top. The tackle is the line that runs through the block. It is used for lifting and pulling. Some block and tackle rigs have a brake that holds the load once it is lifted; others do not. The types that do not have a brake require continuous pull on the hauling line, or the hauling line must be tied off to hold the load. There are two types of block and tackle rigs: simple and compound.

8.1.0 Simple Block and Tackle

A simple block and tackle consists of one sheave and a single line. It is used to lift or pull very light loads. The line hook is attached to the load, and the line is pulled by hand to lift the load. The load capacity of this type of block and tackle is equal to the capacity of the load line. The block must be attached to a building structure or other support by a method that provides adequate load capacity to support the load and the tackle (*Figure 32*).

8.2.0 Compound Block and Tackle

A compound block and tackle uses more than one block. It has an upper, fixed block that is attached to the building structure or other support and a lower, movable block that is attached to the load. Each block may have one or more sheaves. The more sheaves the blocks have, the more parts of line the block and tackle has, and the higher the lifting capacity. For example, two-part blocks

Figure 32 Simple block and tackle.

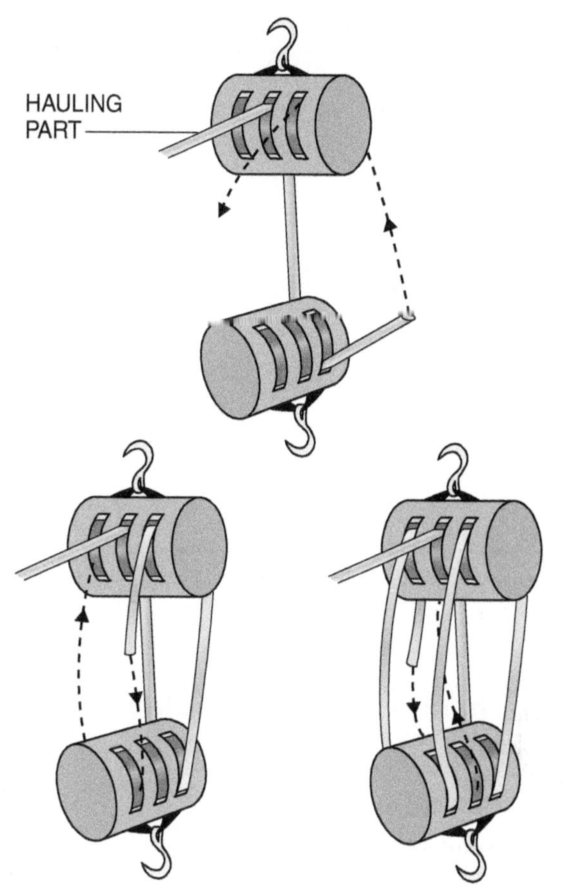

Figure 33 Reeving a three-part block and tackle.

(two sheaves per block) provide a mechanical advantage of 4 to 1. Using three-part blocks raises the mechanical advantage to 6 to 1. The compound block and tackle is capable of handling heavier loads than the simple block and tackle and requires less effort to raise the load.

8.3.0 Reeving a Block and Tackle

Reeving is a term commonly used by riggers. It simply means passing a line through a block. The two blocks should be placed at right angles to each other (one on its face and the other on its edge), with the hooks facing in opposite directions. The lifting line should first pass through the center sheave(s) of the upper block. This will keep the block balanced during the lift and prevent damage to the lifting line. *Figure 33* shows the reeving pattern for a three-part block and tackle. At the end, secure the dead end of the line to the becket on the upper block using a becket hitch. An eye splice would be used in place of the becket hitch for a permanent setup.

9.0.0 Guidelines for Unloading and Yarding Materials

Reinforcing bars and other structural steel components are ordered from the fabricator. Delivery of these materials is usually scheduled to coincide with the construction schedule. Structural and reinforcing steel usually arrives at the job site on flatbed trucks or tractor trailers. If a spur track is available at the job site, shipments may be received by flat or gondola railroad cars. Where delivery of steel is scheduled to meet daily placement requirements, truckloads are delivered to the points of placement. Workers should be aware of safety factors that are necessary to achieve safe, efficient handling, storage, and hoisting of structural materials.

Charts are available for most types of materials so that riggers can calculate the weight of a load. *Table 2* is an example of such a chart. By knowing the number of reinforcing bars and their

Table 2 ASTM Reinforcing Bar Properties

Bar Size		Nominal Characteristics*					
		Diameter		Cross-Sectional Area		Weight	
Metric	[in-lb]	mm	[in]	mm	[in]	kg/m	[lbs/ft]
#10	[#3]	9.5	[0.375]	71	[0.11]	0.560	[0.376]
#13	[#4]	12.7	[0.500]	129	[0.20]	0.944	[0.668]
#16	[#5]	15.9	[0.625]	199	[0.31]	1.552	[1.043]
#19	[#6]	19.1	[0.750]	284	[0.44]	2.235	[1.502]
#22	[#7]	22.2	[0.875]	387	[0.60]	3.042	[2.044]
#25	[#8]	25.4	[1.000]	510	[0.79]	3.973	[2.670]
#29	[#9]	28.7	[1.128]	645	[1.00]	5.060	[3.400]
#32	[#10]	32.3	[1.270]	819	[1.27]	6.404	[4.303]
#36	[#11]	35.8	[1.410]	1006	[1.56]	7.907	[5.313]
#43	[#14]	43.0	[1.693]	1452	[2.25]	11.38	[7.65]
#57	[#18]	57.3	[2.257]	2581	[4.00]	20.24	[13.60]

*The equivalent nominal characteristics of inch-pound bars are the values enclosed within the brackets.

49110-11_T02.EPS

lengths, the rigger can easily calculate the weight of the load.

9.1.0 Unloading

Trucks should be promptly unloaded to prevent job-site clutter and potential safety hazards. Structural materials that are stored on the ground must be placed on timbers or other suitable blocking to keep them free from mud and to allow safe and easy rehandling.

Bars and other structural stock are normally stored by size and shape, and by length within each size. Tags are kept at the same end of each piece for easy identification. When a bundle is opened and part of the stock removed, the unit with the tag should remain with the bundle, and a new number indicating the remaining inventory should be inserted. Machine oil or grease that collects on the stock must be removed using a solvent. All mud must be washed off before placement.

A spreader bar may be necessary when unloading bundles of bars or similar stock. They are usually made from a fabricated truss, a piece of heavy-duty pipe, or an I-beam.

Tag lines are required when bundles are being hoisted. Tag lines guide the load as it is being hoisted. Safe hoisting requires a skilled crane operator and an authorized signal person. Operators take their directions from the signal person using hand signals, radios, or telephone headsets.

9.2.0 Using Slings

Never pick up a bundle by the wire wrappings that are used to tie the bundle together. Slings must be used to lift bundles. Choose a sling appropriate to the material being lifted. Check the weight of each bundle on the bill of lading, and ensure that the sizes of the wire rope slings, hooks, and shackles are safe. Chokers and slings of sufficient strength must be selected to lift the load.

The load on each sling depends on the number of slings or chokers, the angle of the choker, and the total load. Although the total weight lifted can be divided between the supporting slings, the force of the load is down. The greater the angle of the sling, the greater the tension is on the sling. If the hoisting line and the slings are the same material and size, the tension on each sling must be less than or equal to the tension in the hoist line. This means that the strength of the hoist line determines the maximum load-lifting capability of the combination.

If two slings are used to lift bundles of structural iron, always thread the loops in the same direction. If one sling is looped from the opposite side, the bundle will twist when being lifted, making it less safe and causing difficulty when the bars are removed from the bundle. If possible, double wrap the slings to help prevent slipping of the slings and items in the bundle.

Summary

One of the important tasks a rigger performs is guiding the crane operator during a lift. In some instances, radio communications will be used. In others, it will be necessary to use the *ASME B30.5 Consensus Standard* hand signals. Both the rigger and operator must know these signals by heart.

The selecting and setting up of hoisting equipment, hooking cables to the load to be lifted or moved, and directing the load into position are all part of the process called rigging. Performing this process safely and efficiently requires selecting the proper equipment for each hoisting job, understanding safety hazards, and knowing how to prevent accidents. Riggers must have the greatest respect for the hardware and tools of their trade because their lives may depend on them working correctly.

The size of a crane and the huge loads it handles create a large potential for danger to anyone in the vicinity of the crane. The crane operator and rigging workers must be vigilant to ensure that the crane avoids power lines and that site personnel are not endangered by the movement of the crane or the load. Because of tipping hazards, it is important to make sure a crane is on level ground with outriggers extended, if applicable, before lifting a load. Although using cranes to lift personnel was once a common practice, it is discouraged by OSHA, and can only be done in special situations, and under special conditions.

Review Questions

1. The main advantage of a hardwired communication system is _____.
   a. low cost
   b. portability
   c. less radio interference
   d. ease of use

2. The action signaled by tapping your fist on your head, then using regular hand signals is _____.
   a. use main hoist
   b. lower
   c. use whip line
   d. raise boom and lower load

3. Which of the following is a mode of nonverbal communication?
   a. Buzzer
   b. Portable radio
   c. Bullhorn
   d. Hardwired radio

4. When the signal person gives the Dog Everything signal, the operator is expected to _____.
   a. turn the machine
   b. travel the machine
   c. retract the boom sections
   d. make no machine movement

5. If it is apparent that the crane operator has misunderstood a hand signal, the signal person should _____.
   a. climb up on the cab and knock on the door
   b. find a supervisor
   c. signal the operator to stop
   d. get away from the crane as quickly as possible

6. A one-handed signal that is given with the fist in front of the body and the thumb pointing toward the signaler's body means _____.
   a. extend the boom
   b. move the crane toward me
   c. swing the load toward me
   d. I'm taking a break

7. When calculating the weight of the load, the rigging and related hardware must also be considered.
   a. True
   b. False

8. To allow controlled rotation of the load for position, use a _____.
   a. tag line
   b. equalizer beam
   c. hoist line
   d. safe-load indicator

9. What diameter of rope is commonly used for tag lines?
   a. ½"
   b. ⅝"
   c. ¾"
   d. 1"

10. When using two or more slings on a load, the rigger must ensure that _____.
    a. the sling angle is greater than 30 degrees relative to the horizontal
    b. all slings are made from the same material
    c. only chain slings are used
    d. only wire rope slings are used

11. The minimum safe crane-operating distance from power lines of up to 50,000 volts (or 50kV) is _____.
    a. 10 feet
    b. 15 feet
    c. 20 feet
    d. 25 feet

12. If a crane boom has come into contact with a power line, the operator should _____.
    a. immediately jump off the crane
    b. stop and turn off the crane
    c. try to back the crane away from the power line
    d. find out if the line is energized

13. The minimum amount of time required before resuming work after seeing lightning or hearing thunder is _____.
    a. 15 minutes
    b. 30 minutes
    c. 45 minutes
    d. 1 hour

14. When cranes are used to lift people, what percentage of a crane's lift capacity is changed as a safety factor?
    a. Increased 10 percent
    b. Increased 20 percent
    c. Decreased 20 percent
    d. Decreased 50 percent

15. Personnel platform rigging using rotation-resistant rope slings must be capable of supporting how much of the maximum load applied?
    a. At least twice
    b. At least 3 times
    c. At least 5 times
    d. At least 10 times

16. When a wire rope bridle sling is used to lift a personnel platform, each leg must be connected to a _____.
    a. positive locking hook
    b. properly rated chain
    c. clevis bolt
    d. master link or bolt-type shackle

17. Bridles and rigging for attaching a personnel platform to the hoist line _____.
    a. may be used for other lifting applications when not in use
    b. may be used only for other lighter lifting applications
    c. must not be used for any other purpose
    d. may be used to lift other loads of equal weight, if necessary

18. Regarding block and tackle, the term *tackle* refers to the _____.
    a. supporting hooks
    b. lifting hooks
    c. lifting/hauling line
    d. braking system

19. A set of blocks containing 3 sheaves has a mechanical advantage of _____.
    a. 3 to 1
    b. 4 to 1
    c. 2 to 1
    d. 6 to 1

20. When using slings for lifting, the greater the angle of the sling, the greater the tension on the sling.
    a. True
    b. False

Trade Terms Quiz

Fill in the blank with the correct term that you learned from your study of this module.

1. Timbers stacked in alternate tiers to support heavy loads are known as ___________.
2. The point at which an object is perpendicular to and balanced in relation to the earth's gravitational field is its ___________.
3. Devices that provide warnings to prevent the lower load block or hook from contacting the upper load block, boom point, or boom point machinery are ___________.
4. Pieces of hardwood used to support and brace equipment are known as ___________.

Trade Terms

Anti-two-blocking devices
Blocking
Center of gravity
Cribbing

Cornerstone of Craftsmanship

Larry Harvey

Supervisor, Substation Construction BGE
(Formerly Baltimore Gas and Electric),
Baltimore, MD

For Larry Harvey, there's tremendous satisfaction in walking onto a bare plot of land and, after all the planning and construction, walking away with an energized substation. The work is never boring.

How did you get started in the construction industry?
I was serving a tool and die maker apprenticeship and applied at the local utility company for a position in their machine shop. They hired me as a helper in the construction department but said once I began working, I could post for a position in the machine shop when one became available. Needless to say I found the construction work interesting and never left.

Who inspired you to enter the industry?
Oh, that was my mother. She thought having a skilled trade would always come in handy. And I needed a job!

What do you enjoy most about your job?
The whole process interests me. There's tremendous satisfaction in being part of the construction of a substation from the ground up. You walk onto a bare plot of land and, after all the planning and construction, walk away with an energized substation. It's never boring. You get involved with all the trades and skill sets. The variety of work involved in construction keeps me stimulated.

Do you think training and education are important in construction?
Definitely! Training and education go hand in hand with the actual work. It's about more than what you learn in the field or on the job. Formal training gives an employee a basic knowledge of not only how to do things, but it provides them with some theory behind why we do things a certain way, and that makes a difference in how you approach a job. With formal training, you have a chance to learn the right way or the safe way, not just the way the person beside you does it. The best workers usually have both experience and training.

How important are NCCER credentials to your career?
I haven't been exposed to NCCER for any length of time, but I can see that having the NCCER credentials can play an important role in a person's future.

How has training/construction impacted your life and your career?
Early in my career, I began taking construction math courses and attending code school at night. Having the extra knowledge helped me move from helper to mechanic. Later, I took construction management courses at a community college, which helped me progress to supervisor.

Would you suggest construction as a career to others?
Yes. Having a skill is an asset in pursuing a career. The work itself is very rewarding. You get a feeling of accomplishment from seeing the end product.

How do you define craftsmanship?
A craftsman is someone who practices a craft with great skill and precision. True craftsmanship comes from knowing your tools and knowing your materials and then applying that knowledge to create a finished product.

Trade Terms Introduced in This Module

Anti-two-blocking devices: Devices that provide warnings and prevent two-blocking from occurring. Two-blocking occurs when the lower load block or hook comes into contact with the upper load block, boom point, or boom point machinery. The likely result is failure of the rope or release of the load or hook block.

Blocking: Pieces of hardwood used to support and brace equipment.

Center of gravity: The point at which an object is perpendicular to and balanced in relation to the earth's gravitational field.

Cribbing: Timbers stacked in alternate tiers. Used to support heavy loads.

Power Line Awareness Permit

POWER LINE AWARENESS PERMIT

Today's Date __________ **Job Number** ____________________________________

Contractor Name			
Job Address			
Telephone Number		**Fax Number**	

Emergency Contact Number	

Survey

Before beginning any project, you must first survey your work area to find power lines at the job site. (See job-site sketch on reverse side.)

Identify

After finding all of the power lines at your site, identify the activities you'll be doing that may put you or your workers at risk. Mark one or more of the following:

- ☐ Cranes (mobile or truck mounted)
- ☐ Drilling rigs
- ☐ Backhoes/excavators
- ☐ Long-handled tools
- ☐ Other tools/high-reaching equipment
- ☐ Concrete pumper
- ☐ Aerial lifts
- ☐ Dump trucks
- ☐ Ladders
- ☐ Material handling & storage
- ☐ Scaffolding
- ☐ Other__________________

Eliminate or Control

After identifying the power line and high-risk activities on the job site, determine how to eliminate or control the risk of electrocution (a successful determination is often reached only after consultation with the utility).

Mark one or more of the following:

- ☐ Move the activity
- ☐ Change the activity
- ☐ Have the utility de-energize the power line
- ☐ Have the utility move the power line
- ☐ Use barrier protection (insulated sleeves)
- ☐ Use an observer
- ☐ Use warning lines with flags
- ☐ Use non-conductive tools
- ☐ Use a protective technology:
  - ☐ Insulated link
  - ☐ Boom cage guard
  - ☐ Proximity device

Always maintain your minimum safe clearance distance from the power line, except when the utility has de-energized and visibly grounded the power line.

Voltages	Distance from Power Line
Less than 50 kV	10 Feet
More than 50 kV	10'+(0.4")(# of kV over 50 kV)

WARNING!
It is unlawful to operate any piece of equipment within 10' of energized lines

CONSTRUCTION SAFETY COUNCIL

49110-11_A01.EPS

Job-Site Sketch

(Draw in location of power lines and their proximity to construction site, including such things as proposed excavations, location of heavy equipment, scaffolding, material storage areas, etc.)

Completed by____________________ **Date**__________

Title____________________

Approved by____________________ **Date**__________

Title____________________

CONSTRUCTION SAFETY COUNCIL

49110-11_A02.EPS

Additional Resources

This module presents thorough resources for task training. The following resource material is suggested for further study.

Bob's Overhead Crane and Rigging Handbook, Pellow Engineering Services, Leawood, KS.

Refer to the Associated General Contractors web site for a list of training materials and video programs on crane safety. www.agc.org.

Occupational Safety and Health Standards for the Construction Industry, 29 CFR Part 1926. Washington, DC: OSHA Department of Labor, U.S. Government Printing Office.

Figure Credits

Bob Groner, Module opener

Jim Mitchem, Figure 1

Link-Belt Construction Equipment Co., SA01 and Figure 21

AEARO Company, Figure 2

Topaz Publications, Inc., Figures 20 and 23 (running bowline, sheet bend, sheep shank, and figure eight)

Alan W. Grogono, M.D., Figure 23 (bowline, clove hitch, and square knot)

Dave Root, Figure 23 (timber hitch)

Engineered Lifting Technologies Co., www.eltlift.com, SA02 and Figure 28

Manitowoc Crane Group, SA03

Lifting Technologies, Figure 29

The Crosby Group, Inc., Figures 30 and 31 (photos)

Wire Rope Technical Board, Figures 30 and 31 (illustrations)

Klein Tools, Inc., Figure 32

Construction Safety Council, Appendix

CONTREN® LEARNING SERIES — USER UPDATE

NCCER makes every effort to keep its textbooks up-to-date and free of technical errors. We appreciate your help in this process. If you find an error, a typographical mistake, or an inaccuracy in NCCER's Contren® materials, please fill out this form (or a photocopy), or complete the online form at www.nccer.org/olf. Be sure to include the exact module number, page number, a detailed description, and your recommended correction. Your input will be brought to the attention of the Authoring Team. Thank you for your assistance.

Instructors – If you have an idea for improving this textbook, or have found that additional materials were necessary to teach this module effectively, please let us know so that we may present your suggestions to the Authoring Team.

NCCER Product Development and Revision

3600 NW 43rd Street, Building G, Gainesville, FL 32606

Fax: 352-334-0932
Email: curriculum@nccer.org
Online: www.nccer.org/olf

☐ Trainee Guide ☐ AIG ☐ Exam ☐ PowerPoints Other ____________________

Craft / Level: ____________________ Copyright Date: ____________________

Module Number / Title: ____________________

Section Number(s): ____________________

Description: ____________________

____

____

____

____

Recommended Correction: ____________________

____

____

____

____

Your Name: ____________________

Address: ____________________

____

Email: ____________________ Phone: ____________________

Setting and Pulling Poles

49111-11

Trainees with successful module completions may be eligible for credentialing through NCCER's National Registry. To learn more, go to **www.nccer.org** or contact us at **1.888.622.3720.** Our website has information on the latest product releases and training, as well as online versions of our *Cornerstone* newsletter and Pearson's Contren® product catalog.

Your feedback is welcome. You may email your comments to **curriculum@nccer.org**, send general comments and inquiries to **info@nccer.org**, or use the User Update form at the back of this module.

V.1 7/11

49111-11

SETTING AND PULLING POLES

Objectives

When you have completed this module, you will be able to do the following:

1. Describe and demonstrate how to load and unload wood poles in preparation for installation.
2. Explain and demonstrate the importance of using the proper hand signals when setting a pole.
3. Describe and demonstrate how to set a wood utility pole using a digger derrick.
4. Describe and demonstrate how to set a wood utility pole by hand.
5. Describe and demonstrate how to pull a wood utility pole from the ground.

Performance Tasks

Under the supervision of the instructor, you should be able to do the following:

1. Load and unload wood poles in preparation for installation.
2. Demonstrate the proper use of ASME hand signals.
3. Pull a wood pole with a hydraulic pole puller.
4. Set a wood pole with a digger derrick.

Trade Terms

Bump board
Cant hook
Digging bar
Drift
Jenny (mule)
Pike
Piking method
Pole grabber
Pole key
Posthole digger
Snatch block
Spoil bag
Spoon
Tamping bar
Tow plate
Two-blocking

Contents

Topics to be presented in this module include:

Figures and Tables

1.0.0 Introduction

Power transmission and distribution system technicians perform a variety of tasks. Among the most common of these tasks is the installation and maintenance of overhead power lines. In the United States, wood poles carry most power lines. New poles have to be installed to bring power to new customers. Poles that have been damaged or weakened by rot have to be replaced. Line workers must be able to perform this routine but very important task in a safe and efficient manner.

It is estimated that there are currently as many as 135,000,000 wood poles installed in the United States. Wood is clearly the material of choice for utility poles. When you install or replace wood utility poles, safety should be the core value. The poles are heavy, increasing the risk of serious injury if they are handled or installed incorrectly. This module describes how to safely install wood utility poles by hand and by using mechanized equipment. Methods for removing damaged poles are also shown.

2.0.0 Safety

Technicians who install wood poles are required to wear the following personal protective equipment (PPE) (*Figure 1*) to prevent on-the-job injuries:

- *Clothing* – Clothing should be made of natural fibers, such as cotton or wool, and be treated to be fire-retardant. Avoid clothing made of synthetic fibers that could burn or melt and cling to skin. Pant legs should extend over the top of shoes or boots. Never wear shorts on the job. Shirts should be long sleeved to protect the skin from the sun, to minimize injuries caused by wood splinters, and to provide protection from chemicals used to treat the wood against rot. On most job sites, workers are required to wear a bright colored shirt or vest to increase their visibility.

> GOING GREEN
>
> **Wood Pole Preservatives**
>
> The chemicals used to treat wood poles to prevent rot are toxic and can damage the environment as they leech from the wood into the surrounding soil. In an effort to reduce this problem, new, less toxic chemicals are now being introduced to treat wood poles.

49111-11_F01.EPS

Figure 1 Line worker wearing PPE.

- *Footwear* – Only safety-toe leather shoes or boots with slip-resistant soles should be worn on the job. High-top boots have the added advantage of greater support for the ankles. Never wear sandals or canvas-type shoes on the job.
- *Eye protection* – Safety glasses with side shields must be worn at all times while on the job. If a power tool such as a chain saw is used on the job, a full face-shield may also have to be worn.
- *Hearing protection* – If power tools are used on the job, or noisy machinery such as an air compressor is running, approved earmuffs or earplugs may also have to be worn.
- *Respirator* – If a wood pole is cut or drilled, a respirator or dust mask may need to be worn. The preservatives in the wood can be harmful and any sawdust made while cutting or drilling should not be inhaled. Leather chaps are required whenever a chain saw is being used.
- *Gloves* – The rigging and tools used to erect poles, and the wood of the poles can be rough on hands. Leather work gloves provide protection from rope burns, wood splinters, and other hazards.
- *Rubber gloves and sleeves* – Poles are often set near energized lines or equipment. Rubber gloves and sleeves protect against electrical shock.
- *Hard hat* – An approved safety helmet made of a non-conducting material such as fiberglass or plastic must be worn at all times while on the job site. The helmet protects the head from injuries caused by bumps and falling objects. It also provides shade from the sun. *ANSI Standard Z89.1-1986* covers requirements for safety headgear worn by electrical workers.

3.0.0 Storage of Wood Poles

Wood poles are shipped to the utility by truck or on rail cars. The poles are then unloaded using a special forklift or front-end loader and placed in storage at the utility until needed. Most poles are treated with a wood preservative so they do not need to be stored under cover. However, in areas where snow builds up, the poles may be kept under cover. The poles are stored on racks that are off the ground (*Figure 2*). Poles of the same length are stored together.

3.1.0 Loading Wood Poles for Transport

When a pole is needed on the job, it is taken from storage and loaded onto a trailer designed to transport poles (*Figure 3*). Before loading the trailer, the trailer should be placed near the pole storage area with the brakes on and wheels chocked to prevent any movement. The pole(s) may be loaded on the trailer using a special forklift or front-end loader. If the line truck used to tow the trailer is equipped with a boom or digger derrick, it can be used to lift and load the poles. Some digger derricks are able to carry small poles (*Figure 4*).

49111-11_F02.EPS

Figure 2 Wood poles in storage.

GOING GREEN

Composite Poles

Fiberglass-reinforced composite poles offer many advantages over wood poles. These advantages include lighter weight, uniform dimensions, and equal strength. In addition, they do not rot. When all factors are considered, composite poles are often no more costly to install than wood poles, without the disadvantages of wood.

49111-11_F03.EPS

Figure 3 Trailer for transporting poles.

49111-11_F04.EPS

Figure 4 Pole carried on a digger derrick.

The wire sling used to lift the pole should be secured near the pole's center of gravity so that the pole is balanced as it is lifted. Attach at least one tag line to one end of the pole to help control its movement during the lift. The tag line(s) should be long enough so that the worker(s) controlling the load can maintain a safe distance from the pole as it is moved into position over the trailer. Once over the trailer, the pole is lowered into position on the trailer.

Trailers are equipped with uprights or chocks to prevent the pole(s) from rolling from side-to-side. Once in position, the pole is secured to the trailer with ratchet straps or chains. If more than one pole is being loaded, position them on the trailer so that the load is stable and balanced before strapping the poles down. Do not overload the trailer.

WARNING!

Poles can shift or roll during loading or unloading. This can pinch or crush any body part that might be in the way. Always be on the lookout for and avoid potential pinch points. A pinch point is defined as any narrow area between two surfaces that is likely to catch or trap objects.

In some locations, it is common practice to use one of the poles being transported as the tongue of the trailer. The pole being used for this should be long enough and strong enough to serve the purpose. The tow plate of the trailer (*Figure 5*) is securely bolted to the end of the pole using chains. Always place a warning flag and/or lights on any poles that hang off the back of the trailer to warn following vehicles. A trailer loaded with poles is a long load and may require the use of a flag person when making tight turns in traffic.

3.2.0 Unloading Wood Poles

Once at the job site, the poles are unloaded from the trailer using a digger derrick or the boom of the line truck. The trailer should be parked with its brakes on and wheels chocked before unloading. The poles should be lifted in a balanced condition and controlled with tag lines in a manner similar to the way they were loaded. The pole may be set right away, or it may be placed in storage at the job site. If poles are stored, they must be blocked or chocked to prevent rolling and placed in an area that does not interfere with other work.

49111-11_F05.EPS

Figure 5 Trailer tow plate.

4.0.0 Setting Wood Poles with Mechanized Equipment

Due to pole weight and length, and for safety reasons, most wood poles today are set using mechanized equipment. The digger derrick is the piece of equipment most widely used for setting poles.

On Site

Pole Handling Machines

Skid loaders equipped with a special pole handling attachment can be used to load and unload poles at the yard, and to install the poles at the job site. Setting poles by hand requires a larger work crew and can be hazardous. Setting a pole by hand is often done when a digger derrick cannot access the site. Today, there are compact pole-setting machines that are able to get into areas that a regular digger derrick cannot. Pole-setting machines are often tracked vehicles. This allows them to go off road and into rough terrain. The use of a pole-setting machine reduces crew size and is safer than setting the pole by hand.

49111-11_SA01.EPS

49111-11_SA02.EPS

4.1.0 Digger Derrick Operation

A digger derrick (*Figure 6*) is basically a large auger installed on the end of a telescoping boom on a truck. The auger is operated by hydraulic pressure and is used to bore the hole in which the pole will be set. Augers come in diameters up to 48". The auger is placed over the center of the hole, and lowered into place for boring the hole. Once the hole is bored, the auger is stowed out of the way on the side of the boom of the truck. The boom is then used to place the pole into the hole. The truck is equipped with outrigger legs that must be deployed before boring a hole or setting a pole. The outriggers increase the stability of the digger derrick.

NOTE

Before any work begins, workers should survey the work area for any hazards that might be present. In addition, the truck must be grounded before any work begins.

WARNING!

Only workers trained and qualified to operate a digger derrick can operate the equipment. Unauthorized operation can be hazardous.

49111-11_F06.EPS

Figure 6 Digger derrick.

5.0.0 Load Control

Safe and efficient load control when setting a pole with a digger derrick involves the following:

- Good communication
- The use of physical load control techniques
- The safe handling of loads
- Taking landing zone precautions
- Following sling disconnect and removal practices

5.1.0 American Society of Mechanical Engineers (ASME) Hand Signals

When a pole is being moved to the hole with the digger derrick, workers on the ground communicate with the digger derrick operator by using radio or hand signals to direct the movement and placement of the pole. In some cases, there may be too much noise on the job site for radio communication to be effective. Using hand signals in those instances ensures a safe placement of the pole. The ASME standard hand signals used for mobile cranes (*Figure 7*) are the recommended signals. The correct use of these hand signals provide clear and definite instructions for the various movements needed to set a pole.

5.2.0 Load-Handling Safety

The safe and effective control of the pole being set involves the rigger's strict observance of load-handling safety requirements.

- If a swing-type crane is used, keep the front and rear swing paths clear during pole placement. Most people watch the pole when it is in motion, and this prevents them from seeing the back end of the crane coming around. Use safety flags or barricades to keep other workers out of the swing path.
- Keep the area where the pole will be placed clear of all workers except those handling tag lines and/or guiding the pole into place.
- Account for the speed and direction of the wind when setting the pole.
- Never stand beneath a pole being lifted or moved.
- If the pole is being placed on the ground, make sure that it is blocked or chocked so that it cannot move or roll.
- Never remove the rigging used to set the pole until the pole is securely in place without further need for the crane or digger derrick.

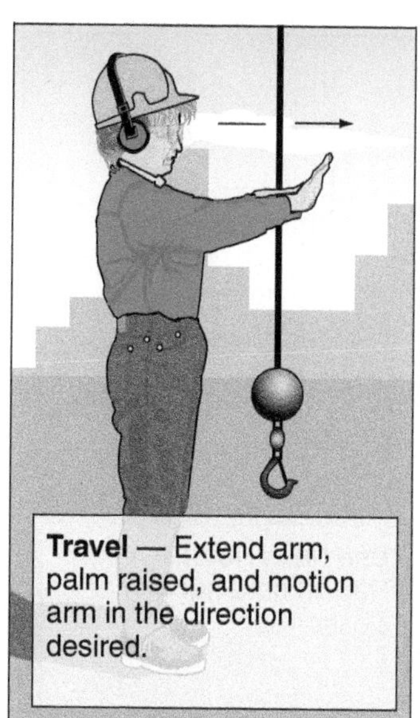

49111-11_F07A.EPS

Figure 7 Standard hand signals. (1 of 2)

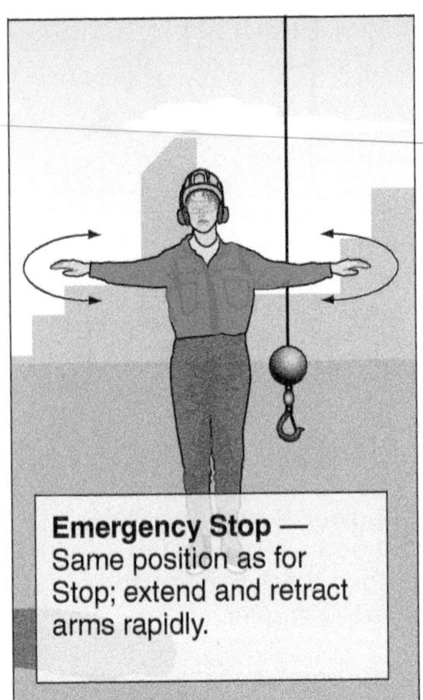

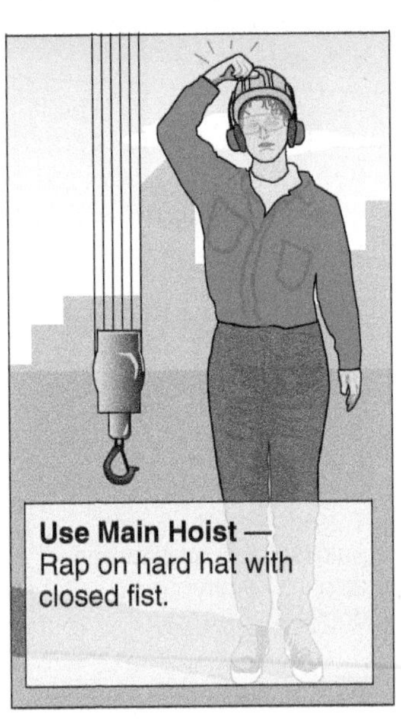

49111-11_F07B.EPS

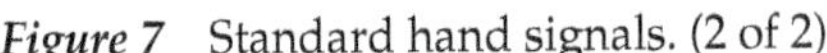

Figure 7 Standard hand signals. (2 of 2)

6.0.0 Site Preparation

Before setting the pole, the site where the pole will be placed must be prepared. If the ground is uneven, is littered with rocks, or is overgrown with brush or trees, a lot of prep work will be needed. This work must be done so that the mechanized equipment can have access to the site.

WARNING!

Before any site preparation work begins, make sure that the local utility companies have surveyed and marked the location of their utilities within or near the planned work area.

In order for the crew to work efficiently, the work area must be cleared to make room for trucks and workers. Small trees and brush can be removed with hand tools such as axes or machetes. A chain saw is required to remove larger trees. Tree limbs hanging over the right of way can also be a hazard, and may need to be removed. Small stumps present a tripping hazard. Larger stumps may interfere with boring the hole. Use the blade of a bulldozer or the bucket of a backhoe to clear smaller stumps and rocks from the site. A chain hoist attached to a line truck can be used to pull larger stumps. The auger on the digger derrick can remove small tree stumps.

On Site

Tree and Stump Removal

Large trees and tree stumps are often not removed by line workers. In many cases, this task is contracted out to a tree removal service. The removal crew may use special stump grinding equipment like the machine shown here to eliminate the stump.

49111-11_SA03.EPS

6.1.0 Placement and Sizing of the Hole

On new projects, location of the hole into which the pole will be set is predetermined and will be marked with a stake. A survey crew or the designer does the location and placement of the stake. In the case of replacement poles, their location is either next to the pole being replaced or in the same hole. The diameter of the hole is determined by the diameter of the pole. Because the diameter of a wood pole is greatest at its base, the diameter at the base is used to determine hole size. The hole must be large enough so that the pole can be moved into a plumb position and to allow room for tools to tamp fill soil around the pole after it is in place. As a general rule, allow at least 3" around the pole. For example, a 12" diameter pole would require an 18" diameter hole (*Figure 8*).

The depth of the hole into which the pole will be set (setting depth) is determined by two factors: the length of the pole being set, and the type of soil in which the hole will be dug. Taller poles require a deeper hole. A pole set in soil requires a deeper hole than the same length pole set in rock (*Table 1*). When setting a pole in soil, use the following general rule to determine setting depth.

10 percent of pole length + 2' = setting depth

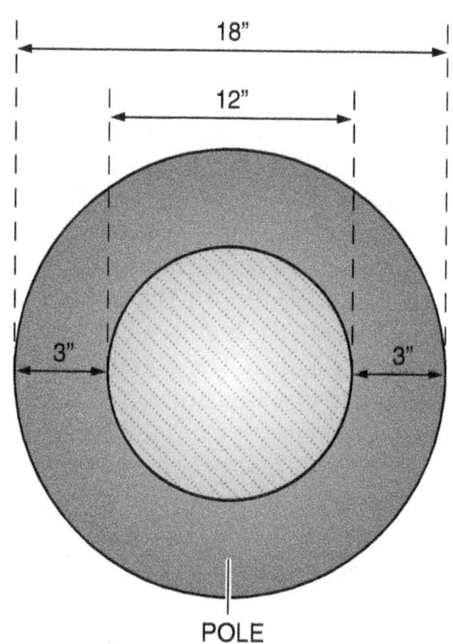

49111-11_F08.EPS

Figure 8 Hole diameter.

Table 1 Pole Height vs. Hole Depth

	Required Hole Depth	
Pole Length	Setting Depth (Soil)	Setting Depth (Rock)
25'	5'	3.5'
40'	6'	4'
50'	7'	4.5'
60'	8'	5'
80'	10'	6.5'

When using this rule, keep in mind that the hole must always be set at least 5' deep. For example, when setting a 50' wood pole in soil, what is the setting depth? Ten percent of 50' is 5'. Add two feet to this value to obtain the setting depth: 5' + 2' = 7'.

Holes can be difficult to dig in very wet or sandy soil because the soil on the sides of the hole tends to cave in as soil is removed. During excavation, shoring in the form of a metal barrel with open ends or a length of corrugated steel culvert can be used to prevent cave-ins.

As the hole is dug, the shoring is forced into the ground where it prevents cave-ins. The end result is a metal-lined hole in which the pole is set. Poles set in wet or sandy soils are generally set deeper and may require other methods such as guy wires or concrete fill to help stabilize the pole. If a pole is set in concrete, the concrete must be given time to cure—sometimes several days—before the pole is used.

A pole can also be stabilized with a device called a pole key (*Figure 9*). The pole key can be made of pressure-treated wood or steel. One or more keys can be used, based on soil conditions. The key is usually placed in the pole hole on the strain side of the pole.

Holes in solid rock do not have to be as deep as holes made in soil. The rock-cutting drill used to make the hole is mounted on the digger derrick like an auger (*Figure 10*). It cuts a circular hole in the rock that leaves a core. The core is then broken off and removed from the hole in one piece. In extreme cases, explosives are used to blast holes in rock. Blasting is dangerous and must only be done by workers trained in the use of explosives.

6.1.1 Pulling a Wood Pole

A pole that is being replaced is cut down with a chain saw with its stump sticking about 4' above ground. This pole stump must be removed using a hydraulic pole puller (*Figure 11*). The pole puller is clamped around the stump of the pole with a chain. A large footpad on the pole puller prevents the puller from sinking into the ground. Hydraulic pressure is applied from the truck hydraulics or from the digger derrick. An engine-driven hydraulic pump attached to the hydraulic cylinder on the pole puller may also be used. The pole puller exerts a pushing force on the ground and a pulling force on the pole to pull it from the ground. As the pole stump is pulled from the hole, the pole puller is repositioned farther down the stump and the process is repeated until the pole stump is removed.

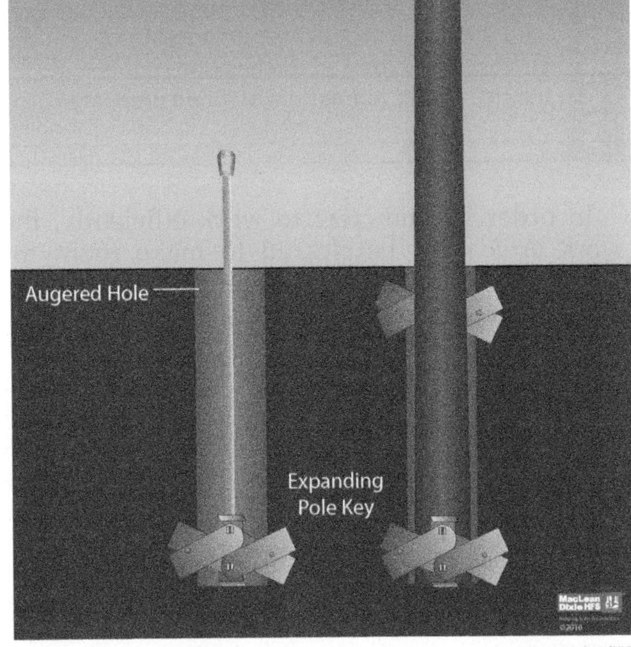

49111-11_F09.EPS

Figure 9 A pole key.

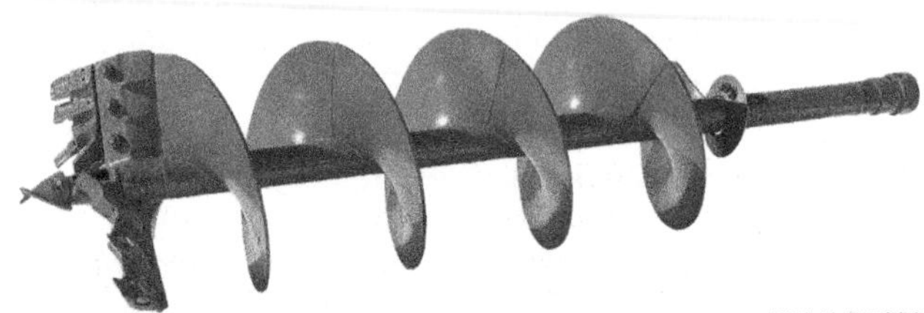

49111-11_F10.EPS

Figure 10 Rock drill.

49111-11_F11.EPS

Figure 11 Hydraulic pole puller.

WARNING!

Never use a digger derrick or the boom/hoist of a line truck to pull a pole from the ground. Personal injury or damage to the boom may occur.

49111-11_F12.EPS

Figure 12 Using the digger derrick auger.

6.2.0 Digging the Hole

CAUTION

Before any digging is done, a survey must be performed to locate any buried utilities such as water and gas pipelines, electrical cables and communications cables. This must be done well in advance because it requires someone to visit the site and mark the locations of all underground utilities with color-coded flags or colored paint.

Whenever possible, the hole is dug with an auger on a digger derrick (*Figure 12*). The auger is positioned above the spot where the pole is located, and is then driven into the earth. The auger is removed from the hole from time to time to clear soil from the blades (*Figure 13*). A shovel or other hand tool may be used for this purpose. When the auger has completed its work, hand tools such as a posthole digger, long-handled digging bar,

49111-11_F13.EPS

Figure 13 Auger ready to be cleaned off.

long-handled spade shovel, or a long-handled shovel called a **spoon** are used to dress the sides of the hole and clean out any loose soil from the bottom of the hole (*Figure 14*). In areas where a digger derrick cannot go, the hole can be dug using the hand tools just described.

Soil removed from the hole should be placed far enough away that it does not drop back in the hole or get in the way of setting the pole. All or most of the soil will later be used for backfill around the upright pole. Excess soil and/or rock can be dealt with in several ways. In some cases, as soil is being removed from the hole, it is placed on a canvas mat called a **spoil bag**. The spoil bag is meant to store and transport clean soil. The spoil bag has lifting rings on its four corners, allowing it to be lifted from the ground with the truck's boom. After the pole is set, any excess soil in the spoil bag is lifted onto the truck and carried away for proper disposal. If the soil is contaminated, it is loaded into barrels and disposed of in accordance with local and federal laws.

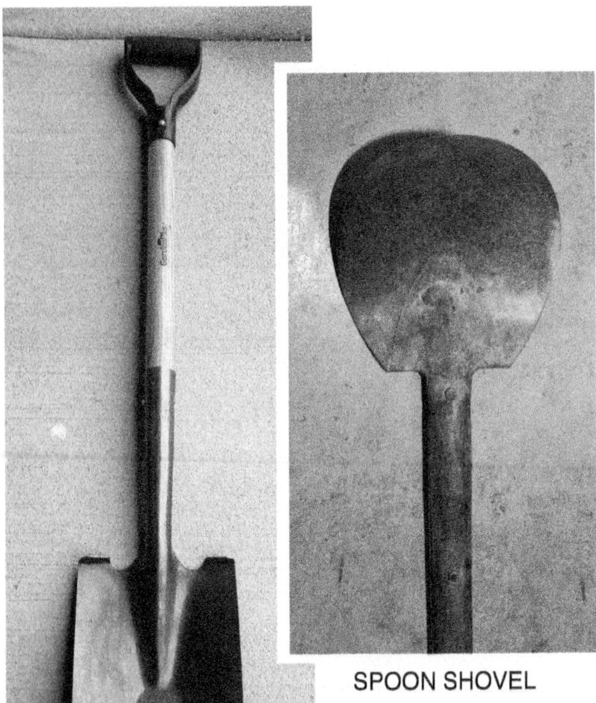

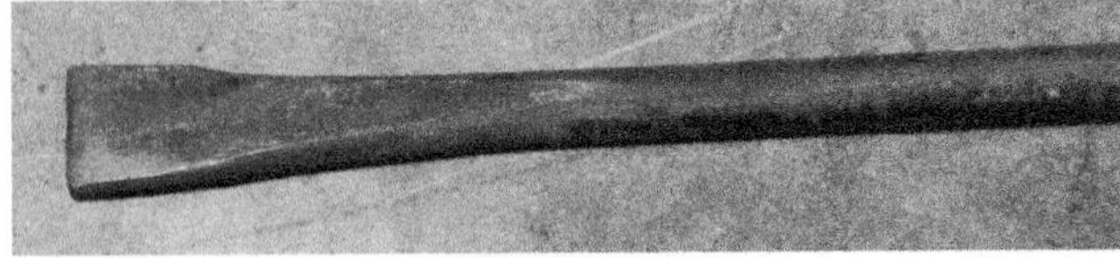

49111-11_F14.EPS

Figure 14 Manual digging tools.

7.0.0 Setting the Pole with a Digger Derrick

Once the pole hole is prepared, the pole can be raised and moved into position (*Figure 15*). Lifting charts that show weight vs. boom angle must be referenced to ensure that the weight of the load will not cause the digger derrick to tip over. The winch line with the hook and approved sling is attached to the pole above its center of gravity so that the pole will be butt-heavy and easier to elevate to an upright position. As the winch line is reeled in, the pole will be raised to the boom. At that point, the **pole grabber** claws are opened to receive the pole and are then closed tightly around the pole (*Figure 16*). The claws help to guide and stabilize the pole while it is being moved into place. Note that the claws are only intended to guide the pole; the lifting must be done by the winch. Tag lines are attached to the pole to help control it as it is lifted. The pole should be in a balanced position, slightly butt-heavy, as it is lifted.

WARNING!

If the pole is being installed near energized overhead lines, place an insulated pole cover over the pole. Also, when setting a pole, never allow yourself to get in a position where you are underneath the pole as it is being lifted into place. Always wear the proper personal protective equipment including rubber gloves and sleeves when setting a pole.

49111-11_F15.EPS

Figure 15 Lifting the pole.

49111-11_F16.EPS

Figure 16 Pole grabber claws secure the pole.

The pole is raised until it is in a near-vertical position with the butt end of the pole clear of the ground. Then the ground worker guides the top of the pole into the pole claws. The digger derrick then moves the pole to the hole and lowers it into the hole. Workers use cant hooks to position the pole (*Figure 17*). Once the pole is vertical in the hole, a plumb bob is used to check that the pole is plumb. When plumb is established in one position, move to a position that is 90 degrees from the original position and check for plumb again (*Figure 18*).

When lifting with a digger derrick, the boom must be extended enough so that when the pole

49111-11_F17.EPS

Figure 17 Positioning the pole.

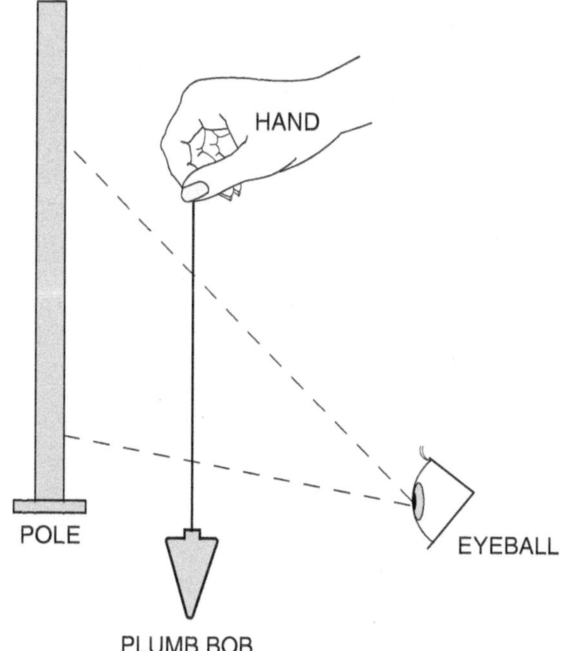

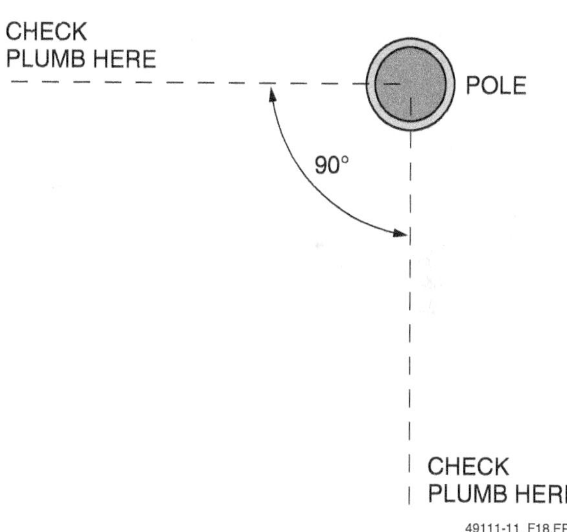

49111-11_F18.EPS

Figure 18 Checking plumb.

is raised into position, the hook at the end of the lifting line does not come into contact with the boom. This condition, which is known as two-blocking, can cause the load to fall. The distance between the hook and the tip of the boom is called drift.

Once the pole is vertical and plumbed, the hole can be backfilled with soil. In areas where soil conditions are poor, the pole may be backfilled with crushed stone or the pole may be set in concrete to increase stability. Proper backfilling is critical to maintaining the stability of the pole in the hole. Start by shoveling a few inches of soil into the hole with a round shovel or spade. After shoveling, tamp the soil down with a hydraulic tamper, or do it by hand with a tamping bar until it is as firm as the soil surrounding the hole (*Figure 19*). Repeat this process several times until the hole is filled with firmly packed soil. Ideally, all of the soil removed when the hole was dug should be used during the backfill and tamping process. It is a good idea to leave 3 to 4 inches of soil heaped up around the pole to account for settling. Any soil left over should be removed from the site with a spoil bag.

NOTE

Backfilling the hole completely before tamping the soil will not allow the soil to compact properly and may result in an unstable pole. If you backfill in increments and tamp a small quantity of soil each time, the soil will be firmly compacted and the pole will be more stable.

49111-11_F19.EPS

Figure 19 Filling-in and tamping the hole.

GOING GREEN

Pole-Setting Foam

Poles can be set using a two-part foam product. The two parts of the foam are mixed and poured into the hole after the pole has been plumbed. The foam hardens and sets the pole within a very short time. The foam also acts as a barrier that prevents wood preservatives in the pole from leaching into the surrounding soil. Pole-setting foam can be used to help keep a pole straight if it tends to lean and is often used to set a pole in wet or sandy soil.

8.0.0 Setting the Pole by Hand (Piking Method)

In places where it is difficult or impossible to use a digger derrick to set a pole, the pole can be set by hand using the piking method. The size of the crew needed to set a pole by hand is based on the length and weight of the pole to be set. The longer and heavier the pole, the larger the crew required for setting the pole (*Table 2*). For example, to set a 40' pole, a piking crew of at least 8 workers is required. Members of an eight-person crew have different tasks. Six workers, called pikers, handle the pike poles (*Figure 20*). One worker, called the jennyman, handles the jenny (mule), a device used to support the pole during the lift (*Figure 21*). One more worker is stationed at the butt of the pole to help guide the pole into the hole. Before the lift is started, the butt of the pole should be placed over the hole and against a bump board (*Figure 22*). The purpose of the bump board is to prevent the pole from gouging the side of the hole and causing soil to drop into the bottom of the hole during the lift. The person at the butt end of the pole uses a cant hook (*Figure 23*) to prevent the pole from rolling during the lift.

NOTE

Before the lift, the pole should be raised off the ground and placed on chocks. This prevents the pole from rolling and it enables the crew doing the lifting to easily grasp the pole.

To start the lift, the pikers at the top end of the pole lift the pole high enough so that the jenny can be placed under it for support. The pikers again lift the pole and the jenny is advanced toward the butt of the pole, supporting the pole as it is raised higher. Once the pole reaches a height where pikes can be used, the lifting team leaves the pole supported by the jenny. Pikers then spread out in a fan-shaped pattern and place the spike end of their pikes into the pole. Lifting and pushing together, the crew pushes the pole higher. At the same time, the jenny is moved to support the pole.

As the pole is lifted to the limit of the length of the pikes, the pikes are repositioned lower on the pole as the pole is held in place by the jenny. This process is repeated until the pole slides into the hole. The bump board is removed once the pole is upright in the hole.

Table 2 Required Crew Size for Piking

Pole Length	Piking Method Crew Size			
	Crew Size	Pike Attendants	Jenny Attendant	Butt Attendant
25'	5	3	1	1
30'	6	4	1	1
40'	8	6	1	1
50'	10	8	1	1

Figure 20 Pike pole.

Figure 21 Jenny.

NOTE

When setting long poles by hand, it is common practice to dig a trench that descends into the pole hole. The trench allows the butt end of the pole to sit lower in the hole and makes lifting easier.

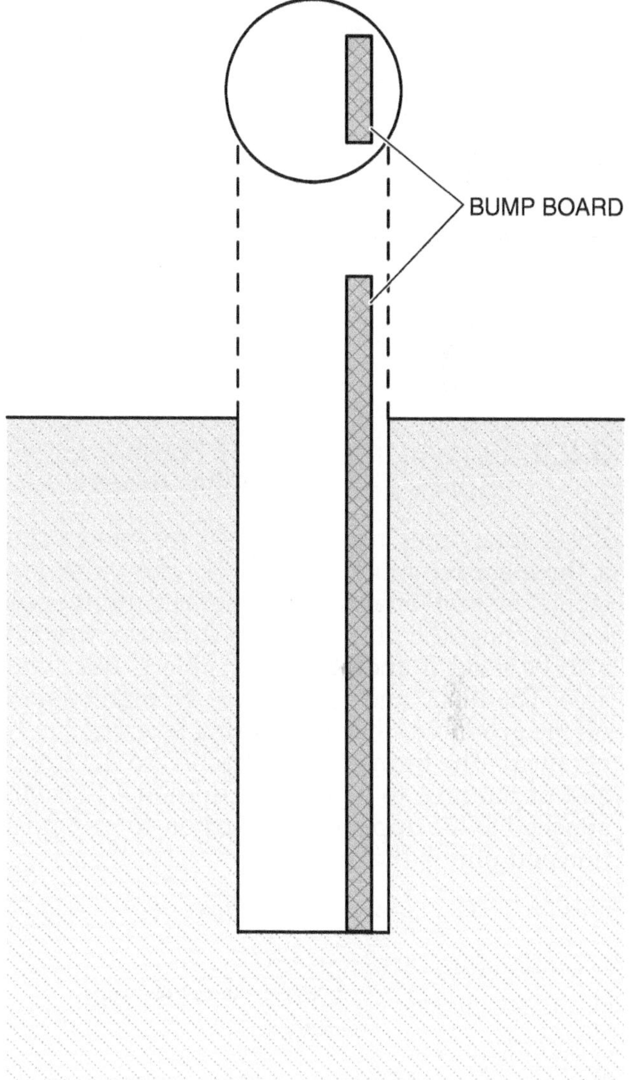

Figure 22 Bump board.

Use a plumb bob to check that the pole is plumb in the same manner as described earlier. The pike poles are used to adjust the position of the pole to obtain plumb. Once the pole is plumbed, backfill and tamp the soil in place around the base of the pole with hand tools.

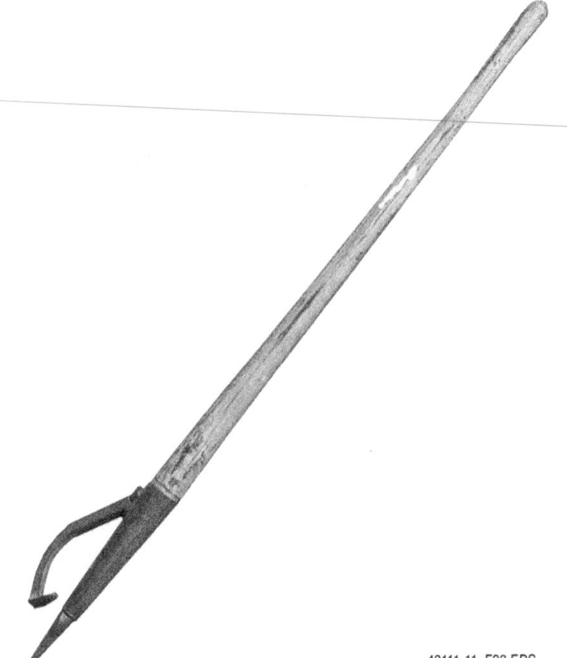
49111-11_F23.EPS

Figure 23 Cant hook.

9.0.0 Setting the Pole by Hand (Block and Tackle Method)

Poles sometimes need to be set by hand using a three-sheave block and tackle. The block and tackle is first attached to another utility pole located near the pole to be set. A nearby tree can also be used if another utility pole is not available. The block and tackle must be high enough off the ground so that good lifting leverage can be obtained. The lifting line is attached to the upper end of the pole to be lifted. Additional lines are attached to the upper end of the pole to help position and plumb the pole.

To start the lift, team members pull on the lifting line through the block and tackle. As the pole is raised, it is positioned over the hole. Once in position over the hole, the pole is lowered into the hole. A **snatch block** (*Figure 24*) is sometimes used to change the direction of the pull line to better align the pole during the lift (*Figure 25*). The snatch block is attached to a utility pole, or other fixed object. When using a block and tackle and/or snatch blocks, the lifting line can be pulled by hand, or the power winch on a line truck can be used. The lifting line is pulled until the pole is upright. Other workers on the positioning lines pull their lines to bring the pole into plumb. Once plumb, the pole is backfilled in the same manner as described earlier.

49111-11_F24.EPS

Figure 24 A snatch block.

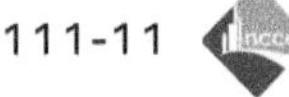

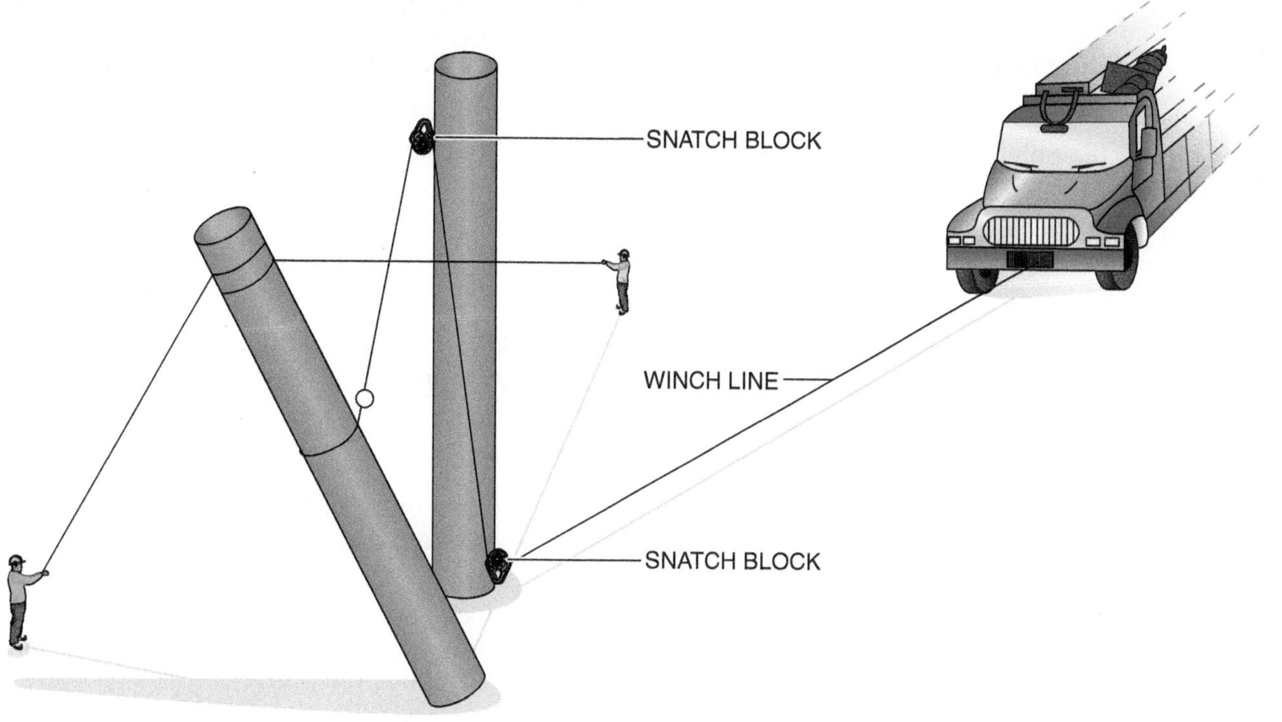

Figure 25 Using a snatch block.

Summary

Wood poles are widely used by electric utilities in the United States to carry overhead power lines. Poles are constantly being installed to replace old or damaged poles and to provide power to new customers. Line workers spend a good amount of time installing wood poles. For that reason, it is important to know how to do the job right while avoiding injuries. When you install or replace wood utility poles, safety should be your highest priority. The poles are heavy, increasing the risk of injury if not handled and/or installed the correct way. Always wear the proper personal protective equipment, and follow all safety precautions.

Most wood poles are installed with mechanized equipment, such as a digger derrick or a pole-setting machine. These machines require fewer workers, take less time to set the pole, and expose the workers to fewer safety hazards. Setting a pole by hand is rarely done today. However, if machinery cannot access a job site, the pole can be set with a crew of workers. A block and tackle or pike poles can be used to raise and set a pole by hand. The size of the work crew needed to set a pole by hand is based on the type of method used and on the length of the pole being installed.

Review Questions

1. The cable used to load a pole on a trailer is attached at the pole's _____.
   a. tip
   b. butt end
   c. center of gravity
   d. lifting eye

2. The auger on a digger derrick is driven by _____.
   a. hydraulic pressure
   b. a belt drive
   c. a chain drive
   d. a gasoline engine

3. Outriggers on a digger derrick are used to provide _____.
   a. increased lifting capacity
   b. greater lifting height
   c. increased pulling capacity
   d. stability

4. A person on the ground signals the digger derrick operator with an arm extended, fist closed, and thumb pointing up. This hand signal indicates _____.
   a. raise the boom
   b. everything is OK
   c. stop the lift
   d. swing the load faster

5. A person on the ground signals the digger derrick operator with arms extended in front of the body, palms down and fists closed, and thumbs pointing inward. This hand signal indicates _____.
   a. raise the boom
   b. retract the boom
   c. start the lift
   d. lower the load

6. What size (diameter) hole is required to set a 10" diameter pole?
   a. 10"
   b. 13"
   c. 16"
   d. 24"

7. What is the depth of the hole required to set a 40' pole in soil?
   a. 4'
   b. 6'
   c. 8'
   d. 10'

8. A pole key is usually placed in the pole hole on the strain side of the pole.
   a. True
   b. False

9. The stump of an old utility pole is typically removed from its hole using _____.
   a. dynamite
   b. a block and tackle
   c. the boom of the digger derrick
   d. a hydraulic pole puller

10. Check the plumb of an installed pole in two positions that are _____.
    a. 180 degrees apart
    b. 120 degrees apart
    c. 90 degrees apart
    d. 45 degrees apart

11. Backfill the hole completely before compacting the soil around the base of the pole.
    a. True
    b. False

12. When setting a pole, drift refers to the _____.
    a. movement of the pole caused by the wind
    b. movement of the derrick boom caused by the wind
    c. distance from the end of the boom to the hook on the end of the lifting line
    d. movement of the butt end of the pole once it is in the hole

13. When setting a pole by hand, what is used to support the pole and prevent it from falling during the lift?
    a. A jenny
    b. The digger derrick boom
    c. A tamping bar
    d. A bump board

14. The purpose of a bump board is to _____.
    a. protect workers from a falling pole
    b. act as a temporary pole support
    c. prevent damage to the hole when setting a pole by hand
    d. help roll the poles off the transport trailer

15. The tool used to prevent rolling of the butt end of the pole when setting a pole by hand is a _____.
    a. jenny
    b. pike pole
    c. pole grabber
    d. cant hook

Trade Terms Quiz

Fill in the blank with the correct term that you learned from your study of this module.

1. Use a ____________ to prevent the walls of the pole hole from being gouged when setting a pole.
2. A ____________ is used to change the direction of a pull to better align a pole as it is being raised.
3. A device used to stabilize a pole that tends to lean is called a ____________.
4. To keep the butt end of a pole from turning as it is being lifted, use a ____________.
5. A hand tool used to compact the soil around the base of an installed pole is called a ____________.
6. When setting a pole by hand, use a ____________ to provide temporary support during the lift.
7. A hand tool used to dig a pole hole is called a ____________.
8. Soil that is left over after setting a pole is removed in ____________.
9. A ____________ is used to break up hard soil when digging a pole hole by hand.
10. A method for setting a pole that requires line workers to raise a pole in steps using long poles is called the ____________.
11. To use a wood pole as the tongue of a trailer that transports poles, you must install a ____________ on the pole.
12. A ____________ attached to the end of the boom is used to hold a pole when it is being raised.
13. The distance between the end of a lifting boom and the hook on the end of the lifting line is called ____________.
14. A hand tool that has a spike on one end and is used to raise poles by hand is called a ____________.
15. A condition that occurs when the load block is pulled up into the boom sheaves is known as ____________.
16. A hand tool used to remove soil from the bottom of a pole hole is called a ____________.

Trade Terms

Bump board
Cant hook
Digging bar
Drift
Jenny (mule)
Pike
Piking method
Pole grabber
Pole key
Posthole digger
Snatch block
Spoil bag
Spoon
Tamping bar
Tow plate
Two-blocking

Cornerstone of Craftsmanship

Jonathan Sacks

Craft Training Manager
Cianbro Corporation

Growing up, Jonathan Sacks never considered learning a construction trade; for the past 34 years, however, he had made a fulfilling career in electrical work. He has worked as an electrical helper, electrician, carpenter, electrical contractor, supervisor for larger companies, and now as an electrical instructor and training manager.

How did you get started in the construction industry?
As a young person, I was not directed to look at the trades as a career. I discovered by accident that it could be very satisfying. After my sophomore year in college, I took a year off. I wanted a job. I met a guy who I thought was a carpenter. I asked if he needed some help and he did. Turns out he was an electrician and I became an electrical helper. At first, I did not think of it for the long term.

As I kept working, I liked the feeling of looking at what you have done at the end of the day. It is very satisfying. I went back to college and got my degree in history and a teaching certificate. But I never really left the trades, working part time when I was in school and full time after I graduated. I realized I really liked electrical work and worked toward getting my license.

Who inspired you to enter the industry?
My first boss was inspirational. I also had many really good role models over the years who inspired me to stay in it. When I was already established in my career as a craftsperson, one individual, an electrical manager, inspired me to continue to learn and grow. He told me that if you can wire a three-way switch, you can wire a power plant. His name was Pete Schien and he helped me understand that you can learn whatever you set your mind to.

What do you enjoy most about your job?
I enjoy the problem solving, troubleshooting, and planning aspects of the work. You look at the work, lay it out, and then do it on schedule. This job is both mental and physical; it is a combination of using your hands and your head. It is great to make a plan and have things work when you are done.

Do you think training and education are important in construction?
Yes, training is essential now, especially for the electrical trade. I got in at a time when prior education was not required. I wished it had been, because I learned some things the hard way. Now a young person needs a formal education when they start out. There is so much to learn and things keep changing. It is impossible to get licensed at this point without formal training.

How important do you think NCCER credentials will be in the future?
I think they will gain in importance in the coming years. NCCER credentials are already important in certain areas and markets.

I think any credentials are important and will only become more important in the future. NCCER is setting the benchmark in many trades. Credentials are especially important in trades that do not have licensing. The NCCER credentials are establishing a benchmark in those trades.

How has training/construction impacted your life and your career?
Both training and construction have been how I have made my living, as an electrical contractor, teacher, and training manager. I have enjoyed it tremendously. I have met and worked for many fine people through those opportunities. It has been very satisfying.

Would you suggest construction as a career to others?
Yes. It a satisfying career if you like it. You can make good wages as a skilled craftsperson. Much of the work is teamwork. Working with other people can be very enjoyable. At the end of the day, you can see what you've done. You can look back 15 years or more and show your grandkids what you accomplished.

Some aspects of construction require travel, which is not for everybody, but some enjoy it. With other types of construction work, you can work close to home everyday. It is a varied field.

How do you define craftsmanship?
Craftsmanship is professional work done by professionals. It is a person's signature—showing pride in their work, as if they were putting their signature on their work. Some work is behind walls or in places that cannot be seen. But a craftsman does that work to the same level as the jobs that can be seen by anyone. A craftsman takes pride in everything they do, from driving a nail or bending conduit, to building a beautiful building. Working safely is also part of being a craftsman. It is how you do the work.

Trade Terms Introduced in This Module

Bump board: A board that is placed in a vertical position in a pole hole. It is used to prevent the pole being installed from gouging the sides of the hole and causing the loosened soil to drop into the hole.

Cant hook: A device used to grip the butt end of a pole. It prevents the pole from rolling during the manual setting of a pole. It is often called a peavey.

Digging bar: A heavy, long-handled tool used to break up soil for removal with a shovel.

Drift: The distance between the hook at the end of the lifting line and the tip of the boom. Enough drift must be maintained so that the digger derrick can lift the pole high enough so that it can be placed in the hole.

Jenny (mule): A device used to hold a pole above the ground when setting a pole by hand. It allows the lifting team to remove and reposition their pike poles.

Pike: A long pole with a spike at the end. It is used to raise and align a wood utility pole by hand.

Piking method: A method for setting a wood utility pole by hand that uses pikes to raise and align the pole.

Pole grabber: A device attached to the end of a boom that grabs and holds the pole as it is being positioned by a digger derrick.

Pole key: A device that is placed in the pole hole next to the pole. Its purpose is to stabilize the pole and prevent the pole from leaning.

Posthole digger: A device used to dig round, deep holes in which a pole will be placed. It can be a manual device, or it can take the form of a hand-held, engine-driven auger.

Snatch block: A type of block (sheave) with an open side that allows a loop of rope or cable to be easily inserted. It can be used to change the direction of a pull.

Spoil bag: A canvas storage bag in which soil removed from the pole hole is placed. It is also used to remove any excess clean soil from the job site.

Spoon: A long-handled shovel with a curved blade resembling a ladle. It is used to remove soil from the bottom of deep holes.

Tamping bar: A hand tool used for packing soil around the base of an installed pole.

Tow plate: A device that is clamped on to the end of a wood utility pole so that the pole can act as the tongue of the trailer used to transport poles.

Two-blocking: A condition that occurs when the load block is pulled up into the boom sheaves. This condition can damage the boom and may cause the load to fall.

Additional Resources

This module presents thorough resources for task training. The following resource material is suggested for further study.

www.diggerderricks.org.

Figure Credits

EZ Spot UR, www.ezspotur.com, 1-877-433-5733, Module opener, SA01, and SA02

Salisbury Electrical Safety, Figure 1

BG&E, Figures 2 and 5

Photo courtesy of Brooks Brothers Trailers, Figure 3

Tony Vazquez, Figures 4 and 6

Vermeer Midwest, SA03

A. B. Chance Co./Hubbell Power Systems, Figure 9 (photo)

Maclean Power Systems, Figure 9 (illustration)

Versalift East, Figure 10

Tiiger, Inc., Figure 11

Bob Groner, Figures 12, 13, 15–17, and 19

Topaz Publications, Inc., Figures 14, 21, and 23

ARB 4x4 Accessories, Figure 24

CONTREN® LEARNING SERIES — USER UPDATE

NCCER makes every effort to keep its textbooks up-to-date and free of technical errors. We appreciate your help in this process. If you find an error, a typographical mistake, or an inaccuracy in NCCER's Contren® materials, please fill out this form (or a photocopy), or complete the online form at www.nccer.org/olf. Be sure to include the exact module number, page number, a detailed description, and your recommended correction. Your input will be brought to the attention of the Authoring Team. Thank you for your assistance.

Instructors – If you have an idea for improving this textbook, or have found that additional materials were necessary to teach this module effectively, please let us know so that we may present your suggestions to the Authoring Team.

NCCER Product Development and Revision

3600 NW 43rd Street, Building G, Gainesville, FL 32606

Fax: 352-334-0932
Email: curriculum@nccer.org
Online: www.nccer.org/olf

☐ Trainee Guide ☐ AIG ☐ Exam ☐ PowerPoints Other ____________

Craft / Level: Copyright Date:

Module Number / Title:

Section Number(s):

Description:

Recommended Correction:

Your Name:

Address:

Email: Phone:

Trenching, Excavating, and Boring Equipment

49112-11

Trainees with successful module completions may be eligible for credentialing through NCCER's National Registry. To learn more, go to **www.nccer.org** or contact us at **1.888.622.3720.** Our website has information on the latest product releases and training, as well as online versions of our *Cornerstone* newsletter and Pearson's Contren® product catalog.

Your feedback is welcome. You may email your comments to **curriculum@nccer.org**, send general comments and inquiries to **info@nccer.org**, or use the User Update form at the back of this module.

V.1 7/11

49112-11

Trenching, Excavating, and Boring Equipment

Objectives

When you have completed this module, you will be able to do the following:

1. Identify the trenching, excavating, and boring safety guidelines.
2. Identify and explain the use and operation of compact and pedestrian trenchers.
3. Identify and explain the use and operation of a backhoe.
4. Identify and explain the use and operation of a horizontal directional drilling machine.

Performance Tasks

This is a knowledge-based module; there are no performance tasks.

Trade Terms

Mobiling
Thrust

Contents

Topics to be presented in this module include:

Figures and Tables

1.0.0 Introduction

Power line workers install and maintain power lines above and below ground. This module focuses on the equipment used to place power lines underground. This equipment includes backhoes, trenchers and boring equipment, commonly referred to as horizontal directional drilling (HDD) equipment.

The type of equipment used depends on the working environment and the type of service being installed, along with time constraints, job specifications, and site resources. The operation of some of this equipment is sometimes contracted out and performed by specially trained operators. However, some of the equipment is used by power line workers and other personnel on the job site. Regardless of who is doing the work, power line crews will be working near the equipment and may be helping the operator. For these reasons, you must understand the safety issues associated with the equipment, and in some cases, know how to safely operate it according to the manufacturer's recommendations. Always follow your company policy regarding operator training, certification, and qualification requirements before operating any type of trenching, excavating, or boring equipment.

2.0.0 Trenching, Excavating, and Boring Safety Guidelines

OSHA provides federally mandated regulations governing excavations in *OSHA 29 CFR 1926, Subpart P – Excavations.* Compliance with the rules and regulations stated in this standard is mandatory during excavation and trenching operations. Failure to comply with these regulations can result in serious injury or death. This module focuses on trenching, excavating, and boring equipment and the safety guidelines that apply to this equipment.

2.1.0 General Safety Guidelines

Each operation has its own safety concerns. The following general safety guidelines apply to all trenching, excavating, and boring equipment:

- Complete proper training and any required certifications and read the operator's manual for the specific equipment you are operating.
- Read and understand all the safety instructions listed in the operator manual and identify and understand all safety decals and warnings on the equipment.
- Wear appropriate PPE at all times.
- Avoid wearing loose-fitting clothing and jewelry and confine long hair, if applicable.
- Inspect the job site and perimeter for any evidence of underground hazards and classify the job site as electric, gas, or other. Evidence of underground hazards to look for include the following:
  - Buried-utility notices
  - Utility facilities (especially those that do not have overhead lines)
  - Gas and water meters
  - Junction boxes
  - Light poles
  - Manholes
  - Trenches
  - Standing or flowing water
  - Embankments
  - Sunken ground
- Review job-site hazards, safety and emergency procedures, and all individual responsibilities with all personnel before work begins.
- Ensure that the equipment is in good operating condition and all safety devices are functioning properly.

2.2.0 Locate Underground Utilities

Any time that the ground is broken for maintenance or installation activities, you must ensure that all measures are taken to identify any existing power, gas, water, sewer, or telecommunications lines that may be present in the work area. Sometimes these utility lines are marked (*Figure 1*), but most times, they are not.

Before starting any digging project, you must call the local One-Call system in your area and any utility company that does not belong to the One-Call system. In most states, the contractor must

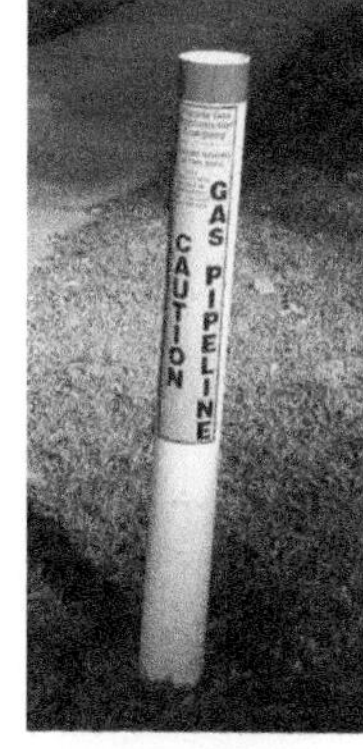

GAS PIPELINE

FIBER-OPTIC CABLE

TELEPHONE CABLE

49112-11_F01.EPS

Figure 1 Examples of marked utilities.

notify the One-Call system 48 hours in advance of the actual dig, but the lead time for marking utility lines may vary from state to state. After calling the local One-Call system, the various utility companies that have buried lines in the area come out to mark their lines. The lines should be marked using the color code system developed by the American Public Works Association (APWA), as follows:

- *Red* – Electrical power lines, cables, conduit, and lighting cables
- *Yellow* – Gas, oil, steam, petroleum, or gaseous materials
- *Orange* – Communication, alarm or signal lines, cables, or conduit
- *Blue* – Potable water
- *Green* – Sewers and drain lines
- *Purple* – Reclaimed water, irrigation, and slurry lines
- *White* – Proposed excavation
- *Pink* – Temporary survey markings

On Site

Safety Manuals

A series of easy-to-read and use equipment safety manuals, such as those shown here, are available from the Association of Equipment Manufacturers (AEM), formally the Equipment Manufacturers Institute (EMI). These manuals were produced by the EMI to provide equipment owners, operators, service personnel, and mechanics with the basic safety procedures and precautions that are pertinent in the day-to-day operation and maintenance of various types of industrial equipment. These safety manuals are generic in nature and are to be used in conjunction with the operator/user manuals provided by manufacturers for a specific piece of equipment.

49112-11_SA01.EPS

2.3.0 In Case of a Rupture

If you come into contact with a buried utility, you must contact the utility company that owns the line. In some instances, as with gas lines, you should call 911. Do not resume the work until the utility company that owns the line has instructed you that it is okay to do so. The following guidelines describe the general actions to take in the event of striking a buried utility. Always follow your company policies and procedures and all local, state, and federal ordinances should a utility line be ruptured.

- *Electrical utility line:*
  - The safest place is on the machine. If you are on the excavator, compact trencher, or boring machine, do not leave the machine. Raise the backhoe, reverse the trencher, or reverse the drilling direction of the boring machine to try and break contact.
  - If you are on the ground operating a pedestrian trencher or boring machine, or assisting the equipment operator, stay where you are and don't move. Do not touch the equipment that has contacted the line or anything connected with the equipment.
  - If you are on an insulated mat, do not leave the mat.
  - Warn others that an electrical strike has occurred and to stay away.
  - Do not resume operations until you have been cleared to do so by the utility company.
- *Gas line:*
  - Warn others that a gas strike has occurred and to stay away.
  - Immediately shut off all engine(s) and remove any ignition sources.
  - Do not smoke or do anything that could cause a spark.
  - If you are on the excavator, backhoe, or boring machine, leave the machine immediately after shutting off the engine. Do not raise the backhoe, reverse the trencher, or reverse the drilling direction of the boring machine to try and break contact, as this may cause a spark.
  - Evacuate and secure the work site so that no one will enter the hazardous area.
  - Call 911 or the utility's emergency number and other local emergency personnel to report the leak.
  - Do not resume operations until you have been cleared to do so by the utility company.

- *Fiber-optic line:*
  - Raise the backhoe, reverse the trencher, or reverse the drilling direction of the boring machine to try and break contact.
  - Do not look into the cut ends of a fiber-optic line or any other unidentified cable. A cut fiber-optic cable can cause severe eye injury if you look into the damaged end of the cable.
  - Warn others that a fiber-optic line rupture has occurred.
  - Contact the owner of the fiber-optic line so that they can make repairs.
- *Water or sewage line:*
  - Warn others that a water or sewage strike has occurred.
  - Try to contain the seeping water if possible and slow down the leak.
  - Do not come in contact with leaking sewage as it may contain pathogens. It may be advisable to seek medical attention for personnel coming in contact with sewage.

3.0.0 Trenchers

Trenchers are used to dig trenches for the installation of underground cables. There are different styles of trenchers available, depending on the scale of the work to be performed. This module discusses two types of trenchers and provides examples of operating procedures for each type. The two styles of trenchers are pedestrian walk-behind trenchers (*Figure 2*) and compact or ride-on trenchers (*Figure 3*). Trenchers use either a gasoline or diesel engine to drive a hydraulic system that provides power to the drive wheels and the digging boom.

49112-11_F02.EPS

Figure 2 Pedestrian trencher.

49112-11_F03.EPS

Figure 3 Compact trencher.

3.1.0 Trencher Assemblies

The two types of trenchers each have a boom. The boom provides the path that the digging chain follows when in operation. The boom is raised, lowered, and moved from side to side using hydraulic cylinders. The size of the boom determines the maximum depth of the trench. Digging chains have replaceable digging teeth that remove material from the ditch. Both trenchers use an engine to power the hydraulic system components and to move the trencher.

The pedestrian machine (*Figure 4*) uses a gasoline engine and the compact trencher (*Figure 5*) has either a gasoline or diesel engine. The engines provide power to move the machines around the work area and to move the trencher forward as the materials are removed from the trench. Both machines have an operator control area. These areas and controls differ because of the way each of the trenchers is operated. The pedestrian machine is used by the operator walking behind the trencher. The compact trencher is a ride-on machine with associated controls for driving the trencher. It is also equipped with a blade for backfilling the trench when work is complete.

Some compact trenchers are also equipped with cable layers and plug-type augers (*Figure 6*). The cable layers will place the cable into the trench

Figure 4 Pedestrian trencher components.

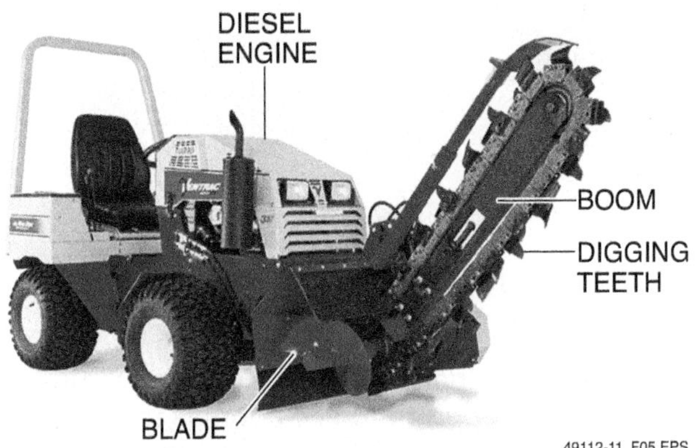

Figure 5 Compact trencher components.

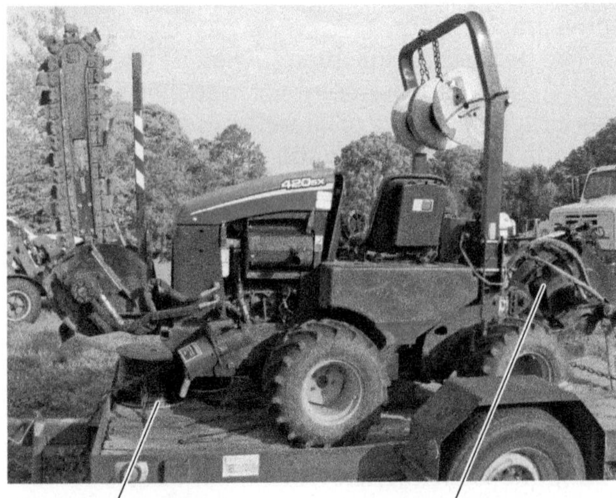

Figure 6 Additional trencher components.

as it is being dug. The plug-type auger is used to install a line underground for a short distance, such as 20 feet. This allows a line to be installed under a driveway, sidewalk, or other obstruction. The trencher hole drill typically uses 8-foot sections of drill pipe to bore into the ground.

3.2.0 Trencher Operator Qualifications

Only trained, authorized persons are allowed to operate a trencher. Operators of trenchers must be mentally and physically qualified in order to operate the equipment safely. Personnel who have not been fully trained in the operation of trenchers may only operate them for the purpose of training. The training must be conducted under the direct supervision of a qualified trainer.

3.3.0 Typical Trencher Controls

This section covers the typical controls associated with pedestrian-style and compact trenchers. This discussion is for explanation only and may not include all controls that could be found on all trenchers. The operator must read and fully understand the operator's manual provided by the equipment manufacturer. The operator's manual contains specific instructions for the operation of the trencher. *Figure 7* shows a typical pedestrian trencher control panel. The panel contains the controls and indicators the operator uses when moving or digging with the trencher. The following text identifies the controls and indicators, including a brief explanation of the purpose of each component:

- *Axle lock control* – This control shifts the trencher in and out of two-wheel drive. Two-wheel drive is used for digging, driving over rough ground, and loading and unloading the trencher from a trailer. It is also used when the trencher is parked or is being left unattended. Trencher maneuverability is reduced in two-wheel drive mode.
- *Boom depth control* – This control raises and lowers the digging boom.
- *Engine oil pressure light* – This light indicates when engine oil pressure has dropped below an acceptable level. If the indicator lights during operation, the engine should be turned off and the oil level checked.
- *Operator presence switch* – This switch must be pressed whenever the speed and direction control is moved from the neutral position or the digging chain control is moved from the stop position. Releasing the switch causes the engine to shut off.

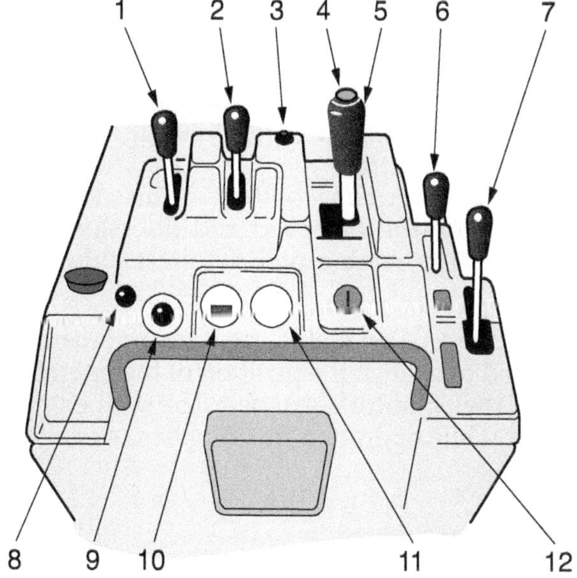

Figure 7 Pedestrian trencher control panel.

- *Speed/direction control* – This control selects the speed and direction of machine travel and provides steering control. The speed/direction control must be in neutral when starting the trencher.
- *Pump clutch* – The pump clutch controls the application of engine power to the hydraulic pump and should be disengaged when starting the engine. The pump clutch must be engaged before using the speed/direction control, boom depth control, or digging chain control.
- *Digging chain control* – This control starts and stops rotation of the digging chain. It must be in the stop position before starting the engine. If it is not in the stop position, the engine will not start.
- *Throttle lever* – The throttle lever regulates engine speed.
- *Choke* – The choke helps to start a cold engine.
- *Hour meter* – The hour meter registers the engine operating time. It is used in the scheduling of lubrication and maintenance.
- *Voltmeter* – The voltmeter shows battery charge level. When the ignition switch is in the on position and the engine not running, it should read near 12V.
- *Ignition switch* – This is generally a three-position (stop/on/start) switch with a removable key.

Figure 8 shows a typical compact trencher control panel. The panel contains the controls and indicators the operator uses when moving or digging with the trencher. The following identifies the controls and indicators, including a brief explanation of the purpose of each component:

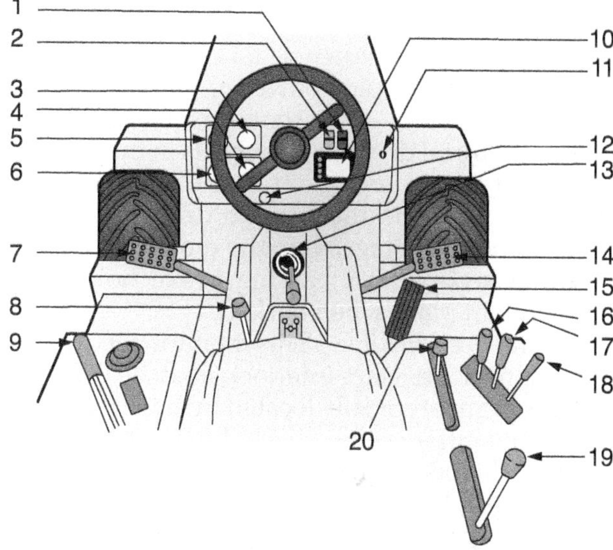

Figure 8 Compact trencher control panel.

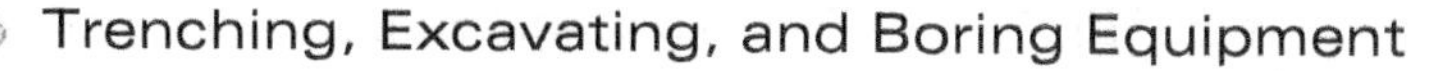

- *Engine high temperature warning light* – This warning light turns on when the engine temperature is too high. Some machines automatically shut down the engine when an engine high temperature condition is reached.
- *Engine low oil pressure warning light* – This warning light turns on when the engine oil pressure falls below a preset limit. If the indicator lights during operation, shut down the engine and check for the cause of the problem.
- *Air filter indicator* – This indicator monitors the condition of the air filter.
- *Hour meter* – This meter shows engine operating time. It is helpful in the scheduling of lubrication and maintenance.
- *Fuel gauge* – This gauge shows the level of fuel in the fuel tank.
- *Voltmeter* – The voltmeter shows battery charge.
- *Clutch pedal* – The clutch pedal is used to engage and disengage engine power.
- *Mobile-dig control* – This control is used to transfer mechanical drive to the wheels in the mobile mode or to the dig chain in the dig mode. It engages the hydraulic ground drive in the dig mode and disengages it in the mobile mode.
- *Parking brake* – The parking brake holds the trencher in place when parked.
- *Start interlock monitor* – This monitor checks the position of the start interlock switches. These interlocks are typically located in the seat, transmission shift, digging chain clutch, and crowd speed/directional control. If any interlocks are not correct, the engine will not turn over.
- *Ignition switch* – This is a three-position switch with a removable key.
- *Hand throttle* – This throttle controls engine revolutions per minute (rpm) for trenching. It must be returned to idle before the foot throttle can be used.
- *Transmission shift lever* – This lever is used to select a gear for travel in the mobile mode or to select the digging chain speed in the dig mode. It prevents the engine from starting unless in neutral.
- *Brake pedal* – The brake pedal is used to stop the trencher and hold it in place while the pedal is pressed. To brake in the mobile mode, push down on the clutch pedal, then press the brake pedal. To brake in the dig mode, position the crowd speed/directional control to neutral and press the brake pedal.
- *Foot throttle* – The foot throttle is used to control engine speed while backfilling or driving.
- *Blade lift* – The blade lift raises and lowers the backfill blade.
- *Blade angle* – The blade angle swings the backfill blade to the left or right.
- *Attachment lift* – This lift raises and lowers the digging boom.
- *Digging chain clutch* – This clutch is used to engage and disengage the digging chain. The engine will not start unless the digging chain clutch is in neutral.
- *Crowd speed/direction control* – This control is used to control the speed and direction when digging. The engine will not start unless this control is in neutral.

Other devices available are a seat adjustment lever used to change the position of the operator's seat and the operator's seat belt. The seatbelt must be worn while operating the trencher.

Case History

Trencher Safety

While following a trencher, and in the process of shoveling dirt back into the partially dug trench, an apprentice fell into the trench and was killed. Witnesses standing at the opposite end of the trench line noticed a rapid movement and something bright thrown into the air (the apprentice's hardhat) near the trencher. They immediately alerted the trencher operator who stopped the trencher and moved it away from the trench. Co-workers went into the trench to aid the victim as one of the witnesses called for emergency assistance. Unfortunately, the victim was pronounced dead at the scene.

The Bottom Line: Since no one actually saw what caused this accident, it was theorized that the apprentice either lost his balance and fell toward the machine or that his clothing got caught in the digging chain, causing him to be pulled into the moving chain and the trench. This accident could have been avoided had the following precautions been taken:

- The apprentice should have maintained a minimum 10' safety zone away from the moving digging chain.
- Before starting work, a job safety analysis for all the tasks involved with digging the trench should have been performed.

Note: Safety decals that alert the trencher operator and others to the dangers associated with trencher machine operation should be installed on all older trenching machines not equipped with such decals.

Source: U.S. Department of Energy.

3.4.0 Trencher Safety Precautions

The safe operation of a trencher is the operator's responsibility. The following are general safety precautions that must be observed when operating a trencher. Remember that the other safety precautions, already discussed in this module, also apply to the operation of a trencher.

- Incorrect procedures could result in death, injury, or property damage. Learn how to use the equipment correctly.
- Digging teeth can kill you or cut off an arm or a leg. Stay away from the boom when it is operating.
- The trencher may move when the chain starts to dig. Allow 3 feet between the end of the chain and any obstacles.
- Keep the digging boom low when operating on a slope.
- If there is any doubt about the classification of a job site, or if the job site could contain electrical utilities, classify and treat the job site as electrical.
- When practical, stay upwind of the trencher. This reduces the amount of dust and dirt blown toward the operator.

3.5.0 Pedestrian Trencher Operating Procedure

This section covers generic operation of a trencher. A power line worker, who may not be the operator, should expect to see these operations take place to ensure a safe work area. Each manufacturer provides detailed operating instructions with their equipment. Refer to the operator's manual provided with the trencher for specific instructions. The general guidelines given here are intended to help you understand the operator controls. Always perform a complete inspection of the trencher and correct any deficiencies before beginning operation. Each manufacturer provides specific inspection procedures in the operator's manual. The checklist normally includes the following types of items:

- Check the engine oil level, hydraulic oil level, fuel level, fuel filter, hoses, and clamps, air filter system, and all of the lubrication points.
- Inspect tires for proper pressure and general condition.
- Check all control function positions.
- Verify that all guards and shields are in place and secure.
- Inspect the condition of drive belts.
- Inspect the digging chain for tension and condition.
- Inspect the condition of the battery and cables.
- Verify that all safety warning signs are in place and readable.

3.5.1 Starting the Trencher

The procedures for starting the trencher are contained in the operator's manual. A typical starting procedure is as follows:

Step 1 Begin by disengaging the pump clutch. Place the speed/directional control in neutral and move the digging chain control to stop. If the engine is cold, use the choke.

Step 2 Insert the key in the ignition switch and turn it to the On position. Turn the key to start. When the engine starts, release the key. Idle the engine five to ten minutes before moving to the job site.

Step 3 Engage the pump clutch slowly while warming up the engine to warm the hydraulic oil.

Step 4 While the engine is idling, check the operation of the speed/directional control, boom depth control, and digging chain control.

3.5.2 General Operation

The procedures for moving a trencher around the job site when not digging, called mobiling the trencher, are contained in the operator's manual. Mobiling normally proceeds as follows:

Step 1 Using the boom depth control, raise the boom to the transport position (*Figure 9*). On rough terrain, keep the boom as low as possible to help with stability.

49112-11_F09.EPS

Figure 9 Boom raised in transport position.

Step 2 Engage and disengage the axle lock to match ground conditions. On level ground, the axle lock should be disengaged to increase maneuverability. On rough ground, the axle lock should be engaged to improve traction. The axle lock should always be in the engaged position when the machine is left unattended.

Step 3 Press the operator presence switch down. Use the speed/directional control to move the machine in the desired direction.

Step 4 Steer the machine with the speed/directional control.

The stop mobiling procedure is also contained in the operator's manual. A typical procedure is as follows:

Step 1 Move the speed/direction control to the neutral position.

Step 2 Release the operator presence switch and place the throttle to the idle position.

Step 3 Using the boom depth control, lower the digging boom to the ground. Engage the axle lock and disengage the pump clutch.

Step 4 Turn the ignition key to Off. If you are leaving the machine, remove the key from the ignition switch.

3.5.3 Trench Digging

The trenching procedure is also contained in the operator's manual. A typical procedure is as follows:

Step 1 Begin digging the trench by slowly engaging the pump clutch. Drive the trencher to a point in line with the intended trench. Digging is done with the trencher moving away from the digging boom (*Figure 10*). Align the trencher with the boom pointing at the starting end of the trench.

Step 2 Place the speed/directional control to neutral. Engage the axle lock. Lower the digging chain to within 1 inch of the ground using the boom depth control.

Step 3 Place the digging chain control to the rotate position. The digging chain will begin to move. Slowly lower the digging boom to the desired digging depth using the boom depth control.

Step 4 When the digging boom reaches the desired depth, adjust the engine throttle for optimum digging speed.

49112-11_F10.EPS

Figure 10 Trenching.

CAUTION

Do not overload the engine at low rpm as this could cause it to stall.

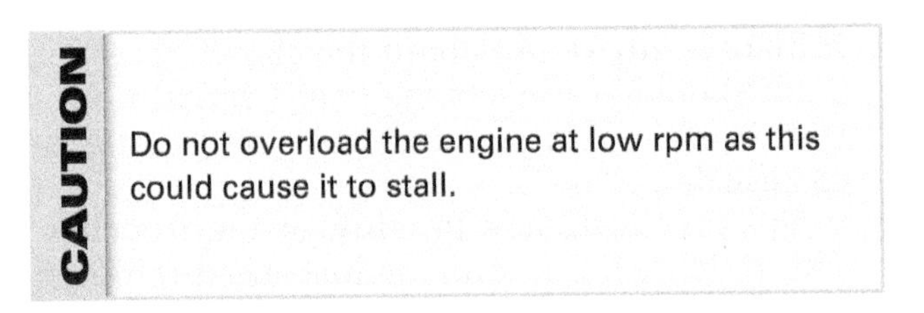

Step 5 Use the speed/directional control to begin trenching and to regulate the ground travel speed. Find the best ground travel speed for digging without causing the engine to lug down. Minor, slow direction changes can be made using the speed and direction control.

The stop trenching procedure is also contained in the operator's manual. A typical procedure is as follows:

Step 1 To stop digging, place the speed/direction control to neutral. Using the boom depth control, raise the boom to ground level.

On Site

Use Your Senses

During normal operation of the trencher, listen for thumps, bumps, rattles, squeaks, squeals, or other unusual sounds. Smell for odors such as burning insulation, hot metal, burning rubber, or hot oil. Feel for any changes in the way the trencher is operating. Look for problems with wiring and cables, hydraulic connections, or other equipment. Correct anything you hear, smell, feel, or see that is different from what is expected, or that seems to be unsafe.

Step 2 Place the digging chain control in the stop position. Raise the boom all the way up using the boom depth control.

Step 3 Use the speed/direction control to drive away from the trench. On level ground, disengage the axle lock and shut down the machine.

3.5.4 Compact Trencher Operations

Some additional requirements must be discussed concerning the operation of a compact trencher. This discussion is limited to the major differences between the two styles of trenchers. A power line worker, who may not be the operator, should expect to see these operations take place to ensure a safe work area. Each manufacturer provides detailed operating instructions with their equipment. Refer to the operator's manual provided with the compact trencher for specific instructions. This section is intended to familiarize you with typical operations associated with a compact trencher. The operator should perform a complete inspection of the trencher and correct any deficiencies before beginning operation. All preoperational checks for the compact trencher are identical to those for the pedestrian trencher.

3.5.5 Summary of Compact Trencher Operational Differences

This section summarizes the operating differences between pedestrian and compact trenchers. Each item is an action the operator is required to perform. These actions are in addition to the actions discussed for the pedestrian trencher. Remember, the biggest difference is that the operator is riding on the compact trencher during operation. Before starting the compact trencher, perform the following:

- Before beginning operation, fasten and adjust the operator's seat belt.
- Verify that the parking brake is set and the crowd speed/directional control, transmission shift, and digging chain clutch are all in neutral.
- Set the parking brake.
- Press the clutch pedal and place the transmission gear shift in neutral.

Before mobiling with the compact trencher, perform the following:

- Before mobiling, fasten and tighten the operator's seat belt and start the engine.
- Raise the boom and backfill blade off the ground.
- Push down on the brake pedal and release the parking brake.
- Press in the clutch pedal and place the mobile/dig control in the mobile position.
- Place the transmission shift into gear.
- Press the foot throttle and slowly release the clutch pedal.
- As the clutch engages, release the brake pedal. The trencher will begin moving.
- Adjust the engine speed with the foot throttle.
- Stop the trencher by depressing both the clutch and brake pedals.

Before beginning to dig with the compact trencher, perform the following:

- Fasten and adjust the operator's seat belt.
- Place the mobile/dig control to the dig position.
- Lower the backfill blade to the ground. This stabilizes the unit when digging is started.
- Press the clutch pedal and put the transmission shift into gear. Select an appropriate gear for the ground and digging conditions.
- Slowly release the clutch pedal to start the digging chain.
- Lift the backfill blade from the ground.
- Release the parking brake and move the crowd speed/direction control forward until the engine is loaded, but is not laboring.
- When the trench is finished, stop the machine by moving the crowd speed/direction control to the neutral position.
- Push down on the clutch pedal.
- Move the transmission shift into neutral.
- Place the mobile-dig control into the mobile position.
- Move the transmission shift into gear and mobile away from trench.

3.5.6 Backfilling with a Compact Trencher

To begin backfilling using a compact trencher, have the engine at idle and place the mobile/dig control in the mobile position. Next, swing the blade to one side with blade angle control and lower the blade to the ground using the blade lift lever. Push down on the clutch pedal and move the transmission shift into gear, then push down on the foot throttle and slowly release the clutch pedal. Begin pushing the soil back into the trench. Adjust the blade height as required to match the ground conditions. Adjust the speed using the foot throttle. After backfilling, straighten and raise the blade all the way up. Mobile away from the covered trench.

3.6.0 Trencher Operator's Maintenance Responsibility

The instruction manual provided with a trencher normally includes a preventive maintenance schedule in chart form. Most charts include a lube column that tells the operator the type of lubricant to use and a diagram that shows the points at which lubricants are applied. Other types of maintenance required include changing the engine oil, lubricating the digging chain clutch, and changing the hydraulic oil filter. The proper maintenance will extend the life and performance of the machine.

4.0.0 Backhoe/Loader

A backhoe/loader (*Figure 11*) is a dual-purpose, highly maneuverable machine used to dig trenches, foundations, and similar excavations. It is also used to move dirt, crushed stone, gravel, and other materials around the job site. Backhoes/loaders are equipped with either a gasoline or diesel engine. The backhoe bucket and loader bucket attachments are hydraulically operated under the control of a single operator who performs all backhoe and loader operations.

4.1.0 Backhoe/Loader Operator Qualifications

Before operating a backhoe/loader, the following requirements must be met:

- The operator must successfully complete a training program that includes actual operation of the backhoe/loader.
- The operator is responsible for reading and understanding the safety manual and operator's manual provided with the equipment.
- The operator must know the safety rules and regulations for the job site.

Figure 11 Backhoe/loader.

Figure 12 Operator controls.

4.2.0 Backhoe/Loader Controls

Depending on the manufacturer and model of backhoe/loader, there can be several different arrangements in the operating controls and their locations. For this reason, always refer to the operator's manual for the specific equipment you are operating in order to become familiar with the different controls and their locations. The basic controls (*Figure 12*) used to operate any backhoe/loader fall into three categories:

- Vehicle engine and movement related controls
- Backhoe bucket and stabilizer controls
- Loader bucket controls

4.2.1 Vehicle Engine and Movement Related Controls

Most of the backhoe/loader engine- and movement-related controls and their functions are similar to those found in other vehicles. These controls include the engine start switch; transmission lock control and direction control lever; steering wheel; steering column tilt control; accelerator pedal; parking brake; and service brake. Gauges used to monitor engine performance can include battery, engine coolant, tachometer, oil pressure, transmission oil temperature, and service-hour meter. Some engine- and movement-related controls and their functions unique to some backhoes/loaders are as follows:

- *Accelerator lever* – Used to control the engine speed between low and high idle during backhoe operation.

- *Differential lock pedal* – Used to engage the differential lock to prevent wheel slippage when moving on soft or wet ground.

Some backhoes/loaders are equipped with an all-wheel drive and/or an all-wheel steering capability. The all-wheel drive capability enables the operator to switch from two-wheel drive to all-wheel drive any time additional traction is needed.

All-wheel steering capability provides three modes of steering to suit various job-site conditions: two-wheel steering, all-wheel steering, and circle steering. Two-wheel steering is normally used to operate the machine on the road or other surfaces when additional maneuvering ability is not needed. In this mode, only the front axle is used to steer the backhoe/loader. All-wheel steering allows the operator to choose independent rear axle operation so that the rear wheels will move the back of the backhoe/loader either to the left or right when the unit is moving forward. This allows the operator to position the front and back wheels in opposite directions or in the same direction. Turning the front and back wheels in opposite directions is typically used when moving the backhoe/loader around tight corners. Turning the wheels in the same direction is done when crab steering is desired. The operator can determine the position of the rear axle by looking at a rear axle position gauge provided for this purpose. Selection of the circle mode reduces the turning radius of the backhoe/loader, allowing for better steering in confined spaces.

NOTE

Before changing from one steering mode to another, always center both the front and rear wheels.

4.2.2 Backhoe Bucket and Stabilizer Controls

The stabilizer controls are used to lower and raise the left and right stabilizer arms. Before using the backhoe bucket, the stabilizer arms must be lowered firmly to the ground in order to provide a firm and level base for backhoe operation. The left and right stabilizers are independently controlled by separate stabilizer levers. Each lever has three positions: down, hold, and up. Moving the lever in the down direction lowers the associated stabilizer. When the stabilizer firmly contacts the ground, lowering it further causes the rear of the backhoe to be raised. Moving the lever in the up position raises the stabilizer, thus lowering the rear of the backhoe. Releasing the stabilizer lever either from the down or up position causes the lever to return to the hold (center) position, stopping stabilizer movement.

Backhoe operation is controlled by the backhoe boom and backhoe stick and bucket controls. The arrangement of these controls may vary from one machine to another. Some machines use a two-lever control scheme; others use a three-lever scheme with a foot swing control. Still others use a four-lever control scheme. The two-lever control scheme is briefly described here.

A backhoe with a two-lever control scheme uses one lever to control the backhoe boom and the other to control the stick and bucket. The backhoe boom has a five-position lever, with two positions to lower and raise the boom and two positions to swing the boom to the left or right. Movement of the boom is stopped when the lever is in the hold (center) position. The lever returns to the hold position when released from any of the other four positions.

The backhoe stick and bucket lever also has a five-position lever. Two positions are used to extend and retract the stick and two positions to load and dump the bucket. Movement of the stick and/or the bucket is stopped when the lever is in the hold (center) position. The lever returns to the hold position when released from any of the other four positions.

A boom lock lever is used to enable or prevent boom operation. When in the lock position, a boom lock is engaged to prevent the backhoe from moving and swinging into objects or into traffic. When in the release position, movement of the backhoe is enabled for digging operations.

4.2.3 Loader Bucket Controls

The loader bucket for the equipment being described here is controlled by an eight-position lever. The control positions are as follows:

- *Float* – Allows the loader bucket to move freely, following the contour of the ground.
- *Lower* – Lowers the loader bucket.
- *Hold* – Stops the movement of the loader bucket. The lever returns to the hold position when it is released from any other position, except for the float position. The lever stays in the float position until it is moved by the operator to the hold position.
- *Raise* – Raises the bucket.
- *Tilt back* – Tilts back the loader bucket.
- *Return-to-dig* – Returns the loader bucket to the dig position. The lever stays in this position until the bucket is level, then it automatically returns to the hold position.

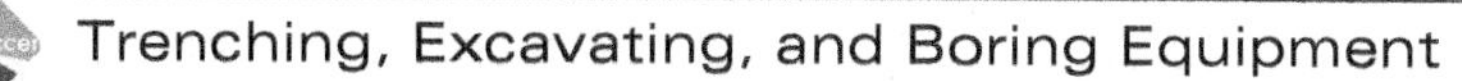

- *Dump* – Used to empty the loader bucket.
- *Quick dump* – Shortens the time required to tilt the loader bucket.

4.3.0 Backhoe/Loader Safety Precautions

The following is a list of general safety rules specific to the operation of a backhoe/loader. Remember that the other safety precautions, already discussed in this module, also apply.

- Mount the backhoe/loader only at locations that are equipped with steps and/or hand-holds.
- Face the backhoe/loader while entering and leaving the operator's compartment.
- Do not mount a backhoe/loader while carrying tools or supplies.
- Do not mount a moving backhoe/loader.
- Do not use controls as handholds when entering the operator's compartment.
- Do not obstruct your vision when traveling or working.
- Never lift, move, or swing a load over a truck cab or over workers.
- When traveling, operate at speeds slow enough so that you have complete control of the backhoe/loader, especially when traveling over rough or slippery ground and on hillsides. Never place the transmission in neutral in order to allow the backhoe/loader to coast.
- Never approach overhead power lines with any part of the backhoe/loader unless you are in strict compliance with clearance restrictions as mandated by local, state, and federal regulations.
- Make sure that you know the underground location of all gas and water pipelines and of all electrical or fiber optic cables.

When operating the backhoe the following safety precautions apply:

- Never enter or allow anyone to enter the backhoe swing pivot area.
- Operate the backhoe from the correct backhoe operating position. Never operate the backhoe controls from the ground. Never allow riders in or on the backhoe.
- Do not dig under the backhoe or its stabilizers. This is necessary in order to prevent cave-ins and the chance of the backhoe falling into the excavation.
- When operating the backhoe on a slope, swing to the uphill side to dump the load, if possible. If necessary to dump downhill, swing only as far as required to dump the bucket.
- Always dump the soil far enough away from the trench to prevent cave-ins.

When operating the loader, the following safety precautions apply:

- Carry the bucket low for maximum stability and visibility.
- Stay in gear when traveling downhill and use the same gear range as you would for traveling up a grade.
- When on a steep slope, drive up or down the slope. Do not drive across the slope. If the bucket is loaded, drive with the bucket facing uphill. If empty, drive with the bucket pointed downhill.
- When operating the loader, make sure that the backhoe is in the transport lock position to prevent backhoe movement.
- When working at the base of a bank or overhang, never undercut a high bank and/or operate the loader close to the edge of an overhang or ditch.

4.4.0 Backhoe/Loader Operating Guidelines

The following section covers some basic operations done with a backhoe/loader. A power line worker, who may not be the operator, should expect to see these operations take place to ensure a safe work area. Each manufacturer provides detailed operating instructions with their equipment. To obtain specific operating instructions, refer to the operator's manual provided with the backhoe/loader being used.

4.4.1 Backhoe/Loader Inspection

Before operating the backhoe/loader, the operator should inspect the equipment and attachments and perform the recommended operator's daily maintenance procedures in accordance with the manufacturer's instructions. Examples of the daily maintenance tasks are given later in this section. A backhoe/loader with defective or missing parts must be repaired before operation.

4.4.2 General Backhoe/Loader Operating Guidelines

Startup and warm-up of the backhoe/loader engine should be done according to the manufacturer's instructions. Remember that the duration of the engine warm-up period is determined by the prevailing temperature; the colder the temperature, the longer the required warm-up period. During the warm-up period, check the gauges and controls for proper indications/operation. Follow these general guidelines pertaining to machine travel:

- During travel, raise all attachments enough to clear any expected obstacles.
- Make changes in the direction of travel by setting the transmission direction control lever to the desired speed and direction. When making changes in direction, the speed of the backhoe/loader should be reduced.
- Select the best gear speed for conditions before starting down a hill, and do not change gears while going down the hill. When going downgrade, the rule of thumb is to use the same gear speed that would be used if going upgrade. Do not allow the engine to overspeed when going downhill. If provided, activate the all-wheel drive function when operating on a hill or when additional traction and/or braking are needed.
- Avoid turning on a slope.
- When the load will be pushing the backhoe/loader, put the transmission lever in the first speed position before you start downhill. Engage the all-wheel drive function, if provided.

CAUTION

Never use the float position to lower a loaded bucket. Damage to the machine can result from the bucket falling too fast.

When the work is finished and the backhoe/loader is to be shut down, lower all the attachments to the ground. Shift the controls to neutral/park-and-lock (if provided) and set the parking brake. Before turning off the engine, allow it to idle for a short time in order to cool down. After turning off the engine, cycle the hydraulic controls. If the backhoe/loader is being parked on an incline, block the wheels.

4.4.3 Backhoe Operating Guidelines

The backhoe must be prepared for operation and operated according to the manufacturer's instructions. When digging with the backhoe, follow these general guidelines:

- When digging, close the bucket slowly and move the stick inward at the same time for maximum performance.
- Keep the bucket teeth at an angle that gives the best earth penetration. This helps prevent the bucket from just scraping the ground.
- Apply downward pressure with the boom to increase bucket penetration in hard-packed ground conditions.
- Move the boom downward to close the bucket completely. Move the stick and the bucket outward from the backhoe slightly while lifting the bucket from the excavation. Then, swing the bucket to the side and dump the load as you approach the pile.
- After the bucket is emptied, the stick and bucket should be operated to the closed dig position and returned to the excavation.
- When backfilling, lift the bucket over the dump pile; then pull the bucket inward and lift the boom evenly.

CAUTION

When digging with the backhoe, do not swing the bucket against the sides of the excavation and/or dump pile. Such actions can cause structural damage to the boom and result in premature wear of the boom pin and bushings.

4.4.4 Loader Operating Guidelines

The loader should be prepared for operation and operated according to the manufacturer's instructions. When using the loader, follow these general guidelines:

- Carry an empty or loaded bucket low to the ground to achieve equipment stability and better vision.
- When carrying a load downhill, travel in reverse. When carrying a load uphill, travel forward.
- To load the bucket, skim the ground as you move the loader forward. Move the control lever in order to lift the bucket and tilt it back. Fully tilt back the bucket to avoid spillage.
- Move the loader as close as possible to the truck before dumping the load. Dump the load in the center of the truck bin. On successive dumps, load the truck starting from the front of the truck bin and working toward the rear.
- When possible, keep the wind to your back when picking up and dumping loads. This helps keep dust and debris from getting on the operator and equipment.

4.5.0 Backhoe/Loader Maintenance

The manufacturer's manual provides a schedule and procedures to be followed when performing maintenance on a backhoe/loader. Before performing any maintenance the operator must be trained, authorized, and have the proper tools to perform any procedures. The following are examples of the types of maintenance an operator performs on a daily basis:

- Clean the radiator core.
- Clean the windows.
- Check the engine fuel level.
- Check the cooling system level.
- Check the engine oil level.
- Check the hydraulic system oil level.
- Check the transmission oil level.
- Check the brake reservoir oil level.
- Check the engine air filter service indicator.
- Check the tire inflation.
- Check the wheel nut torque.
- Check for the proper installation of access covers and the guards.
- Drain the fuel system water separator.
- Lubricate the backhoe boom, stick, bucket, and cylinder bearings.
- Lubricate the front kingpin bearings.
- Lubricate the loader bucket, cylinder, and linkage bearings.
- Lubricate the stabilizer and cylinder bearings.
- Lubricate the swing frame and cylinder bearings.

5.0.0 Directional Boring Equipment

Directional boring, also known as horizontal directional drilling (HDD), is a trenchless method of installing underground pipes, electrical conduit, and cables. It is used when traditional trenching methods are not practical or when trenching would be too disruptive to the area where the line needs to be buried. It is suitable for most types of soil and is primarily used for crossing beneath roads, rivers, railroads, landscapes, congested areas, and environmentally sensitive areas. HDD is also used to place installations under small bodies of water (*Figure 13*). The advantages of using directional boring include lower cost, less traffic disruption, deeper and longer installations, shorter completion times. In many cases, it is also safer for the environment.

While HDD equipment simplifies underground installation projects, it also presents some additional safety hazards. In addition to having the knowledge and ability to recognize and avoid common job-site hazards on HDD projects, workers must also recognize and avoid the hazards specific to HDD. These hazards apply to anyone working near the boring equipment or assisting the machine operator.

HDD jobs can vary from installing pipes or cables that are as small as 2 inches in diameter up to those with diameters as large as 48 inches. For this reason, directional boring machines (*Figure 14*) are available in various sizes and configurations. Most are rated by the amount of thrust, pullback force, and rotary pressure. Thrust and pullback are typically measured in the amount of pounds of force, while rotational power is measured in foot-pounds of pressure.

5.1.0 Directional Boring Process

The directional boring machine is set up at the entry point and two power stake-down augers (*Figure 15*) are drilled into the ground at the front of the drilling machine. These stabilize the machine and prevent movement as boring operations begin.

49112-11_F14.EPS

Figure 14 Horizontal directional boring machine.

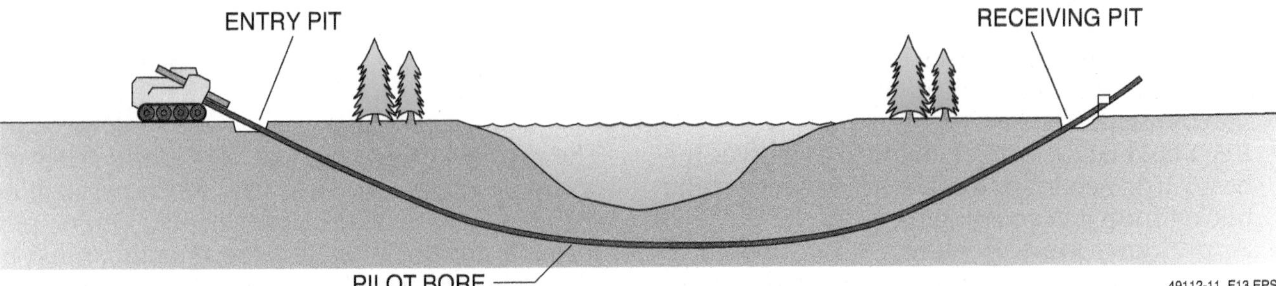

49112-11_F13.EPS

Figure 13 HDD installation under a river bed.

49112-11_F15.EPS

Figure 15 Power stake-down augers.

49112-11_F16.EPS

Figure 16 Drill cutting heads.

49112-11_F17.EPS

Figure 17 Fluid tank truck.

The boring machine drills a drill string into the ground along the planned path of the cable or conduit to be buried. This drill string is made up of several sections of pipe or conduit called drill rods or drill string. A cutting head (*Figure 16*) is attached to the first length of drill rod. The first length of this drill rod is bored into the ground at the entry point of the planned burial. As the rod is inserted into the ground, additional rods are attached to the end of the previous rod to continue the dig.

This initial boring is known as drilling the pilot hole. As the pilot hole is dug, lubricating fluid is pumped through the drill rods to lubricate the cutting tip of the drill head. The fluid tank is typically transported on a separate trailer or flatbed truck and parked close to the drilling machine (*Figure 17*).

The drill head is guided using aboveground tracking systems that monitor the location of the drill head (*Figure 18*). These are handheld devices known as drilling tool locators that the ground assistant monitors and provides feedback to the boring machine operator on the direction the drill head needs to move.

Upon reaching the exit point, the drill bit is detached and the end of the drill pipe is attached to a back-reamer (*Figure 19*), if the bore hole must be enlarged. The reamer is pulled back through the hole while rotating the drill pipe to enlarge the bore hole.

Additional drill pipe is added behind the reamer so that there is always pipe in the bore hole. When the bore hole is about 25 percent larger than the pipe, conduit, or cable to be installed, the reamer is connected to a swivel that is connected to the product pipe or cable (*Figure 20*). The product pipe or cable is then pulled back through the drill hole while the drill pipe and reamer are rotated. The swivel prevents rotation of the product pipe or cable.

For some telecommunications or power cable projects, the drill pipe itself becomes the conduit and is left in the ground upon reaching the

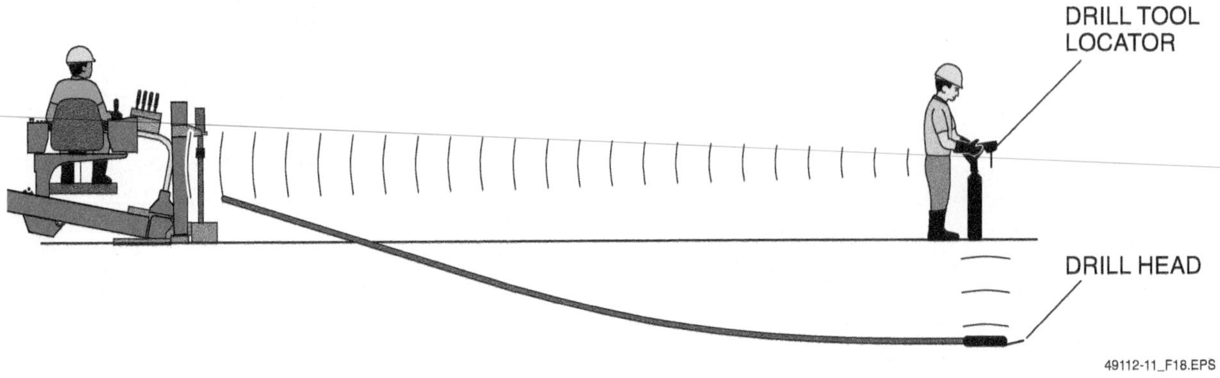

Figure 18 Tracking drill head.

exit point. This type of installation is known as drill and leave. Once the pipe, conduit, or underground cable is installed, the exit and entry points are excavated as necessary.

Figure 19 Back-reamer.

Figure 20 Back-reamer and swivel.

5.2.0 Directional Boring Safety Precautions

Due to the nature of this work, boring equipment has some unique safety features that do not typically apply to trenchers and excavators. Always follow the guidelines included on the warning and danger tags posted on the equipment (*Figure 21*). Manufacturers of this equipment also recommend additional safety practices to help

Figure 21 Warnings and dangers.

protect the operator and other personnel at the job site assisting the operator.

5.2.1 Personal Protective Equipment

You must wear protective equipment when operating the boring machine (*Figure 22*). This equipment typically includes a hard hat, wrap-around safety goggles, hearing protection, and electrically insulated boots. If you are working in a high traffic area, you should also wear a reflective vest.

The operator of a boring machine with an operator seat is not typically required to wear electrically insulated gloves while operating the machine, but should wear electrically insulated boots. This will protect the operator against an electrical shock in case he or she inadvertently steps off the machine during an electrical strike.

The operator of a smaller boring machine that requires the operator to stand on the ground should always stand on an equal-voltage insulated mat and wear electrically insulated boots and gloves while operating the machine. These types of boring machines should also be set up with the tracks on an insulated mat.

Anyone installing the electric strike system or assisting the operator on the ground during the bore must wear electrically insulated boots and gloves. The ground assistant holding the drilling tool locator must also wear electrically insulated boots. These should be high-top boots with pant legs tucked completely inside. Boots must meet local and federal electrical hazard protection guidelines. These guidelines typically require testing for adequate protection at 14,000 volts. At a minimum, gloves must have protection ratings at 17,000 AC maximum use voltage. If working around higher voltages, use gloves and boots with higher ratings.

49112-11_F22.EPS

Figure 22 Personal protective equipment.

5.2.2 Radio Communications

The distance between the boring machine and the ground assistant monitoring the drill head locator may prevent visual contact and direct voice communication. The boring machine operator and the ground assistant should use two-way radios to communicate with each other. Always use good quality radios with sufficient communication range and test the radios before operation begins to ensure that communications can be heard above background noise. The radio used by the ground assistant along the bore path and at the exit location must be assigned to one designated person. This person will be the only person communicating with the boring machine operator. During these communications, always begin the communications by stating your name and the name of the person you are talking to so there is no question as to who should be speaking on the radio. After receiving instructions, always confirm receipt of the instructions by repeating them. This will ensure clear communications between the boring operator and the ground assistant. If no acknowledgement is received, send the message again until an acknowledgement is returned.

5.2.3 Removing Components

When the drill head emerges at the exit point, specific lockout procedures must be followed before removing the drill head from the drill pipe.

> **WARNING!**
>
> Drill rod rotation and thrust can cause serious injury or death if rotation is started unexpectedly. If rotation starts, you could be struck by any tools attached to the drilling rod or entangled in rotating parts. Always follow the manufacturer's lockout procedure before removing or installing any components to the drill pipe.

Newer designs use connections that do not require the use of pipe wrenches or tongs to remove the drilling head from the drill pipe. These designs include Spinelok® connections and hex collar connections that secure components to the pipe using pins and sleeves. These designs help prevent injury from long pipe wrenches being left on the pipe when it is started. Always follow the specific manufacturer's instructions when removing and installing drill pipe components.

If threaded connections are used, special tools must be used to uncouple connections. Always follow the manufacturer's recommendations and instructions for using these tools safely.

WARNING!

When attaching a hex-coupled or direct-threaded component to the drill rod, never hold the tool or component. Use the machine-powered rotation of the drill rod to connect the component.

Regardless of the system used to remove these components, a lockout procedure must be followed that will typically include the following steps:

Step 1 Shut off the boring machine ignition and remove the key.

Step 2 Carry the key to the exit location and hand the key to the ground assistant holding the other radio. The key must remain in this person's possession at all times while work is completed on the drill pipe.

Step 3 After components are removed and/or installed, the key is returned to the boring machine operator.

Step 4 The machine must not be restarted until the area is clear and the ground assistant radios the boring operator that it is safe to resume operations.

5.2.4 Equipment Grounding and Strike Alert System

The power stake-down augers used to stabilize the boring machine during operation are sufficient grounding rods when they are driven into the ground at full depth. If they are not driven to full depth, an optional ground rod must be driven into the ground away from the boring machine and connected with the grounding cable provided.

Electrical strike alert systems come standard on boring machines. Anytime a boring machine is operated, an electrical strike system must be properly set up, tested, and used during the entire drilling process. When used properly, they alert (warn) the operator of the boring machine that the drill pipe has come in contact with an electrical source. This information is relayed back to the machine sensor and an alarm is sounded.

WARNING!

The strike system does not prevent electric strikes or detect strikes before they occur. It is a warning device and not a protective device.

Case History

Loose Connection

While conducting drilling on a site, the drill string was being pulled from the hole. The driller loosened the top connection on an 8' drill rod where it was connected to the drive head. He then proceeded to raise the drive head and drill string to break the bottom connection to the drill rod. After unscrewing the bottom connection, the driller started to raise the drive head and attached drill rod in order to swing the drill rod out so that the driller's helper could remove it and place it on the pipe rack. As the drive head was raised, the drill rod (weighing approximately 70 pounds) came loose from the top connection. The helper, who was standing by to receive the drill rod, saw it come loose from the drive head and was able to step out of the way and let it fall to the drill deck. That was quick thinking on his part.

During an investigation of the incident it was determined that, although the threads on the drive head subassembly showed signs of wear, they were not worn enough to be entirely responsible for the rod falling. It was determined that the driller had loosened the top connection between the drill rod and the top drive subassembly too much. This compromised the holding power of the tapered threads, allowing the drill rod to come loose and fall. Fortunately, no one was hurt when this happened.

The Bottom Line: It is important to remember that your carelessness not only affects you, it affects those around you. Make sure all connections are tightened properly.

Source: U.S. Department of Energy.

If the alarm is activated, a strike has already occurred and all precautions must be taken to prevent the operator or ground assistants from being electrocuted. The electrical strike system (*Figure 23*) typically consists of the following components:

- *Voltage stake* – The voltage stake is a copper rod that is driven into the ground 6 feet or more away from the boring machine, but not over the drill path.
- *Alarm* – The alarm is mounted on the boring machine and will sound in the event of an electrical strike.
- *Flashing warning light* – A warning light will flash in the event of an electrical strike.

Figure 23 Typical electrical strike alert system.

- *Voltage test button* – The voltage test button is a pushbutton used to test the voltage sensing circuit. The audible alarm should sound and the flashing warning light should illuminate when this is pressed.
- *Current test button* – The current test button is a pushbutton used to test the current sensing circuit. The audible alarm should sound and the flashing warning light should illuminate when this is pressed.
- *Reset button* – The reset pushbutton will stop the alarm and reset the system after testing.
- *Current sensing coil* – The current sensing coil senses electrical current and is mounted to the front of the boring machine.
- *Electronic module* – The electronic module that monitors the system is typically mounted in or near the operator station on the boring machine.

The strike alert system detects electrical strikes by sensing voltage differences between the high voltage of the machine and the low voltage at the voltage stake. It also senses electrical current flow through the current sensing coil, indicating an electrical strike has occurred. Always follow manufacturer's recommendations when setting up and testing the electric strike system. The system should be tested before each dig. Do not operate the boring machine if the electric strike alert system fails to operate properly during the test.

Case History

Amputation Caused by Lack of Training

On Friday, November 16, 2001, a hollow-stem auger drill rig operator lost a finger in a pinch-point injury. The middle and index fingers of the same hand were broken at the knuckle. He also suffered a large cut on the back of his hand beginning at the knuckle of the middle finger, crossing the index finger, and extending toward the thumb. This cut severed the tendon to his index finger.

An investigation of the accident found that manually aligning extension rod sections to the threaded rod cap was a practice that the operator considered a routine and accepted risk. It is not! This action regularly exposed the operator's hand into the pinch-point hazard. That's why he lost his finger. During the investigation, it was discovered that the drilling subcontractor never provided written procedures or training to this worker.

The Bottom Line: Make sure you know how to do your job safely and that you get the training you need.

Source: U.S. Department of Energy.

Summary

As a power line worker, it is important that you be familiar with the different types of trenching, excavating, and boring equipment that you are likely to see on the job. This equipment includes trenchers, backhoes, and horizontal directional drilling (boring) machines. There will be many occasions when you will be working in close proximity to these machines. In some cases, you will be required to assist the operator. You may also be trained to operate some of this equipment. For all these reasons, it is important that you recognize the equipment, understand the operation of each machine, and recognize the safety hazards associated with each. This knowledge will help you perform your work safely and more efficiently.

Safety has been a major discussion point throughout this module. The safe operation of any trenching, excavating, and boring equipment is key to the health and well-being of yourself and your co-workers.

Review Questions

1. During excavation and trenching operations, compliance with the rules and regulations stated in *OSHA 29 CFR 1926, Subpart P – Excavations* is _____.
   a. optional
   b. voluntary
   c. mandatory
   d. not applicable

2. Using the color code system developed by the American Public Works Association red markings designate _____.
   a. electrical lines
   b. gas lines
   c. water lines
   d. sewage lines

3. If you are on the ground operating a pedestrian trencher or boring machine, and an electrical line is struck, _____.
   a. stay where you are and don't move
   b. leave the area immediately
   c. saturate the ground with water
   d. climb onto the machine

4. It is dangerous to look into the cut end of a(n) _____.
   a. water line
   b. fiber-optic cable
   c. drill string
   d. telephone line

5. The trencher may move when the chain starts to dig. The minimum clearance you should allow between the end of the chain and any obstacles is _____.
   a. 6 inches
   b. 1 foot
   c. 2 feet
   d. 3 feet

6. To raise the trencher boom to the transport position, use the _____.
   a. chain depth control
   b. boom depth control
   c. blade lift
   d. attachment lift

7. Digging is done with the trencher moving _____.
   a. away from the digging boom
   b. toward the digging boom
   c. faster than the digging boom
   d. slower than the digging boom

8. To begin backfilling using a compact trencher, have the engine at idle and place the mobile/dig control in the _____.
   a. neutral position
   b. off position
   c. dig position
   d. mobile position

9. The loader bucket control that allows the loader bucket to move freely following the contour of the ground is the _____.
   a. float control
   b. lower control
   c. hold control
   d. tilt back control

10. For some telecommunications or power cable projects, the drill pipe itself becomes the conduit and is left in the ground upon reaching the exit point. This type of installation is known as _____.
    a. permanent drilling
    b. drill and leave
    c. pit drilling
    d. back-reaming

11. Anyone installing the electric strike system or assisting the operator on the ground while the boring machine is being used must wear _____.
    a. metal mesh gloves
    b. ear plugs
    c. fall protection
    d. insulated boots and gloves

12. When using two-way radios during boring, always begin the communication by_____.
    a. blowing into the radio mouthpiece
    b. saying, "test, test"
    c. stating your name and the name of the person you are talking to
    d. tapping on the radio mouthpiece and saying, "come in"

13. When the boring drill head emerges at the exit point, the first step is to _____.
    a. check the condition of the drill head
    b. dig around the drill head to completely free it for easy access
    c. disconnect the drill head from the drill pipe
    d. perform specific lockout procedures

14. The electrical strike alert system _____.
    a. protects the worker from being electrocuted
    b. alerts the operator that the drill pipe is about to come into contact with an electrical source
    c. alerts the operator that the drill pipe has come in contact with an electrical source
    d. serves as an equipment ground for the machine

15. On boring machines, the strike alert system detects voltage differences between the high voltage of the machine and the _____.
    a. high voltage of any buried lines in the boring path
    b. low voltage at the current sensing coil
    c. low voltage at the voltage stake
    d. presence of voltage leakage in the ground

Trade Terms Quiz

Fill in the blank with the correct term that you learned from your study of this module.

1. The amount of force required to push or drive an object, normally measured in pounds of force, is known as ___________.

2. A term used to describe driving a trencher around the job site when not digging is ___________.

Trade Terms

Mobiling
Thrust

Cornerstone of Craftsmanship

Tony Vazquez

President/CEO VisionQuest-Academy,
Ocala, FL

Tony Vazquez came from a family of furniture-makers that emigrated from Cuba to the United States. Influenced by his father and grandfather, he entered the construction industry. It wasn't long before he and his father opened their own company, but the need to find good craftsmen led him to training. As the need for quality training became more apparent, he eventually opened his own construction training school. Tony enjoys sharing his love of building and encouraging others to excel in the industry that's made him a success.

How did you get started in the construction industry and who inspired you?

I was raised in a second-generation, construction/craft, working family. I was inspired as a boy by my father and grandfather back in Cuba, where they hand-made furniture. Later, here in the U.S., my father continued his craft as a construction worker and I, as a teenager, often went with him on weekends. After I left high school, my father and I started a construction company. The hard work that came with the job was outweighed by the gratification of seeing a well-built project that would last for many years. In 1990, while looking for trained construction workers, I ended up at the Miami Job Corps. I was impressed with the training going on in our industry. Later that year I volunteered my time to help with their program, and a few years later when the instructor retired, I was asked to take over. I returned to college and obtained my instructor/teacher certification in the industry and have enjoyed training others ever since.

What do you enjoy most about your job?

I still love to build. However, today I get the most satisfaction from teaching others the art of craftsmanship. I have taught at both the secondary (high school) and post-secondary (adult) levels for years. For decades the basics of construction have been taught at a few high schools (shop class) and a few post-secondary training programs scattered throughout the country. As our industry becomes more united in providing an industry-driven curriculum and nationally recognized training, it is revitalizing the art of construction craftsmanship. It has come during a much-needed time as our older workforce is preparing to retire and we have need of new blood and craftsmanship in our industry.

Do you think training and education are important in construction?

Yes, training and education are at a most influential time. Tough times today require specialized training. In today's way of building, each trade has become specialized, where those with the most training, experience, and credentials get the job. As our industry moves forward in creating nationally recognized credentials and certifications, so too have specialized training centers and schools at both secondary and post-secondary levels.

How important are NCCER credentials to your career?

Extremely important, in two distinct and different ways. First, by establishing standard trainer credentials, the industry ensures that training taking place across the country is similar and of the same quality. Second, completion points and credentials given to trainees are also similar and of the same quality across the country, thus making it possible to have national certification that is valid anywhere. At a personal level, having and acquiring multiple credentials allows me to excel. Reaching Master Trainer and Subject Matter Expert status with NCCER in the construction and green job trade has given me the opportunity to create my own construction training academy.

How has training/construction impacted your life and your career?
Construction training has impacted my life since my teen years, both informally and formally. In the early years it provided the technical background for the work I was performing. Years later, training became more formal in specialized areas. Currently, the combination of field work/training and formal classroom training has provided me with a good salary and lifestyle. Furthermore, it has provided me the opportunity to open VisionQuest-Academy, a full-service, post-secondary educational/training facility in Ocala, Florida. The academy is dedicated to providing cutting-edge construction and green job training, and courses are offered in English and Spanish. NCCER has been a key to this opportunity; with its support, many of our county's adults/working population now have the opportunity to obtain national industry credentials.

Would you suggest construction as a career to others?
Yes, absolutely. As I tell my students, both youth and adults, it's never too late for training and education, either informal or formal. However, a well-planned training strategy is like a well-planned trip that includes a road map to your destination. NCCER has that covered for the construction and green job industry along with pathways to success. Ultimately, the construction and green job industry is especially rewarding in that, at the end of the day, you have something tangible to look back on.

How do you define craftsmanship?
Craftsmanship can be defined in a multitude of ways. For the most part, in the construction industry, craftsmanship is the final product of a job, task, or project completed within the quality standards set for that industry. However, I define it as follows: the skills, expertise, ability, and technique used by a person to shape, mold, transform, and convey an idea into products of value to others. A craftsman's level of work and performance can only be achieved by a combination of hard work, informal and formal training, time, and mostly a love of the craft.

Trade Terms Introduced in This Module

Mobiling: Term used when driving a trencher around the job site when not digging.

Thrust: The amount of force required to push or drive an object, normally measured in pounds of force.

Additional Resources

This module presents thorough resources for task training. The following resource material is suggested for further study.

Construction Equipment Guide. New York, NY: John Wiley & Sons.

Engineering Trenchless Solutions, Mesa, CO. www.hdd-consult.com.

North American Society for Trenchless Technology. Liverpool, NY. www.nastt.org.

Figure Credits

The Charles Machine Works, Inc., Manufacturer of Ditch Witch products, Module opener, Figures 2, 3, 9, 10, 14, 18, and 19

Topaz Publications, Inc., Figures 1, SA01, 6, 15, 17, 21, and 23

Vermeer Manufacturing Company, Figures 4, 16, and 22

Ventrac by Venture Products, Inc., Figure 5

http://www.vectorsite.net/gfxpxv_08.html by Greg Goebel, Figure 11

Caterpillar Inc., Figure 12

Igo Ditching & Backhoe, Figure 20

CONTREN® LEARNING SERIES — USER UPDATE

NCCER makes every effort to keep its textbooks up-to-date and free of technical errors. We appreciate your help in this process. If you find an error, a typographical mistake, or an inaccuracy in NCCER's Contren® materials, please fill out this form (or a photocopy), or complete the online form at www.nccer.org/olf. Be sure to include the exact module number, page number, a detailed description, and your recommended correction. Your input will be brought to the attention of the Authoring Team. Thank you for your assistance.

Instructors – If you have an idea for improving this textbook, or have found that additional materials were necessary to teach this module effectively, please let us know so that we may present your suggestions to the Authoring Team.

NCCER Product Development and Revision
3600 NW 43rd Street, Building G, Gainesville, FL 32606

Fax: 352-334-0932
Email: curriculum@nccer.org
Online: www.nccer.org/olf

☐ Trainee Guide ☐ AIG ☐ Exam ☐ PowerPoints Other ______________

Craft / Level: Copyright Date:

Module Number / Title:

Section Number(s):

Description:

Recommended Correction:

Your Name:

Address:

Email: Phone:

Introduction to Electrical Test Equipment

49113-11

Trainees with successful module completions may be eligible for credentialing through NCCER's National Registry. To learn more, go to **www.nccer.org** or contact us at **1.888.622.3720.** Our website has information on the latest product releases and training, as well as online versions of our *Cornerstone* newsletter and Pearson's Contren® product catalog.

Your feedback is welcome. You may email your comments to **curriculum@nccer.org**, send general comments and inquiries to **info@nccer.org**, or use the User Update form at the back of this module.

V.1 7/11

49113-11

Introduction to Electrical Test Equipment

Objectives

When you have completed this module, you will be able to do the following:

1. Describe the following pieces of test equipment and explain their purpose:
    - Voltmeter
    - Ohmmeter
    - Clamp-on ammeter
    - Multimeter
    - Megohmmeter
    - Hi-pot tester (dielectric strength tester)
    - Motor and phase rotation testers
    - Recording instruments
    - High-voltage detector
    - Phasing sticks
2. Select the appropriate meter for a given work environment based on category ratings.
3. Identify the safety hazards associated with various types of test equipment.

Performance Task

Under the supervision of the instructor, you should be able to do the following:

1. Select the appropriate meter for a given work environment based on category ratings.

Trade Terms

Coil
Continuity
D'Arsonval meter movement
Frequency

Contents

Topics to be presented in this module include:

Figures and Tables

1.0.0 INTRODUCTION

The use of electronic test instruments and meters generally involves troubleshooting electrical/ electronic circuits and equipment. It also has to do with verifying the proper operation of instruments and equipment.

The test equipment selected for a task depends on the type of measurement required and the level of accuracy needed. This module focuses on some of the test equipment used by power transmission and distribution system technicians. After completing this module, you should be able to select the correct test equipment for an application and identify the related safety hazards.

2.0.0 METERS

In 1882, a Frenchman named Arsene d'Arsonval invented the galvanometer. This meter used a stationary permanent magnet and a moving coil to indicate current flow on a calibrated scale. The early galvanometer was very accurate, but it could only measure very small currents. Over time, many improvements were made that extended the range of the meter and made it more rugged. The D'Arsonval meter movement (*Figure 1*) is the basis for analog meters.

A moving-coil meter movement works on the electromagnetic principle. In its simplest form, the moving-coil meter uses a coil of very fine wire wound on a light aluminum frame. A permanent magnet surrounds the coil. The aluminum frame is mounted on pivots. The pivots allow the frame and the coil to rotate freely between the poles of the permanent magnet. When current flows through the coil, it becomes magnetized, and the polarity of the coil is repelled by the field of the permanent magnet. This causes the coil frame to rotate on its pivots. The distance the frame rotates is determined by the amount of current that flows through the coil.

The amount of current flowing through the meter can be measured by attaching a pointer to the coil frame and adding a calibrated scale. Multiplier resistors are used to extend the range of the meter movement for voltage measurements. To extend the range of the meter movement for current measurements, shunt resistors are used.

Most meters in use today are solid-state digital systems. They are easier to read than mechanical (analog) meters and have no meter movement or moving parts.

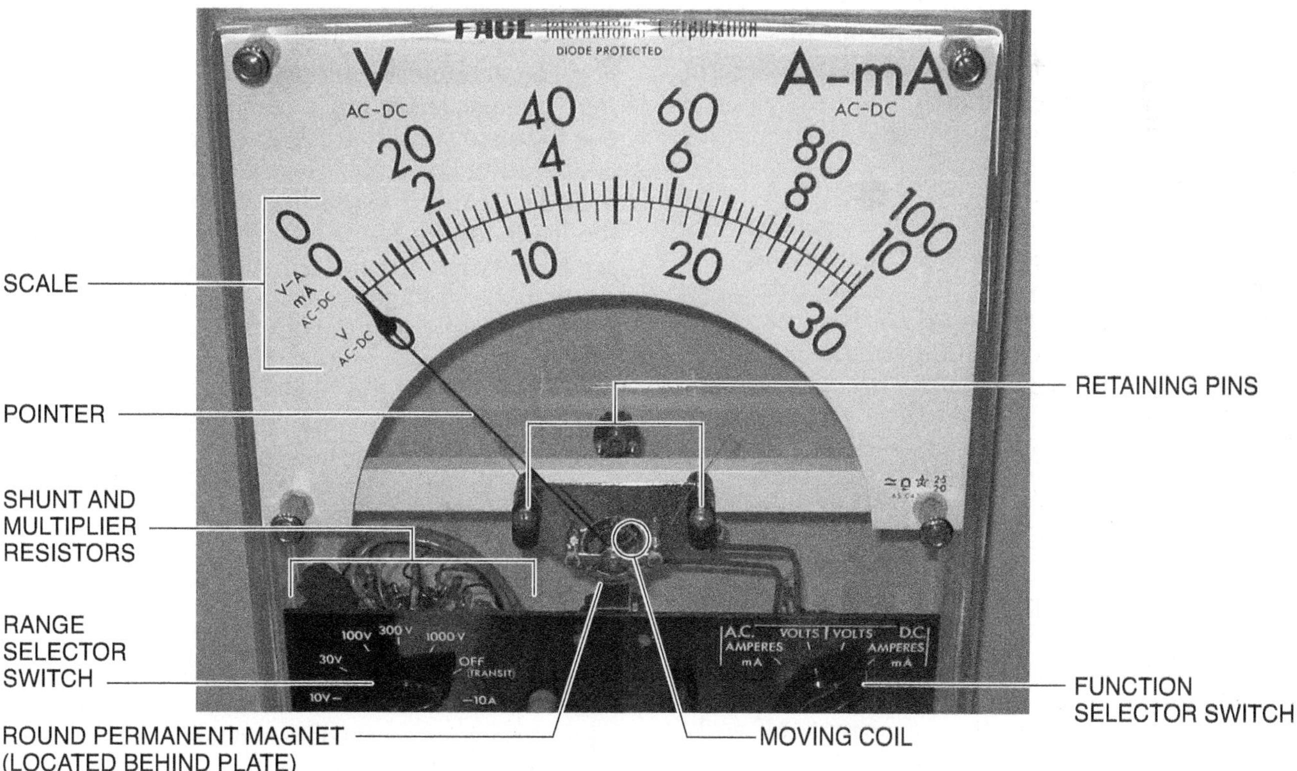

Figure 1 D'Arsonval meter movement.

On Site

Phantom Readings

The sensitivity of a digital meter can sometimes produce a low reading known as a phantom, or ghost, reading. This low reading is due to the induction from the electrical field around the energized conductors in close proximity to the meter.

2.1.0 Voltmeter

A voltmeter measures voltage, which is also known as potential difference or electromotive force (emf). A voltmeter is connected in parallel with the circuit or component being measured. An analog meter uses the basic D'Arsonval meter movement with internally switched resistors to measure different voltage ranges. A digital meter uses an analog-to-digital converter chip to convert the sensed values into a digital or graphic display.

Many digital voltmeters are autoranging. This means that the meter automatically searches for the correct scale. When using a voltmeter that is not autoranging, always start with the highest voltage range. Then work down until the indication reads somewhere between half and three-quarter scale. This will provide a more accurate reading and prevent damage to the meter. On many meters, a DC value is indicated by a straight line with three dashes beneath it. An AC value is often shown by a sine wave.

WARNING!

Measuring voltages above 50V exposes the technician to potentially life-threatening hazards. Follow all applicable safety procedures as found in *NFPA 70E®*, OSHA standards, and company and institutional policies and standards.

A voltmeter is used when the exact value of the voltage is required. Often, technicians only need to know if voltage is present, and if so, the general range of the voltage. In other words, is the source energized? If it is energized, is it at 120V, 240V, or 480V? In these cases, a voltage tester is used. The range of voltage and the type of current (AC and/or DC) that the voltage tester is designed to measure are usually shown on the scales that display the reading (*Figure 2*).

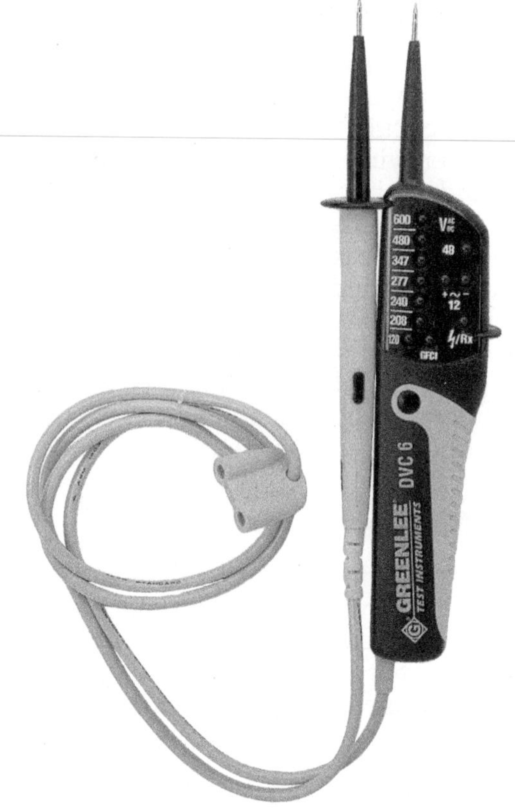

49113-11_F02.EPS

Figure 2 Voltage tester.

Advanced voltage testers offer more features, such as a digital readout, GFCI test capability, and the ability to switch between use as a contact and non-contact detector (*Figure 3*).

A voltage tester must be checked before each use to make sure that it is in good condition and is operating correctly. The external check should include a careful inspection of the insulation on the leads for cracks or frayed areas. Faulty leads are a safety hazard and must be replaced. To verify that the voltage tester is operating correctly, the probes of the tester are first connected to a known energized source. The voltage indicated on the tester should match the voltage of the source. If there is no indication, the voltage tester is not working properly and must be repaired or replaced. Repair or replacement is also required if the tester indicates a voltage that is different from the known voltage of the source.

CAUTION

Care should be taken when placing the probes of the tester across the voltage source. Some voltage testers are designed to take quick readings. They may become damaged if there is longer contact with the voltage source.

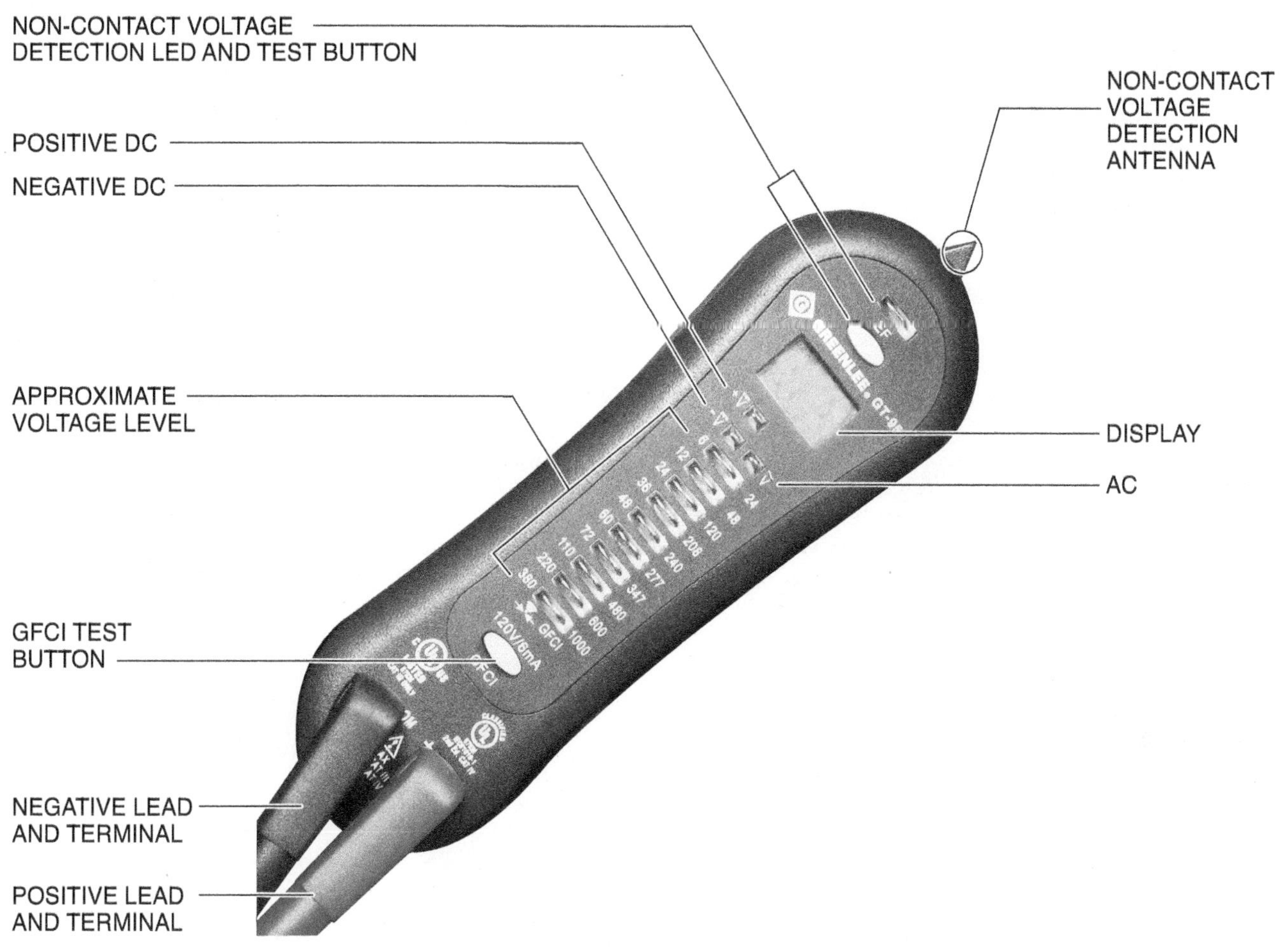

Figure 3 Multi-function voltage tester.

Voltage testers are used to ensure that voltage is available when it is needed and that power has been cut off when it should have been. When troubleshooting, it may be necessary to confirm that power is available in order to verify that the lack of power is not the issue. For example, a problem could exist with a power tool, such as a drill. A voltage tester might be used to ensure that power is available to the drill. Another possible use of a voltage tester would be to verify that power is available to a three-phase motor that will not start.

> **WARNING!**
>
> When testing for voltage to verify that a circuit is de-energized as part of an electrical lockout, always perform a live-dead-live test. First verify the operation of the test equipment on a known energized (live) source. Next, de-energize and test the target circuit to ensure it is dead. Then test the known energized (live) source again before making contact with the de-energized circuit. This test is an OSHA requirement for systems above 600V. However, it is a good practice at all voltage levels.

2.2.0 High-Voltage Detectors

Voltages in a power transmission and distribution system are often well over 1,000 volts. In some distribution systems, the voltage can be as high as 765,000 volts (765kV). Technicians must be very careful when checking these voltages to avoid electric shock. The voltmeters and voltage testers discussed earlier are not suitable for safely measuring very high voltages. A non-contact high-voltage detector (*Figure 4*) is a type of voltmeter that can be safely used to measure high voltages. The voltage detector is typically placed on the end of an insulated pole, called a hot stick. The detector is positioned near (not touching) the conductor being tested. The presence of high voltage is indicated by a light and/or buzzer.

Other high-voltage detectors (*Figure 5*) are designed to physically contact the high-voltage

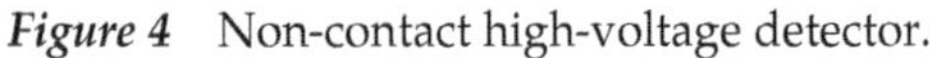

Figure 4 Non-contact high-voltage detector.

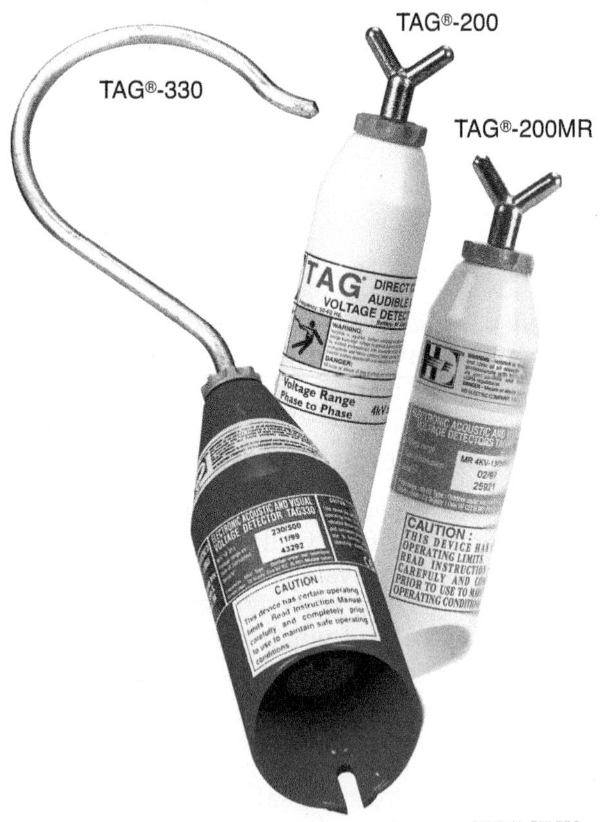

Figure 5 Contact-type high-voltage detectors.

conductor under test. Like the non-contact voltage detector, the contact-type detector is attached on the end of a hot stick for safety. The presence of voltage is indicated by a light and/or buzzer.

On Site

Voltage Detectors

Simple non-contact (proximity) voltage detectors can also be used to show the presence of voltage within their specified range rating. However, these devices do not discriminate between ranges of values in the same way as a voltage tester. They are handy for quickly scanning for the presence of voltage in junction boxes or termination cabinets. They can even be used to trace circuits through walls. The voltage detector shown here glows in the presence of voltages between 50V and 1,000V.

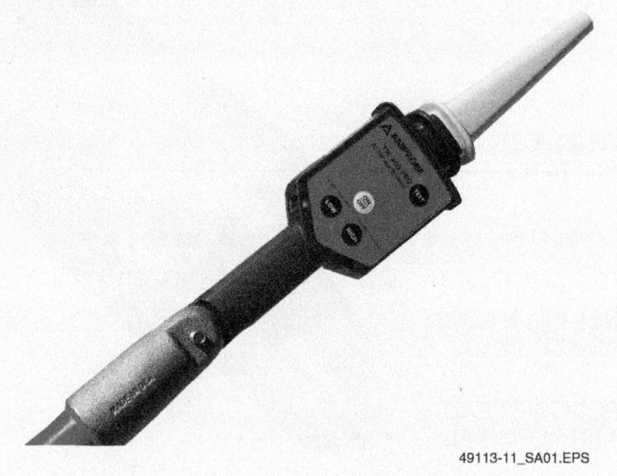

2.3.0 Ohmmeter

An ohmmeter measures the resistance of a circuit or component. It can also be used to locate open circuits or shorted circuits. An ohmmeter consists of a DC current meter movement, a low-voltage DC power source (usually a battery), and current-limiting resistors. All are connected in series with the meter (*Figure 6*).

Before measuring the resistance of an unknown resistor or electrical circuit, connect the test leads together. This zeroes out or nulls the resistance of the leads. Some analog ohmmeters have a zero adjustment knob. With the leads connected together, turn the adjustment knob until the meter registers zero ohms. This adjustment must be made each time a different range is selected.

Many digital ohmmeters are autoranging. The correct scale is internally selected and the reading will indicate the range (ohms, K ohms, or M ohms). In analog ohmmeters, the desired range must be selected. Most digital meters also have an audible tone when the measured value is very low or at zero ohms. This indicates a closed circuit,

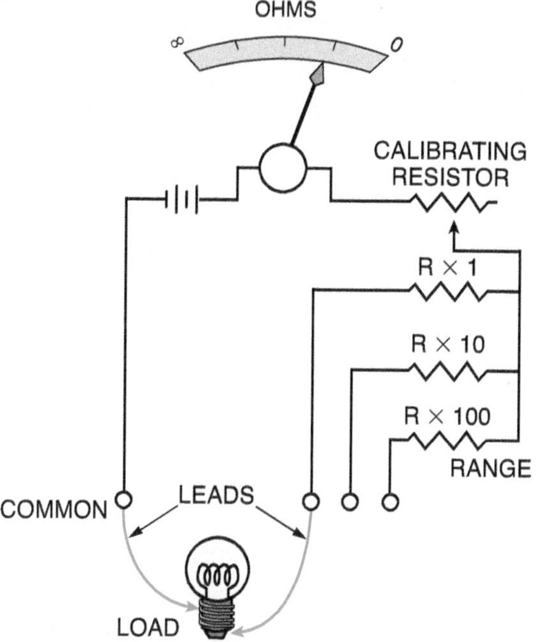

Figure 6 Ohmmeter schematic.

which is useful when using the meter as a continuity tester. A continuity test is used to determine if a circuit is complete.

WARNING!

Before taking a reading with an ohmmeter, use a voltmeter to verify that both sides of the circuit are de-energized. If the circuit proves to be energized, its voltage could cause a damaging current to flow through the ohmmeter. This could damage the meter and/or circuit and cause personal injury.

When making resistance measurements in circuits, each single component in the circuit can be tested by removing it from the circuit and connecting the ohmmeter leads across it. However, the component does not have to be totally removed from the circuit. Usually, the part can be effectively isolated by disconnecting one of its leads from the circuit. Note that this method can still take some time.

Think About It

Resistance

When measuring resistance, why does the resistance vary when you hold the resistor by pinching the meter leads against the resistor with your fingers versus measuring it while you hold it in clips?

2.4.0 Ammeter

A clamp-on ammeter, or clamp meter, can measure current without having to make contact with uninsulated wires (*Figure 7*). This meter works by sensing the strength of the electromagnetic field around the wires.

Clamp-on ammeters measure current by using simple transformer principles. The conductor(s) being measured would be the primary and the jaws (clamp) of the meter would be the secondary. The current in the primary winding induces a current in the secondary winding. If the ratio of the primary winding to the secondary winding is 1,000, then the secondary current is 1/1000 of the current flowing in the primary. The smaller secondary current is connected to the meter's input. For example, a 1A current in the conductor will produce 0.001A (1mA) in the meter.

To measure the current, open the jaws of the meter and close them around the conductors to be measured. Make sure that the jaws are clean and close tightly. Then read the magnitude of the current on the meter display.

Many meters have a Hold function. This is useful in tight locations where it may be hard to read a meter that is clamped around the conductors. Just press the Hold button when the value is measured, and then remove and read the meter.

CAUTION

When using a clamp-on ammeter, make sure that the range of the meter is at least as high as the current to be measured. If the meter is digital and the current is too high, the meter will become overloaded. The display will read OL (overload). If the meter is analog, the indicator needle will peg (move) above the maximum limit on the scale. This might damage the meter.

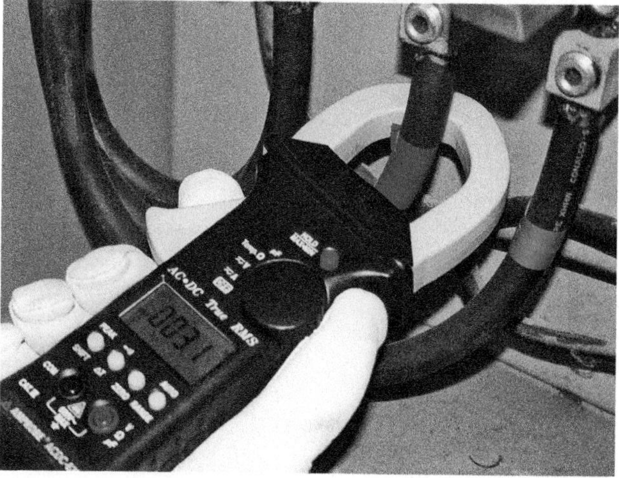

Figure 7 Clamp-on ammeter.

Think About It

Current Measurements

What happens when you loop the conductor so that the meter measures two turns instead of one?

Some meters have a Min/Max or Peak function. This allows the technician to record the maximum inrush, as with a motor start.

WARNING!

Using a clamp-on ammeter may expose you to energized systems and equipment. Never use this type of meter unless you are qualified and are following NFPA, OSHA, and company/institutional safety procedures.

2.5.0 Multimeter

The multimeter is also known as a volt-ohm-milliammeter (VOM). An analog multimeter is shown in *Figure 8*. It is a multi-purpose instrument that combines features of the voltmeter,

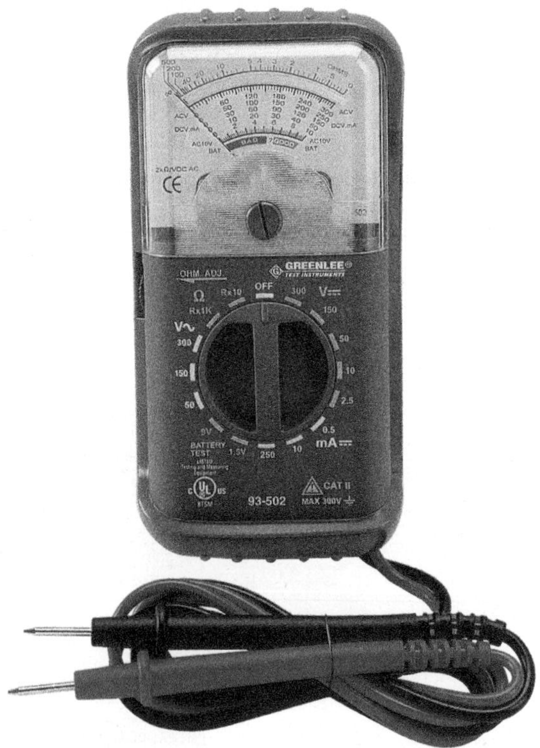

49113-11_F08.EPS

Figure 8 Analog VOM.

On Site

High-Voltage Ammeters

In power transmission and distribution systems, the voltages are too high to safely measure current with a conventional clamp-on ammeter. Instead, a special high-voltage clamp-on ammeter is used. The clamp or jaws of the meter are often in the form of a hook or fork and are mounted on the end of a hot stick. This allows the meter to be safely applied on a high-voltage conductor.

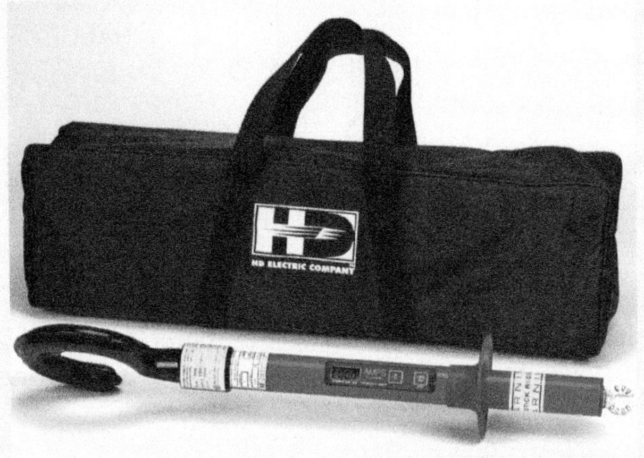

49113-11_SA02.EPS

49113-11_SA03.EPS

ohmmeter, and ammeter. When using an analog meter, you must select the proper voltage (DC or AC) and range (in volts, amps, or ohms). When using a digital VOM (*Figure 9*), you must select the proper voltage. Most digital VOMs have an auto-ranging feature. Unless the VOM is specifically designed for it, do not use the meter to measure voltages above 600V.

WARNING!

Some multimeters have dedicated jacks into which the leads must be plugged to get the correct reading. Failure to plug the meter leads into the correct jack can damage the meter or cause personal injury.

Before use, a VOM must be checked on a known power source.

Using a VOM is the same as using the individual voltmeter, ohmmeter, and clamp-on ammeter. Current clamps can be used with most multimeters for measuring AC and DC currents above the milliamp level. Depending on the current clamp, these can be plugged into either the amp or voltage test lead connections. Always refer to the manufacturer's instructions for the device in use. Current clamps are available in various current ranges, from 50 amps to several thousand amps. The jaws are available in different shapes and sizes for various applications. Examples include round, rectangular, and flexible jaws.

Some multimeters can measure the frequency of an AC waveform. This function is useful for diagnosing harmonic problems in an electrical distribution system.

Another added feature is the Min/Max memory function. It records the minimum and maximum readings over the time period selected. Other common multimeter functions include capacitance measurement, diode and transistor testing, temperature measurement, and true RMS measurement for accurate voltage and current readings at different frequencies. Always refer to the manufacturer's instructions for the meter in use.

In addition to the current clamps used with standard multimeters, clamp-on multimeters are also available (*Figure 10*). They are used to measure AC current, AC and DC voltage, resistance, and other values.

2.6.0 Megohmmeter

An ordinary ohmmeter cannot be used for measuring resistances of several million ohms. Such resistances are found in conductor insulation or between motor or transformer windings. The instrument used to measure very high resistances is known as a megohmmeter. Megohmmeters are also called meggers, or insulation resistance testers. They can be powered by alternating current, battery (*Figure 11*), or hand cranking (*Figure 12*).

When using a megohmmeter, the following minimum safety precautions must always be followed to avoid personal injury or damage to the equipment:

- High voltages are present when using a megohmmeter. For example, in 600V class systems, applied megohmmeter voltages are typically 500V and 1,000V. Only qualified persons may use this equipment. Always wear the correct personal protective equipment when approaching energized parts.
- De-energize the circuit, and verify that it has been de-energized before connecting the meter. Make sure that all capacitors are discharged.

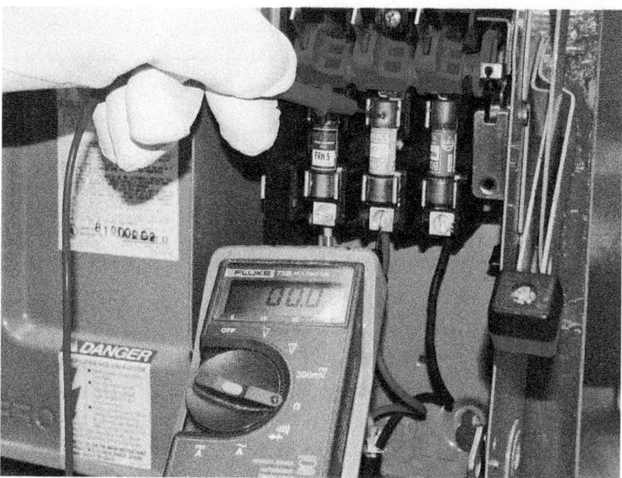

49113-11_F09.EPS

Figure 9 Digital VOM.

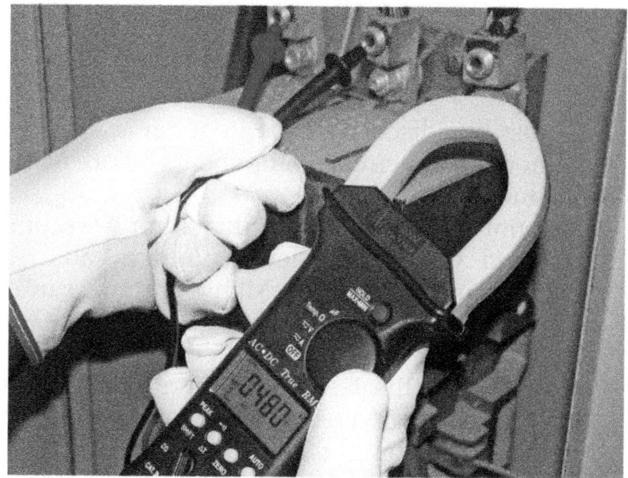

49113-11_F10.EPS

Figure 10 Clamp-on multimeter.

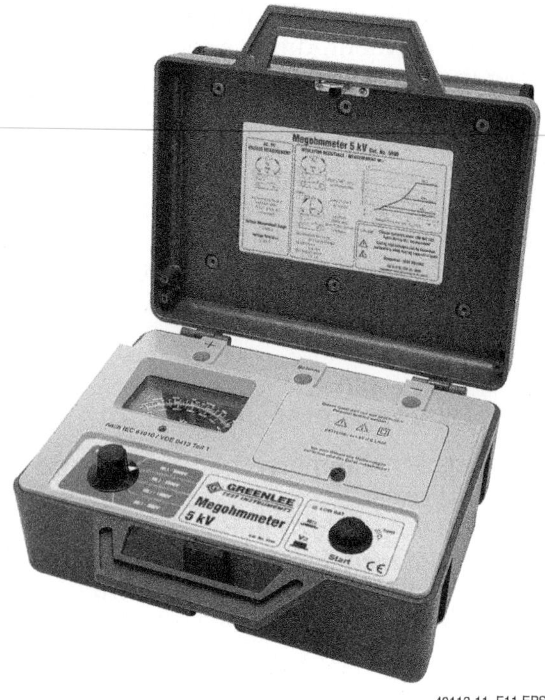
49113-11_F11.EPS

Figure 11 Battery-operated megohmmeter.

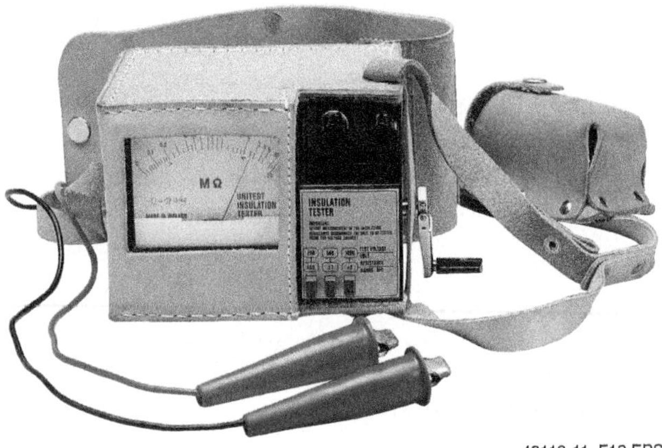
49113-11_F12.EPS

Figure 12 Hand-crank megohmmeter.

- If possible, disconnect the item being checked from the other circuit components before using the meter.
- Do not exceed the manufacturer's recommended voltage test levels for the cable or equipment under test. Many manufacturers have different test levels based on the age of the cable or equipment being tested.
- Never touch the test leads when the meter is energized or powered. Meggers generate high voltage. Touching the leads could result in injury or electrical shock.
- After the test, discharge any energy that may be left in the circuit by grounding the conductor or equipment for a period of time equal to the duration of the test.
- When meggering cables or busducts and you have exposed parts that are remote from your testing position, safely secure or barricade the exposed end to protect others from accidental contact with the test voltage.

CAUTION

If a megohmmeter is used to test switchgear, all of the electronics must be disconnected prior to testing the switchgear. The voltage produced by the megohmmeter may damage electronic equipment.

Meter manufacturers supply detailed manuals for testing various devices and equipment. Always follow the instructions.

2.7.0 Motor and Phase Rotation Testers

Before connecting a three-phase motor to a circuit, first match the legs or windings of the motor (T1, T2, and T3) to the phases of the circuit (L1, L2, and L3). This ensures that the motor rotates in the proper direction. A motor rotation tester can be used to identify the legs of the motor (*Figure 13*). A phase rotation tester can be used to detect the phases of the circuit (*Figure 14*).

WARNING!

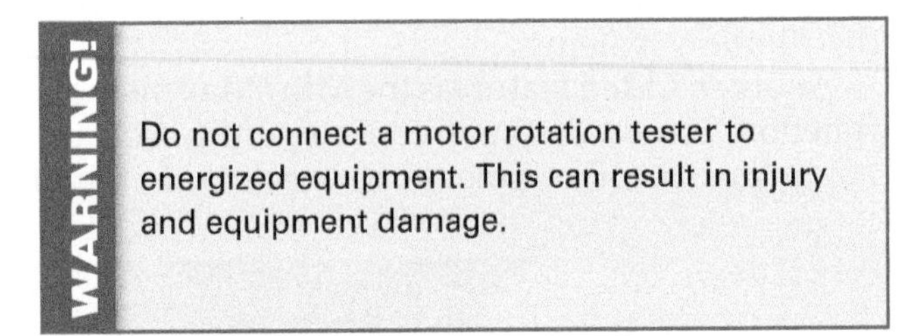

Do not connect a motor rotation tester to energized equipment. This can result in injury and equipment damage.

A motor rotation tester is a passive device. It operates on residual magnetism present in the motor after it has been run or tested by the manufacturer before shipping. To use a motor rotation tester, connect the three motor wires to the T1, T2, and T3 leads on the tester. Then rotate the motor shaft a half-turn while pressing the Test button (the direction of rotation depends on the tester in use; always follow the manufacturer's instructions). Either the clockwise or counterclockwise LED will light up. If the required rotation is clockwise and the clockwise LED lights up,

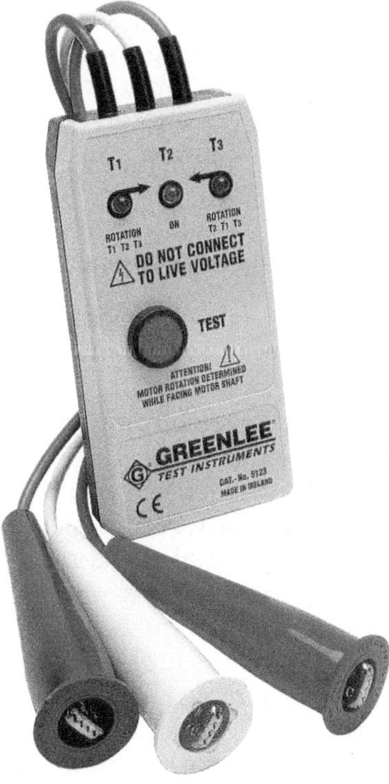

49113-11_F13.EPS

Figure 13 Motor rotation tester.

tag the motor wires to correspond to the motor rotation leads. If the required rotation is clockwise and the counterclockwise LED lights up, switch a pair of leads and retest.

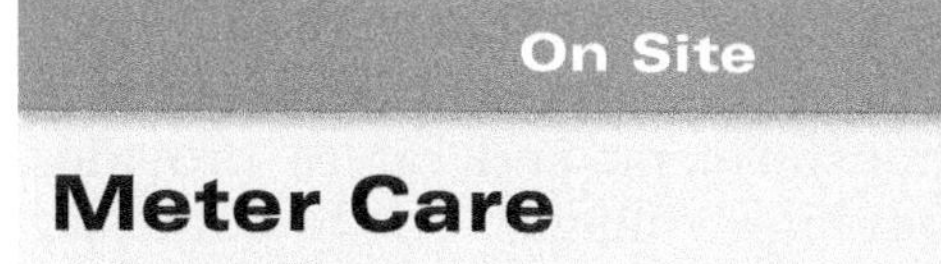

On Site

Meter Care

Like all meters, a megohmmeter is a sensitive instrument. Treat it with care and keep it in its case when not in use.

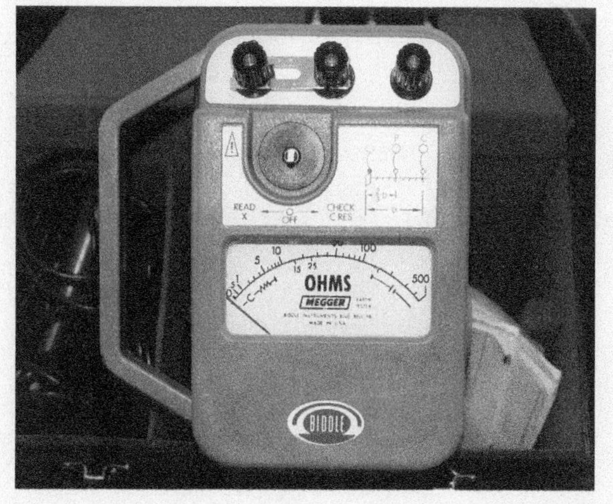

49113-11_SA04.EPS

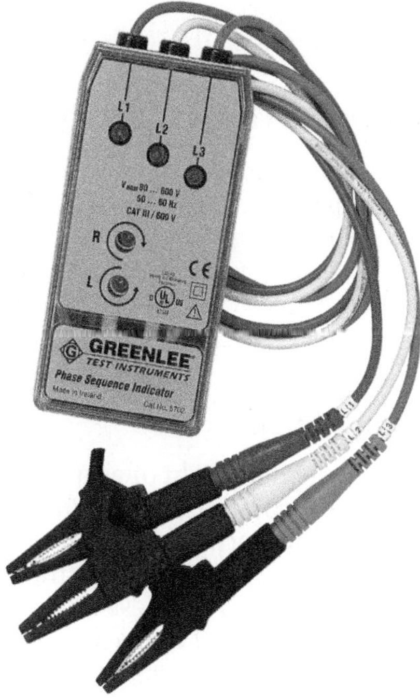

49113-11_F14.EPS

Figure 14 Phase rotation tester.

A phase rotation tester, or phase sequence indicator, is used on three-phase electrical systems to indicate the phase sequence rotation of the voltages. These testers typically have LEDs to show the phase rotation. A phase sequence is measured as clockwise or counterclockwise rotation.

WARNING!

Phase rotation testers may only be used by qualified persons. Always wear appropriate personal protective equipment when near energized parts.

A phase rotation tester is used to ensure the same phase rotation throughout a facility, when this is required. To test phase rotation, de-energize and lock out power to the circuit. Then connect the three leads of the tester to the phase conductors in the circuit. Next, safely energize the circuit and observe the meter. Make note of both the color scheme of the connected leads to the system and the phase sequences as indicated on the meter. This is needed to ensure that the added equipment follows the same phase rotation. De-energize and lock out the circuit before disconnecting the leads.

On Site

Phase Rotation Tester

The leads of a phase rotation tester may not correspond to typical circuit color coding. It is a good idea to put phase tape on the meter leads that corresponds to the circuit color coding. This will help to ensure correct connections.

2.8.0 Phasing Sticks

Phasing sticks can be thought of as a high-voltage version of a phase rotation tester. Phasing sticks are typically used to verify that any two conductors in a three-phase power distribution system are the same phase. Both wireless and wired versions are available (*Figure 15*). Because of the high voltages involved, phasing sticks are mounted on a hot stick for safety. The wired version consists of a voltmeter and two hot sticks connected together with a conductor. Wireless versions are designed to be mounted on the end of a hot stick. As the name implies, they are not connected together. Both types are used on energized high-voltage conductors. Extreme caution must be exercised when using either type.

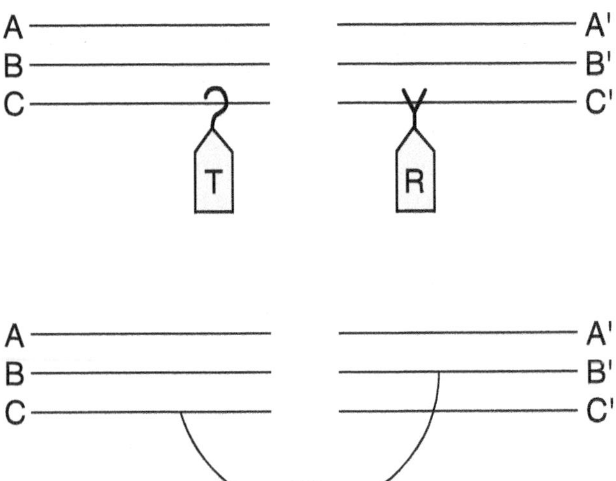

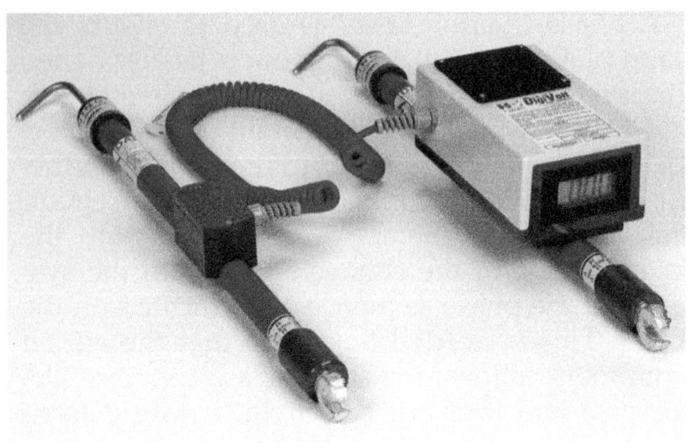

49113-11_F15.EPS

Figure 15 Phasing equipment applications.

2.9.0 Recording Instruments

Recording instruments include many devices that make a permanent record of measured values over a period of time. Recording instruments use either a paper strip or electronic memory. Those using electronic memory are usually called data loggers. They record electrical quantities, including voltage, current, power, resistance, and frequency. Data loggers can also record nonelectrical quantities by electrical means. An example is a temperature recorder that uses a potentiometer system to record thermocouple output.

Technicians must often know the conditions that exist in an electrical circuit over a period of time. This assists them in determining such things as peak loads and voltage fluctuations. An automatic recording instrument can be set up to take readings at specified intervals for later review and analysis. Some meters can upload data to a PC for real-time data logging and graphing (*Figure 16*).

2.10.0 Hi-pot Tester

Hi-pot is short for high potential. A hi-pot tester (*Figure 17*) is used to verify that the insulation on motors, transformers, and cables provides protection from high voltages. The hi-pot tester is used to conduct insulation resistance tests to verify that the insulation will not fail under high-voltage conditions. A hi-pot test can indicate

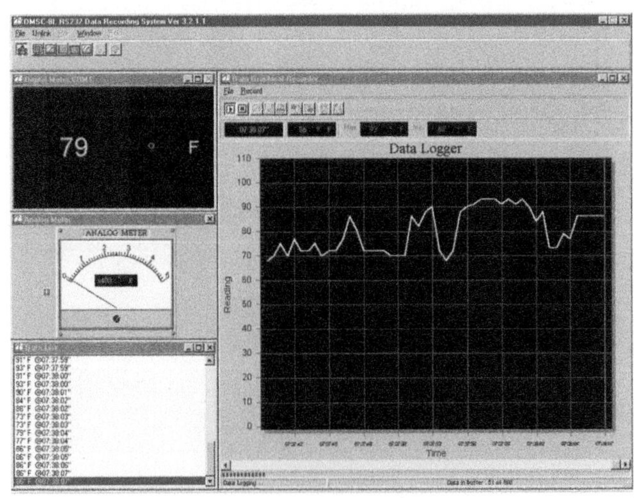

49113-11_F16.EPS

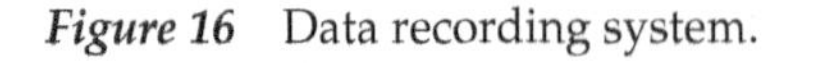

Figure 16 Data recording system.

49113-11_F17.EPS

Figure 17 Hi-pot tester.

Table 1 Overvoltage Installation Categories

Overvoltage Category	Installation Examples
CAT I	Electronic equipment and circuitry
CAT II	Single-phase loads such as small appliances and tools, outlets at more than 30 feet from a CAT III source or 60 feet from a CAT IV source
CAT III	Three-phase motors, single phase commercial or industrial lighting, switchgear, busduct, and feeders in industrial plants
CAT IV	Three-phase power at meter, service-entrance, or utility connection, any outdoor conductors

damaged insulation, loose wires, and other defects that would otherwise not be recognized.

3.0.0 Category Ratings

Distribution systems and loads are becoming more complex. This increases the risk of transient power spikes. Lightning strikes on outdoor transmission lines and switching surges from normal switching operations can also produce dangerous high-energy transients. Motors, capacitors, variable speed drives, and power conversion equipment can also generate power spikes.

Safety systems are built into test equipment to protect technicians from transient power spikes. The International Electrotechnical Commission (IEC) developed a new safety standard, *IEC 1010*, for test equipment. This new safety standard was adapted as *UL Standard UL3111-1*. These standards define four overvoltage installation categories, often abbreviated as CAT I, CAT II, CAT III, and CAT IV (*Table 1*). These categories identify the hazards posed by transients. The higher the category number, the greater is the risk to the technician. A higher category number refers to an installation that has higher power available and higher-energy transients.

When selecting a meter, choose one rated for the highest category you will be working in. Then select the appropriate voltage level. In addition, make sure that your test leads are rated as high as your meter. Choose meters that are tested and certified by UL, CSA, or another recognized testing organization. Certified meters are marked with the category rating on the meter housing (*Figure 18*).

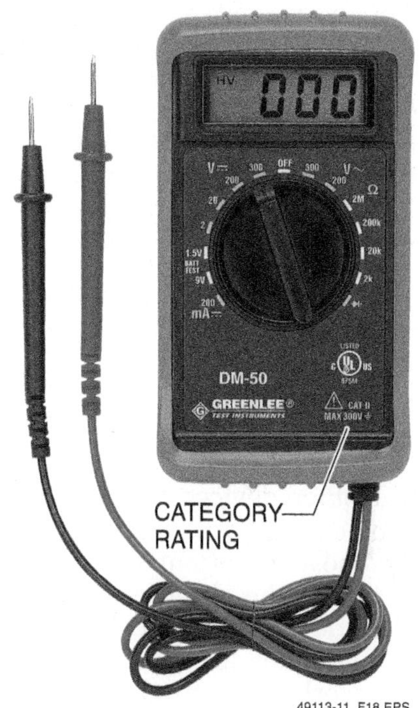

49113-11_F18.EPS

Figure 18 Category rating on a typical meter.

4.0.0 Safety

Safety must be the primary responsibility of all personnel on a job site. The safe installation, maintenance, and operation of electrical equipment requires strict adherence to local and national codes and safety standards, as well as facility and company safety policies. Carelessness can result in serious injury or death due to electrical shock, burns, falls, flying objects, or other causes. After an accident, investigation almost always shows that it could have been prevented by following simple safety precautions and procedures. It is your personal responsibility to identify and eliminate unsafe conditions and unsafe behaviors that cause accidents.

Bear in mind that de-energizing main supply circuits by opening supply switches does not necessarily de-energize all circuits in a given piece of equipment. A source of danger that is often neglected or ignored, sometimes with tragic results, is the input to electrical equipment from other sources, such as backfeeds. The rescue of a victim shocked by the power input from a backfeed is often hampered because of the time required to determine the source of power and isolate it. Always turn off all power inputs before working on equipment, and lock out and tag. Then check with an operating voltage tester to be sure that the equipment is safe to work on.

WARNING!

When performing lockout/tagout procedures, remember that other forms of energy may be present and must also be locked out and tagged. These include water pressure, steam, springs, gravity, and other forms of energy.

Keep in mind that the common 120V power supply voltage is not a low, relatively harmless voltage. It is a voltage that has caused more deaths than any other.

Safety can never be stressed enough. There are times when your life literally depends on it. Always observe the following precautions:

- Thoroughly inspect all test equipment before each use. Check for broken leads or knobs, damaged plugs, or frayed cords. Do not use equipment that is wet or damaged.
- Make sure that the rating of any leads or accessories meets or exceeds the rating of the meter.
- Do not work with energized equipment unless you are both qualified and approved by your supervisor.
- Never take shortcuts with safety. Strictly adhere to all energized work policies and procedures.
- When testing circuits, test at higher ranges first, then work your way down to lower ranges.
- Always have a standby person present during hot work. That person should know how to disconnect the power and whom to contact in case of an emergency.

SUMMARY

Meters and other devices are used to test and troubleshoot circuits and electrical equipment. One of the most important tests you will perform is verifying the absence of voltage before working on a device or circuit. This is often done using a voltage tester. In power transmission and distribution work, specialized equipment is used for this purpose because of the increased level of hazard from extremely high voltages. Other common test equipment includes devices such as multimeters, clamp-on ammeters, megohmmeters, and motor and phase rotation testers. You must understand the operation of and the safety precautions related to each piece of test equipment. You must also be able to select the proper test equipment based on the task and the category rating of the environment in which the work will be done. Always inspect and verify the operation of all test equipment before using it. Your life may depend on it.

Review Questions

1. The difference between a digital multimeter and an analog multimeter is that analog multimeters have a(n) _____.
   a. red lead wire
   b. battery
   c. meter movement
   d. auto-ranging setting

2. A voltmeter is used to measure _____.
   a. exact voltages
   b. voltage ranges
   c. power
   d. sine waves

3. Before measuring low voltages, you should first test for the presence of _____.
   a. resistance
   b. current
   c. vibration
   d. higher voltages

4. What is likely to happen if an ohmmeter is used in an energized circuit?
   a. The battery will lose its charge.
   b. The meter indication will not be accurate.
   c. The meter will read voltage instead of resistance.
   d. The meter will be damaged and you may be injured.

5. An ammeter is used to measure _____.
   a. current
   b. voltage
   c. resistance
   d. insulation value

6. Clamp-on ammeters operate by _____.
   a. sensing the strength of the electromagnetic field around the wires
   b. measuring the high resistance end of a power transformer
   c. measuring sine wave amplitude
   d. using a resistive shunt

7. An insulation resistance tester is another name for a(n) _____.
   a. ammeter
   b. multimeter
   c. phasing stick
   d. megohmmeter

8. A motor rotation tester _____.
   a. tests an energized motor
   b. works on an energized circuit
   c. works on residual magnetism
   d. works only on clockwise rotation

9. A phasing stick is used to _____.
   a. measure current
   b. verify phase relationships
   c. measure single-phase power levels
   d. ground de-energized high-voltage conductors

10. Category ratings relate to the ability of a meter to withstand _____.
    a. physical damage
    b. electrical noise
    c. transient voltage spikes
    d. ground faults

Trade Terms Quiz

Fill in the blank with the correct term that you learned from your study of this module.

1. A number of turns of wire used for electromagnetic effects or for providing electrical resistance is a _____.
2. A permanent magnet and moving coil arrangement are used to move a pointer across a scale in a _____.
3. Usually expressed in hertz, the number of cycles completed each second by a given AC voltage is called _____.
4. A term used to describe an uninterrupted electrical path for current flow is _____.

Trade Terms

Coil
Continuity
D'Arsonval meter movement
Frequency

Trade Terms Introduced in This Module

Coil: A number of turns of wire, especially in spiral form, used for electromagnetic effects or for providing electrical resistance.

Continuity: An electrical term used to describe a complete (unbroken) circuit that is capable of conducting current. Such a circuit is also said to be closed.

D'Arsonval meter movement: A meter movement that uses a permanent magnet and moving coil arrangement to move a pointer across a scale.

Frequency: The number of cycles completed each second by a given AC voltage, usually expressed in hertz. One hertz equals one cycle per second.

Additional Resources

This module presents thorough resources for task training. The following resource material is suggested for further study.

ABCs of Multimeter Safety: Multimeter Safety and You. Everett, WA: Fluke Corporation, 2000.

ABCs of DMMs: Multimeter Features and Functions Explained. Everett, WA: Fluke Corporation, 2006.

Clamp Meter ABCs. Everett, WA: Fluke Corporation.

Electronics Fundamentals: Circuits, Devices, and Applications. 8th ed. Thomas L. Floyd. Upper Saddle River, NJ: Prentice Hall, 2009.

Fluke Corporation: www.fluke.com.

Principles of Electric Circuits: Electron Flow Version. 8th ed. Thomas L. Floyd. Upper Saddle River, NJ: Prentice Hall, 2006.

Figure Credits

Tony Vazquez, Module opener

Tim Dean, Figures 1, 7, 9, and 10

Greenlee Textron, Inc., a subsidiary of Textron Inc., Figures 2, 3, 8, 11–14, 16, and 18

Photograph provided by AEMC® Instruments, Figure 4

HD Electric Co., Figure 5, SA02, and Figure 15

Mitchell Instrument Co., SA01

A. B. Chance Co./Hubbell Power Systems, SA03

Topaz Publications, Inc., SA04

Advanced Test Equipment Corp., Figure 17

CONTREN® LEARNING SERIES — USER UPDATE

NCCER makes every effort to keep its textbooks up-to-date and free of technical errors. We appreciate your help in this process. If you find an error, a typographical mistake, or an inaccuracy in NCCER's Contren® materials, please fill out this form (or a photocopy), or complete the online form at www.nccer.org/olf. Be sure to include the exact module number, page number, a detailed description, and your recommended correction. Your input will be brought to the attention of the Authoring Team. Thank you for your assistance.

Instructors – If you have an idea for improving this textbook, or have found that additional materials were necessary to teach this module effectively, please let us know so that we may present your suggestions to the Authoring Team.

NCCER Product Development and Revision

3600 NW 43rd Street, Building G, Gainesville, FL 32606

Fax: 352-334-0932
Email: curriculum@nccer.org
Online: www.nccer.org/olf

☐ Trainee Guide ☐ AIG ☐ Exam ☐ PowerPoints Other ____________

Craft / Level: Copyright Date:

Module Number / Title:

Section Number(s):

Description:

Recommended Correction:

Your Name:

Address:

Email: Phone:

Glossary

Aerial device: Any vehicle mounted device, telescoping or articulating, or both, which is used to raise and position personnel to work at heights.

Ammeter: An instrument for measuring electrical current.

Ampere (A): A unit of electrical current. For example, one volt across one ohm of resistance causes a current flow of one ampere.

Anchor point: The location of attachments on a structure for all types of climbing and/or rigging systems.

Anchorage point: A place for properly securing a lanyard.

Angle of repose: The greatest angle above the horizontal plane at which a material will adjust without sliding.

Anti-two-blocking devices: Devices that provide warnings and prevent two-blocking from occurring. Two-blocking occurs when the lower load block or hook comes into contact with the upper load block, boom point, or boom point machinery. The likely result is failure of the rope or release of the load or hook block.

Arc blast: An explosion similar to the detonation of dynamite that occurs during an arc flash incident.

Arc fault: A high-energy discharge between two or more conductors.

Arc flash: A dangerous condition caused by the enormous release of thermal energy in an electric arc, usually associated with electrical distribution equipment.

Arc rating: The maximum incident energy resistance demonstrated by a material (or a layered system of materials) prior to material breakdown, or at the onset of a second-degree skin burn. Expressed in joules/cm^2 or calories/cm^2.

Arc thermal performance value (ATPV): The incident energy limit that a flame-resistant material can withstand before it breaks down and loses its ability to protect the wearer. Expressed in joules/cm^2 or calories/cm^2.

Articulating boom: An aerial device with two or more hinged boom sections. They are designed to reach up and over obstacles.

Atmospheric hazard: A potential danger in the air or a condition of poor air quality.

Atom: The smallest particle into which an element may be divided and still retain the properties of that element.

Battery: A DC voltage source consisting of two or more cells that convert chemical energy into electrical energy.

Benching: A method of protecting workers from cave-ins by excavating the sides of an excavation to form one or a series of horizontal levels or steps, usually with vertical or near-vertical surfaces between levels.

Blast hazard: The explosive expansion of air and metal in an arc path. Arc blasts are characterized by the release of a high-pressure wave accompanied by shrapnel, molten metal, and deafening sound levels.

Blocking: Pieces of hardwood used to support and brace equipment.

Body belt: A belt that functions as the anchor point for the fall restraint safety belt and as a utility belt for carrying all necessary tools.

Body harness: A system of straps and rings worn on the body with the intent of distributing weight and force applied evenly across the shoulders, chest, waist, thighs, and pelvic area. The body harness must not be confused with a body belt, which is simply worn around the waist and is not approved as a fall arrest device.

Bonding: A means of interconnecting conductive metal objects that are not intended to carry current in order to protect people from electric shock. Bonding causes all the objects to be at the same potential.

Boom: A spar or beam projecting from the mast of a derrick for supporting and guiding an aerial platform to be lifted.

Boring: The process of drilling a hole into the ground.

Bullwheel: A motorized piece of equipment used to provide consistent tension and braking to a conductor supply during pulling and installation.

Bump board: A board that is placed in a vertical position in a pole hole. It is used to prevent the pole being installed from gouging the sides of the hole and causing the loosened soil to drop into the hole.

Cable grabs: Components which ride along the length of a cable smoothly when moved casually, but lock onto the cable sharply when movement becomes too rapid or violent.

Cant hook: A device used to grip the butt end of a pole. It prevents the pole from rolling during the manual setting of a pole. It is often called a peavey.

Carabiner: A chain-link shaped device which can be opened on one side for the insertion of a line, then closed securely. They are usually rated in Newtons, since they are tested by impact to determine their strength; a connecting ring for belts and straps, which was first designed and used for mountain climbing.

Center of gravity: The point at which an object is perpendicular to and balanced in relation to the earth's gravitational field.

Chicago-style grip: A wire grip used for securing guy wires during tensioning. This type of grip cannot be used for gripping conductors.

Chinese finger: A term to describe wire mesh grips and splices that cinch around a conductor when tension is applied. This type of grip will not damage conductors.

Chuck: A specialized type of clamp used to hold an object.

Circuit: A complete path for current flow.

Circular mil: A unit of area equal to the area of a circle with a diameter of one mil. A mil is one thousandth of an inch. A circular mil is a unit for referring to the area of the cross section of a wire or cable with a circular cross section.

Coil: A number of turns of wire, especially in spiral form, used for electromagnetic effects or for providing electrical resistance.

Combustible: Air or materials that can explode and cause a fire.

Conductor: A material allowing the flow of electric current; a material through which it is relatively easy to maintain an electric current. A wire constructed line used to convey electric power from the point of generation to an end user.

Connecting devices: Devices used to connect the PFAS and positioning belts to anchor points and positioning points.

Continuity: An electrical term used to describe a complete (unbroken) circuit that is capable of conducting current. Such a circuit is also said to be closed.

Coulomb: An electrical charge equal to 6.25×10^{18} electrons or 6,250,000,000,000,000,000 electrons. A coulomb is the common unit of quantity used for specifying the size of a given charge.

Cribbing: Timbers stacked in alternate tiers. Used to support heavy loads.

Cross-arm: A pole mounted structure used to support conductors and components.

Current: The movement, or flow, of electrons in a circuit. Current (I) is measured in amperes.

Cutout: A term that refers to a combination fuse and knife switch used on power poles. An automatic fused safety device located in the line between a primary conductor and a secondary conductor or other components. An unexpected and unintentional dislodging of a gaff from the pole.

D-ring: The metal connecting ring used as the anchor point for straps and lanyards. On the body belt, the fall restraint strap is connected to the D-rings.

D-size: The distance between the heels of the D-rings on a body belt. D-size is the critical measurement for correctly sizing body belts.

D'Arsonval meter movement: A meter movement that uses a permanent magnet and moving coil arrangement to move a pointer across a scale.

Dead-end: A term used to identify the terminating arrangement at the end of a run of primary conductors.

Dielectric: A material with very little electrical conductivity.

Digging bar: A heavy, long-handled tool used to break up soil for removal with a shovel.

Distribution capacitors: A bank of components connected to distribution lines to stabilize power where fluctuations could occur.

Drift: The distance between the hook at the end of the lifting line and the tip of the boom. Enough drift must be maintained so that the digger derrick can lift the pole high enough so that it can be placed in the hole.

Drill string: The length of pipe that connects the boring machine to the drill head.

Electrical power: The rate of doing electrical work. Electrical power is measured in watts (W).

Electrically safe work condition: A state in which the conductor or circuit part to be worked on or near has been disconnected from energized parts, locked/tagged in accordance with established standards, tested to ensure the absence of voltage, and grounded if necessary.

Electron: A negatively charged particle that orbits the nucleus of an atom.

Energy of break-open threshold (E_{BT}): The incident energy limit that a flame-resistant material can withstand before the formation of one or more holes that would allow flames to penetrate the material. Expressed in joules/cm^2 or calories/cm^2.

Equipotential plane: An area where wire mesh or other conductive elements are embedded in or placed under concrete, bonded to all metal structures and fixed non-electrical equipment that may become energized, and connected to the electrical grounding system to prevent a difference in voltage from developing within the plane.

Equipotential region: A region of constant voltage that contains no electric field. A charged particle in an equipotential region experiences no electric force, so there will be no electric current between two points of equal voltage because there is no force to drive the electrons.

Fall arrest: A means of stopping or controlling a fall in progress without injury to the climber.

Fall arrest belt: A belt that provides support for the climber and that will stop a fall in progress.

Fall restraint: A means to prevent a fall from occurring.

Fall restraint belt: A belt that only functions as a support for the climber; it will not stop or arrest a fall in progress.

Flame-resistant (FR): The property of a material whereby combustion is prevented, terminated, or inhibited following the application of a flaming or non-flaming source of ignition, with or without subsequent removal of the ignition source.

Flash hazard: A dangerous condition associated with the release of energy caused by an electric arc.

Flash hazard analysis: A study investigating a worker's potential exposure to arc flash energy, conducted for the purpose of injury prevention and the determination of safe work practices and appropriate levels of PPE.

Flash protection boundary: An approach limit at a distance from exposed energized electrical conductors or circuit parts within which a person could receive a second-degree burn if an electrical arc flash were to occur.

Frequency: The number of cycles completed each second by a given AC voltage, usually expressed in hertz. One hertz equals one cycle per second.

Gaff gauge: A manufacturer-specific measuring tool used for inspecting the condition of the climbers and gaffs.

Gaff guard: A device that safeguards the gaffs and preserves their integrity and sharpness.

Gaff protectors: Equipment that protects the climber's ankles and lower legs from sharp gaffs.

Grounding: Connecting to ground or to a conductive body that extends the ground connection.

Grounding mat: A mesh mat that is placed underneath equipment to provide a single electrical path to ground should the equipment come in contact with an energized conductor during a utility strike.

Guy anchor: A device well secured in the ground to provide an anchor point for a guy wire.

Guy rod: A length of fiberglass rod connected at the top of guy wires that extend into the primary conductor level on a pole.

Guys: Various types of supports used to maintain the vertical stability of poles.

Hitchhiking: The term for ascending the pole with the fall restraint belt connected.

Hypothermia: A life-threatening condition caused by exposure to very cold temperatures.

Insulated: A material that resists the flow of electric current.

Insulated aerial device: An aerial device designed for work in close vicinity of energized lines and apparatus.

Insulator: A ceramic or polymer component used to support and isolate energized conductors from the pole structure.

Insulator: A material through which it is difficult to conduct an electric current.

Jenny (mule): A device used to hold a pole above the ground when setting a pole by hand. It allows the lifting team to remove and reposition their pike poles.

Jib: An extension added to the boom and mounted to the boom tip, which may be in line with the boom or offset from it.

Joule (J): A unit of measurement that represents one newton-meter (Nm), which is a unit of measure for doing work.

Jumper: A length of wire used to connect two electrical distribution sources together.

Kilo: A prefix used to indicate one thousand. For example, one kilowatt is equal to one thousand watts.

Kirchhoff's current law: The statement that the total amount of current flowing through a parallel circuit is equal to the sum of the amounts of current flowing through each current path.

Kirchhoff's voltage law: The statement that the sum of all the voltage drops in a circuit is equal to the source voltage of the circuit.

Lanyard: Any of various small cords or ropes used for securing or suspending.

Limited approach boundary: An approach limit at a distance from an exposed energized electrical conductor or circuit part within which a shock hazard exists.

Matter: Any substance that has mass and occupies space.

Mega: A prefix used to indicate one million. For example, one megawatt is equal to one million watts.

Minimum approach distance (MAD): The distance from energized electrical conductors or circuit parts that a qualified person may approach without wearing rubber insulated PPE.

Mobiling: Term used when driving a trencher around the job site when not digging.

Neutrons: Electrically neutral particles (neither positive nor negative). Neutrons have the same mass as a proton and are found in the nucleus of an atom.

Newtons: A measure of force applied, equal to the amount of force required to accelerate a mass of one kilogram at a rate of one meter per second, per second. One pound of force = 4.45 Newtons.

Nucleus: The center of an atom, which contains protons and neutrons.

Ohm (Ω): The basic unit of measurement for resistance.

Ohm's law: A statement of the relationships among current, voltage, and resistance in an electrical circuit. Current (I) equals voltage (E) divided by resistance (R). This is generally expressed as a mathematical formula: $I = E/R$.

Ohmmeter: An instrument used for measuring resistance.

Operator presence sensing system (seat switch): A device that is used to stop the HDD rig when the operator leaves the seat with the drill turned on.

Outrigger: An extension that projects from the main chassis of the utility truck to add stability and support.

Oxygen-deficient: An atmosphere in which there is not enough oxygen. This is usually considered less than 19.5 percent oxygen by volume.

Parallel circuits: Circuits containing two or more parallel paths through which current can flow.

Pathogen: A bacteria, chemical, or virus that is known to cause disease.

Pike: A long pole with a spike at the end. It is used to raise and align a wood utility pole by hand.

Piking method: A method for setting a wood utility pole by hand that uses pikes to raise and align the pole.

Platform: Any personnel-carrying device, basket, or bucket, which is a component of an aerial device.

Point of daylight: The point where anchor assemblies or tower components meet the soil and they are no longer exposed to daylight.

Pole climbing gaffs: Pole climbing gaffs, also called climbers, are sturdy, narrow metal bars that extend down the lower leg, flare out slightly around the ankle, and form a stirrup that curves under the work boot just in front of the heel. A strong and durable V-shaped gaff (spike) that is securely attached at the lower curve of the device serves as the foot hold for the climber.

Pole grabber: A device attached to the end of a boom that grabs and holds the pole as it is being positioned by a digger derrick.

Pole key: A device that is placed in the pole hole next to the pole. Its purpose is to stabilize the pole and prevent the pole from leaning.

Pole-top rescue: The rescue of an injured, incapacitated, or unconscious worker, who must be secured and safely lowered to the ground. All climbers must be properly trained and able to perform the pole-top rescue of an injured worker.

Positioning lanyard: A device used as a secondary safety belt when climbing over obstructions.

Posthole digger: A device used to dig round, deep holes in which a pole will be placed. It can be a manual device, or it can take the form of a hand-held, engine-driven auger.

Power: The rate of doing work, or the rate at which energy is used or dissipated.

Prohibited approach boundary: An approach limit at a distance from an exposed energized electrical conductor or circuit part within which work is considered the same as making contact with the energized conductor or part.

Protons: The smallest positively charged particles of an atom, found in the nucleus.

Qualified worker: One who has the skills and knowledge related to the construction and operation of the electrical equipment and installations, and has received safety training to recognize and avoid the hazards involved.

Rappel: Descent of a vertical surface by sliding down a rope, typically while facing the surface and performing a series of short backward leaps to control the descent.

Relay: An electromechanical device consisting of a coil and one or more sets of contacts. Used as a switching device.

Rescue line: A properly rated rope of at least ½ inch in diameter that is used to secure and lower an injured worker during a pole-top rescue. The line, which is an important part of the rescue kit, must be twice the length of the climbing height plus 10 feet.

Resistance: An electrical property that opposes the flow of current through a circuit. Resistance (R) is measured in ohms.

Resistor: Any device in a circuit that resists the flow of electrons.

Restricted approach boundary: An approach limit at a distance from an exposed energized electrical conductor or circuit part within which there is an increased risk of electrical shock.

Sag: The amount of drop in a span of mounted conductors.

Schematic: A type of drawing in which symbols are used to represent the components in a system.

Series circuit: A circuit with only one path for current flow.

Series-parallel circuits: Circuits that contain both series and parallel current paths.

Shield: A structure that is able to withstand the forces imposed on it by a cave-in and thereby protect employees within the structure. Shields can be permanent structures or can be designed to be portable and moved along as work progresses. Additionally, shields can be either pre-manufactured or job-built in accordance with *29 CFR 1926.652 (c)(3) or (c)(4)*.

Snap ring: A connecting ring used to secure belts and lanyards.

Snatch block: A type of block (sheave) with an open side that allows a loop of rope or cable to be easily inserted. It can be used to change the direction of a pull.

Solenoid: An electromagnetic coil used to control a mechanical device, such as a valve.

Splice: A device used to securely join two conductor ends together.

Spoil bag: A canvas storage bag in which soil removed from the pole hole is placed. It is also used to remove any excess clean soil from the job site.

Spoon: A long-handled shovel with a curved blade resembling a ladle. It is used to remove soil from the bottom of deep holes.

Step potential: The voltage between the feet (usually about 1m in length) of a person standing near an energized grounded object.

Stringing block: A pulley-style device used to smoothly guide conductors between poles during installation.

Subsidence: A depression in the earth that is caused by unbalanced stresses in the soil surrounding an excavation.

Swing zone: The area in space where the momentum and inertia of a fall would cause the body or protected object to swing until a center of gravity is stabilized (hanging straight down).

Tamping bar: A hand tool used for packing soil around the base of an installed pole.

Telescopic boom: A boom that consists of straight shafts that fit together and extend to raise an aerial platform. They are designed to reach vertically or horizontally.

Temporary grounding device (TGD): A device, such as a shorting cable, that places ground potential on points of a de-energized circuit. It is used to protect personnel working on a de-energized circuit from electrical shock hazards should the circuit somehow become energized.

Thrust: The amount of force required to push or drive an object, normally measured in pounds of force.

Touch potential: The voltage between the energized object being touched and the ground.

Tow plate: A device that is clamped on to the end of a wood utility pole so that the pole can act as the tongue of the trailer used to transport poles.

Transformer: A device consisting of one or more coils of wire wrapped around a common core. Transformers are commonly used to step voltage up or down.

Two-blocking: A condition that occurs when the load block is pulled up into the boom sheaves. This condition can damage the boom and may cause the load to fall.

V-shaped stance: The term for the stance commonly taught to apprentice climbers in which their heels are close together and their toes turned outwards. This stance correctly orients the gaff angle to the pole.

Valence shell: The outermost ring of electrons that orbit about the nucleus of an atom.

Volt (V): The unit of measurement for voltage, also known as electromotive force (emf) or difference of potential. One volt is the force required to produce a current of one ampere through a resistance of one ohm. In addition, one volt is the potential difference between two points for which one coulomb of electricity will do one joule (J) of work.

Voltage: The driving force that makes current flow in a circuit. Voltage (E) is also referred to as electromotive force (emf) or difference of potential.

Voltage drop: The change in voltage across a component caused by the current flowing through it and the amount of resistance opposing it.

Voltmeter: An instrument for measuring voltage. The resistance of a voltmeter is fixed. When the voltmeter is connected to a circuit, the current passing through it is directly proportional to the voltage at the connection points.

Watt (W): The basic unit of measurement for electrical power.

Index

C

D

E

F

G

H

I

J

K

L

P

T

U

V

W

Y

Z